OCEANOGRAPHY

OCEAN

third edition

OCRAPHY:

A View of the Earth

M. GRANT GROSS

PRENTICE-HALL, Inc., Englewood Cliffs, New Jersey 07632

Library of Congress Cataloging in Publication Data

Gross, M. Grant (Meredith Grant), (date)
    Oceanography, a view of the earth.

    Includes bibliographies
    and index.
        1. Oceanography.   I. Title.
GC16.G7 1982   551.46   81-15336
ISBN 0-13-629683-1   AACR2

OCEANOGRAPHY:
A View of the Earth, third edition
M. Grant Gross

© 1982, 1977, 1972 by Prentice-Hall, Inc.,
Englewood Cliffs, N.J. 07632

Frontispiece:

The submersible Alvin is operated by the Woods Hole Oceanographic Institution for the American Oceanography Research Community. The sub takes a pilot and two scientific observers to depths of 4000 meters for sampling, photography, and observation in the deep ocean. (Photograph courtesy of John Porteous, Woods Hole Oceanographic Institution.)

Editorial and production supervision by Maria McKinnon
Interior design by Lee Cohen and Judy Winthrop
Cover design by Lee Cohen
Cover photograph courtesy F. Grehan, The Image Bank
Page layout by Mary Greey
Manufacturing buyer: John Hall

Printed in the United States of America

ISBN 0-13-629683-1

10 9 8 7 6 5 4 3 2 1

Prentice-Hall International, Inc., *London*
Prentice-Hall of Australia Pty. Limited, *Sydney*
Prentice-Hall of Canada, Ltd., *Toronto*
Pentice-Hall of India Private Limited, *New Delhi*
Prentice-Hall of Japan, Inc., *Tokyo*
Prentice-Hall of Southeast Asia Pte. Ltd., *Singapore*
Whitehall Books Limited, *Wellington, New Zealand*

CONTENTS

WAVES   210

TIDES AND
TIDAL CURRENTS   236

ESTUARIES AND
THE COASTAL OCEAN   260

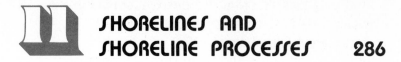

PREFACE

I have been extremely pleased with the enthusiastic reception of the first two editions of *Oceanography*. This new edition retains those features that have proven to be most useful to students and teachers. But it also incorporates numerous changes suggested by users to make it more useful in the classroom.

My first objective in the new edition is to commmunicate the exciting developments of the 1970s that have so dramatically changed our view of the ocean. The new ways of studying the ocean, new ships, and new instruments are incorporated into the text. New illustrations from cameras and other sensors on earth-orbiting satellites are used to demonstrate the importance of the ocean. Some of these new illustrations show previously unsuspected surface-ocean processes such as the complicated mixing processes between coastal and open ocean. Other new illustrations from manned and unmanned submersibles show dramatic examples of new deep-ocean processes seen for the first time—chimney-like structures discharging superheated waters surrounded by newly discovered worms 8 feet long, and clams and other organisms never seen before. Virtually every chapter contains some new materials as a demonstration of the impact of these discoveries on our understanding of the ocean and ocean processes.

The second objective is to present the new role of the ocean in human affairs. Increased importance of ocean resources has led to radical changes in the legal status of the ocean. For more than 300 years, the ocean was free to all nations. Now the United Nations Conferences on the Law of the Sea have partitioned large ocean areas, giving jurisdiction to the coastal states. Thus the ocean as a frontier area has begun to disappear. Various aspects of the role of ocean processes in human affairs are incorporated throughout the book.

This edition remains at the same level of difficulty as previous editions—an introductory, nonmathematical text for nonscience majors which emphasizes the familiar and important aspects of the ocean. For example, the emphasis is retained on beaches and coastal processes. Problems of using and protecting the ocean are extensively discussed. Many examples tie ocean processes to discussions of important topics such as weather, climate, fish production, and mineral resource recovery.

The distinctive illustrations of the late Roger Hayward are retained, although presented in a different format. Some drawings were modified to illustrate more clearly the important points for teaching purposes. Many new illustrations were added to provide wider geographic coverage, especially along the Pacific and Gulf coast regions.

A concerted effort was made to make the book more useful to students, following suggestions from both students and teachers familiar with the book. Review questions were added to each chapter to help students identify major concepts in the chapter and master the major points of the book. Summary outlines and annotated reading lists in each chapter were retained from previous editions.

The book was designed to be used in a one semester course. Various chapters can be dropped without disrupting the flow of material. No college-level pre-requisites are assumed; only high school level science are required.

The many revisions in the third edition have arisen from recent developments in the science and from the suggestions of many colleagues who have used the book. I am especially grateful to E. Stroup and R. Moberly, University of Hawaii; C. Drake, Dartmouth College; D. Folger, United States Geological Survey; R. Heath and R. Smith, Oregon State University; Y. Hsueh, Florida State University; R. Dugdale, University of Southern California; R. Barber, Duke University; G. Fryxell and R. Merrill, Texas A&M University; R. Fine and G. Ostlund, University of Miami; J. Walsh, Brookhaven National Laboratory; W. Boicourt and J. Taft, Johns Hopkins University; A. Alldredge and J. Childress, University of California, Santa Barbara, and M. J. Perry, University of Washington.

My associates at the National Science Foundation deserve special thanks for their untiring efforts to educate me, especially C. Collins, B. Malfait, D. Heinrichs, E. Dahvin, N. Andersen, R. Baier, D. Frankenberg, R. Wall, and R. Carney.

V. Cullin, Woods Hole Oceanographic Institution and J. Jahnke, Scripps Institution of Oceanography were extremely helpful in locating photographs.

Ann Georgilas typed the final revised manuscript.

To all of you, I am deeply grateful.

M. GRANT GROSS

OCEANOGRAPHY

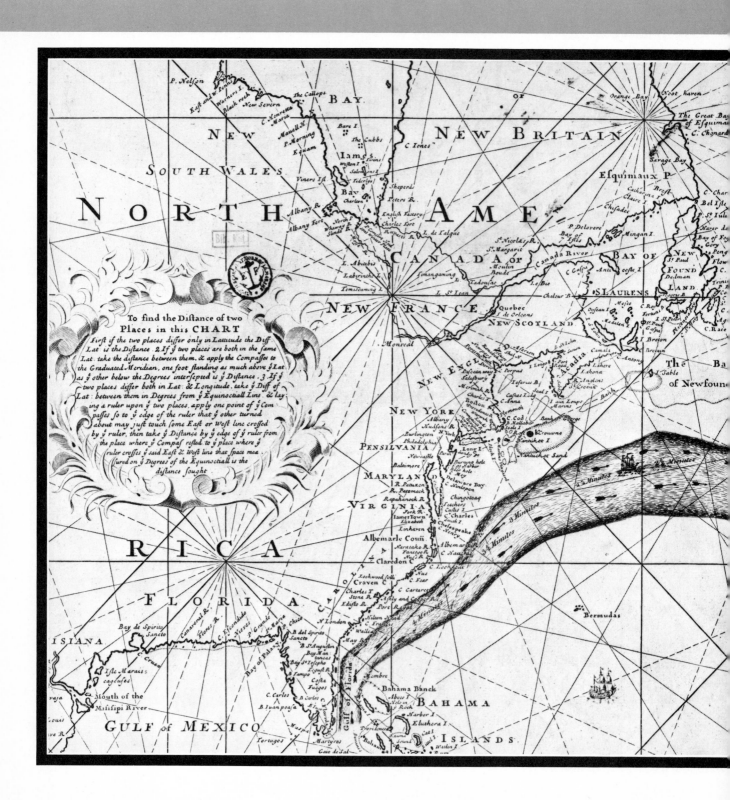

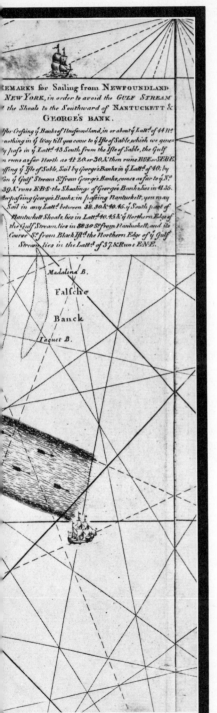

Benjamin Franklin published the first good map of the Gulf Stream in 1769 or 1770. Location of the Gulf Stream was sketched by Timothy Folger, a Nantucket ship captain. (Photograph courtesy of Philip Richardson, Woods Hole Oceanographic Institution.)

The ocean plays a central role in our lives and has done so for a long time. Much that has been learned about the ocean and a great deal that is now being studied come from perceived needs to understand the ocean better and to be able to predict how it will behave in the future, especially after absorbing the various impacts of human activities. In this chapter we review the history of ocean science with particular emphasis on the period since the end of World War II, when governments in many countries made available enough money to support ocean research on a scale comparable to the processes under study.

EARLY HISTORY OF OCEANOGRAPHY

Since antiquity the study of the ocean contributed to solving problems and locating new resources. Fishermen, merchants, and traders explored the ocean, searching for new trade routes or fishing grounds.

Because of these explorations, the basic principles of oceanography were understood by ancient thinkers. Aristotle, for instance, noted in the fourth century B.C. that the sea neither dries up nor overflows; thus the amount of rainfall equals evaporation over the earth. The astronomer Ptolemy (A.D. 150) developed a scheme of latitude and longitude for charting the known world and made surprisingly accurate maps.

The Venerable Bede (673–735) knew that the moon controls the tides and in the fifteenth century Cardinal Nicholas of Cues devised a method of calculating the sea's depth. He suggested measuring the rate of rise of a buoyant object in "diverse waters" of known depth and then dragging such an object to the bottom of the ocean, releasing it, and timing its return. During the later Renaissance explorers circled the earth and made charts as they went. And the seventeenth-century scientist Sir Robert Boyle (1627–1691) studied ocean temperatures, water pressure, sea salts, and the effects of storm waves. In short, study of the ocean is an ancient activity.

Working at sea is expensive and time consuming. As a result, most major advances have been sponsored by governments or, rarely, by wealthy individuals. Governments have long supported ocean exploration. The voyages of Vasco da Gama, Columbus, Balboa, Magellan, and Captain Cook mapped the ocean basin. Much remained unknown until the 1800s. Polar regions were still inaccessible to ships. Oceanic depths had never been charted. Only surface marine life was known. Open ocean phenomena were largely the province of myth and mystery.

4

Ocean exploration led to studies of ocean processes. One of the most successful early explorer–scientists was a British naval officer, Sir James Clark Ross (1800–1862). As a young man, he sailed under his uncle, Sir John Ross, searching for a Northwest Passage from the Atlantic to the Pacific, and tried to reach the North Pole. At 29, on another Arctic expedition, he reached the magnetic pole. Later he was assigned by the British Admiralty to head an Antarctic expedition that made him famous. From 1839 to 1843 his ships *Erebus* and *Terror* made three separate voyages, circling Antarctica and charting its coast. Scientific parties aboard collected biological specimens from great depths, although most of the samples were either lost or allowed to deteriorate through inexpert handling. The scientists also dredged to then-incredible depths of 5000 meters and tried to measure temperatures; thermometers of that time, however, could not give accurate readings at great pressures and so the data were wrong.

The primitive early nineteenth-century measuring and sampling techniques caused serious misconceptions about deep-ocean conditions. Often such misconceptions caused scientists to pursue fruitless studies. It was believed, for instance, that the deepest ocean waters did not circulate but lay stagnant at the bottom of ocean basins. Consequently, no supply of incoming oxygen would be available to support life there. Furthermore, a respected naturalist, Edward Forbes (1815–1854), claimed to have proved the absence of life below 600 meters. He had done some collecting in the Aegean Sea to compare findings with those of Aristotle nearly 2500 years earlier and had noticed that fewer and fewer species were caught as he let his nets sink farther below the surface. He concluded that most of the ocean was *azoic* or devoid of animal life. This idea gained wide acceptance— so much so that occasional reports of animals taken from below 600 meters were ignored or discounted.

Forbes's azoic theory was not refuted until the 1860s when ocean basin topography was studied in connection with the laying of transoceanic telegraph cables. A cable from 2000-meters depth in the Mediterranean Sea was raised for repairs and on it were living corals. This discovery came at a time when interest in deep-sea exploration had been aroused in connection with Charles Darwin's (1809–1882) theories about the origin of species through evolution. Scientists became interested in finding stable environments in which conditions had not changed over long time periods. It was hoped that the deep-ocean floor might be inhabited by ancient species or "old generic types," the ancestors of modern organisms. Primitive animals on stalks, called *crinoids,* had recently been discovered by a Norwegian vicar and amateur scientist, Michael Sars, which seemed a promising sign.

CHALLENGER EXPEDITION

By the 1860s, then, there were several compelling reasons for studying the deep ocean. Widespread interest existed in setting up large-scale research projects for which the personal funds of even a wealthy scientist would be inadequate. Fortunately members of the Royal Society of London, a group of eminent citizens devoted to philosophical and scientific programs, provided financial support. The British Admiralty also provided two ships for North Atlantic deep-sea studies during the late 1860s. As a result of these research efforts, some of the old prejudices and misconceptions disappeared. It was demonstrated that waters can move through deep-ocean basins. Scientists dredged to depths of 1200 meters and discovered new organisms.

Finally, the Royal Society was persuaded to sponsor the most ambitious and innovative project that had ever been attempted. This expedition (1872–1876) aboard *H.M.S. Challenger* established a tradition of large-scale, interdisciplinary cooperation that has been an important component of major efforts in oceanography ever since.

The *Challenger* was a full-rigged corvette with an auxiliary 1200-horsepower steam engine (Fig. 1-1). The scientific party under Sir Wyville Thomson (1830–1882) had been assigned to investigate "everything about the sea"; they planned to study physical and biological conditions in every ocean, recording whatever might influence the distribution of marine organisms. This process included taking water samples and temperature measurements of both deep and surface waters, recording currents and barometric pressures, and collecting bottom samples (see Fig. 1-2) in order to study sediments and identify new species. Fishes were caught in nets dragged behind the ship.

Before returning to England, the *Challenger* traveled 68,000 miles, hauled on board some 133 loads of rock and sediment from the ocean floor, and took 492 soundings, using a weighted hemp line. All ocean basins except the Arctic had been sampled and 4717 new animal species had been collected and classified. Circling the globe, the ship brought back data that eventually filled 50 large volumes. The reports, written by 76 authors over a 23-year period, are a watershed in ocean science.

Many delusions about the ocean's mysteries were swept away by the *Challenger* results. The most famous, perhaps, was the debunking of *Bathybius*, a supposed "primordial slime" that had been thought to represent preevolutionary life on the ocean floor. Bathybius had been "discovered" by Thomas H. Huxley (1825–1895), a well-known biologist whose staunch support of Charles Darwin had helped make

Figure 1-1

The *Challenger* crew attempting to land small boats at St. Paul's Rocks in the mid-Atlantic. Strong currents and high waves made it a difficult and dangerous operation.

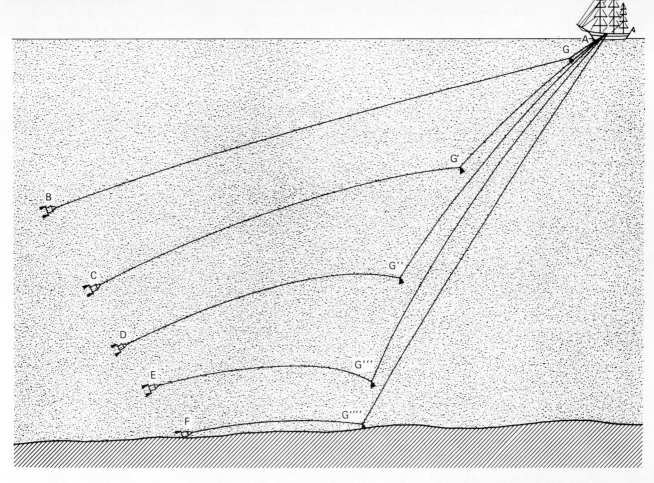

Figure 1-2

This sketch shows a deep-sea dredge aboard the *Challenger* being lowered by stages to the bottom. A weight slid down the line (G, G′, G″, etc.), drawing the heavy-chain dredge bag downward to land on its side. This step put the dredge in position to scoop up samples as the ship dragged it along.

evolution respectable. In 1868 Huxley found a slimy substance in a jar containing a biological specimen several years old. He may have been overly anxious to provide new evidence for his friend's theory, which was then under attack from supporters of the biblical creation. At any rate, Huxley concluded that he had identified the protoplasmic substance from which life originated in the deep ocean. It was a fortuitous idea in that Bathybius stimulated much interest in marine research. But while at sea a *Challenger* chemist revealed that the slime was made from calcium sulfate, precipitated by mixing preservative alcohol with seawater. Fortunately the theory of evolution survived this setback.

A British government department, the *Challenger* Expedition Commission, was established to analyze and publish the results of the voyage. It was to have been headed by Wyville Thomson, but when he died suddenly, that position went to Sir John Murray (1841–1914), a Canadian-born geographer and naturalist who had served as Thomson's assistant. Murray was a brilliant scientist whose career did not end with publication of the *Challenger* Reports. He is generally credited with laying the foundations for modern submarine geology and authored an important study on marine sediments. He also did valuable work on coral reefs and plankton, the microscopic drifting organisms that are the ocean's major food source. After persuading the British government to annex uninhabited Christmas Island in the Indian Ocean, Murray made his fortune there in phosphate mining. This wealth he used in part to promote oceanographic research; for example, he paid the expenses of a Norwegian expedition in the North

Atlantic aboard the *Michael Sars* in 1910. In collaboration with the director of that expedition, Dr. Johann Hjort (1869–1948), Murray wrote his most important work, *The Depths of the Ocean*, which was for many years the most widely read and authoritative oceanography text available in English in the early twentieth century.

EARLY AMERICAN OCEANOGRAPHY

In the United States governmental support for oceanography also grew out of a need to solve practical problems. These included ensuring the safety of passengers and goods aboard U.S. ships, maintaining adequate coastal defenses, and developing marine resources, such as fisheries in U.S. waters. In 1830 the Navy created a Depot of Charts and Instruments to supervise use of navigational instruments on government vessels. Matthew Fontaine Maury (1806–1874), a naval officer, was appointed Superintendent in 1842.

Maury (Fig. 1-3) wanted to make ocean transportation safer and speedier and he had written a navigation textbook before coming to the Depot. Once in charge, he organized the vast amount of wind, current, and seasonal weather data stored in ships' logbooks. To gather further information, Maury furnished ships' crews with blank charts so that they could make daily records of weather and ocean conditions, showing the time and place of each instrument reading. The Wind and Current Charts he compiled, in this way, revolutionized navigation and cut weeks off transoceanic runs by U.S. clipperships. In addition, Maury's sounding and bottom-sampling projects were used to make maps of the Atlantic Ocean floor.

In 1853 Lieutenant Maury organized the first international meteorological conference, which led to international cooperation in collecting weather information at sea. His popular and influential book, *The Physical Geography of the Sea* (1855), was the first major oceanographic work in English. It stimulated widespread public interest in ocean currents, although many scientists were skeptical of his rather florid writing style as well as his often unsupported theories about the causes of oceanic phenomena. He is sometimes called the father of physical oceanography even though today he would be categorized as a gifted navigator, a popularizer of science, and a dynamic administrator.

Figure 1-3

Mathew Fontain Maury grew up on frontier farms in Virginia and Tennessee. At 19 he was appointed a midshipman and within the next 10 years he had sailed around the world on Navy vessels. When a stagecoach injury ended his seagoing career, Maury embarked on the administrative duties that established a foundation for government-sponsored oceanographic research.

While the Navy was responsible for safeguarding American lives and property on the high seas, the Treasury Department was interested in maritime commerce. Charting of harbors, ports, and coastal waters was urgently needed, and Thomas Jefferson created a civilian agency within the Treasury Department to take charge of that program. But in 1807 scientific research under government sponsorship was a new idea. Internal disorganization kept the newly formed Coast Survey (now the National Ocean Survey) from undertaking significant mapping or charting projects during its first 25 years.

Finally, Alexander Bache (1806–1867), who was trained in physics and chemistry, took over the Survey in 1843 and set up a comprehensive hydrographic program. His people worked in East Coast waters, mapping hazards to navigation, taking depth measurements, collecting bottom samples, measuring currents, and monitoring tides. They also made pioneer studies of the Gulf Stream, charting its temperatures, its depths, and changes in its velocity. Actually, Bache was carrying on the work of his great-grandfather, Benjamin Franklin, who had mapped the Gulf Stream in the 1770s (Fig. 1-4). Another of his innovative ideas was to invite scientists to join Coast Survey cruises. This policy greatly expanded the scope of federally sponsored research and provided support for basic oceanographic science at little public expense.

Figure 1-4

Prior to the American Revolution Benjamin Franklin noted that transatlantic mail ships made the voyage to England more speedily than the voyage home. His cousin Timothy Folger, a whaling ship captain, pointed out that a current flowing northward along the U.S. coast and eastward into the North Atlantic was responsible for the difference in sailing time. In 1770 Franklin published this map, with instructions on how "to avoid the Gulph Stream, on one hand, and on the other the Shoals that lie to the Southwest of Nantucket and St. George's Bank." (Map courtesy Woods Hole Oceanographic Institution.)

Lt. Charles Wilkes in 1836 was selected to command the U.S. *Exploring* Expedition to "extend the bounds of human knowledge." During the 4-year expedition studies were made in the Antarctic, the central Pacific Islands, and the western coast of the Americas. The report of the expedition was the greatest scientific publishing program undertaken by the U.S. government before the Civil War. Some of the charts prepared from the Wilkes expedition were the best available in 1943 for military landings in the southwest Pacific.

Another significant naval expedition, the U.S. Surveying and Exploring Expedition, went to the southwest Pacific in 1853, returning by way of the Arctic in 1855. Deep-water bottom samples were taken and some useful hydrographic data were published, but there was no money for a complete report of the expedition. And neither expedition left a major mark on ocean science.

In 1871 a Fish Commission was established, with a distinguished ornithologist and natural history professor as unpaid Commissioner. Spencer Fullerton Baird (1823–1887) was more interested in establishing a center for basic research in coastal ecology than in regulating commercial fish stocks. For his laboratory, he chose a tiny fishing village on Cape Cod—Woods Hole near Falmouth, Massachusetts, where university-based scientists could come to work during the sum-

Figure 1-5

Woods Hole Station, about 1890. The Fish Commission's laboratory is at center, the residence to the right, and the *Albatross* is moored at the left. (Photograph courtesy National Marine Fisheries Service.)

mer. It soon became an important center for biological research; eventually the Marine Biological Laboratory and the Woods Hole Oceanographic Institution were set up there (Fig. 1-5). Thus two of America's foremost marine science institutions resulted from Baird's dedication and foresight and the need to understand fisheries better in order to protect known stocks and explore for new ones.

The first specially equipped U.S. research ship, the 234-foot steamer *Albatross*, was built in 1882 under Baird's direction. Research parties studied ways in which changing water conditions—temperatures, salinity, and currents—affected the abundance of fishes in an area. At the same time scientists collected specimens and conducted their own studies on the way fishes and invertebrate animals grow and develop in marine environments.

Scientists were invited to participate on Fish Commission cruises. The most famous was a biologist, Alexander Agassiz (1835–1910), whose father, Louis Agassiz, had sailed with Alexander Bache. A wealthy man, Agassiz, Jr., personally funded more than one Pacific cruise, outfitted the *Albatross* with special equipment, and subsidized oceanographic research in other ways. Late in the nineteenth century a record depth of 7632 meters was dredged under his direction, on the Pacific floor near Japan. It was not surpassed until 1951. In addition, Agassiz maintained a private laboratory at Newport, Rhode Island, where he made comparative studies of corals and other specimens collected during his travels.

NATIONAL EXPEDITIONS, 1900–1950

Between the *Challenger* Expedition and World War II, small but important oceanographic expeditions were sent out from various countries. Notable among them was the work of Prince Albert I of Monaco (1848–1922). This diligent investigator was rich and had a life-long interest in the ocean. He equipped four private schooners as research vessels. In one study Prince Albert set adrift over 1500 bottles and barrels containing messages in ten languages. Enough of them were recovered to explain the relation of the Gulf Stream to the North Atlantic current system. Using cleverly designed equipment, he collected many new kinds of mid-depth fishes, including several specimens vomited by a dying whale. Later the Prince established a museum and an oceanographic institute in Monaco that continue today.

The Scandinavian countries, whose economies are so dependent on the sea, have long been leaders in ocean research. As investigators began to realize that the abundance of commercial fishing stock is

10

affected by currents and the distribution of water masses, research in these areas received increasing government support. Improved equipment was developed to study movements of ocean waters, both near the surface and at great depths.

Scandinavian scientists realized that temperature and salinity differences drive currents, a fact that had already been suspected by Maury and others earlier in the century. The result led to rapid developments in the field of chemical, as well as physical, oceanography. A surge of interest in theoretical studies combined with the development of better deep-sea *reversing thermometers* (Fig. 1-6) and methods of determining salt concentrations. Presently another Scandinavian, Bjorn Helland–Hansen (1877–1957), developed techniques to calculate current direction and speed from precise observations of temperature and salinity. This was a particularly outstanding contribution to modern oceanographic technique. Another advance was made by V. W. Ekman (1847–1954), who showed how the earth's rotation affects wind-driven currents.

Figure 1-6

Reversing thermometers were used for decades to obtain precise temperature measurements in the ocean. More precise methods of determining salinity and temperature have been developed and have replaced these thermometers on most ships.

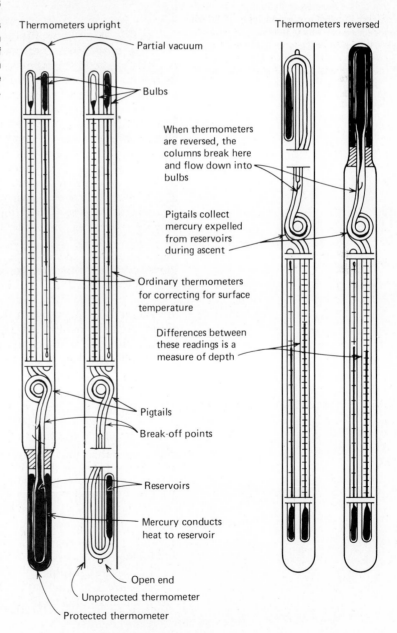

Plankton studies (Fig. 1-7) followed the discovery that the physical and chemical properties of water masses determine which organisms can live in them. Soon techniques were developed for identifying water masses by their biological communities. Because North Sea circulation is tied to that of the Atlantic as a whole, local studies inevitably led to more extensive work that could not be effective if carried on by one nation alone. A Swedish scientist, Otto Pettersson (1848–1941), correctly foresaw that measurements made by several vessels simultaneously, at exactly predetermined stations, would be the most efficient method for studying ocean properties on a large scale. He enlisted support from Sweden's King Oscar II and thus the plan for an International Council for the Exploration of the Sea (ICES) was born.

Figure 1-7

Plankton (microscopic marine plants and animals) are collected by towing a cone-shaped net of fine-meshed cloth through seawater. Organisms unable to swim faster than the net are captured for study. (Photograph courtesy Woods Hole Oceanographic Institution.)

Germany, Russia, Great Britain, Holland, and the Scandinavian countries participated in ICES from the start. The Council was founded in 1902 to foster international cooperation in learning how best to utilize marine resources rationally, especially fisheries. It has functioned since then as a coordinating body and as a sponsor for joint research enterprises.

Scientific teams working within ICES followed Pettersson and Ekman in studying currents and water masses that move with them. Others tagged fishes to study their life cycles and the unpredictable migrations that had frustrated generations of fishermen. Johann Hjort demonstrated that fish abundance is largely determined by the amount and kind of food available to them just after they hatch. This finding was highly significant in fisheries research. Victor Hensen (1835–1924) followed with important studies on how larval and juvenile (planktonic) fishes are nourished. This factor turned out to depend again on ocean water movements, demonstrating that physical and biological processes were invariably linked within the total framework of oceanographic research. Always, the pooling of knowledge and resources at international levels promoted communication between scientists and their cooperation in solving common problems.

Word of ICES methods reached a young Harvard zoologist, Henry Bryant Bigelow (1879–1967), who had originally trained as a naturalist under Alexander Agassiz. Bigelow, encouraged by the visiting John Murray, undertook a comprehensive study of biological and physical oceanography in the Gulf of Maine. Beginning in 1912, he spent 15 years working on detailed, comprehensive investigations, using methods pioneered by Swedish and Norwegian scientists. His work provided incentive for other Americans to plan their research around the new techniques. It meant studying the geology, biology, and chemistry of a single body of water over a long enough period to get a complete picture of its properties and their variability.

Nationally sponsored expeditions during the early part of the twentieth century produced outstanding results. Already mentioned was the 1910 Murray and Hjort collaboration for biological and physical studies on the *Michael Sars*. Another highlight was the German Antarctic Expedition of 1911–1912, which charted temperature and salinity distributions from surface to ocean bottom and from north to south across the entire Atlantic Ocean. Studies of Antarctic waters completed the expedition. In 1925–1927 the German research vessel *Meteor* crossed the Atlantic 14 times between 20°N and 65°S, charting temperature–salinity profiles and mapping the ocean floor (Fig. 1-8). This was the most comprehensive study of water masses ever made over such a vast area. Such studies were finally beginning to provide a total oceanographic picture of the great ocean basins.

Two other government-sponsored projects deserve mention. Sailing under the Danish flag, Johannes Schmidt spent 20 years trying to locate the spawning grounds of American and European eels. Finally, after being interrupted by World War I, he identified the birthplaces of these long-distance swimmers far out in the mid-Atlantic (Fig. 1-9). In 1929–1930, aboard the *Willibrord Snellius*, Dutch scientists explored Indonesian seas. Invention of the *Piggot gun* explosive bottom sampler enabled them to bring a "core" of sediment 6 feet long aboard the ship. Analyzing its successive layers of sediment revealed the history of an ocean basin.

In a class by itself was the remarkable polar voyage of the Norwegian scientist and adventurer Fridtjof Nansen (1861–1930). Convinced that winds and current drift caused pack ice to move between the North Pole and Franz Josef Land, from the Siberian Arctic to

Figure 1-8

South Atlantic crossings made during the *Meteor* expedition (1925–1927). Each tiny mark represents a station at which temperature and salinity measurements were made at various levels to get a profile of conditions throughout the entire basin.

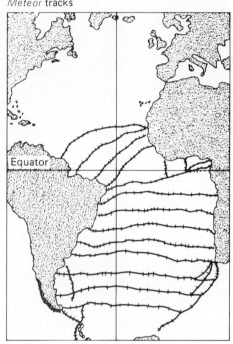

Meteor tracks

Figure 1-9

The anguillid "freshwater" eels, 0.5 meter or longer, inhabit both sides of the Atlantic. They live in rivers and lakes, but at sexual maturity (5–8 years old) they begin a migration to the Sargasso Sea. After a journey of thousands of miles and many months without food, they spawn at depths of around 500 meters and then die. Their eggs float at the surface and soon hatch into tiny larvae that resemble little fish. Each begins the long journey to freshwater first described by Johannes Schmidt. European eels take 3 years to complete the journey, but the American variety requires only 1 year. There is a physical transformation at that time; the slender young eels ("elvers"), about 7 centimeters long, travel up the streams to live there until maturity.

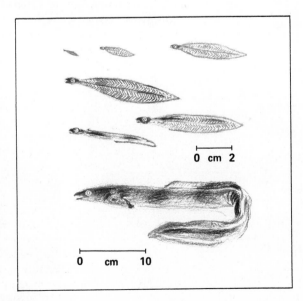

eastern Greenland, Nansen persuaded his government and the British Royal Geographical Society to help him test his theory. A special ship, the *Fram*, was built to carry a scientific party eastward from the Barents Sea toward the North Pole, frozen in an ice floe (Fig. 1-10). This expedition was an outstanding success. From September 1893 to August 1896 the *Fram* drifted with the ice, from 70°50′ to an extreme north latitude of 85°57′, closer to the pole than any surface ship has ever been before or since. The crew then broke her free during warm weather and sailed home. Nansen, with a companion, left the ship at 84°N and set out by dogsled toward the pole. The two men spent 14 months alone on the Arctic ice but made only 86°14′ before movements of the floating ice pack forced them to turn back toward Franz Josef Land.

The remainder of Nansen's career was equally distinguished. He worked on current theory with Ekman and Helland–Hansen and published several books on North Sea oceanography. He also designed a standard water-sampling bottle that bears his name (Fig. 1-11). From 1905 onward Nansen was active in politics. He received the Nobel Peace Prize in 1923 for his contributions to the relief and repatriation of World War I refugees; later he became an expert on Soviet agricultural economics. A lifelong athlete and famous skier, Nansen did much to popularize winter sports in Switzerland and elsewhere outside Scandinavia.

One important series of research projects followed from the tragic sinking of the *Titanic*. After she hit an unusually southerly iceberg in 1912, the International Ice Patrol was set up to guard against future mishaps. Its primary function is to monitor the hundreds of icebergs that break from the Greenland ice cap every spring to create a hazard to North Atlantic shipping. The patrol soon made additional contributions to oceanography, however, by demonstrating that the best way to predict current movements is by the so-called dynamic technique worked out by Bjorn Helland–Hansen. The method had never before been tested on a large scale. Its successful application was a triumph for those who believed that theoretical studies could be the key to increasing the mastery of the ocean.

New research fields opened up in the early 1920s. Efforts at submarine detection led to development of the *echo sounder*, a device

Figure 1-10

Scenes from the voyage of the *Fram*. (a) A crew member tests the depth of a melted pool on the ice pack. (b) Fridtjof Nansen at the helm.

(a)

(b)

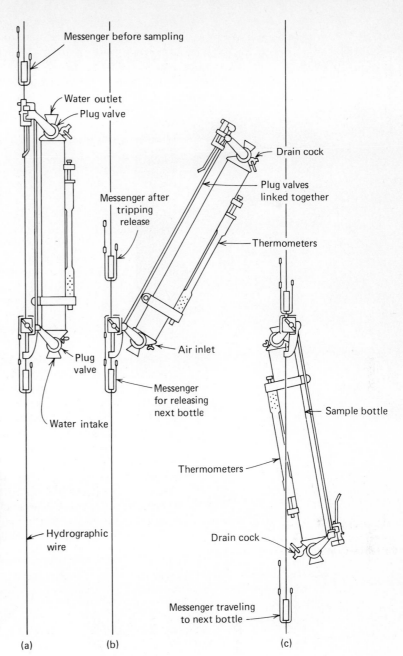

Figure 1-11

Nansen bottles attached to a wire line for lowering into the water. The messenger (a brass weight on the line above the bottle) releases the trigger, permitting the bottle to invert and take a sample. Precision reversing thermometers in the two cases mounted on the bottle record the temperature at the time that the sample is taken.

that emits sounds toward a solid object, such as a ship, an iceberg, or the sea floor. It then records the time elapsed before the returning echo is detected. Earlier ocean depths had been sounded by using a weighted wire; the echo sounder made continuous depth readings possible and ocean basin mapping took a giant step forward (Fig. 1-12). The *Meteor* Expedition was one of the first to take full advantage of this method.

By World War II foundations for modern American ocean research had been laid. A new source of funds, coming from nonprofit institutions set up by industrial millionaires, started to provide some solid financial support. Several million dollars from the Rockefeller Foundation supported the Woods Hole Oceanographic Institution. The Oceanographic's ketch *Atlantis* (Fig. 1-13) provided new data on the Gulf Stream and western Atlantic Ocean. From Scripps Institution of Oceanography, also a recipient of Rockefeller money, *R/V E. W.*

15

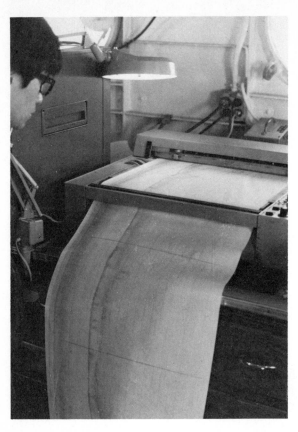

(a)

Figure 1-12 (above)

A depth recorder provides a continuous record of water depths below the ship's track.

Figure 1-13 (right and below)

(a) The Atlantis sailing under the auxiliary power. (b) Atlantis II, a vessel designed for oceanographic research. (Photographs courtesy of Woods Hole Oceanographic Institution.)

(b)

Scripps explored the nearby California Current and Gulf of California. *Catalyst,* owned by the University of Washington, worked off the northwest coast and in the Gulf of Alaska. For more distant studies, scientists from various institutions cruised on Coast Guard and Navy ships in the Bering Sea, through the North Pacific, and off Central America.

INTERNATIONAL OCEANOGRAPHY

Modern oceanography dates from the 1950s. Most civilian ocean research came to a sudden halt with the outbreak of World War II. Scientific resources were mobilized for the war effort and previously neglected research topics assumed new importance. As usual, projects initiated to attack specific problems led to theoretical studies and basic science was advanced in many cases. Acoustical studies were needed to help track submarines by their sounds transmitted through water. Because temperature affects the acoustic properties of seawater, this led to new studies of temperature distribution in the ocean. The *bathythermograph,* which records a profile of temperatures as it is lowered through the water, was perfected in the 1930s and widely used during the 1940s. Wave research was needed to plan safer amphibious troop landings. In the same way, coral reef studies resulted from the war in the Pacific because Allied troops needed the information for landings. Sea-floor mapping was done to locate safe routes for submarines through narrow straits and other hazards.

U.S. government funding for such projects was generous compared to the meager support Congress had given to prewar research. But, for scientists, possibly the most significant outcome of the war was an increased confidence in their potential for contributing to the nation's welfare. After the war, as defense-related research was followed up by further studies and prewar efforts were resumed, government contracts and grants became available for projects of generally recognized importance. Such is the pattern for much research funding today.

Defense-related projects have been mainly underwritten by the Office of Naval Research and basic science by the National Science Foundation. The Departments of Energy and Commerce, as well as the National Aeronautics and Space Administration, have funded marine science, outside of government laboratories.

National security, mineral and petroleum resources, living resources, such as fisheries, and environmental protection are major areas of publicly sponsored research. Basic research has also been well supported. During the postwar decades several major projects undertaken by international research teams have revolutionized our understanding not just of ocean processes but of the entire Earth's structure.

Following directly from wartime research and postwar nuclear weapons testing, a comprehensive study was undertaken of two Pacific atolls, Bikini and Eniwetok. The study included scientists from government, universities, and major oceanographic research institutions in the United States. The islands were to be used as test sites for atomic bombs directed at ships and researchers wanted to study what such operations might do to the islands themselves.

By then many scientists and government officials realized that small, nationally based seagoing programs could probably no longer make major scientific contributions. Large-scale international cooperation was clearly the way of the future. A major step in this direction was taken by the International Council of Scientific Unions, which organized the International Geophysical Year (IGY) of 1957–1958.

The IGY was carried out by loosely coordinated scientific groups from 67 nations. These scientists examined many kinds of physical phenomena during a period of intense sunspot activity. In oceanography, several areas were emphasized. For mapping currents, the Swallow Float was used to study water movements at any desired depth. Preset to maintain its position in water of given density, this device emits sounds by which it may be tracked from a ship or from listening stations on islands or ashore. Subsurface currents can be mapped by following the float's path and averaging its movements over many days.

Tide gages were used to study variations in average sea level, which may be due to seasonal effects, temperature, or winds. Other investigations included one on carbon dioxide content of deep-ocean waters, in connection with atmospheric pollution, and another on movements of deep-ocean waters, using radioactive carbon-14 as a tracer.

But the most exciting oceanographic result of the IGY turned out to be exploration of the sea floor. Intensive work on the topography of ocean basins had already been going on for some years. At Columbia University's Lamont–Doherty Geological Observatory, for instance, geologists mapped the ocean basins. With Lamont's director, Maurice Ewing (1906–1974) (Fig. 1-14), they developed theories about the great submarine mountain ranges, called midocean ridges, that occur in major ocean basins. This work was compared with important new research on the earth's magnetism, its gravitational field, and the heat that flows from its interior. Data collected from these studies led to a problem that had intrigued scientists for hundreds of years—the possibility that continents have changed their relative positions during geologic time, as suggested by the apparent "picture-puzzle" fit of the African continent against Central and South America.

In 1962 Harry H. Hess (1906–1969) of Princeton (Fig. 1-15) theorized that movements in the earth's mantle, a plastic layer below its rocky crust, caused continents to move. Data from many fields of geological science, often previously confusing and apparently con-

Figure 1-14

Dr. W. Maurice Ewing with Lamont's research vessel, *Vema*. (Photograph courtesy Lamont–Doherty Geological Observatory of Columbia University.)

flicting, have been bound to fit this theory, known as sea-floor spreading or plate tectonics. It states that the ocean floor is constantly but slowly moving with the underlying mantle. Oceanic crust is continually being formed by volcanic action at midocean ridges and destroyed at deep trenches near ocean basin margins.

One way to test the validity of this theory was to drill into the ocean floor for samples of deep fossils and crustal rock and determine the age of the material recovered. Scientists from Woods Hole, Lamont, Scripps, and the University of Miami agreed to cooperate on such a project, called JOIDES for Joint Oceanographic Institution for Deep Earth Sampling. Later other groups joined them. The National Science Foundation provided funds and a specially built vessel, the *Glomar Challenger,* did the drilling (Fig. 1-16).

By dating the sediments that had accumulated on newly formed oceanic crust, the JOIDES scientists confirmed that the newest crust is next to midocean ridges and that its age increases with distance from them. Since the 1960s many modifications have been made in details of this theory, but basically it is still similar to what Hess proposed.

Figure 1-15

As navigation officer aboard a World War II troop transport, Dr. Harry Hess took echo soundings across large regions of the Pacific Ocean floor and discovered numerous flat-topped volcanoes. Later, as he continued to study these ancient "guyots," data on their age helped convince him that the sea floor moves and is continuously consumed at the trenches. (Photograph courtesy Department of Geology, Princeton University.)

Figure 1-16

The *Glomar Challenger* was specially designed to maintain position and to drill in very deep water. Position is maintained by means of a sonar beacon; dropped to the ocean floor, its emitted signals are picked up by hydrophones below the ship's hull. Computers calculate the ship's position with respect to the beacon and automatically hold her on station while drilling is in progress. Samples of sediment and rock have been recovered from more than 1000 meters below the ocean bottom and the vessel can operate in waters nearly 7 kilometers deep. Analyses of cores from this deep-sea drilling project have provided important evidence supporting the theory of sea-floor spreading. Absence of sediment over midocean ridges and progressive thickening of sediment toward ocean basin margins suggested that ridges are sites of new-crust formation. Increasing age of the crust in direct proportion to its distance from the ridge is indicated by its increasingly thick sediment cover. (Photograph courtesy of Global Marine, Inc.)

THE OCEAN AND CLIMATE CHANGE

The importance of ocean processes in controlling climate has been studied since the 1960s. Interest in large-scale climatic variation was accelerated by threats of widespread food shortages. Global food supplies could be seriously disrupted by fluctuations in temperature and rainfall and so accurate prediction of future climatic variations was recognized as a vital need. Consequently, the World Meteorological Organization's GARP (Global Atmospheric Research Programme) studies processes that control weather and cause long-term climate changes.

Oceans and atmosphere interact, forming a single worldwide climate control system (Fig. 1-17). To learn the most important causes of climatic variation, scientists construct a continuous picture of global atmospheric conditions, using weather satellites, deep-sea buoys, and computers to collect and correlate the data. Past climatic variations are analyzed statistically, using deep-ocean fossil records up to hundreds of thousands of years old to map ocean surface temperatures during the last Ice Age and earlier. From an understanding of past variations, scientists formulate numerical models of simulated climatic conditions to test hypotheses and predict how future climates will develop. Increasingly powerful computers permit numerical models to simulate interactions between ocean and atmosphere under glacial and nonglacial conditions.

Figure 1-17

Schematic illustration of the components of the coupled atmosphere–ocean–ice earth climate system. Heat, moisture, and momentum are exchanged at the sea surface; transport of heat and moisture occurs in atmospheric circulation and ocean currents. Changes in any of their relationships may cause significant variation in other parts of the system. Mathematical models are constructed to explore these adjustment mechanisms quantitatively. (From U.S. Committee for the Global Atmosphere Research Program, 1975. *Understanding Climate Change: A Program for Action.* National Academy of Science, Washington, D.C.)

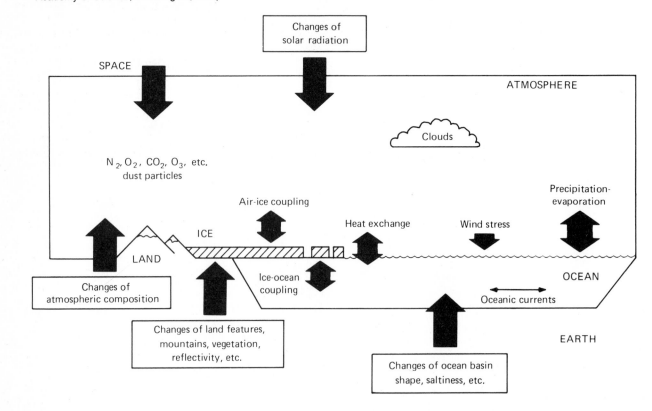

FUTURE TRENDS IN OCEAN SCIENCES

Interest in solving practical problems continues to dominate the ocean sciences in the 1980s. New instruments available to oceanographers have greatly increased capabilities to investigate ocean processes by direct observation. The need to picture large areas of the ocean nearly instantaneously, for example, has significantly changed the practice of oceanography. Oceanographers always relied on ships to take samples and measure ocean properties. But ships move slowly, 10 to 20 kilometers per hour. So only small areas could be studied or sampled at a time. And often the ocean was changing too rapidly to study this way. The problem is partially solved by having several ships work together and by anchoring floating meters to measure currents while the ships are engaged elsewhere. A carefully coordinated field program, or experiment, was designed to detect a circular current pattern called an *eddy*. Thus in the 1970s oceanographers were able to demonstrate that the ocean has highly energetic, nearly circular current patterns that resemble storms in the atmosphere. A single ship would not be able to carry out such an experiment. Such experiments have changed the way oceanographers work at sea.

Research submarines (see Fig. 1-18) have also changed our concept of the ocean floor and its processes. Submersibles permit scientists to observe directly small-scale processes on the ocean floor. In 1977 oceanographers diving on the submersible *Alvin* found entirely unexpected marine organisms growing on the midocean ridge near vents discharging hot waters. Waters are heated by circulating through recently erupted volcanic rocks. These waters support bacteria that contribute food for dense growths of large, gutless worms, up to 3 meters long, blind crabs, fishes, and clams as big as dinner plates. None of the details of the relationships of the vents and organisms could have been determined by working solely from surface ships. Furthermore, submersibles permit oceanographers to work at great depths and to put geophysical instruments and experimental devices on the ocean floor, just as astronauts placed instruments on the moon's surface. Detailed accurate measurements are necessary to work out the details of ocean floor processes. Improvements in remotely controlled samples and better underwater television systems will probably permit greater use of remote-controlled vehicles for ocean floor studies.

Remote sensing of the ocean surface from aircraft and satellites dramatically improved the capability to map surface ocean currents and water masses. Although satellites and aircraft have difficulty observing very far below the ocean surface (seawater is only relatively transparent to light), instruments on earth-orbiting satellites can meas-

Figure 1-18

Alvin, the U.S. submersible research vessel used to explore the deep-ocean floor.

ure ocean surface temperature, map surface waves and winds, measure ocean surface elevations, and detect color changes caused by the growth of phytoplankton (minute floating plants) and benthic plants in shallow water and by suspended sediment in river waters flowing into the sea. These surface and near-surface measurements provide snapshots of ocean processes on a scale that cannot be obtained by observations from surface ships.

Other remote-sensing techniques use sound to study the depths of the ocean below the surface layers visible to satellites and aircraft. Floats can be adjusted to remain at predetermined depths below the ocean surface and to move with the currents. A "pinger" on the float gives off distinctive signals that can be picked up by microphones on the ocean floor at island stations. It is possible to locate these floats and to map their movements through time, thereby deducing the movements of subsurface waters.

Other techniques also take advantage of seawater's transparency to sound. Sound sources transmit pulses that are detected by arrays of distant microphones. The paths followed by these pulses moving from the source to the receiver are modified by the temperature and salinity (salt content) of the waters between them. Using computers, oceanographers can calculate distributions and variability of temperature and salinity within the ocean waters traversed by these sound pulses. From such observations maps of the ocean waters can be prepared, showing the ocean much like a weather map represents the atmosphere.

The production of petroleum and natural gas, as well as sand and gravel, has increasingly moved offshore. Major oil fields are now producing in the hostile ocean environments of the North Sea between Great Britain and Norway and oil has been discovered in the Arctic Ocean off Alaska and Canada. Engineers require information about ocean conditions, such as waves, tidal currents, and floating ice, in order to design drilling and production platforms, plus the pipelines needed for offshore oil and oil fields.

Because gas and oil discoveries are being made in deeper waters, new production platforms taller than the Empire State Building are needed. These platforms, sitting on the ocean bottom, raise questions about the stability of the ocean floor as a foundation. Platforms have failed because of massive submarine slides and oil spills have occurred because of lack of knowledge about the ocean bottom. Such engineering concerns are dictating more detailed studies of ocean floor processes.

Finally, the largest volumes of potentially oil-bearing sediment deposits lie at the edges of the continents in waters 2 to 4 kilometers deep. Exploration—and possibly production—of these deposits requires new technologies and places increasing demands on ocean scientists.

LAW OF THE SEA CONFERENCE

Since European exploration and colonization of the newly discovered continents in the sixteenth century, the ocean has been largely free to all nations. Hugo Grotius (1583–1645) formulated the legal basis in the doctrine of *mare liberum* in 1635. For more than 300 years the ocean, outside of a narrow territorial sea extending 3 nautical miles (5 kilometers) from the shore, was freely used by all nations.

National interest in the potential of immense quantities of petroleum on the continental margins and metal-rich manganese nodules from the deep-ocean floor challenged this concept. Uncertainties about the legal status of these deposits greatly inhibited the investments necessary to exploit them.

In 1974 the United Nations convened a Conference on the Law of the Sea. The nations of the world negotiated their many conflicting needs in order to develop a consensus, often fuzzy, to divide the ocean among them. The treaty was completed in 1981. Its implementation will markedly change the legal status of the ocean.

First, the territorial sea was extended to 12 nautical miles (22 kilometers) from the shore. Within this zone the coastal state has rights and responsibilities equal to those it has over the land. Consequently, some major straits through which ships must pass in going from one ocean to another came within the territories of coastal states. Negotiations were necessary to ensure the rights of ships to navigate these areas.

More dramatic was the recognition of a 200-nautical mile (370 kilometers) Exclusive Economic Zone. In this zone the coastal state regulates fisheries and resource exploration and exploitation. About one-third of the ocean falls in the Exclusive Economic Zone of some coastal state. And many important seas—North Sea, Mediterranean, Gulf of Mexico, Caribbean—are totally divided among the coastal states.

Enormous areas of ocean thus come under national jurisdictions. The United States, for instance, assumed responsibility for an ocean area approximately 80% of its land area (see Fig. 1-19). New programs were required to regulate the fisheries on a much larger scale than ever tried before. And new—largely unknown—mineral resources were opened for future exploitation as their legal status was clarified, thus encouraging investment.

Figure 1-19

The 200-mile Exclusive Economic Zone of the United States is approximately 80% the size of the land area of the nation. It includes large areas of open ocean in the central Pacific.

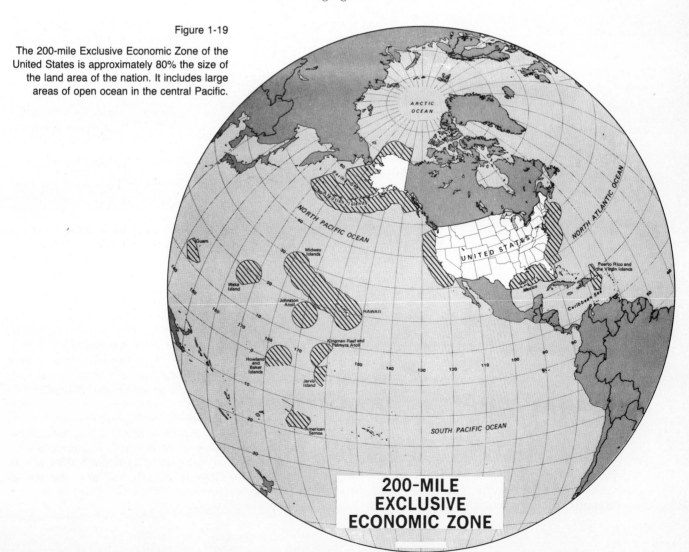

200-MILE EXCLUSIVE ECONOMIC ZONE

Many details were intentionally left fuzzy because no agreement was possible even after many prolonged sessions. Yet the legal regime of the world ocean has been radically changed. And the result will be widely felt not only by ocean scientists but also by those even remotely involved with the ocean.

REVIEW QUESTIONS

1. Explain the importance of the *Challenger* Expedition to modern oceanography.
2. List the principal contributions of M. F. Maury, H. H. Hess, and Benjamin Franklin.
3. Describe the principal contributions of Fridtjof Nansen to oceanography.
4. How has drilling the deep-ocean floor contributed to our knowledge of ocean basin history?
5. What changes has the UN Conference on the Law of the Sea caused in the legal status of the ocean near the continents?
6. List some instruments that have improved observations of ocean processes.

SUMMARY OUTLINE

History of oceanography—ocean studied to solve problems
   Explorers made charts of the ocean

Foundations of modern oceanography—early explorers knew little about the ocean
   James Clark Ross—explorer–scientist; charted waters near Antarctica
   Misconceptions delayed scientific progress—for example, deep waters stagnant; no life below 600 meters
   Laying of transatlantic cable led to discoveries
   Voyage of the *Challenger*—Sir Wyville Thomson; beginning of modern oceanography
      Studied marine life, ocean floors, ocean water properties
   Sir John Murray—published *Challenger* Reports, studied submarine geology

Early American oceanography—need to protect trade, passengers, aid shipping
   Matthew Fontaine Maury—Dept. of Charts and Instruments
      Began program of Wind and Current Charts
   Alexander Bache—Coast Survey
      Studied U.S. coastal waters, Gulf Stream
   Spencer F. Baird—Fish Commission
      Founded Woods Hole laboratory; *Albatross* made scientific cruises; studied fish populations

National expeditions
   Prince Albert of Monaco—used personal fortune for oceanographic research; established oceanographic institute in Paris

Scandinavian research—studied currents, temperature and salinity of water masses, in connection with fisheries research
International Council for the Exploration of the Seas—studies of plankton, currents, water masses
ICES-inspired research by Henry Bryant Bigelow—oceanography of Gulf of Maine
German Antarctic Expedition of 1911–1912; later *Meteor* Expedition
   Studied temperature, salinity of Atlantic Ocean; sounded ocean floor
Fridtjof Nansen drifted with *Fram* in pack ice
International Ice Patrol—established after sinking of *Titanic*
American research promoted with support from foundations—for instance, Rockefeller

International oceanography
   War-related research in 1940s—development of bathythermograph, research on waves and coastal conditions, sea-floor mapping
   Government sponsored research for national security, mineral and fuel resources, fisheries
   Bikini–Eniwetok studies of coral atoll structure and formation
   International Geophysical Year—mapped currents, studied tides
   Sea-floor spreading theory—coordinated studies of earth's heat flow, gravity, magnetic field, ocean basin topography
   Deep-Sea Drilling project drilled deep-ocean sediments, corroborated sea-floor spreading theory of moving continents, continuous formation of ocean floor

Future trends
New observing techniques and instruments—submersibles, satellites, computers, acoustic remote sensing
Offshore petroleum and mineral production

Law of the Sea Conference
Changed legal status of ocean areas
New boundaries, up to 200 nautical miles offshore

SELECTED REFERENCES

DEACON, MARGARET. 1971. *Scientists and the Sea, 1650–1900.* Academic Press, London. 445 pp. History of oceanography up to twentieth century.

DIETRICH, GUNTER, and others. 1980. *General Oceanography: an Introduction,* 2nd ed. Interscience, London. 626 pp. Comprehensive, technical-level treatise, emphasizing physical oceanography.

DRAKE, C. L., JOHN IMBRIE, JOHN A. KNAUSS, and KARL K. TUREKIAN. 1978. *Oceanography.* Holt, Rinehart, and Winston, New York. 447 pp. Intermediate level.

NELSON, S. B. 1971. *Oceanographic Ships: Fore and Aft.* U.S. Government Printing Office, Washington, D.C. 240 pp. Pictures and brief histories of important oceanographic ships, instruments, and oceanographers in the United States.

PICKARD, GEORGE L. 1979. *Descriptive Physical Oceanography: An Introduction,* 3rd ed. Pergamon Press, Elmsford, N.Y. 233 pp. Elementary-level text, emphasizing physical oceanography.

SCHLEE, SUSAN. 1973. *The Edge of an Unfamiliar World: A History of Oceanography.* Dutton, New York. 398 pp. A readable history of oceanography since the nineteenth century, with emphasis on American oceanography.

SCIENTIFIC AMERICAN. 1971. *Oceanography.* W. H. Freeman, San Francisco. 417 pp. Selected articles from *Scientific American.*

SHORE, ELIZABETH NOBLE. 1978. *Scripps Institution of Oceanography: Probing the Oceans 1936 to 1976.* Tofua Press, San Diego, CA. 502 pp. History of ocean research at one of the world's foremost oceanographic institutions.

SVERDRUP, H. V., MARTIN W. JOHNSON, and RICHARD H. FLEMING. 1942. *The Oceans: Their Physics, Chemistry, and General Biology.* Prentice-Hall, Englewood Cliffs, N.J. 1087 pp. Classic oceanography text; detailed and scholarly.

WEIHOUPT, JOHN G. 1979. *Exploration of the Oceans.* Macmillan, New York. 588 pp. General coverage, stressing geological features and history. Intermediate in difficulty.

2 THE OCEAN BASINS

New Horizon, a 52-meter (170-foot) oceanographic research vessel, was built for the Scripps Institution of Oceanography, University of California, San Diego, at a cost of 3.33 million dollars. This marks the first new ship to be added to the institution's fleet since acquisition of the 75-meter (245-foot) *Melville* in 1969. Scripps operates a fleet of four vessels and two research platforms.

R/V *New Horizon* is capable of performing a large variety of experiments in all fields of modern oceanography and in any part of the ocean except polar seas. Much of her work will be for marine fisheries and environmental studies in the eastern Pacific Ocean between northern California and the tip of Baja California. Her range is about 7,000 nautical miles (at 10.5 knots), and her endurance is 30 days. She carries 13 scientists and a crew of 12.

Funds to build the new ship were provided by the state of California, the first time the state has funded construction of a Scripps vessel. She was designed by Florida naval architect Rudolph F. Matzer 7 Associates, Inc., Jacksonville, and constructed by Atlantic Marine, Inc., Fort George Island, Florida. (Photograph courtesy of Scripps Institution of Oceanography, University of California, San Diego.)

The water-filled ocean basin is Earth's most distinctive feature. Most of the land is in the Northern Hemisphere (Fig. 2-1, Table 2-1), with the Pacific, Atlantic, and Indian ocean basins extending northward as three "gulfs" from the ocean-dominated area around Antarctica. Sometimes a fourth—the Antarctic or Southern ocean basin—is designated. Its northern limit is set at 40°S latitude, where opposing currents cause a downward flow of surface waters known as the *subtropical convergence*.

Relationships between oceanic depths and land elevations can be visualized as a *hypsographic curve* (Fig. 2-2) that shows two distinct levels of the earth's surface. On the average, land projects about 840 meters above sea level and ocean basins lie an average 3800 meters below it. But despite the great depth of the oceans and the enormous height of the mountains compared to the height of a human being, differences in elevation or depth for the earth as a whole are truly insignificant when compared to the dimensions of the earth. If the earth were a smooth sphere with the land planed off to fill the basins, it would be covered with water to a depth of 2430 meters. Viewed in this way, the ocean is a thin film of water on a nearly smooth globe.

Figure 2-1

Land and ocean are unequally distributed; most land occurs in the Northern Hemisphere while the Southern Hemisphere is dominated by ocean, as seen in these views of the earth from space. One is a view from above Antarctica. The other view is from a point above Europe.

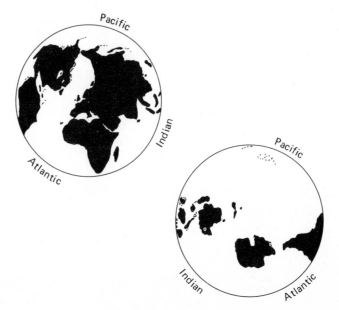

TABLE 2-1
Useful Data About the Earth and the Ocean

Size and Shape of the Earth

DIMENSIONS	MILES	KILOMETERS
Equatorial radius	3963	6378
Polar radius	3950	6357
Average radius	3956	6371
Equatorial circumference	24,902	40,077

Areas of the Earth, Land, and Ocean

PART OF EARTH	MILLIONS OF SQUARE MILES	MILLIONS OF SQUARE KILOMETERS
Land (29.22%)	57.5	149
Ice sheets and glaciers	6	15.6
Oceans and seas (70.78%)	139.4	361
Land plus continental shelf	68.5	177.4
Oceans and seas minus continental shelf	128.4	332.6
Total area of the earth	196.9	510.0

Distribution of Land and Water On the Earth's Surface*

HEMISPHERE	LAND (percent)	OCEAN (percent)
Northern	39.3	60.7
Southern	19.1	80.9

*After Sverdrup, Johnson, and Fleming, 1942. p. 13.

Volume, Density, and Mass of the Earth and Its Parts*

PART OF EARTH	AVERAGE THICKNESS OR RADIUS (km)	VOLUME ($\times 10^6$ km³)	MEAN DENSITY (g/cm³)	MASS ($\times 10^{24}$g)	RELATIVE ABUNDANCE (percent)
Atmosphere	–	–	–	0.005	0.00008
Oceans and seas	3.8	1370	1.03	1.41	0.023
Ice sheets and glaciers	1.6	25	0.90	0.023	0.0004
Continental crust†	35	6210	2.8	17.39	0.29
Oceanic crust‡	8	2660	2.9	7.71	0.13
Mantle	2881	898,000	4.53	4068	68.1
Core	3473	175,500	10.72	1881	31.5
Whole earth	6371	1,083,230	5.517	5976	

*From Holmes, 1965.
†Including continental shelves.
‡Excluding continental shelves.

Heights and Depths of the Earth's Surface

Land HEIGHT	FEET	METERS	Oceans and Seas DEPTH	FEET	METERS
Greatest known height: Mt. Everest	29,028	8848	Greatest known depth: Mariana Trench	36,200	11,035
Average height	2757	840	Average depth	12,460	3800

Figure 2-2

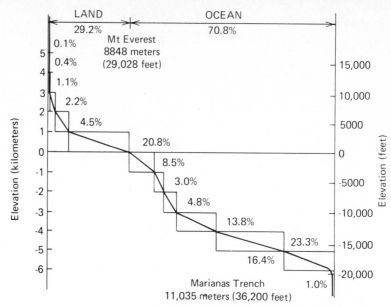

A hypsographic curve shows the relative proportions of the earth's surface at elevations above and below sea level. The length of each bar shows the relative proportion within each depth zone. (From Arthur N. Strahler, 1960. *Physical Geography*, 2nd ed. John Wiley & Sons, London.)

MAJOR OCEAN BASINS

Stretching about 17,000 kilometers (10,000 miles) across from Ecuador to Indonesia, the Pacific Ocean occupies more than one-third of the earth's surface (Fig. 2-4) and contains more than one-half of the earth's free water (see Fig. 2-3). Mountain building dominates Pacific shorelines, the mountains creating barriers between inland areas and ocean. Along the Pacific's entire eastern margin (the western coasts of the Americas from Alaska to Peru), rugged young mountain ranges parallel the coast.

The mountains inhibit direct access of rivers to the Pacific; consequently, relatively little freshwater or sediment is carried to it from the interiors of the continents. Continental margins are narrow because the steep slope of the mountains tends to continue below sea level. As a result, the central Pacific is little affected by the continents around it.

The *Pacific* is the deepest, the coldest, and by far the largest of the oceans (see Table 2-2). Paradoxically, it is also the least salty (see Table 2-3) even though it receives the least river water of any major ocean (see Table 2-4 and Fig. 2-4). Some of the Asian rivers, discharging into the marginal seas along the western side, are among the largest sources of sediment carried to the ocean (see Fig. 2-4).

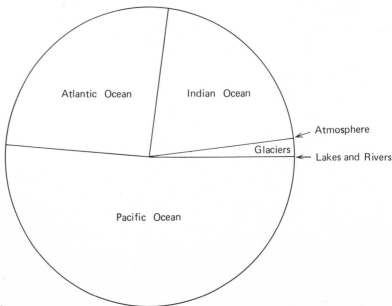

Figure 2-3

Mass of free water (in tons) on the earth's surface. Note the minute fraction of the total that occurs as freshwater lakes and rivers at any instant.

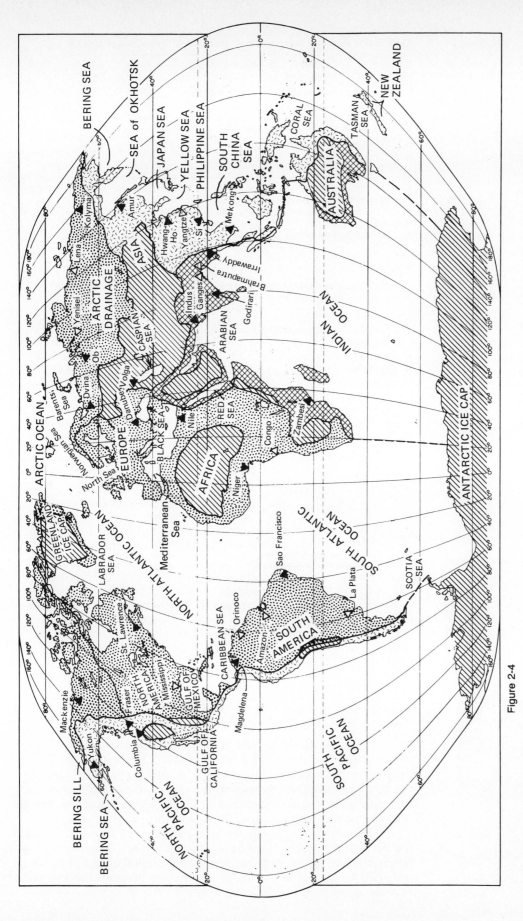

Figure 2-4

Map of the earth's surface showing the arbitrary boundaries drawn for the major oceans. Areas of the continents whose rivers drain into each of the ocean basins are shown, as well as some of the major rivers. Note the large area that drains into the Atlantic and Arctic oceans.

▽ Rivers discharging more than 15,000 cubic meters (525,000 cubic feet) per second

▼ Rivers discharging more than 3000 cubic meters (100,000 cubic feet) per second

Runoff to Atlantic Ocean and Arctic Sea

Runoff to Pacific Ocean

Runoff to Indian Ocean

No runoff to oceans

31

TABLE 2-2

*Surface and Drainage Areas of Ocean Basins and Their Average Depths**

OCEAN†	OCEAN AREA (millions of square kilometers)	LAND AREA DRAINED‡ (millions of square kilometers)	RATIO OF OCEAN AREA TO DRAINAGE AREA	AVERAGE DEPTH† (meters)
Pacific	180	19	11	3940
Atlantic	107	69	1.5	3310
Indian	74	13	5.7	3840

*From H. W. Menard and S. M. Smith, 1966. Hypsometry of ocean basin provinces. *Journal of Geophysical Research* **71**:4305.
†Includes adjacent seas. Arctic, Mediterranean, and Black seas included in the Atlantic Ocean.
‡Excludes Antarctica and continental areas with no exterior drainage.

TABLE 2-3

*Average Temperatures and Salinity of the Oceans, Excluding Adjacent Seas**

	TEMPERATURE (°C)	SALINITY (parts per thousand)
Pacific (total)	3.14	34.60
North Pacific	3.13	34.57
South Pacific	3.50	34.63
Indian (total)	3.88	34.78
Atlantic (total)	3.99	34.92
North Atlantic	5.08	35.09
South Atlantic	3.81	34.84
Southern Ocean†	0.71	34.65
World ocean (total)	3.51	34.72

After L. V. Worthington, 1981. *The water masses of the world ocean: Some results of a fine-scale census*, pp. 42–69. In B. A. Warren and C. Wunsch (Eds.). *Evolution of Physical Oceanography*. MIT Press, Cambridge, MA. 623 pp.
†Ocean area surrounding Antarctica, south of 55°S.

TABLE 2-4

*Water Sources for the Major Ocean Basins (centimeters per year)**

OCEAN	PRECIPITATION	RUNOFF FROM ADJOINING LAND AREAS	EVAPORATION	WATER EXCHANGE WITH OTHER OCEANS
Atlantic	78	20	104	6
Arctic	24	23	12	35
Indian	101	7	138	30
Pacific	121	6	114	13

*From M. I. Budyko, 1958. *The Heat Balance of the Earth's Surface,* Trans. N. A. Stepanova, Office of Technical Services. Department of Commerce, Washington.

The large Pacific islands, occurring along the western margin of the basin, are continental in structure. They extend from New Zealand in the south to Japan in the north, the largest being New Guinea. Consisting of continental-type rocks, these large islands are generally richer in industrial minerals and other natural resources than the smaller volcanic or carbonate islands, in the open ocean.

The *Atlantic Ocean* was the first explored by Europeans and it has been studied more thoroughly than any other. This ocean is relatively narrow, about 5000 kilometers wide (3000 miles). It forms an S-shaped basin that is the major connection between the two polar regions of the world ocean (Fig. 2-5).

Figure 2-5

Distribution of land and water in each 5° latitude belt. (Redrawn from M. Grant Gross, 1980. *Oceanography,* 4th. ed. Charles E. Merrill, Columbus, Ohio.)

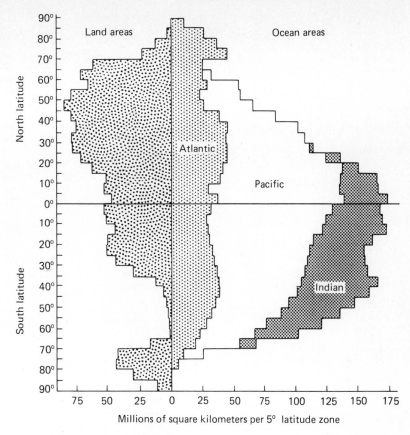

The boundary between the Atlantic and Indian oceans is arbitrarily drawn from the Cape of Good Hope at the southern tip of Africa and extends southward along longitude 20°E to Antarctica. The boundary between the Atlantic and Pacific oceans is drawn between Cape Horn at the lower tip of South America to the northern end of the Antarctic Peninsula. When the Arctic Ocean is included as part of the Atlantic, the northern boundaries of both the Atlantic and the Pacific are made to meet at the Bering Strait between Alaska and Siberia—a shallow area known as the Bering sill forms an easily recognizable boundary between the two oceans. (Although oceanographers draw such boundaries with ease, the ocean pays little attention to them. There is substantial movement of water across all these artificial boundaries, including the Bering sill.) Defined this way, the Atlantic Ocean extends from the shore of Antarctica northward across the North Pole to about 65°N on the other side of the earth. Neither of the other two major oceans extends so far in a north–south direction (Fig. 2-5).

The Atlantic is relatively shallow, having an average depth of only 3310 meters (10,800 feet), the result of an abundance of continental shelf in the ocean, shallow marginal seas, and the presence of the shallow Mid-Atlantic Ridge. The Atlantic Ocean receives large amounts of riverborne sediment. There are relatively few islands in the Atlantic Ocean. Greenland, the world's largest island, is actually a part of the North American continent, isolated by the relatively narrow channel formed by Davis Strait and Baffin Bay. Most other Atlantic islands are tops of volcanoes, many of them located on the Mid-Atlantic Ridge.

The Atlantic Ocean receives large amounts of freshwater from rivers. The Amazon and the Congo, the world's two largest rivers (see Fig. 2-4 and Table 2-5), flow into the equatorial Atlantic, together

33

TABLE 2-5

*River Discharges to Ocean Basins**

RIVER	DISCHARGE (10⁹ cubic meters per year)	(cubic miles per year)	OCEAN BASIN
Amazon	5550	1330	Atlantic
Congo	1250	300	Atlantic
Yangtze	688	165	Pacific
Ganges	590	141	Indian
Yenisei	550	130	Arctic
Mississippi	550	180	Atlantic
Lena	490	118	Arctic
St. Lawrence	446	107	Atlantic
Mekong	350	84	Pacific
Columbia	178	42.7	Pacific
Yukon	165	39.6	Pacific
Nile	117	28	Atlantic
Colorado	5	1.2	Pacific
Total world	30,000	7200	

*Data from U.S. Geological Survey, 1964.

they carry about one-quarter of the world's river discharge to the ocean. Other large rivers flow into the marginal seas that are common around the North Atlantic. The Mississippi River, draining much of North America, for example, discharges into the Gulf of Mexico and the St. Lawrence flows into another small gulf. Also, freshwater running from several large rivers into the Arctic Ocean eventually enters the Atlantic.

Despite its connections to both polar regions and the large amount of freshwater discharged into it, the Atlantic is the warmest (average temperature 3.99°C) and saltiest (average salinity 34.92 parts per thousand, usually written 34.92‰) of the world ocean basins (average ocean temperature is 3.51°C, average salinity 34.71‰). This apparent paradox is a result of the large number of marginal seas adjoining the Atlantic. The Mediterranean Sea, in particular, exerts a strong influence on the entire Atlantic. Mediterranean waters are warmed and exposed to dry atmospheric conditions. Substantial amounts of water are lost by evaporation from the sea surface so that the seawater becomes more saline. This warm, salty water flows out into the Atlantic and is detectable at mid-depths over thousands of kilometers from the Strait of Gibraltar (see Chapter 6). And the entire North Atlantic Ocean has the highest average temperature (5.08°C) and salinity. (35.09‰) of any major ocean basin.

The *Arctic Ocean* is a landlocked arm of the Atlantic including all waters north of Eurasia and the North American continent. It is unique in several respects: shallowly submerged continental margins form one-third of its floor, it is almost completely surrounded by land, and—especially characteristic—much of it is covered by sea ice during part or all of the year.

The Arctic as a whole is quite shallow, largely because of its wide continental shelves. On North America, between Alaska and Greenland, the shelf is only 100 to 200 kilometers (60 to 300 miles) wide, but offshore from western Alaska and Siberia it is much wider, 500 to 1700 kilometers (300 to 1000 miles) across. During much of the year it is unnavigable because its surface is covered by *pack ice* (sea ice in thick chunks that freeze together, break apart, and freeze again to form a solid surface) to a depth of 3 to 4 meters (10 feet or more).

In the central Arctic this cover is permanent, although it melts to a minimum thickness of about 2 meters by the end of August.

Being almost entirely surrounded by land, the Arctic Ocean is greatly influenced by river discharge and has low salinity. Large amounts of sediment on the bottom of Arctic basins are the result of extensive erosion of surrounding continents by glaciers during the past several million years. Much of this material was deposited as glaciers melted and lost the sediment they accumulated while advancing over the land. Rivers discharge considerable sediment into the Arctic region.

The *Indian Ocean* lies primarily in the Southern Hemisphere, extending from Antarctica to the Asian mainland at about 20°N (see Fig. 2-4). Its Pacific boundary runs through Indonesia and extends from Australia through Tasmania and southward to Antarctica along longitude 150°E. It is the smallest of the three major ocean basins, triangular in shape, with a maximum width of about 15,000 kilometers between South Africa and New Zealand. The continental shelves around this ocean are relatively narrow. The large amount of sediment that has been deposited in the northern part of the basin makes it intermediate in depth.

Three of the world's major rivers (Ganges, Indus, Brahmaputra) discharge large amounts of water and sediment into the Indian Ocean (see Table 2-5) along its northern margin, also the location of its only marginal seas. Of these the Red Sea and Arabian Gulf have the greatest effect on the Indian Ocean as a whole because they are areas of intensive evaporation. Most of the water discharged into the Indian Ocean drains from the Indian subcontinent; little comes from Africa.

There are relatively few islands in the Indian Ocean—Madagascar being the largest one. It is continental in composition, as are several shallow banks in the ocean basin—all apparently fragments of an ancient continental block. A few volcanic islands and several groups of carbonate islands, including atolls, exist.

OCEANIC RIDGES AND RISES

The deep-ocean floor begins at about 4 kilometers below sea level and is shown very simply in Fig. 2-6. Midocean ridges and rises stand out clearly above the deep basins, covering 23.1% of the earth's surface (Table 2-6). Compare this figure with the land above sea level, which covers 29.2% of the earth. The midoceanic ridge and rise system includes high-relief segments (shown schematically in Fig. 2-7) and low-relief segments (Fig. 2-8).

The rugged high-relief *Mid-Atlantic Ridge* stands 1 to 3 kilometers above the nearby basin floor. It is 1500 to 2000 kilometers across and has a prominent steep-sided rift valley (Fig. 2-9). The valley is usually 25 to 50 kilometers wide and 1 to 2 kilometers deep. It is bordered by steep mountains, also formed by faulted blocks; these peaks come to within 2 kilometers of the ocean's surface. The closest terrestrial analog to conditions on the Mid-Atlantic Ridge is found in the East African Rift Valley shown in Fig. 2-10. Both apparently have similar origins.

Away from the central valley of the Mid-Atlantic Ridge, bottom topography is still quite rugged. There is a great deal of faulting and insufficient sediment accumulation to bury the irregular terrain. Faults cut across the ridge and offset it in a series of segments, as indicated in Fig. 2-11. One of these transverse faults is the Romanche Trench, an important link for the flow of deep-ocean water from the western to the eastern Atlantic basin. There are numerous and frequent shallow-focus earthquakes (occurring within 70 kilometers of the earth's surface) along the Mid-Atlantic Ridge as well as volcanoes and volcanic islands, including the Azores, Iceland, Ascension, and

Figure 2-6

Ocean bottom provinces showing relationships among submerged continental margins, midoceanic rises, ocean basins, island arcs, and volcanic (aseismic) ridges. Note that the trenches, the deepest parts of the ocean bottom, occur along the basin margins.

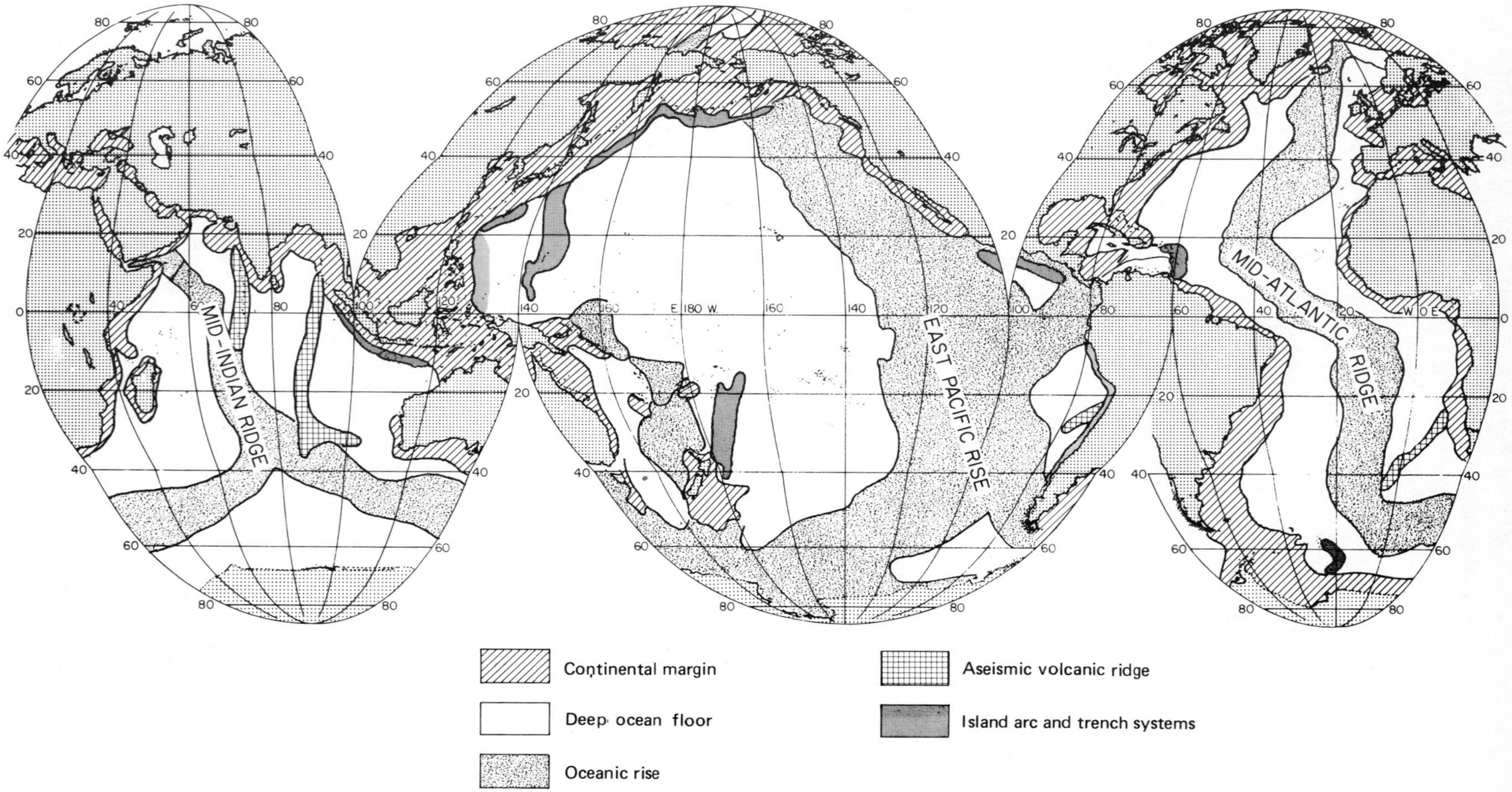

MID-ATLANTIC RIDGE

EAST PACIFIC RISE

MID-INDIAN RIDGE

Continental margin

Deep ocean floor

Oceanic rise

Aseismic volcanic ridge

Island arc and trench systems

TABLE 2-6

*Physiographic Provinces of the Ocean**

OCEAN†	SHELF AND SLOPE (percent)	CONTINENTAL RISE (percent)	DEEP-OCEAN FLOOR (percent)	VOLCANOES AND VOLCANIC RIDGES (percent)	RISE AND RIDGE (percent)	TRENCHES (percent)
Pacific	13.1	2.7	43.0	2.5	35.9	2.9
Atlantic	19.4	8.5	38.0	2.1	31.2	0.7
Indian	9.1	5.7	49.2	5.4	30.2	0.3
World ocean	15.3	5.3	41.8	3.1	32.7	1.7
Earth's surface	10.8	3.7	29.5	2.2	23.1	1.2

*After H. W. Menard and S. M. Smith, 1966. Hypsometry of ocean basin provinces. *Journal of Geophysical Research* 71:4305.
†Includes adjacent seas—for example, Arctic Sea included in Atlantic Ocean.

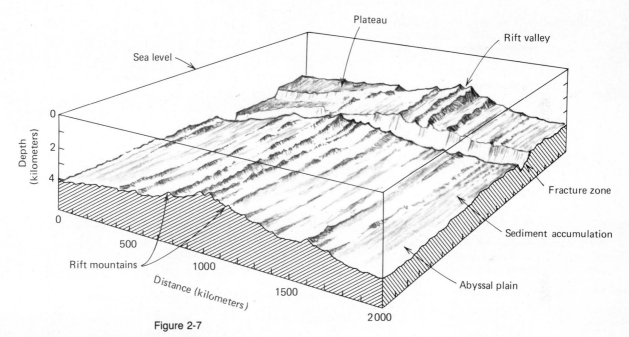

Figure 2-7

Schematic representation of ocean bottom topography of a portion of the mid-Atlantic ridge, typical of a rugged, faulted, midocean rise. Note the rough topography along the fracture zone where two segments of the rise are offset. The fault along which movement occurred is a transform fault. Sediment forming the abyssal plain is deposited on the margin of the rise.

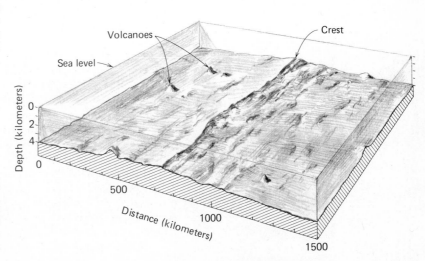

Figure 2-8

Schematic representation of ocean-bottom topography along a portion of the East Pacific rise, an example of a broad, low rise with relatively subdued relief.

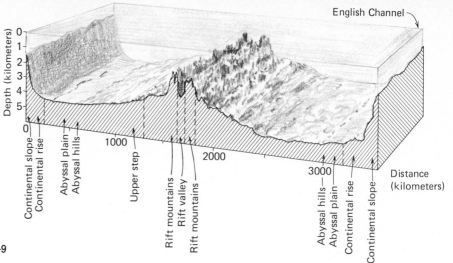

Figure 2-9

Diagrammatic view of the North Atlantic's ocean floor between Canada and Great Britain. Note the great exaggeration in vertical relief.

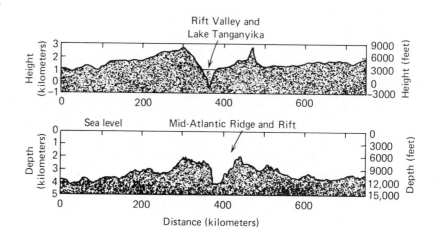

Figure 2-10

Profiles of the rift valley of the mid-Atlantic ridge and the Tanganyika rift of East Africa. Note the similarity in form and size of the two features. (After Holmes, 1965.)

Tristan de Cunha; the distribution of these features is indicated in Fig. 2-12. The Mid-Atlantic Ridge extends into the Arctic Ocean.

The equivalent to the Mid-Atlantic Ridge, the *East Pacific Rise*, lies near the eastern margin of the Pacific basin. Unlike the Mid-Atlantic Ridge, it is a vast, low bulge on the ocean floor, approximately equal in size to North and South America combined. The rise stands about 2 to 4 kilometers above the adjacent ocean bottom and varies from 2000 to 4000 kilometers in width. Intersecting the North American continent in the Gulf of California, its continuation reappears off the Oregon coast and extends into the Gulf of Alaska. The two segments of the rise system are connected by the San Andreas fault in California. Movement along this fault caused the San Francisco earthquake of 1906.

In the Indian Ocean, the high-relief *Mid-Indian Ridge* intersects Africa–Asia in the Red Sea area (see Fig. 2-6) and the active system apparently extends under the East African Rift Valley. On the south, the Mid-Indian Ridge system branches to join the system of active ridges and rises that circle Antarctica. Near Madagascar, a low section of the ridge system permits deep water to move between the deeper parts of the Indian Ocean and the Atlantic Ocean. The Mid-Indian Ridge separates two distinctly different ocean areas.

A narrow belt of frequent shallow-focus earthquakes runs along

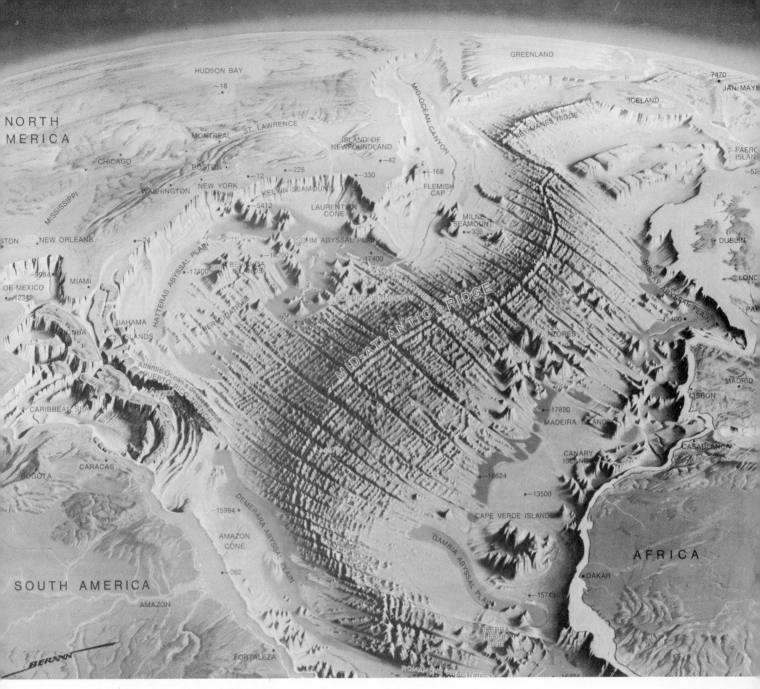

Figure 2-11

Artist's rendition of the bottom of the North Atlantic Ocean. Depths are noted in feet. (Painting by Heinrich Berann; courtesy Aluminum Corporation of America.)

the crests of oceanic rises (Fig. 2-12). In the North Atlantic, where the ridge is about 1500 kilometers wide, the belt of earthquake *epicenters* (point on the earth's surface above the earthquake focus) is about 150 kilometers wide. Detailed surveys show that these earthquakes coincide closely with the axis of the rift valley. Indeed, earthquakes have been used to pinpoint active ridge systems in little-known ocean areas. Such information is especially useful when other information is lacking or when relationships are unclear, as in the Arctic Ocean.

Midoceanic ridge systems are also sites of high heat flow by conduction from the earth's interior—about 1 to 3 microcalories per square centimeter per second (cal cm^{-2} sec^{-1}). These areas of high heat flow do not correspond as closely to the location of rift valleys as do the sites of earthquakes. Instead areas of unusually high heat

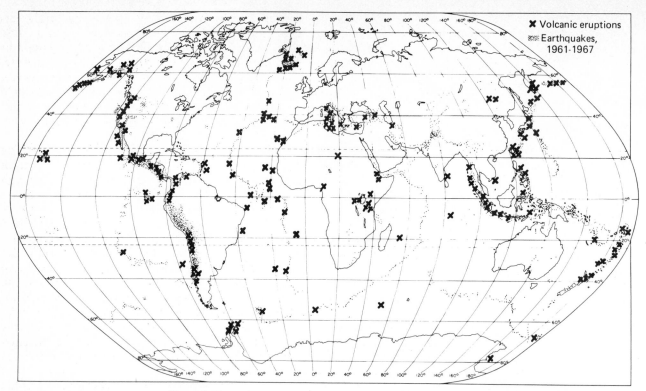

Figure 2-12

Distribution of earthquakes (1961–1967) and active volcanoes. Note the correspondence between earthquake epicenters and the boundaries of the large crustal blocks shown in Fig. 2-17. (Compiled from Bryan Isacks, J. Oliver, and L. R. Sykes, 1968. Seismology and the new global tectonics. *Journal of Geophysical Research* 73:5855–5900; and from A. Holmes, 1965.)

flow (up to 8 cal cm^{-2} sec^{-1}) occur in narrow bands, caused by narrow intrusions of molten rocks. Hydrothermal vents discharge hot waters in areas of recent volcanic activity.

The shallowness of midocean areas affects sediment accumulation. Distinctive types of carbonate-rich sediment, not found in deep-ocean basins, accumulate in shallow areas, particularly in the small, fault-bounded basins of the midoceanic rises. Furthermore, the presence of barriers to the movement of sediment along the bottom prevents continental sediment from reaching many ocean bottom areas. The segment of the East Pacific Rise off the coast of Washington and Oregon, for instance, blocks the movement of sediment along the bottom so that none of it reaches the deep North Pacific ocean floor. Thus the accumulation of sediment on the landward side of the ridge has caused it to be hundreds of meters shallower than the seaward flank of the rise.

DEEP-OCEAN FLOOR   The *deep-ocean floor* occupies about 29.5% (Table 2-6) of the earth's surface. It is covered by sediment deposits hundreds of meters thick, much of which has accumulated particle by particle so that sediment layers are draped over preexisting topography just as a snowfall blankets and blurs the sharp features of the land. Near continents, on the other hand, thick turbidity current deposits can completely bury original topography, forming smooth abyssal plains.

Oceanic features persist much longer than land features because there is no deep-ocean analog to the vigorous erosion by glaciers, wind, or rivers that takes place on land. Nor is there the chemical alteration of rock by water, atmospheric gases, and plant action of a terrestrial environment. Submarine volcanoes formed more than 25 million years ago are still very much in evidence—for instance, as foundation for atolls at the northwestern end of the Hawaiian Islands chain. Rocks have been recovered from submarine volcanoes that apparently formed about 80 million years ago.

Ocean bottom topography is dominated by the fracture zones that extend for hundreds or thousands of kilometers on either side of midocean ridges and rises. In the Pacific Ocean these bands of volcanoes and mountainous terrain are 100 to 200 kilometers wide. Each consists of individual ridges and troughs that are several hundred kilometers long and a few tens of kilometers wide. Some have cliffs or ridges up to a few kilometers above the nearby ocean floor. The greatest depths in the central Pacific Ocean (more than 6 kilometers) occur along these fracture zones.

In some areas, fracture zones offset continental margins. The sudden bends in the coasts of Africa and South America are thought to be caused by a major fracture zone. This zone can be traced through the deeper ocean basin where it is marked by the Romanche Fracture Zone.

Low *abyssal hills* (less than 1000 meters high) cover about 80% of the Pacific deep-ocean floor and about 50% of the Atlantic, including portions of the Mid-Atlantic Ridge. They are probably also abundant over much of the Indian Ocean basin, although data there are more sketchy. These hills have an average relief of about 200 meters, with diameters of about 6 kilometers, and are the most common topographic feature on the earth's surface. Some appear to be small volcanoes covered with a thin sediment layer that is slightly thicker in the valleys than on the hills themselves. Many apparently form by faulting of oceanic crust caused by movements of the sea floor at the midocean ridges.

Some immense areas of exceedingly flat ocean bottom, called *abyssal plains,* lie near continents (Fig. 2-13). These areas are among the flattest portions of the earth's surface. They have slopes of less than 1 in 1000, equivalent to a slope of 1 meter per kilometer. Areas of comparable flatness are found in the High Plains area of the midcontinental United States, where the surface slopes about 1.5 in 1000. Abyssal plains commonly occur at the seaward margin of the deep-sea fans that make up the continental rise with the adjacent abyssal plain.

Most abyssal plains are covered with thick sediment deposits derived from the continents, to which we attribute their nearly flat topography. Deep-sea channels a few meters deep, with leveelike margins, cut across them. These channels are extensions of deep canyons across continental margins through which sediment is carried into the deep ocean. For instance, Cascadia Channel in the northeast Pacific extends more than 500 kilometers from the base of the continental slope near the mouth of the Columbia River and crosses through a gap in the midoceanic rise, finally extending out onto the

Figure 2-13

Block diagram showing the Atlantic Ocean floor in the North American basin.

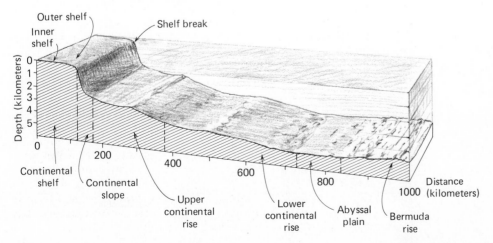

adjacent Tufts Abyssal Plain. Similar channels have been described in the Atlantic and Indian oceans.

Abyssal plains are especially common in marginal seas, such as the Gulf of Mexico, Caribbean, and the many marginal basins of the Pacific borderland. The Sigsbee Abyssal Plain in the Gulf of Mexico has been explored extensively; it lies at the base of the Mississippi Cone, built by sediment coming from the Mississippi River. Sediment that escapes deposition within the river delta or on the cone is deposited on the abyssal plain along with a minor amount coming from the Campeche Bank to the south. During the last glacial (Wisconsin) stage sediment accumulated on the plain at the rate of about 60 centimeters per thousand years. In the past 10,000 years (since the retreat of the ice) sediment has accumulated more slowly, about 8 centimeters per thousand years.

The greatest ocean basin depths occur in *trenches* (Table 2-7). Deepest is Challenger Deep, a part of the Marianas Trench (see Fig. 2-5), where a depth of 1035 meters has been recorded. Obtaining accurate depth measurements in trenches is not easy and depth figures are subject to revision as new techniques are developed or old ones improved. In general, the trenches are tens of kilometers wide and about 3 to 4 kilometers deeper than the surrounding ocean bottom. Individual trenches are thousands of kilometers long.

Trench sides consist of a series of steep-sided steps, suggesting that extensive faulting has occurred and caused blocks of the earth's crust to subside. The result is a narrow, deep, V-shaped trench that is almost invariably flat bottomed due to sediment deposits. The largest negative gravity anomalies on the earth's surface are associated with trenches and result from the downbuckling of the crust beneath them.

Other troughs that are generally less deep than the trenches and not connected with island arcs occur on the ocean bottom. Some deep areas are apparently associated with fracture zones that cut the mid-ocean rises and ridges, such as the Romanche Trench in the Atlantic.

TABLE 2-7

*Characteristics of Trenches**

TRENCH	DEPTH (kilometers)	LENGTH (kilometers)	AVERAGE WIDTH (kilometers)
Pacific Ocean			
Kurile–Kamchatka Trench	10.5	2200	120
Japan Trench	8.4	800	100
Bonin Trench	9.8	800	90
Marianas Trench	11.0	2550	70
Philippine Trench	10.5	1400	60
Tonga Trench	10.8	1400	55
Kermadec Trench	10.0	1500	40
Aleutian Trench	7.7	3700	50
Middle America Trench	6.7	2800	40
Peru–Chile Trench	8.1	5900	100
Indian Ocean			
Java Trench	7.5	4500	80
Atlantic Ocean			
Puerto Rico Trench	8.4	1550	120
South Sandwich Trench	8.4	1450	90
Romanche Trench	7.9	300	60

*After R. W. Fairbridge, 1966. Trenches and related deep sea troughs. In R. W. Fairbridge (Ed.), *The Encyclopedia of Oceanography,* pp. 929–938. Reinhold Publishing Corporation, New York.

MARGINAL OCEAN BASINS

Marginal ocean basins are large depressions in the ocean bottom lying near continents. They are separated from the open ocean by submarine ridges, islands, or parts of continents. Despite their proximity to continents, marginal basins are usually more than 2 kilometers deep and so their bottom waters are partially isolated from ocean waters at comparable depths outside the basin.

These small seas are, in a sense, transitional between continents and deep-ocean basins, as their location would suggest. Studies of their crustal structure has shown that they commonly have a typical oceanic crust overlain by thick sediment deposits. Being near the coast, they receive the discharge of many large rivers. The sediment brought by these rivers is deposited in the basins, resulting in smooth sediment deposits many kilometers thick. These deposits contain about one-sixth of all known ocean sediments.

Nearly all marginal ocean basins are located in areas where crustal deformation or mountain building is active. They may eventually be changed and incorporated into the adjacent continental block. Three kinds of associations are typical of marginal ocean basins. The most common is that of a basin associated with island arcs and submarine volcanic ridges that partially isolate surface waters and may completely isolate the subsurface waters. The many small basins around Indonesia and along the Pacific margin of Asia are examples of this association.

A second type of marginal sea lies between continental blocks—for instance, the Mediterranean Sea between Europe and Africa, the Black Sea (now almost completely landlocked), and the Gulf of Mexico and Caribbean Sea between North and South America (see Fig. 2-4). The Arctic Ocean (Fig. 2-14) may be considered a marginal sea.

The third type is the long, narrow marginal sea formed where oceanic rises are breaking up continental blocks. The Red Sea (shown in Fig. 2-15) and the Gulf of California are examples of such basins.

Figure 2-14

Map of the Arctic Ocean showing the edges of the continental shelves (200-meter contour) and the edges of the continents (2000 meters). The 4000-meter contour outlines the deep-ocean floor.

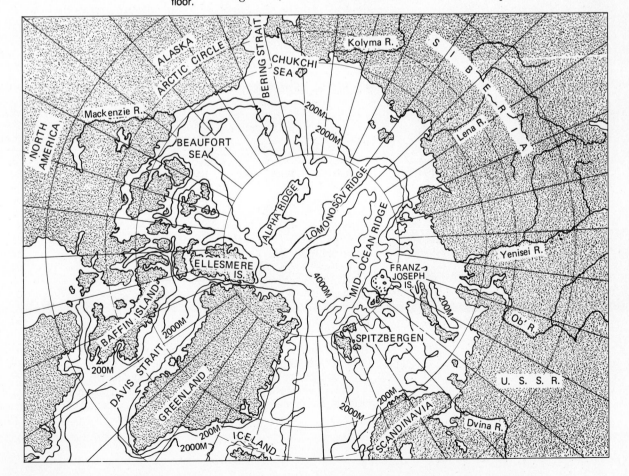

Figure 2-15

The Red Sea with the Gulf of Aqaba and the Gulf of Suez. The dark band of the right is the Nile Valley and the river can be seen as the light meandering line. Note the straight coastlines on either side of the Red Sea. The absence of clouds is an indication of the region's aridity. The photograph was taken from Gemini XII at an altitude of 280 kilometers (175 miles); portions of the spacecraft are visible on the right side. (Photograph courtesy NASA.)

Because of the proximity of large landmasses, marginal seas are subjected to more changeable weather conditions than occur in the open ocean. In high latitudes, such as the Sea of Okhotsk, marginal seas are cold enough during the winter months to freeze over. In midlatitudes, where the world's deserts are located, marginal seas like the Mediterranean Sea and Arabian Gulf are strongly evaporated so that their waters are much warmer and saltier than the world ocean average.

Because of their marked response to seasonal weather changes, water circulating through marginal seas has a recognizable influence on adjacent ocean basins. As noted, this is especially true for the Atlantic Ocean, which has many marginal seas and is a relatively small ocean basin.

In Europe and North America shallow seas occupy depressed areas at the edge of continental blocks, such as Hudson Bay in Canada. The North Sea of northern Europe is another such flooded continental area, leaving only England and her associated islands above sea level at the present time.

CONTINENTS AND CONTINENTAL MARGINS

The continents are four great landmasses—Eurasia–Africa, the Americas, Antarctica, and Australia—isolated in the midst of the world ocean. Although joined by "land bridges," such connections tend to be narrow and easily flooded by even small changes in sea level. For example, land connected Alaska to Siberia during the last Ice Age more than 20,000 years ago, but the connection was broken when melting of the glaciers freed the North American and Eurasian continents of ice and raised sea level so that continental margins were flooded.

The continental margin is the submerged edge of a continental block and consists of the *continental shelf, slope,* and *rise* (see Fig. 2-16). The continental shelf is the submerged top of a continent's outer edge and begins at the shoreline. This part of the continent has been shaped by the ocean except where mountains occur at the edge of the continent. There the shaping force has been the processes that formed the mountains. From the shoreline the shelf slopes gently toward the *continental shelf break* at an average depth of about 130 meters. Off Antarctica, however, the shelf break is more than 350 meters deep. At the shelf break, the gently sloping shelf gives rise to the much steeper continental slope.

On the average, the continental shelf is about 70 kilometers wide. Around much of the Pacific it is relatively narrow, only a few tens of

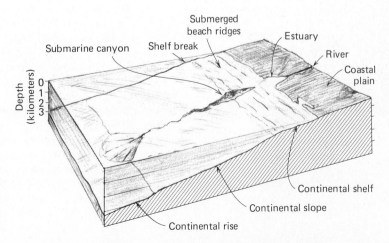

Figure 2-16

Schematic representation of the continental margin showing the continental shelf and its relationship to the coastal plain, the continental slope, and continental rise. Note that a submarine canyon cuts across the shelf break and extends to the base of the continental slope, where it joins the continental rise.

kilometers across. The widest shelves occur in the Arctic Ocean. A broad continental shelf consists of the sculptured sediment deposits shed by a continent, particularly where there has been no major mountain building at the margin over hundreds of millions of years—in other words a passive margin.

The *coastal plain* is the equivalent of the continental shelf. Where the coastal plain is absent and mountains form the coastline, as in southern California, the continental margin is rugged, broken by submarine ridges, banks, and basins. Continental shelves near glaciated coasts are also typically rugged. These shelves tend to be wide and contain basins or deep troughs more than 200 meters deep, generally paralleling the coast. Broad ridges often divide these basins. The Gulf of Maine off the New England coast is such an area.

Some troughs cut across the shelf and connect with deep, flooded indentations called *fjords*. Such troughs can be quite large, such as the trough at the entrance to the Strait of Juan de Fuca connecting the Pacific Ocean with Puget Sound and the Strait of Georgia. Topography of these shelf areas was formed by glaciers during periods of lowered sea level and has not been modified greatly by presently active marine processes.

On unglaciated continental shelves, the bottom topography tends to be much smoother, gently rolling and subdued as a rule. Elongate low sand ridges (sometimes containing gravel) generally parallel the shore. The ridges, formed by infrequent violent storms that stir waters to the bottom and redistribute sediments, are up to 10 meters thick.

The continental shelf break is quite uniform in depth, typically about 130 meters over most of the ocean. The shelf break is thought to have formed when sea level stood at its lowest during Pleistocene glacial times. At that time the shoreline was at the edge of the present continental shelf, which was then a coastal plain. Estuaries and lagoons occurred along the edge, the estuaries occupying the sites where present submarine canyons indent the shelf break. Traces of ancient lagoons are not readily discernible, although they must have been common in many areas. After that time, as glaciers melted, the ocean began to rise. The present level was reached about 3000 years ago; since then there have been relatively minor changes in sea level. In some formerly glaciated areas, the land has risen in response to the melting of glaciers.

Relatively shallow, canyonlike shelf valleys extend from the mouths of estuaries across the continental shelf except in areas where sediment discharged from the estuary has obliterated them. These valleys, a few tens of meters deep, were formed by tidal currents that flowed in and out of the estuary during the migration of sea level.

In tropical oceans there are often coral reefs on the continental shelf. The coral reefs commonly occur on the outer portion of the shelf and are separated from the land by a channel paralleling the coast.

Continental slopes are the relatively narrow, steeply inclined submerged edges of continental blocks. On an ocean-free earth they would be the most conspicuous boundaries on the earth's surface. Although their total relief is substantial, ranging from 1 to 10 kilometers (0.6 to 6 miles high), continental slopes are not precipitous. They average about 4°, ranging from 1 to 10° of slope, either quite straight or gently curving. The most spectacular slopes occur where a continental block borders a deep trench. On the western coast of South America the peaks of the Andes Mountains reach elevations of about 7 kilometers whereas the adjacent Peru–Chile Trench is 8 kilometers deep. This adds up to a total relief of 15 kilometers (9 miles) within a few hundred kilometers. Continental slopes with trenches at their base are common in the Pacific and account for about half the world's continental slopes. They are the active margins and do not have thick sediment deposits.

Faulting on the slope often forms small basins that trap sediment, changing the local topography from a complex of steep-walled basins to small flat-floored basins that gradually become filled and smoothed over, eventually disappearing. With accumulation of a sufficient thickness of sediment, the original outlines of the continental slope are completely buried. Margins of many continental blocks have numerous buried ridges and ancient coral reefs, now covered by sediment deposits. Many of these filled basins are prime locations for petroleum accumulation.

Thick sediment accumulations cover most of the world's passive continental slopes except along the Pacific Ocean. When sediment deposits become too thick and the slopes too steep, the deposits may slump. Consequently, large, spoon-shaped scars form on the edge of the slope, plus a mass of deformed sediment at the base of the slump. This situation probably is fairly common and may account for many of the hills on the lower parts of continental rises.

Submarine canyons are large features commonly found on continental slopes. Typically they are V-shaped, follow curved paths, and have tributary channels leading into them. Some are as large as the Grand Canyon of the Colorado River. A few submarine canyons are obviously associated with large rivers on the continent—for instance, offshore from the Hudson and Congo rivers. In these cases, the upper portions of the valley could have been cut during times of lowered sea level when a river flowed through the canyon; so their origin posed no problem. How, then, explain the portion of a canyon that swept out to the edge of the continental slope where it merged into much shallower channels with low levees on each side? These channels often run for tens or hundreds of kilometers across the fan of sediment that forms the base of the slope. Many canyons have no obvious connection to present-day rivers, although they may have been associated with rivers no longer in the region. Some canyons were cut through sedimentary rocks that are relatively soft. Others were found to be cut through granite. Some sediment commonly occurs on canyon floors.

These canyons are usually cut by *turbidity currents,* flows of sediment-laden water that move along the ocean bottom carrying material from the continental shelf to the deep-ocean floor. The process of canyon cutting is still active in many areas, especially where the continental shelf is relatively narrow and sediment can be readily transported to the head of the canyon. Such canyons have been studied closely in southern California, where their upper portions lie quite close to the shore.

Continental rises (Fig. 2-16) are accumulations of sediment at the base of continental slopes. They cover the transition between the continent and the ocean basin. Low hills are present, perhaps formed by slumping of sediment from the continental slope. Channels a few meters deep, apparently routes of sediment movement, cross the rise. These channels are often connected to submarine canyons.

Continental rises are apparently formed by sediments carried by turbidity currents and deposited at the base of the slope. Deposition is caused by the reduction in current speed when it flows out onto the gently sloping ocean floor. Initially each canyon has its fan-shaped deposit of sediment at its mouth, similar to mountain streams flowing onto a flat desert floor. Eventually individual fans build sideway and coalesce to form a continuous sediment wedge. Sediments are also eroded and redeposited by strong currents, called *contour currents,* that flow along the ocean bottom across the continental rise.

Typically the continental rise is made of sediment deposits several kilometers thick and tens of kilometers wide. One estimate of the total volume of sediment in the world ocean gave the following results:

Continental shelf 1.2 billion cubic kilometers (33%)
Continental slope and rise 1.6 billion cubic kilometers (45%)
Deep-ocean bottom 0.8 billion cubic kilometer (22%)

These data indicate that about 45% by volume of the sediment in the ocean lies in the continental rises at the base of continental slopes. Altogether deposits on the continental margin account for about 78% of the total recognized sediment deposits in the world ocean.

VOLCANOES

Volcanoes and volcanic islands are among the most conspicuous features of the ocean floor. Projecting usually 1 kilometer or more above the surrounding sea floor, they have a variety of shapes, as on land. It is estimated that there are at least 10,000 volcanoes in the Pacific alone, often grouped into provinces covering up to 10 million square miles. Some, like the Hawaiian Islands, form long chains; other groups are more circular. Considering that the Pacific is about half the world ocean, there may be 20,000 or more volcanoes scattered across the ocean bottom. We have no data to indicate how many former volcanoes are now buried by later flows of volcanic material or covered by sediment.

Once formed, volcanoes persist for millions of years on the ocean floor. They are alternately active for extended periods, during which a volcanic cone builds up, and then dormant. Between eruptions volcanoes often subside slightly as crust and mantle adjust to the added load of volcanic rock. Periods of activity last for millions to tens of millions of years. Certain areas of the southwestern Pacific Ocean appear to have undergone an extended period of volcanic activity during which immense volumes of lava poured out onto the ocean floor, beginning about 70 million years ago and lasting perhaps 30 million years. After cessation of volcanic activity, the region gradually sank. In such areas of subsidence, there is often a compensating low rise of the ocean bottom beyond the depression. Many extinct volcanoes are known to occur in other ocean basins—for instance, the Bermuda Islands (Fig. 2-17).

The andesitic island arcs associated with subduction zones are sometimes relatively simple and sometimes quite complex. The Aleutian Islands southwest of Alaska, for instance, are a single chain with a trench on the outward, or seaward side, as shown in Fig. 2-18. West of the Aleutians, the Kamchatka Peninsula consists of a group of similar volcanoes, but here a platform of older, greatly altered rocks is exposed above sea level, as it is in the Japanese Islands to the south. Probably such older, deformed rock underlies many island arcs, but it is submerged and not directly observable.

The complexity of some island-arc systems is illustrated by the Indonesian area (Fig. 2-19), where several chains of active or recently active volcanoes form complexly twisted arcs. The Java (Indonesian) Trench has several associated island groups, including the island of Timor, one of the few places on Earth where ancient deep-ocean sediments are exposed on land. The West Indies of the western Atlantic is another example of a complex island-arc system. Deep-ocean sediments are exposed on the West Indian island of Barbados.

The Fiji Islands east of Australia provide a case history of changes in volcanic island composition that occur after an island arc has moved away from a trench system. The Fiji group was formerly

Figure 2-17

The Bermuda Islands are solidified sand dunes built on a wave-eroded extinct volcano. The shallow top of the volcano (now forming a bank) was exposed during periods of lowered sea level and carbonate sands were blown across the bank, forming the dune, now changed to soft, crumbly limestone. Reefs built primarily of calcareous algae can be seen in the lower portion of the photograph. (Photograph courtesy Bermuda News Bureau.)

Figure 2-18

Aleutian Islands, a relatively simple island-arc system with trench on the convex, seaward side. These islands are built of volcanic rock from ancient eruptions. Some of the volcanoes are still active.

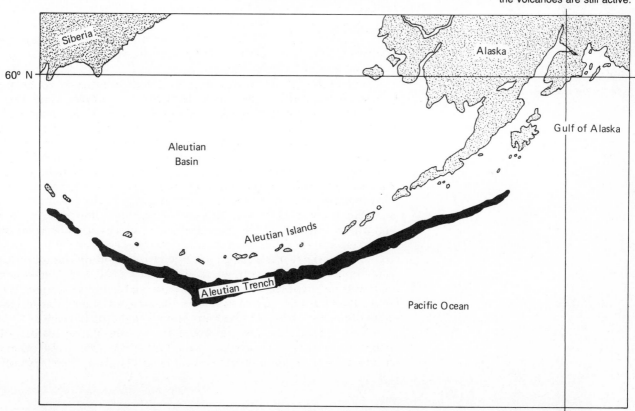

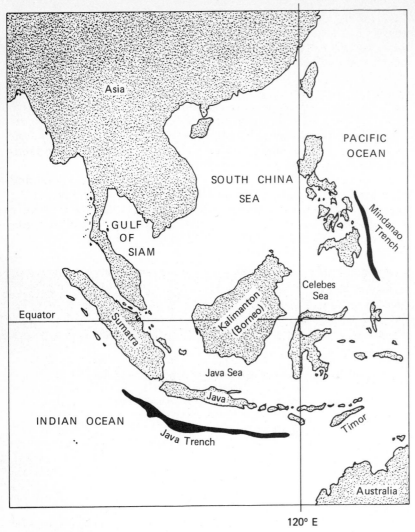

Figure 2-19

Indonesia, a complex island-arc system. Active
volcanoes occur on the larger islands. Some
smaller islands near the trench are built of
deformed sediment. The island of Timor is one
of the few places where deep-sea sediment is
exposed on land.

associated with the Kermadec–Tonga trench system and hence was
andesitic in composition. Subsequently the islands moved westward,
away from the subduction zone. Volcanism since then has become
increasingly basaltic, more like that of an ocean island chain.

Volcanic lava commonly erupts at temperatures between 900
and 1200°C. It flows for a while and then solidifies, first at the surface
and finally throughout the flow. Further cooling turns the lava into
volcanic rock.

There are two modes of volcanic eruption in the ocean. Some
lavas form tranquil flows in which the surface cools first, forming an
insulated cover for the still-molten material in the interior of the flow.
When this type of lava flows into the ocean, it forms *pillow lavas* (Fig.
2-20), rounded masses of volcanic rock. High pressure at oceanic
depths seems to favor such tranquil flows, for they appear to be quite
common features of the deep ocean. The other type of marine volcanic eruption is explosive, producing large amounts of volcanic ash,

small pieces of volcanic rock, and glass formed by quenching of molten rock (Fig. 2-21).

Not all volcanic activity forms volcanoes. Large areas on the ocean bottom are smooth plains of volcanic rock. Apparently lava flowed out in large volumes, covering previous topography over extensive areas of the ocean bottom and leaving a bare rock surface, now slightly covered by sediment. Probably these lavas were quite fluid at first. There may have been several centers of eruption, perhaps long breaks forming linear fissure eruptions. These *archipelagic plains* (or *aprons*) surround volcanic groups or islands and extend tens or hundreds of kilometers from the volcanoes themselves.

Numerous ancient Pacific islands were eroded to sea level and then submerged to depths of 1 to 2 kilometers by subsidence of the ocean floor. They are known as *guyots*. The fact that they were once near the sea surface is demonstrated by the remains of shallow-water organisms and wave-rounded cobbles on their flat tops.

Figure 2-20

"Toothpaste" and "pillow" lava photographed at 1015 fathoms by cameras aboard the submersible *Alvin* in the axial valley of the mid-Atlantic ridge. These lava formations are probably less than 10,000 years old. The remotely controllable sampling arm is picking up a rock sample. (Photograph courtesy Woods Hole Oceanographic Institution.)

(a)

(b)

Figure 2-21

Eruption of Kovachi, a submarine volcano in the Solomon Islands, South Pacific, in October 1969, caused shock waves and water eruptions approximately 30 seconds apart (a). Explosions ejected water and steam 60 to 90 meters into the air. The sea surface was discolored for 130 kilometers (80 miles). (Photograph courtesy Dr. R. B. M. Thompson and Smithsonian Institution Center for Short-Lived Phenomena.) A new island (b) was built by Kovachi in March 1970. A similar island formed by an eruption in 1961 has since been eroded by waves and is now completely submerged. (Photograph courtesy Smithsonian Institution Center for Short-Lived Phenomena.)

REVIEW QUESTIONS

1. Draw a cross section showing the continental margin, the deep-ocean floor, and the adjacent continent. Label and indicate typical depths for major features.

2. Draw a cross section of a midocean ridge. Indicate where volcanic activity and hydrothermal vents are most likely to occur. Label significant features.

3. On an outline map of the world, sketch the locations of earthquake belts, active volcanoes, midocean ridges, and subduction zones. Discuss any correlations you find.

4. Briefly describe the geographic relations of the three major ocean basins. Indicate how and where they connect and where and by what they are separated.

5. Discuss the origins and potential economic significance of marginal ocean basins.

SUMMARY OUTLINE

Ocean—three gulfs; extending northward from Antarctica

Hypsographic curve—two levels: continents and ocean basins

Major ocean basins
    Pacific Ocean—largest basin, contains more than 50% of surface water
        Bordered by mountains
        Many marginal seas; volcanoes and volcanic islands; large continental islands
    Atlantic Ocean—long, narrow; connects North and South polar regions
        Relatively shallow, many marginal seas
        Receives about 68% of world's drainage
    Arctic Ocean—nearly landlocked arm of the Atlantic
        Shallow, wide continental shelves
        Low salinity, covered by ice most of the year
    Indian Ocean—Southern Hemisphere; smallest of three major basins
        Major rivers discharge into northern section

Oceanic rises—a single system throughout all ocean basins
    Mid-Atlantic Ridge—rugged, faulted, with rift valleys
    East Pacific Rise—low bulge, little faulting, no rift valleys
    Mid-Indian Ridge—resembles Mid-Atlantic Ridge
    Characteristics of oceanic rises:
        Sites of oceanic crustal formation
        Comparable in size to continents
        Many volcanoes and shallow-focus earthquakes
        High heat flow, usually in narrow bands

Deep-ocean floor
    Sediment cover typically accumulated particle by particle, covering older topography
    Fracture zones cross ocean basins
    Abyssal hills abundant

Abyssal plains—among flattest portions of earth's crust
    Occur at seaward margin of continental rise
    Crossed by deep-sea channels, avenues of sediment dispersal
    Common in marginal seas—for instance, Gulf of Mexico, Caribbean
Salt domes—in former isolated areas of evaporation of seawater; may trap fossil fuels
Trenches—greatest depths in ocean
    Negative gravity anomalies
    Faulted sides, sediment in bottom
    Sites of crustal subduction

Marginal ocean basins
    Formed by tectonic processes
    Usually have distinct oceanographic characteristics; highly variable
    Many have thick sediment deposits

Continents and continental margins
    Continental shelf—average 70 kilometers wide; rough if region was glaciated, smooth if adjacent to smooth coastal plain
    Formed by migrations of shoreline, due to changing sea level
        Shelf break averages 130 meters depth
    Continental slope—narrow, steeply inclined edge of continents
        Formed by faulting and slumping of rocks in many places
        Cut by submarine canyons probably formed by turbidity currents
    Continental rise—sediment accumulations at base of continental slope contain about 45% of identified sediment in ocean

Volcanoes
    Volcanoes—1 kilometer or more in height; tops may form islands
    Andesitic island arcs—simple or complex
    Archipelagic plains—flat, smooth lava flows; 80% of volcanic rock in Pacific
    Guyots—subsided, flat-topped seamounts

SELECTED REFERENCES

BALLARD, R. D., and JAMES G. MOORE. 1977. *Photographic Atlas of the Mid-Atlantic Ridge Rift Valley*. Springer-Verlag, New York. 114 pp. Photographs of submersible photographs of submarine volcanic features.

HOLMES, ARTHUR. 1965. *Principles of Physical Geology*, 2nd ed. Ronald Press, New York. 1288 pp. Advanced treatment of geological principles.

MacDONALD, GORDON A. 1972. *Volcanoes*. Prentice-Hall, Englewood Cliffs, N.J. 510 pp. Comprehensive study of volcanic activity.

SHEPARD, FRANCIS P. 1973. *Submarine Geology*, 3rd ed. Harper & Row, New York. 517 pp. Advanced discussion of marine geology.

3 THE EARTH AND THE OCEAN BASIN

An overhead view of the Skylab space station cluster in Earth orbit as photographed from the Skylab 4 Command and Service Modules (CSM), during the final "fly-around" by the CSM before the return home. The space station is contrasted against a cloud-covered Earth. Note the solar shield which was deployed by the second crew of Skylab and which shades the Orbital Workshop (OWS) in the area from which a micrometeroid shield has been missing since the cluster was launched on May 14, 1973. The OWS solar panel on the left side was also lost on workshop launch day. Inside the CM when this picture was made were Astronaut Gerald P. Carr, commander; Scientist-Astronaut Edward G. Gibson, science pilot; and Astronaut William R. Pogue, pilot. The crew used a 70mm hand-held Hasselblad camera to take the photo. (Photograph courtesy of National Aeronautics and Space Administration.

The earth's surface—both continents and ocean floor—is slowly but constantly changing. New oceanic crust forms by volcanic eruptions along the midocean ridges and slowly moves away. Over hundreds of millions of years the crust cools and moves toward the trenches, where it slips under the adjacent crust and is assimilated into the underlying mantle. In the process, the continents are moved and ocean basins change their shape. Thus the distribution of ocean basins and continents described in Chapter 2 is a snapshot of a slowly changing earth's surface. In Chapter 3 we examine the structure of the earth and study the processes that cause these profound changes.

STRUCTURE OF THE EARTH

Water covers two-thirds of the earth's surface but accounts for only a small fraction of its total mass. Basically the planet consists of several concentric spheres (Fig. 3-1).

Core is rich in iron and nickel and is very dense. The inner core is solid, the outer core liquid, with a transition zone between.

Mantle is a less dense layer of rock. The lower mantle is thought to be essentially rigid, but the upper mantle or *asthenosphere* is softer and can flow very slowly.

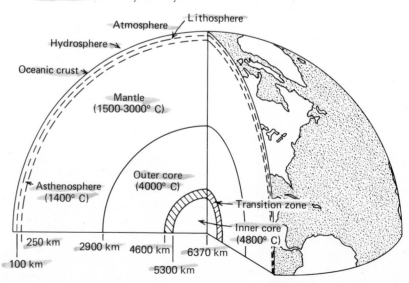

Figure 3-1

Simplified view of the earth's layered structure. [After P. J. Wyllie, 1975. The earth's mantle. *Scientific American* 232(3):50–63.]

Lithosphere is a less dense, rigid outer shell, including granitic *continental crust* as well as basaltic *oceanic crust*. The crust is underlain by and apparently fused with a layer of heavier rock, presumably solidified upper mantle material. These layers are separated at a boundary known as the *Mohorovičić discontinuity* (or Moho) that occurs under continents at a depth of about 25 to 40 kilometers and under ocean basins at 5 to 10 kilometers, typically 7 kilometers (Fig. 3-2).

Hydrosphere is all free water, including the water that has been removed from the ocean for millions of years by incorporation in sediments, sedimentary rocks, and glaciers. Eventually this water also returns to the ocean, which contains about 98% of the earth's free water (Table 3-1).

Atmosphere is primarily nitrogen, oxygen, and inert gases, mixed with variable quantities of water vapor and carbon dioxide.

TABLE 3-1

*Volume of Major Water Reservoirs of Earth's Crust and Mantle**

RESERVOIR	VOLUME	
	(millions of cubic kilometers)	(percent)
Earth's surface		
Ocean	1370	98.0
Sedimentary rocks (waters actively exchanged)	4	0.3
Glaciers	24	1.7
Lakes	0.23	0.02
Atmospheric water vapor	0.014	0.001
Rivers	0.0012	0.00009
Total	1398.245	100
Mantle (0.5% water)	13,000	

*After M. I. Lvovich, 1973. *The World's Water*. Mir Publishers, Moscow.

Continental crust forms large blocks, which do not end at shorelines (see Fig. 3-2). The submerged portion of continental shelf underlies a narrow coastal ocean zone adjacent to the continents (Fig. 3-3). Beyond there is a sharp dropoff known as the continental slope, ending at about 2 kilometers depth (Fig. 3-2), the approximate outer limit of continental blocks. The juncture between continental mass and ocean floor is covered by thick deposits of sediment eroded from the surface of the continent. This is the continental rise, which slopes gently to about 4 kilometers below sea level. Most deep ocean basins range in depth from about 4 to 6 kilometers, with some narrow trenches near Pacific ocean basin margins dropping to as much as 11 kilometers below sea level.

Continents are thick accumulations (35 to 40 kilometers) of granitic rocks containing abundant silicon and aluminum and having a mean density of about 2.8 grams per cubic centimeter. Because of their relatively low density, they tend to remain above sea level. Some continental rocks are nearly 4 billion years old. Wind and water constantly erode continental surfaces, transporting soils and silts to the ocean floor, where they accumulate as sediment deposits. Mountain building continually renews the continents.

Oceanic crustal rocks contain more iron and magnesium than do the continents; some of its basaltic rocks have densities on the average of about 3 grams per cubic centimeter. A layer about 7 kilometers thick underlies the ocean basins. Direct sampling has been limited by the hundreds of meters of sediment and thousands of

Figure 3-2

Schematic representation of the relationships among continental blocks, oceanic crust, lithosphere, and upper mantle on a passive, Atlantic-type margin. Note that the more buoyant continental mass is supported by a thickened lithosphere below and a corresponding depression of the more dense but plastic mantle layer beneath.

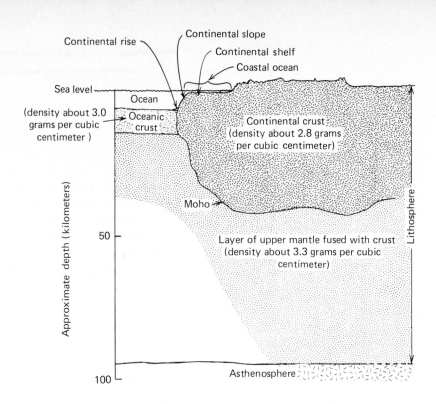

Figure 3-3

Map showing relationship of landmasses to the 2000-meter depth contour that marks the approximate outer limits of the continental blocks and the 4000-meter contour that generally outlines the deep-ocean floor and the midocean ridge system. [From J. E. Williams (Ed.), 1963. *Prentice-Hall World Atlas,* 2nd ed. Prentice-Hall, Englewood Cliffs, N.J.]

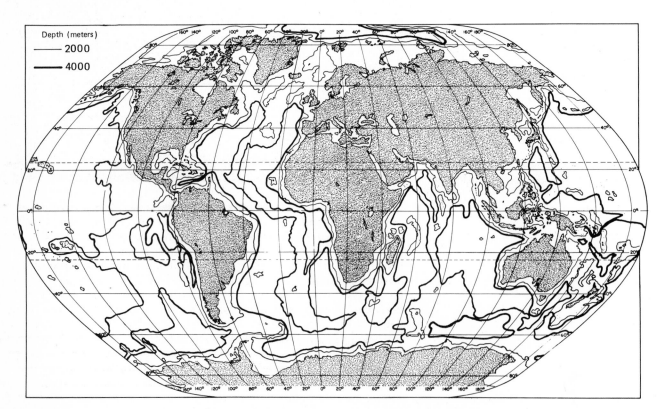

meters of seawater that overlie the oceanic crust. Deep-ocean drilling, however, has supplied valuable information; samples have been taken from considerable depths below the ocean floor (Fig. 3-4). Long cores of oceanic sediment provide information on ocean basin history and composition comparable to that available to geologists working on land.

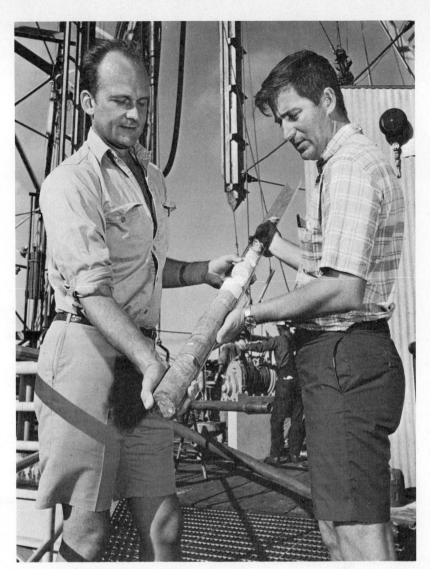

Figure 3-4

A core of altered igneous rock (solidified from a melt) overlain by white marble (an altered sedimentary rock). The core was obtained by drilling operations in water 4700 meters deep in the western North Atlantic Ocean. This rock sample, taken after drilling 450 meters of sediment almost 100 million years old, provides a direct means of investigating ocean-bottom structure and ocean-basin history. (Photograph courtesy Deep-Drilling Project, Scripps Institution of Oceanography, under contract to the National Science Foundation.)

The lithosphere "floats" in the less rigid asthenosphere below (see Fig. 3-2). The surface of the continental blocks extends above the oceanic crust because granite is less dense than basalt. Continents have their "roots" deeply submerged in the heavier upper mantle, just as an iceberg floats in equilibrium with the slightly denser water medium by having most of its bulk below the waterline. The earth's crust behaves like a series of large blocks, each of which "floats" in isostatic equilibrium with the plastic or deformable deeper layer, a concept known as *isostacy*. Therefore the shape of the Moho below a continent is a mirror image of the surface relief, exaggerated several times. Furthermore, isostatic adjustment takes place whenever a portion of the crust becomes heavier or lighter. If an appreciable mass is removed from part of a continent or ocean basin, that part rises to maintain a constant ratio of exposed to submerged material. For example, parts of Scandinavia and Canada are rising as much as 1 meter per century because of melting of glaciers within the past 10,000 years that had previously weighted down the crust. Conversely, whenever a volcano builds a mountain on the ocean floor, its great mass achieves isostatic equilibrium by sinking deeper into the underlying material. Many volcanoes that once formed islands have sunk below sea level through that process.

The earth's interior is studied primarily by indirect techniques. One of the most informative is the study (seismology) of the speed and direction of waves generated by earthquakes and explosions as they change (refract) by passing through the earth's layers. Waves change their speed and direction according to the density, elasticity, and flow properties of the rocks through which they pass. Certain types of waves *(compressional waves)* cause the medium to compress and expand; such waves pass through liquids and solids. Others, known as *shear waves,* distort the shape of a material but do not change its volume. These waves can pass through high-strength, nondeformable substances but not through liquids. Both types of waves may be slowed when passing through rocks of lowered density, through hotter layers, or through low-strength materials that can flow or be very slowly deformed. High pressures increase wave speed, but tension—the stress that occurs when materials are being pulled—lowers it. The Moho, for example, was recognized from seismic observations; compressional waves from the earthquakes abruptly change speed there, from 6.1 to 6.7 kilometers per second in the crust to about 8.1 kilometers per second in the lithosphere below.

Waves are partly bounced back or *reflected* [Fig. 3-5(a)] from a *discontinuity* (boundary marked by a change in seismic properties) between layers and partly transmitted. If wave velocity is different in the new layer, the waves are bent or *refracted* [Fig. 3-5(b)].

Two other physical properties are used to study the earth's structure and composition—its gravity and the flow of heat from its interior. Variations of mass anywhere on the earth cause disturbances in the normal gravity field known as *gravity anomalies.* More dense materials cause a locally greater gravitational force whereas less dense rocks weaken the gravity above them. The size and depth of a subsurface mass anomaly can be calculated if its density is known.

Heat flows from continental and ocean crust at an average of about 1.5 microcalories per square centimeter per second (about 1/3000 the average amount of energy from the sun at the ocean surface).

Figure 3-5

(a) In seismic-reflection surveys, a ship drops an explosive charge at some preset point. Sound pulses reflected from different layers are picked up by sensitive microphones—*hydrophones*—towed by a ship. These signals are recorded and analyzed to determine depths to the reflecting rock units.
(b) Seismic-refraction studies commonly use two widely separated ships—one to release the explosive charge, the second to receive and record the signals. Sound waves from the source travel along the layers and their boundaries. They move up through the overlying layers to be received at a second ship.

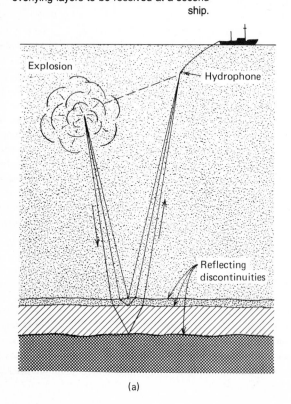

(a)

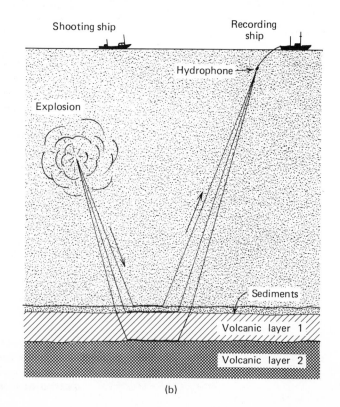

(b)

TABLE 3-2

Heat Flow by Conduction Through the Earth's Crust*

AREA	HEAT FLOW (microcalories per square centimeter per second)
Continents and continental shelves	
Old continental areas (older than 1700 million years)	1.0
Younger continental areas (younger than 250 million years)	1.8
Oceanic areas	
Old oceanic crust (older than 110 million years)	1.2
Young oceanic crust (younger than 4 million years)	3.6

*Data from J. G. Sclater, C. Jaupart, and D. Galson, 1980. The heat flow through oceanic and continental crust and the heat loss of the earth. *Reviews of Geophysics and Space Physics* 18(1):269–312.

Heat flow by conduction is determined by inserting sensitive thermometers at several depths below the surface and measuring the increase in temperature with depth. This value, multiplied by the thermal conductivity of the rock, gives the outward heat flow per unit area and unit time. Heat is produced in rocks by the decay of radioactive isotopes of thorium, uranium, and potassium, which contributes to the earth's high interior temperatures. And some of the heat may remain from the time of the earth's formation, 4.5 billion years ago.

Note, however, that heat flow is highly variable (Table 3-2). It is greatest near midocean ridges and lowest in the oldest oceanic crust near the deep trenches that border the Pacific basin. Molten rock rising through the asthenosphere from deep within the mantle causes locally high heat flows—for instance, at midocean ridges. Very high flows can be explained by the presence of pockets of molten rock (called *magma chambers*) close beneath the ocean floor. Much of the heat dispersed by cooling molten rock is released by seawater circulating through the newly formed crust. Nearly half the heat is removed by circulating seawater. The hot waters are discharged by hydrothermal vents that have been observed in a few locations on active midocean ridges.

Clues as to when and how a surface feature formed are found in the iron-rich minerals of rocks, which become magnetized in the direction of the earth's magnetic field at the time of their formation. Oceanic crust, for example, is formed when iron-rich lavas flow upward from the earth's interior and solidify at the surface. As molten rock cools below the 600 to 700°C temperature range, iron minerals become permanently aligned with magnetic north at that time.

The earth's magnetic field, created by circulation of core materials, has reversed itself many times. This process has occurred at intervals on the order of half a million years (Fig. 3-6), with each reversal probably taking a few thousand years. During periods of so-called normal magnetic orientation the north-seeking end of a compass needle would behave as it does now. But during periods of reversed magnetism the north-seeking end would point south. The permanent magnetism of a single rock can be determined in the laboratory. On a larger scale, instruments are towed by aircraft or ship to map the earth's magnetic field over wide areas.

Such a survey over a portion of the Atlantic Ocean basin revealed the striped pattern shown in Fig. 3-7: Each stripe corresponds to a period of normal or reversed magnetism, as illustrated on the chart

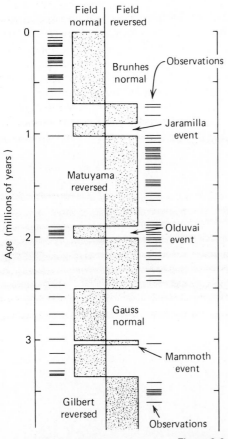

Figure 3-6

Time scale for reversals of the earth's magnetic field. The polarity (direction of the magnetic field) is shown for the past 4 million years. (After A. Cox, R. R. Doell, and G. B. Dairymple, 1964. Reversals of the earth's magnetic field. *Science* 144:1537–1543.)

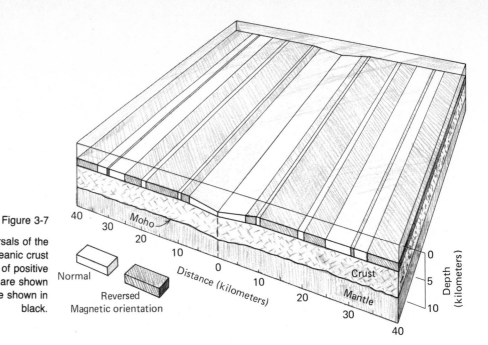

Figure 3-7

A striped pattern caused by reversals of the earth's magnetic field as the oceanic crust forms at the midocean ridge. Areas of positive anomalies (unusually high intensity) are shown in white; areas of low intensity are shown in black.

in Fig. 3-6 and each stripe on the ocean floor can be dated according to the chart. Thus the ocean floor acts like a gigantic tape recorder, the tape being the newly formed crust and the signal being the earth's magnetic field at the time of crustal formation. Data for interpretation of the pattern have been extracted from lavas of known ages on the continents. Comparison of the pattern in the ocean's magnetic stripes with this dated pattern in terrestrial lavas permits ages of various parts of the ocean bottom to be estimated. Estimates based on this evidence show that about 50% of the deep-ocean bottom has formed during the past 70 million years.

SEA-FLOOR SPREADING

The lithosphere and soft mantle below are linked and slowly moving to shape continents and ocean basins. The lithosphere may be compared to an eggshell cracked into six major plates and several smaller ones (Fig. 3-8). Divisions between plates occur at three kinds of boundaries [Figs. 3-9 and 3-10(a)]: at *midocean ridges* and *rises* (see Fig. 3-9); at *trenches* that ring the Pacific Ocean basin; and at major cracks in the crust known as *faults*. The plates move away from each other (*divergence*), toward each other (*convergence*), or past each other, at rates shown in Fig. 3-9, in response to very slow movements in the mantle below.

Figure 3-8

Artist's views of the earth showing its major crustal plates and how the boundaries cut across ocean basins and continents: (1) Pacific plate, (2) Antarctic, (3) Nazca, (4) South American, (5) African, (6) North American, (7) Cocos, (8) Caribbean, (9) Eurasian, (10) China, (11) Australian.

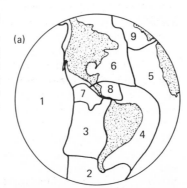

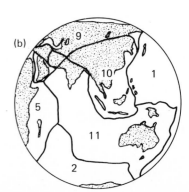

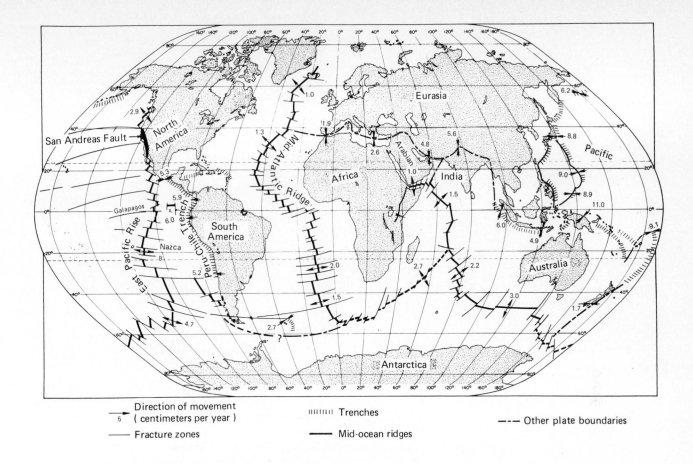

→	Direction of movement (centimeters per year)	IIIIIIII	Trenches	– – –	Other plate boundaries
5					
——	Fracture zones	———	Mid-ocean ridges		

Figure 3-9

Connected oceanic rises form a midocean ridge system that generally bisects ocean basins. Deep trenches occur mainly around Pacific margins. Lithospheric plates, including continental and oceanic crust, generally move away from midocean ridges and rises and toward trenches where subduction takes place. Faults occur where plates move past each other.

New crust is formed when molten rock or *magma* from the earth's interior reaches the surface at midocean ridges, typically at depths of about 2.5 kilometers. Through continuing volcanic activity, about 12 cubic kilometers of new oceanic crust form each year as magma fills the cracks formed when plates are pulled or pushed apart. Reduced pressure near faults or cracks permits rocks to melt at lower temperatures and the faults provide routes for lava to move to the surface.

Age of the crust increases with distance from the ridge crest (Fig. 3-11). Rate of crustal generation is determined from the amount of material created per unit time. The process does not proceed at a uniform rate at all points along a ridge; consequently, the sections of the rise are offset at intervals (see Fig. 3-9). Where sections have moved past each other, a *transform fault* connects the ends of the offset [Fig. 3-10(a)].

Since the earth is not expanding, oceanic crust must be destroyed at about the same rate as it is formed. Crust is destroyed by being *subducted*, or drawn down, and resorbed into the mantle at the trenches. Continental crust as such is not formed at midocean ridges and rises, nor is it subducted at trenches, but the relative position of continental masses changes as a result of lithospheric plate movements. Over the entire earth, the average rate of ocean floor movement from ridges and rises toward trenches is about 2.5 centimeters per year. The Atlantic Ocean, for instance, has only two small trench systems but an extensive midocean ridge. Consequently, it is growing wider by about 5 to 10 centimeters per year. In the process, Europe, Africa, and the Americas are being dragged farther from the Mid-Atlantic Ridge. Plates move at rates as fast as 10 centimeters per year in the Pacific basin and as slow as 1 to 2 centimeters per year in the Atlantic (Fig. 3-9). The Mediterranean Sea is slowly disappearing as

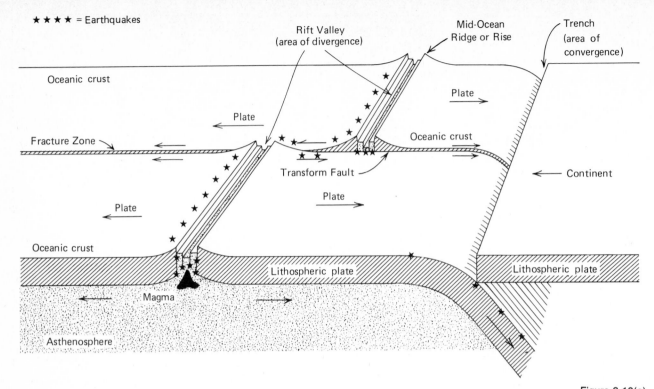

Figure 3-10(a)

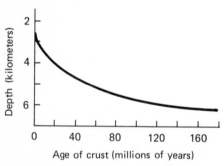

Figure 3-10(b)

The depth of the oceanic crust increases with age. (After J. G. Sclater, R. N. Anderson, and M. L. Bell, 1971. The elevation of ridges and evolution of the central eastern Pacific. *Journal of Geophysical Research.* 76:7888–7915.)

Three types of plate boundaries. (a) The midocean ridge system, here shown with steep central rift valley from which magma issues. The typical *oceanic rise* is gentler in relief and typically not topped by a pronounced rift. These areas of spreading or divergence are offset at intervals. (b) Beyond the offset segments extends a type of fault known as a *fracture zone*, on each side of which material may be moving in the same direction but is nevertheless discontinuous, not having been formed at the same time. The two parts of the ridge crest are connected by a *transform fault* where newly formed crust is moving in opposite directions. Note that the direction of a fault indicates the relative motion of adjacent plates. (c) A trench occurs where two plates converge. Continental crust cannot be easily subducted; so the plate whose leading edge is of oceanic crust descends beneath the continent. [After D. L. Anderson, 1971. The San Andreas Fault. *Scientific American* 225(11).]

Africa collides with Europe. The Pacific grows smaller as the Atlantic widens.

As plates split apart to form new oceans, the nearly parallel shorelines on either side of the new ocean basin still show how they previously fit together. This feature is graphically demonstrated in the Red Sea and Gulf of Aden between Africa and Eurasia (see Fig. 3-12).

No part of the present ocean floor is older than 200 million years (Fig. 3-13). But there is no reason to assume that movement of crustal plates has occurred only in the past 200 million years. Evidence indicates that an ancient Atlantic Ocean separated Europe and America during the early Paleozoic era, more than 450 million years ago, and

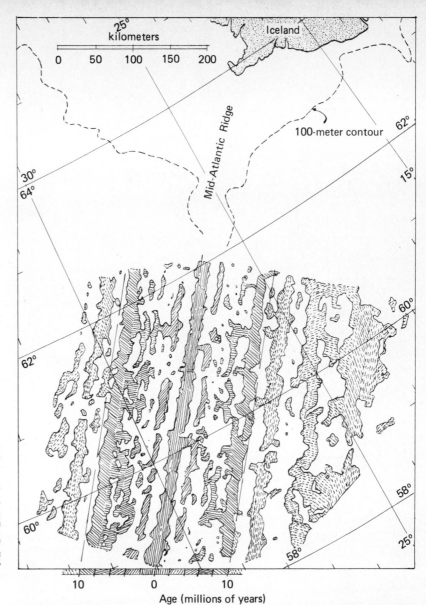

Figure 3-11

Diagrammatic representation of the magnetic field at the mid-Atlantic ridge showing areas of different magnetic orientations of the crust, resulting from crustal formation when the earth's magnetic field is normal and reversed. A constant rate of spreading is assumed in estimated crustal ages given at the bottom of the figure. The patterns indicate the approximate ages of the different magnetic "stripes."

Age (millions of years)

Figure 3-12

The Red Sea (on the left) and the Gulf of Aden extending into the Indian Ocean on the upper right. This seaway formed about 20 million years ago when Africa (bottom) split from Eurasia (upper left) and subsequently drifted apart. Note the parallelism of the coasts of the Gulf of Aden. Clouds occur over the land, especially just above the spacecraft in the lower portion of the photograph. (Photograph courtesy NASA.)

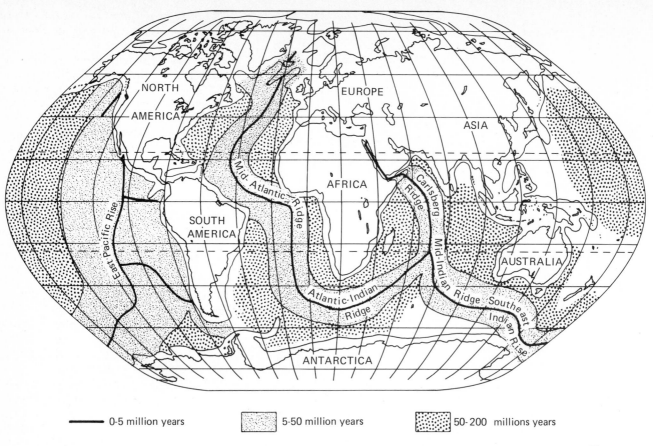

──── 0-5 million years   ⬚ 5-50 million years   ⬚ 50-200 millions years

Figure 3-13

The ages of ocean floors were determined on the basis of magnetic anomaly profiles, made by towing magnometers across ocean basins behind ships or airplanes. Points in the magnetic reversal sequence were then dated by Deep-Sea Drilling Project teams aboard D/V *Glomar Challenger* and found to coincide closely with the magnetic data. Note that in the Atlantic the edge of the ocean floor is symmetrical on either side of the mid-Atlantic ridge but that a large area of the eastern Pacific has been subducted. Compare with Fig. 3-19, where rates of motion are indicated, and with Fig. 3-14. There is a recent rift at the Galapagos volcanic center in the equatorial Pacific. In the Indian Ocean three spreading centers join to form an inverted Y whose northern branch causes rifting the Gulf of Aden and the Red Sea.

was later closed, only to be replaced by the present Atlantic. Many other major rearrangements of landmasses must also have taken place in the more than 3 billion years that continental material has floated at the earth's surface. Each continent contains rocks that were originally parts of other continents. For example, the coast of Maine was apparently part of Europe until about 300 million years ago. Vancouver, British Columbia, was on a tropical island until about 200 million years ago. Florida may once have been part of Africa. And Alaska contains many such ancient crustal fragments.

CAUSES OF SEA-FLOOR SPREADING What causes sea-floor spreading? It seems probable that heat from the core and inner mantle causes convective movements within the asthenosphere. Hot materials rise in some locations and spread out across the top of the plastic layer. As this moving layer cools, it becomes dense and eventually sinks. Some geophysicists believe that as the edge of a lithospheric plate sinks at a trench, it pulls the rest of the plate along. This process is analogous to a towel floating on water, one end of which has lost the surface tension that kept it afloat so that it sinks and drags the rest of the towel after it.

Another possibility is that upward-moving, hotter, and hence less dense material from deep layers pushes up the lithospheric plates at midocean ridges and rises. It would explain the buoyancy of the midocean ridge system, for if the rocks beneath had the same density as ordinary oceanic crust, presumably the huge mountain range system would attain isostatic equilibrium by sinking or flattening out to a depth uniform with the rest of the deep-ocean floor. At the leading edges of lithospheric plates, on the other hand, the thick accumulations of land- and ocean-generated sediments that lie adjacent to continents may depress the plates. According to this model, young crust is elevated relative to older portions because it is hotter and therefore less dense. As the oceanic crust cools and becomes denser, it also subsides. In other words, the youngest crust is the shallowest and the oldest is the deepest [Fig. 3-10(b)]. Gravity acts to drag the lower, colder and hence denser edges of plates "downhill" to sink at the trenches that occur in the deepest parts of ocean basins.

Another line of reasoning suggests that a "plume" of magma rises from deep within the mantle at each major center of prolonged volcanic activity and spreads radially beneath the lithosphere. Many such volcanic centers occur at or near the present ridge system. Other active areas may mark plumes whose force has been insufficient to crack the crust in a line along which magma could come to the surface. When large-scale rifting of plates occurs, a new ridge system forms. The sea-floor spreading continues but with plate movements occurring in different directions.

Evidence for this theory is found in the chains of volcanic seamounts illustrated in Fig. 3-14. In each chain the islands become progressively older with increasing distance from the active volcanic center. To picture their formation, go back about 70 million years. Imagine a "hot spot" or plume of rising molten basaltic rock in the mantle under Hawaii, under Easter Island, and under the Macdonald Seamount, a submerged mountain (Fig. 3-14). Then picture the Pacific plate moving almost due north over those "hot spots" for tens of millions of years. During this time volcanic eruptions built several submerged volcanic chains—the Emperor Seamounts, the Line Island chain, and the Marshall–Ellice islands (Fig. 3-15). But during that period, and subsequently, erosion wore them down above sea level and isostatic adjustment caused them to sink. Now only the tops of some of the largest volcanoes appear as tiny islands.

After these northward-trending island chains were built, imagine that the movements of the Pacific plate shifted direction. Since then it has moved generally northwestward, as indicated by the arrows in Fig. 3-9, perhaps because of tension where the corners of plates collide. Subsequently, the Hawaiian, Tuamoto, and Austral islands formed. Figure 3-15 shows these chains in detail, plus a smaller group that arises from the Cobb Seamount thermal center off the coast of Washington. Spreading beneath the East Pacific Rise, which causes Pacific plate motion, results in earth movements along California's well-known San Andreas Fault (Fig. 3-16). Baja California, being part of the Pacific plate, will eventually be drawn away from the California mainland as the Gulf of California extends northward to join the Pacific Ocean (Fig. 3-14).

Volcanic ridges are another prominent feature of deep ocean basins, also attributable to volcanic activity. Figure 3-14 shows the Rio Grande Rise and Walvis Ridge extending on either side of the Mid-Atlantic Ridge at Tristan da Cunha. Another such feature centers at Iceland and a third trends northwestward from Amsterdam Island;

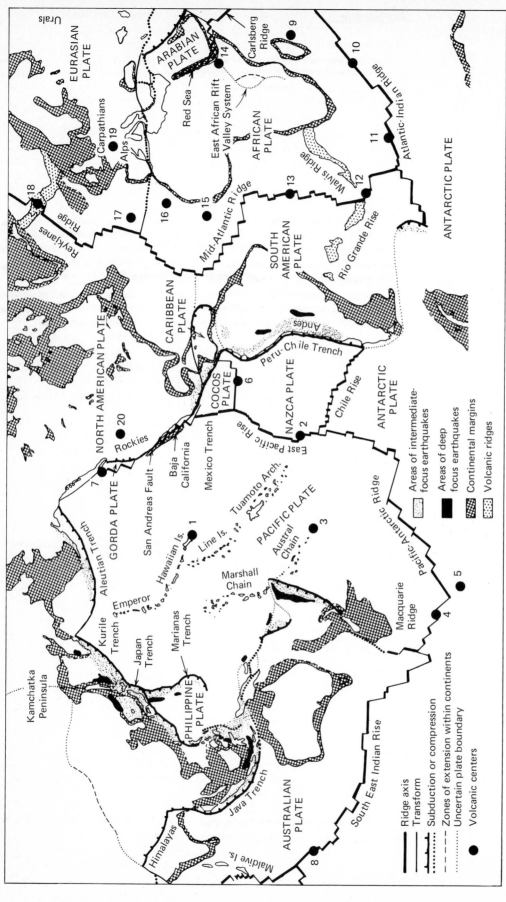

Figure 3-14

Worldwide lithospheric plate system showing some centers of volcanic activity. Three prominent Pacific seamount chains, most of which do not now appear above sea level, demonstrate Pacific plate movement away from thermal centers or "hot spots." Submarine volcanic ridges, extending northeast and northwest from Iceland and Tristan da Cunha, and northwest from Reunion Island, were produced by long-continued volcanic eruptions at the "hot spots" now marked by those islands. Note the transform faults along which the plates move past each other and are perpendicular to ridge axes. Deep and intermediate-focus earthquakes are associated with trenches. [After J. F. Dewey, 1972. Plate tectonics. *Scientific American* 226(5):56–58.]

Legend:

1. Hawaii
2. Easter Island
3. Macdonald Seamount
4. Bellany Island
5. Mt. Erebus (Antarctica)
6. Galapagos Islands
7. Cobb Seamount
8. Amsterdam Island
9. Reunion Island
10. Prince Edward Island
11. Bouvet Island
12. Tristan de Cunha
13. St. Helena
14. Afar
15. Cape Verde
16. Canary Islands
17. Azores
18. Iceland
19. Eifel
20. Yellowstone

Figure 3-15

Pacific seamount and island chains. Most of the volcanoes do not reach sea surface; only a few form islands. Many volcanoes in these chains occur in pairs. This evidence suggests that the processes causing the volcanic center of plume affect an area some 300 kilometers in diameter. [After G. B. Dalrymple, E. A. Silver, and E. D. Jack, 1973. Origin of the Hawaiian Islands. *American Scientist* 6(13):294–308.]

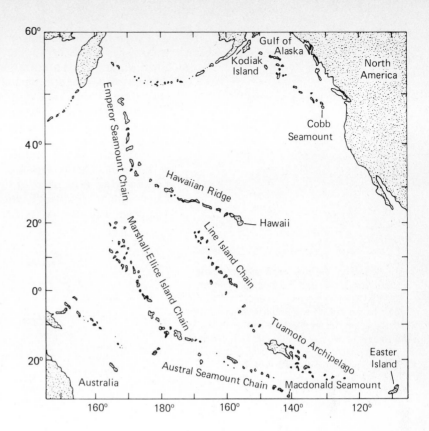

Figure 3-16

Radar image of the San Andreas fault (California). Along the California coast, the Pacific plate (see Fig. 3-14) moves northwestward at its boundary with the westward-moving North American plate. Baja (lower) California is part of the Pacific plate and is being dragged away from the rest of the continent. Strain builds up steadily along the San Andreas; eventually one side moves with respect to the other and an earthquake results. This fault is an example of a *transcurrent fault*. (Photograph courtesy U.S. Geological Survey.)

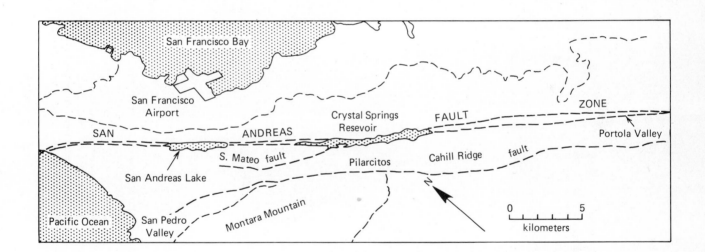

its extension on the extreme left of the figure, the Maldive Island group, lies south of India. As in Pacific island chains, these ridges form when a great amount of lava erupts from a thermal center. The difference is that submarine ridges originate at centers of sea-floor spreading, where lava erupts fairly continuously and in large quantities instead of sporadically as with volcanic islands. Figure 3-17 shows diagrammatically how ridges or island chains develop through time and how they record the direction of plate movements.

Volcanic ridges and midocean rises separate the deep ocean into several small basins (Fig. 3-18) so that bottom waters cannot flow freely between different parts of the world ocean. In the southeast Atlantic Ocean the Walvis Ridge extends from the Tristan da Cunha–Gough Island area to southwest Africa. This sharp ridge comes to within 2

Figure 3-17

Formation of a submarine ridge or volcanic chain. The Walvis and Rio Grande ridges have probably been developing for about 140 million years as the African and South American plates moved apart. Apparently there has been little or no change of direction during that time. [After R. S. Dietz and J. C. Holden, 1970. The breakup of Pangaea. *Scientific American* 224(10).]

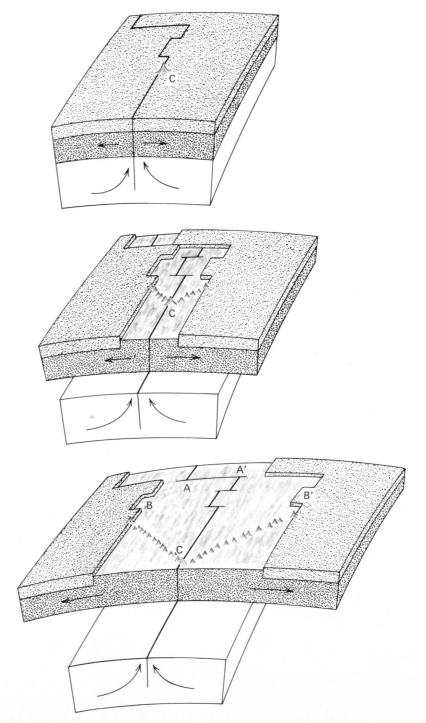

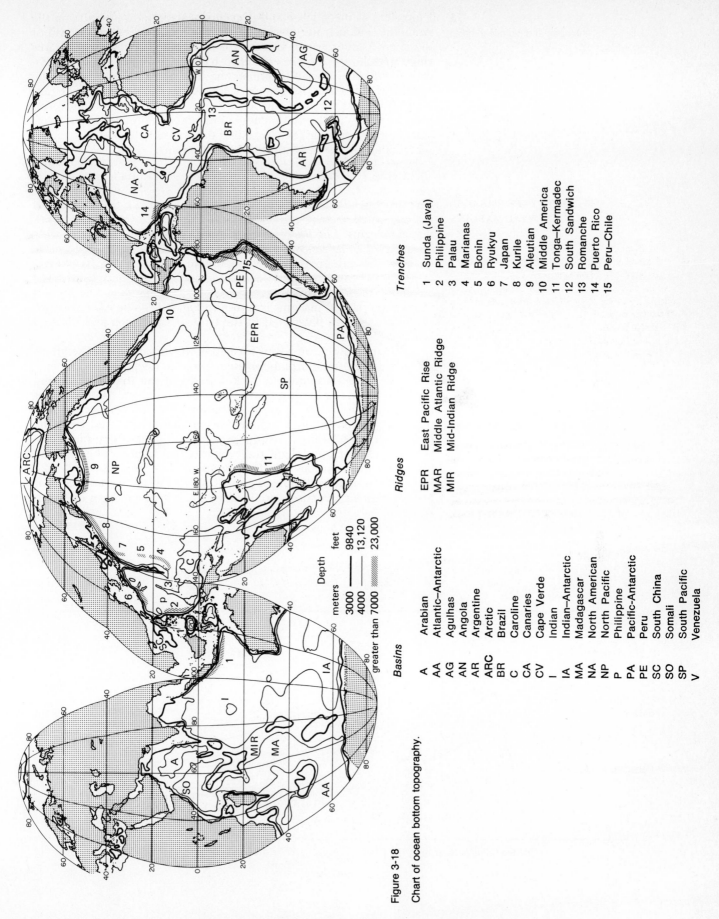

Figure 3-18

Chart of ocean bottom topography.

Depth

meters	feet
—— 3000	9840
—— 4000	13,120
////// greater than 7000	23,000

Basins

A	Arabian
AA	Atlantic–Antarctic
AG	Agulhas
AN	Angola
AR	Argentine
ARC	Arctic
BR	Brazil
C	Caroline
CA	Canaries
CV	Cape Verde
I	Indian
IA	Indian–Antarctic
MA	Madagascar
NA	North American
NP	North Pacific
P	Philippine
PA	Pacific–Antarctic
PE	Peru
SC	South China
SO	Somali
SP	South Pacific
V	Venezuela

Ridges

EPR	East Pacific Rise
MAR	Middle Atlantic Ridge
MIR	Mid-Indian Ridge

Trenches

1	Sunda (Java)
2	Philippine
3	Palau
4	Marianas
5	Bonin
6	Ryukyu
7	Japan
8	Kurile
9	Aleutian
10	Middle America
11	Tonga–Kermadec
12	South Sandwich
13	Romanche
14	Puerto Rico
15	Peru–Chile

71

kilometers of the surface and prevents near-bottom waters from the Antarctic region from flowing northward into the eastern portion of the deep Atlantic basin. In the North Atlantic the massive volcanic ridge stretching from Greenland to Iceland and eastward to the Faeroe and Shetland islands inhibits the flow of Arctic waters into the North Atlantic basin. East–west circulation in the deep ocean is generally inhibited because the location of continents and midocean ridges favors north–south circulation. In certain areas, however, east–west trending ridges also restrict the flow of near-bottom water—for instance, in the Atlantic and Indian oceans.

ATLANTIC-TYPE (PASSIVE) CONTINENTAL MARGINS

The processes that form the ocean basins also create the continental margins as they move the continents. There are two types of continental margins. The *Pacific-type margins* are tectonically active, usually bordered by trenches, and display volcanic activity and frequent earthquakes. They are areas of active mountain building. The *Atlantic-type* or *passive margins* (see Fig. 3-19) are formed when a continental block is split during the early stages of ocean basin formation. The process of forming passive margins is shown in Fig. 3-17. The earliest stages of the process are illustrated in the Gulf of California and the Red Sea (see Fig. 3-12).

The narrow ocean basins formed in the initial stages of sea-floor spreading are often partially isolated from the deep ocean. The narrow basins often accumulate thick deposits of salt that form during the times when the basin is nearly completely isolated. At other times thick accumulations of carbonate rocks and massive reefs form. Later they are covered by thick sediment accumulations derived from the continents. As Chapter 15 shows, the resulting salt domes and carbonate rocks form excellent reservoirs for petroleum. The large oil fields in Saudia Arabia and Iran occur in such deposits, as do the large accumulations of oil and gas off Mexico in the western Gulf of Mexico. Similar structures occur along the Atlantic margin of North America.

Figure 3-19

Typical features of passive or Atlantic-type continental margins. Note that thick sediment deposits cover the transition between continental and oceanic crust. In this schematic diagram, the boundary between the two types of crust is shown to be a fault.

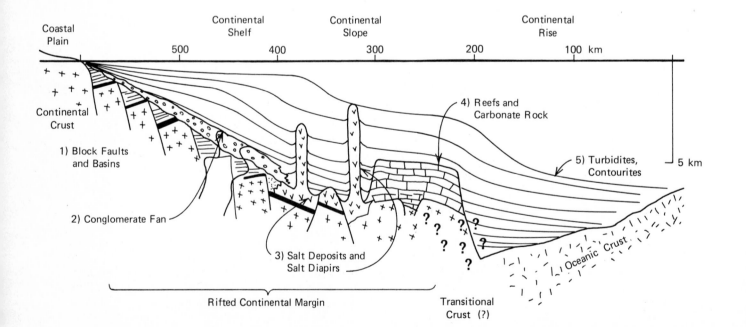

PACIFIC-TYPE (ACTIVE) MARGINS

Pacific-type margins are especially diverse and complicated because of their active tectonic processes. Many of the major processes involved in sea-floor spreading are well exhibited on Pacific-type margins.

Earthquakes at plate margins result from strains that gradually build up in rocks and are suddenly released when the rocks finally break. These earthquakes seldom happen below 70 kilometers depth along ridges and faults, but at subduction zones intermediate and deep-focus earthquakes typically originate at depths from 100 to 700 kilometers (see Fig. 3-20). This situation occurs because the plate being subducted descends under the overriding plate for hundreds of kilometers before losing its identity as a result of extreme heat and pressure. Deep earthquake foci are distributed along inclined planes known as *Benioff zones*, as shown in Fig. 3-10.

Studies of earthquake wave behavior indicate that lithospheric plates are under tension where they bend at trench margins, but pressure on them increases as they descend through the asthenosphere. This factor suggests that plates meet increased resistance as they penetrate farther into the mantle. Earthquakes do not occur below 700 kilometers depth, indicating either that lithospheric plates do not descend below this level or that they become too soft from heating to sustain the stresses that generate earthquakes.

Figure 3-20

Schematic diagram showing the relationships between subducting oceanic crust, forearc basin, volcanic island arc, and the back-arc basin and its spreading center. Back-arc spreading is well developed in the Marianas and Scotia (South Sandwich) basins. See Fig. 3-18 for locations.

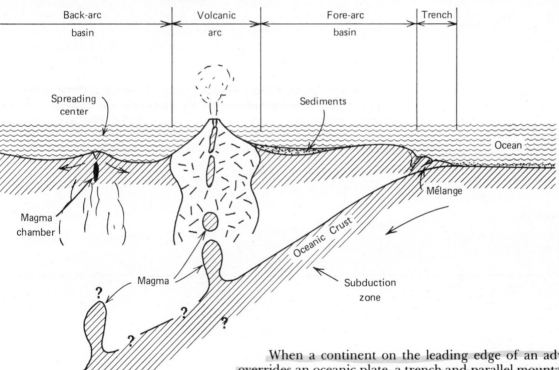

When a continent on the leading edge of an advancing plate overrides an oceanic plate, a trench and parallel mountain ranges are formed. An example is the Peru–Chile Trench, which parallels the western coast of South America. Close beneath it lies an earthquake zone that extends under the continent at an angle, to a depth of about 600 kilometers. About 300 kilometers inland is a range of volcanic mountains, the Andes, which parallel the west coast of South America for more than 6000 kilometers. The Andes and other trench-related mountains are made of *andesite,* a type of volcanic rock intermediate in silica and iron-magnesium content between granites and basalts. Thick sediment layers overlying the ocean floor near a continental margin travel into the trench and are reworked, along with the oceanic crust itself. Altered by the heat and pressure of the melting region, and possibly also incorporating partially molten mantle rocks, both plate and sediment contribute material to the building of andesitic coastal ranges. The Sierra Nevada in California is another example.

In the western Pacific relationships are more complex. Trenches occur several hundred kilometers from the Asian continent and beyond them are arcs of volcanic islands. Shallow marginal seas, filled with sediment washed from the adjacent continent, separate these island-arc systems from the mainland (Fig. 2-23). The Aleutians, Indonesia, and Japan (Fig. 3-14) are examples of island arcs.

Island arcs are built by much the same mechanism as andesitic coastal ranges. Basaltic sea floor, together with a thick layer of wet, silica-rich sediment, descends diagonally and melts, forming an andesitic magma as it reaches the asthenosphere. Made buoyant by heat and the pressure of expanding volatile materials, magma rises and erupts as gases and lava at an overlying volcanic arc and sometimes at a back-arc spreading center. Across a large island, such as Honshu in Japan, volcanoes have sometimes formed in a line from the trench of ocean side to the marginal sea (Fig. 3-21). The shallowest seem to originate at about 80 kilometers depths, where low-density constituents in the rocks begin to melt and move upward. As the lithospheric plate descends to greater depths with increased distance from the trench, other materials begin to melt and undergo *metamorphosis* or changes in mineral composition and texture. Potassium content of the extruded lava, for example, consistently rises with increasing depth of melting. Consequently, changes in potassium content of volcanic rocks can be used as "footprints" to mark the direction of a descending plate, even in continental mountain ranges formed during earlier eras of subduction and volcanism.

Continental crust, being less dense and hence more buoyant than oceanic basalt, is not subducted at trenches. Instead continental margins are deformed by the intense pressure that builds up when continental blocks collide. This situation happens, for instance, if a trench beside one continental mass is approached by another continent, after

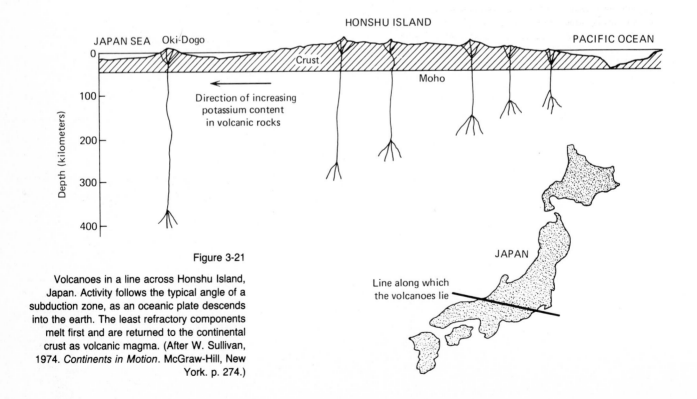

Figure 3-21

Volcanoes in a line across Honshu Island, Japan. Activity follows the typical angle of a subduction zone, as an oceanic plate descends into the earth. The least refractory components melt first and are returned to the continental crust as volcanic magma. (After W. Sullivan, 1974. *Continents in Motion.* McGraw-Hill, New York. p. 274.)

the oceanic crust that separated the continents has been completely subducted. Collisions between continental masses result in folding, thickening, or thrusting of the continental crust, as shown in Fig. 3-22. Tectonic plate movements have been occurring perhaps 15 times longer than the age of present ocean basins and the same continental materials must have collided many times, in different relative positions. Complex continental structures, such as gigantic, upthrust, folded mountain ranges containing mile-high sections of former ocean floor, are formed by billions of years of deformation.

The Himalaya Mountains were formed about 50 million years ago when [see Fig. 3-12(c)] India underthrust the Asian landmass, creating a double continental crust about 70 kilometers thick (Fig. 3-23). A former continental slope, the Himalayan plateau rises sharply to a height of 4 kilometers above the region south of it. Folded mountains extend as far northward as the Russia–China border in central Asia (Fig. 3-14). Drilling in the Indian Ocean has provided data in-

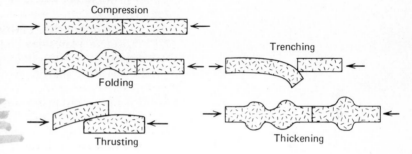

Figure 3-22

Mountains form as continents are pressed together. Most mountains were formed by a combination of these processes, often acting together with volcanism.

Figure 3-23

About 200 million years ago the northern edge of the Indian subcontinent began to underthrust Asia, eventually creating a double continental thickness with folding of the crust that forms the Himalayas, highest mountains on Earth. (After J. Weisberg and Howard Parish, 1974. *Introductory Oceanography.* McGraw-Hill, New York.)

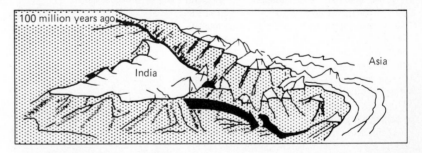

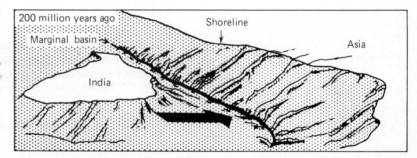

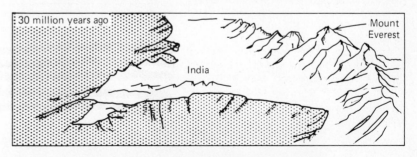

dicating more than 50 million years of pressure at that contact. The Himalayas are still adjusting, as evidenced by their many earthquakes.

Continents grow when materials from the ocean floor are added to their margins. Evidence of this process can often be observed in mountains built over a present or former trench area. They often contain *ophiolitic suites,* or complexly faulted assemblages of deep-sea sediment, submarine lava flows, and oceanic crust that were once part of the sea floor. Under extreme pressure, ophiolites have broken off a descending plate and been thrust to great heights. These layers of ancient ocean floor—volcanic basalts and the sedimentary rocks that formed on top of them through compression over millions of years on the sea floor—are wedged high over much younger rocks in the world's greatest mountain ranges. Ophiolites, for instance, occur along the axis of the Urals (Fig. 3-14), far from any present ocean, and in the mountains near San Francisco Bay. A clearly defined series of bands marks the junction of Switzerland, Italy, southern Yugoslavia, Albania, and Greece with the rest of Europe. Mountains were formed in these regions when the northern portion of what is now Africa collided with present northern Europe, closing the ocean that formerly separated the continental masses.

As oceanic plates descend into a trench, thick layers of sediment eroded from the continent may be scraped off against the trench wall by the overriding plate. These sediments are folded, sliced, and altered by extreme pressure while being carried to depths of 30 kilometers or more. But instead of descending farther toward the mantle, this relatively buoyant material is sometimes returned to the surface. This type of a deposit, known as a *mélange,* is found in present and ancient subduction zones. It typically contains ophiolitic suites and additional layers of sediment originally eroded from a continental mass and deposited at the base of a continental slope. Mélanges commonly occur seaward of andesitic mountain chains or island arcs that have been built by volcanic processes associated with a descending plate (Fig. 3-24).

As continents grow by progressively adding material, ocean basins gradually become deeper. It is estimated that the oceans have deepened at a rate of about 1 meter per million years and the continental blocks have accumulated at a similar rate.

Figure 3-24

An island arc or coastal mountain range is commonly formed from reworked marine sediments and basalts in areas of subduction. Seaward from the volcanic feature there may be mélange deposit that consists of sediment originating on the continent and deposited nearshore, plus deep-ocean sediment carried toward the trench of a descending plate, and also incorporating ophiolitic suites of sedimentary and basaltic rock broken from the ocean floor during subduction. The drawing is not to scale. (After W. R. Dickinson, 1971. Plate tectonics in geologic history. *Science* 174:107.)

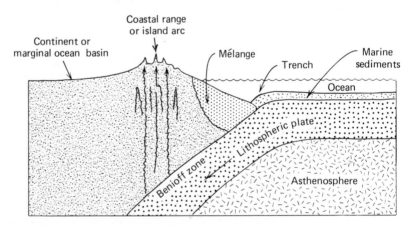

HISTORY OF OCEAN BASINS

Using the concepts of plate movements, the recent history of ocean basin formation can be reconstructed. About 225 million years ago there was only a single landmass or supercontinent, called *Pangaea* after the Greek earth goddess [Fig. 3-25(a)]. The fit of the continents against one another is excellent, like pieces of a jigsaw puzzle; the few

225 my

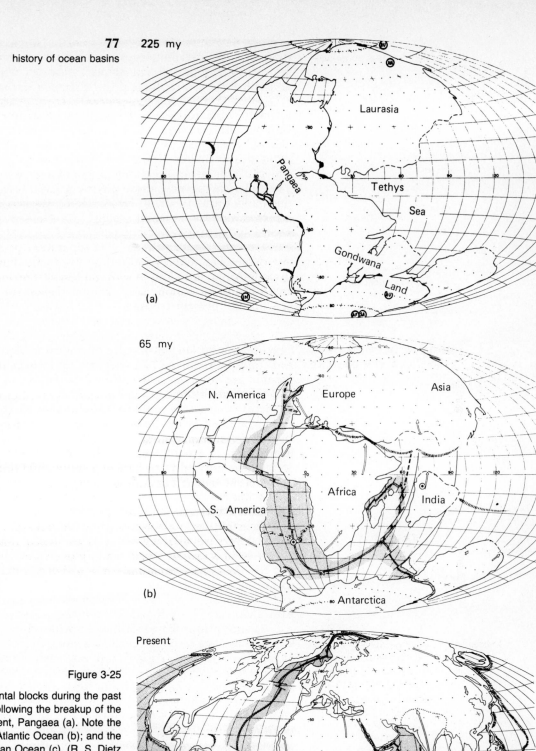

(a)

65 my

(b)

Present

Figure 3-25

Movement of continental blocks during the past 225 million years, following the breakup of the ancient continent, Pangaea (a). Note the opening of the Atlantic Ocean (b); and the formation of the Indian Ocean (c). (R. S. Dietz and J. C. Holden, 1970. "Reconstruction of Pangaea; Breakup and Dispersion of Continents, Permian to Present," *Journal of Geophysical Research,* 75, 4939–56.)

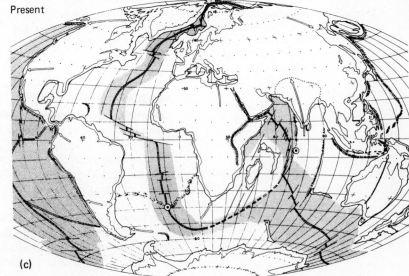

(c)

overlaps or gaps are attributable to well-understood processes. Furthermore, the distribution of certain types of rocks in South Africa appears to be identical with similar rocks of comparable ages found in South America.

About 180 million years ago *Pangaea* began to break up, initially forming a long, thin east–west trending basin, like the Red Sea or the Gulf of California [Fig. 3-12(b)]. Two subcontinents, designated *Laurasia* and *Gondwana,* were separated by this equatorial Sea of Tethys. Around 100 million years ago North America, followed by South America, split off by the development of a north–south ridge system. The Americas moved westward as the Atlantic Ocean opened between them and Eurasia–Africa. Finally, India separated from Africa and moved rapidly northward to collide with southern Asia. This collision caused India to move under part of Asia, buckling the crust above and creating the Himalayas. The separation of Australia and Antarctica [see Fig. 3-25(b) and (c)] permitted the major ocean currents to circle Antarctica. This new current pattern may have been a major factor in the formation of continental glaciers on Antarctica and the onset of the present Ice Age.

ORIGIN OF OCEAN WATERS AND ATMOSPHERE

The ocean is an ancient feature of the earth. But how ocean waters and atmosphere formed, at what rate, and what controlled their composition are questions still actively debated.

The age of the earth is determined from radiometric dating of the decay products of uranium and thorium isotopes that accumulate after billions of years. Some evidence shows that the earth formed about 4.6 billion years ago, but there is no known record of the first billion years of Earth's history. The earth probably formed from an accumulation of cold particles, especially silicon compounds and oxides of iron and magnesium. The heaviest materials fell toward the center and the mass became increasingly compressed. This compaction caused heating and finally melting. Heat released by the decay of radioactive materials also raised interior temperatures. Three concentric layers were segregated: the heaviest material, primarily iron and nickel, became the core; heavy silicate rocks formed the mantle; and the lightest materials rose to form the surfacial crust. Ancient rocks containing earliest evidence of a solid crust on the earth are 3.5 to 3.6 billion years old. The oldest known organisms—microfossils in Australian rocks—are dated as 3.5 billion years old.

If the primordial earth had an atmosphere, it apparently was lost, perhaps as a result of the same heating that caused the separation of core, mantle, and crust. At any rate, there is relatively much less rare noble gas (krypton, neon, xenon) in the earth's atmosphere than in the universe as a whole. This factor indicates that these gases were not retained in the abundance that would be expected from the retention of an ancient atmosphere. Thus the present ocean and atmosphere must have been derived from later alteration of the earth rather than from cosmic sources. Both the water gases must have been combined in rocks, for they were not lost in the early escape of gas from the earth's surface.

As the earth became hotter, parts of the upper mantle or crust melted and the more volatile, water-rich portions of these molten rocks escaped as lava from volcanic eruptions or intruded overlying rocks. In the process, a part of their original water and gas content was lost to the atmosphere. In time the water condensed to fill depressions in the crust—the ocean basins—and the gases remained in the atmosphere. This process apparently continues today, although the amount of new water is too small to be detected with confidence.

The earliest evidence for the existence of the ocean is the presence of recognizable water-laid sediments formed about 2.5 to 3 billion years ago. Although such sediment indicates the presence of abundant water on the earth's surface, it does not tell us about the chemical composition of this primordial ocean. Carbonate sediments about 2 billion years old suggest that ocean water has been salty for at least half of the earth's history. Probably the major constituents of sea salt did not differ greatly from those now present in the ocean.

Evidence on the composition of the early atmosphere is mostly indirect. It probably contained water vapor, hydrogen, hydrogen chloride, carbon monoxide, carbon dioxide, and nitrogen. Free oxygen almost certainly did not occur. This fact is evident because some ancient sediments originally formed in the ocean contain minerals that are not stable in the presence of oxygen. Further evidence is provided by the existence of iron-rich deposits that have been extensively mined as iron ore. They could not have been formed in the past 2 billion years because iron is highly insoluble in seawater containing dissolved oxygen and today does not form large-scale marine sedimentary deposits. Ancient ocean waters were apparently devoid of dissolved oxygen and so large quantities of iron could be transported and precipitated to form extensive deposits.

Substantial amounts of free oxygen were apparently not introduced to the atmosphere, nor to the ocean, until after the evolution of *photosynthesis.* It is the process by which green plants, such as algae, combine carbon dioxide and water in the presence of sunlight to form organic carbon compounds and free oxygen. Fossil evidence suggests that this may have begun to happen about 3 billion years ago. The subsequent accumulation of oxygen in air and ocean permitted the evolution of *respiration,* whereby animals combine oxygen with organic carbon compounds to release energy.

Small amounts of free oxygen apparently began to be used for respiration about 2.7 billion years ago. But oxygen and organic carbon levels continued to build up until about 500 million years ago, when higher animals evolved. The oxygen accumulated in ocean and atmosphere; and the organic carbon was incorporated in sediments from which a small fraction is now recovered as fossil fuel. Relationships between oxygen and carbon in photosynthesis and respiration, showing the former excess production of organic carbon and oxygen, are given in Fig. 3-26.

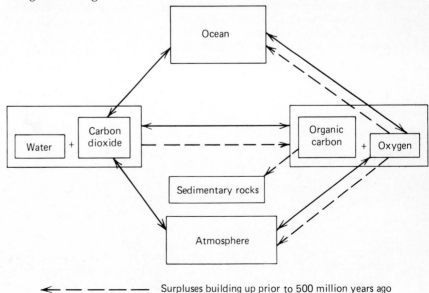

Figure 3-26

Before about 500 million years ago, an excess of photosynthesis over respiration built up surplus organic carbon and oxygen. The former was retained in sedimentary rocks, and the latter in atmosphere and ocean. Today oxygen and organic carbon are used by animals in respiration at about the same rate as they are produced by plants in photosynthesis.

HYDROTHERMAL CIRCULATION AND THE COMPOSITION OF SEAWATER

The composition of sea salts and deep-ocean sediments is influenced by seawater circulating through hot oceanic crust, formed by volcanic eruptions. Oceanic crust is created by volcanic eruptions or emplacement of molten basalt at temperatures around 1200°C. Some of this heat is conducted through solid rock to the overlying waters. But one-half to two-thirds is removed by seawater circulating through the crust. Seawater enters the oceanic crust through faults and crevices and moves along open fractures and permeable zones, penetrating deeply. Circulating waters even reach the vicinity of magma chambers below the spreading center where molten rock collects before eruption.

The hot water flows out into the ocean through vents on the ocean floor and quickly mixes with overlying waters. This cooling process takes nearly 100 million years to complete and involves the circulation of immense volumes of seawater. It is estimated that the flow is about 0.5% of the annual river flow. Put another way, the volume of the entire ocean circulates through the oceanic crust every 5 to 10 million years.

Reactions between hot basalts and seawater play a major role in regulating the chemical composition of salts dissolved in seawater. Magnesium and sulfate, for example, are removed from seawater while other elements—lithium, rubidium, calcium—are added. These processes are discussed in Chapter 5.

Hydrothermal vents discharge hot waters in areas of recent volcanic activity. Only a few parts of the midocean ridge are actively erupting at any time. Volcanic eruptions at any location on a ridge are separated by thousands to tens of thousands of years.

Three types of active vents have been observed discharging heated waters. The most spectacular are the *black smokers*, which discharge superheated waters between 300 to 400°C (water at the ocean surface boils at 100°C). Because of their high temperatures, these waters have a density around 0.7 gram per cubic centimeter compared to the surrounding bottom waters with densities of about 1.025 gram per cubic centimeter (see Fig. 3-27). The heated waters, discharged

Figure 3-27

Sketch of "black smokers" discharging hot waters containing precipitating metal sulfides. The chimneys are fairly fragile structures, up to 10 meters tall and about 50 meters across at the base. (After CNEXO, 1980, Birth of an Ocean: The Crest of the East Pacific Rise, Centre National pour l'Exploitation des Oceans, Brest, France. 84 p.)

at velocities of several meters per second, rise, forming a plume that appears black—hence the name black smoker. The dark color apparently results from the turbulence caused by the firehoselike discharge and from the sulfide minerals that precipitate in the sulfide-rich hot waters as they mix with oxygenated bottom waters. Cooler vents (called *white smokers*) also occur; some vents discharge waters between 25 and 250°C because the hot waters have mixed extensively with bottom waters around 0°C before discharge. In these vents the sulfide minerals are removed by chemical reactions before they reach the surface.

The third type of hydrothermal discharge is through cracks and fissures in the volcanic rocks. These waters are relatively cool, typically a few tens of degrees warmer than surrounding bottom waters. All three types of vent discharges support extensive growths of filter-feeding organisms around them.

Vents form large, irregular, chimneylike, Halloween-colored mounds up to 10 meters high on the volcanic terrain, illustrated in Fig. 3-28. The mounds are made of porous silica, native sulfur, and metallic sulfide minerals that color the vents yellow, black, brown, red, and white. Active vent chimneys discharging cooler waters are coated with the tubes of encrusting worms. The surrounding ocean bottom supports a rich marine life that depends on sulfur-oxidizing bacteria growing in the discharged sulfide-rich waters. (More about these distinctive marine organisms in Chapter 14.) The freshly precipitated sulfide minerals settle out near the vents, causing nearby sediments to be metal rich.

Chimneys and their surrounding deposits resemble valuable ore deposits mined on land for copper and other metals. Such deposits occur worldwide. Examples are the copper mines of Cyprus, worked since Roman times, and sulfide deposits in the Philippine Islands, Japan, and the coast ranges of North and South America and Turkey. Copper- and zinc-rich muds at the bottom of the Red Sea may have been formed by similar processes.

Figure 3-28

Sketch of recently active vents on the midocean ridge. The chimneys through which the vents discharge their waters are made of silica, iron-oxide, and metal sulfides. Encrusting worms and other organisms grow on the cool outer surfaces of such vents. (After CNEXO, 1980, *Birth of an Ocean: the Crest of the East Pacific Rise.* Centre National pour l'Exploitation des Oceans, Brest, France. 84 p.)

1. List the five primary layers, or spheres, that make up the earth and give the distinguishing features of each. Indicate which fraction of the earth's mass is contained in each.

2. Explain how reversals in the earth's magnetic field are used to determine the ages of specific depth zones in sediment cores and of areas of the oceanic crust.

3. Diagram the primary processes involved in sea-floor spreading and illustrate the three types of crustal plate boundaries.

4. Explain how a volcanic center ("hot spot") under an ocean basin forms a chain of volcanic islands.

5. Draw a diagram of an active (Pacific-type) continental margin, labeling the most significant features.

6. Draw a diagram of a passive (Atlantic-type) continental margin, labeling the most significant features.

SUMMARY OUTLINE

Structure of the earth
Water is a small fraction of total mass of Earth
Earth is three concentric layers: core, mantle, lithosphere
Core—dense, rich in iron and nickel
Mantle—outer asthenosphere is capable of flow
Lithosphere, 60 to 100 kilometers thick—includes continental, oceanic crust fused with layer of rigid material derived from mantle
Mohorovičić discontinuity separates crust from rigid mantle layer
Hydrosphere—includes water in sedimentary rocks; most water in ocean
Atmosphere—includes water vapor
Continental crust—35 to 40 kilometers thick, granitic; extends out to 2000-meter-depth contour
Oceanic crust—7 to 10 kilometers thick
Lithosphere floats on asthenosphere: isostatic equilibrium of crust maintained by rising or sinking with respect to deeper layers

Methods of probing the earth—indirect techniques
Behavior of earthquake waves demonstrates layered Earth structure
Gravity anomalies indicate differences in mass
Heat flow varies—high from ocean basins, especially midocean ridges; low at trenches
Magnetic "striping" reveals age of ocean floor, due to periodic reversals of Earth's magnetic field that caused minerals in molten rocks to align themselves with magnetic north at the time of formation

Sea-floor spreading
Lithosphere divided into plates—boundaries at midocean ridges, trenches, transform faults
Generation of crust by magma rising at ridges and rises; transform faults connect offset sections of ridges

Subduction at trenches destroys crust
Pangaea—ancient supercontinent; broke up 200 million years ago to form present continents, Atlantic basin

Causes of sea-floor spreading
Sea-floor spreading caused by convection in mantle; younger crust elevated relative to older crust due to buoyancy
Plumes of magma rise at volcanic centers; seamounts form in linear progression as crust moves; volcanic ridges form similarly, divide ocean into basins

Atlantic-type (passive) continental margins
Form when continents split
Fault controlled
Thick sediment deposits
Often salt deposits and reefs or other carbonate rocks
Major petroleum deposits

Pacific-type (active) continental margins
Subduction zones, often active trenches
Deep-focus earthquakes along inclined plane
Andesitic mountain chains, island arcs form from reworked sediments, plate, mantle
Continental crust deformed when continents converge—mountains built by folding, thickening of crust; contain layers from ocean basins
Mélange deposits add to continental margins

Origin of ocean water and atmosphere
Primordial atmosphere lost when Earth heated
Water and gas released from interior during volcanic activity; free oxygen absent
Oxygen produced during photosynthesis during past 3 billion years
Organic carbon stored—some now used for fuels
Respiration, photosynthesis now maintain oxygen, organic carbon balance

Hydrothermal circulation and the composition of seawater
    Seawater circulates through hot oceanic crust and cools it
    Chemical reactions influence composition of sea salts

Sulfides precipitated from discharged waters form large deposits
Support growth of unusual organisms around vents

SELECTED REFERENCES

HOLLAND, H. D. 1978. *The Chemistry of the Atmosphere and Oceans.* Wiley-Interscience, New York. 351 pp. Advanced technical discussion.

JUDSON, SHELDON, KENNETH S. DEFEYES, and ROBERT B. HARGRAVES. 1976. *Physical Geology.* Prentice-Hall, Englewood Cliffs, N.J. 551 pp. Elementary; includes discussion of sea-floor spreading.

PRESS, FRANK, and RAYMOND SIEVER. 1978. *Earth,* 2nd ed. W. H. Freeman, San Francisco. 649 pp. Comprehensive treatment of earth science.

Scientific American. 1972. *Continents Adrift.* W. H. Freeman, San Francisco. 172 pp. Reprinted articles from *Scientific American.* Elementary and intermediate in difficulty.

SULLIVAN, WALTER. 1974. *Continents in Motion, The New Earth Debate.* McGraw-Hill, New York. 399 pp. Well-written treatment of development of plate tectonic theory.

UYEDA, SEIYA, 1978. *The New View of the Earth.* W. H. Freeman, and Company, San Francisco. 217 pp. Detailed discussion of plate tectonic theory.

WYLLIE, PETER J. 1976. *The Way the Earth Works:* An Introduction to the New Global Geology and its Revolutionary Development. John Wiley & Sons, New York. 296 pp. Elementary, well-written introduction to plate tectonics.

4 SEDIMENTS

A sediment sampler provides a geologist with an undisturbed mud sample from the surface of the ocean floor. (Photograph courtesy of Woods Hole Oceanographic Institution.)

Sediment deposits—accumulations of particles brought in by rivers, glaciers, and winds and mixed with shells and skeletons of marine organisms—cover most of the ocean bottom. Only a few areas of ocean bottom are devoid of sediments. Some are swept clean by strong currents; other areas are newly formed and have not accumulated sediment deposits. Sediment is usually thickest (a few kilometers) near continents, especially in marginal ocean basins, where ridges interfere with sediment transport into the main ocean basin, and thinnest near newly formed crust at midocean ridges.

The ocean has always been the final repository for debris from the land. Oceanic sediment deposits record Earth history during the past 200 million years. Earlier deposits were apparently destroyed by subduction or incorporated into continental blocks. Deep-sea deposits contain a record of ocean history, including effects of climatic changes on oceanic circulation. On land, studies of oceanic sediments and the fossils in them provide a time dimension for ocean studies. In this chapter we study the processes that form oceanic sediment deposits.

ORIGIN AND CLASSIFICATION OF SEDIMENTS

Sediments are classified in two different ways—by origin and by particle size. Classified by origin, sediment particles are divided into the following groups.

Biogenous (derived from organisms): isoluble remains of bones, teeth, or shells of marine organisms. Some typical particles are illustrated in Figs. 4-2 and 4-3.

Hydrogenous (derived from water): formed by chemical reactions occurring in seawater or within the sediments; manganese nodules illustrated in Fig. 4-7 are examples.

Lithogenous (derived from rocks): primarily silicate mineral grains (listed in Table 4-1) released by the breakdown of silicate rocks on the continents during weathering and soil formation; in the open ocean, volcanoes are locally important sources of lithogenous particles.

Sediment particles are also classified according to size, as in Fig. 4-1 and Table 4-2. Particle size is an important property of sediments because it determines the primary mode of transport and how far particles travel before settling to the bottom. Typical particle sizes from various sources are indicated in Fig. 4-1.

TABLE 4-1

Common Minerals in Marine Sediments and Their Chemical Composition

MINERAL*	CHEMICAL COMPOSITION
Silicates—primarily lithogenous, derived from silicate rocks:	
Quartz	SiO_2
Opal	$SiO_2 \cdot H_2O$, glasslike substance formed by diatoms, radiolaria
Feldspars	Alumino-silicates with Na, K, and Ca in variable proportions
Micas	Primarily muscovite (a K, Al hydrous silicate)
Clays	Similar to micas but with more variable composition and structure
Illite	Primarily altered, fine-grained muscovite
Montmorillonite	A group of minerals generally formed from volcanic glass
Chlorite	Fe, Mg, Al hydrous silicate, commonly formed during weathering
Kaolinite	Al silicate formed during tropical weathering
Gibbsite	Al oxide, formed during tropical weathering
Zeolites	Hydrous alumino-silicates, similar to feldspars in structure
Oxide, hydroxides—occur primarily in Fe–Mn nodules:	
Hematite	Fe_2O_3, probably red coloring on deep-sea clays
Goethite	$FeO(OH)$, constituent of Fe–Mn nodules
Lepidocrocite	$FeO(OH)$, another mineral form of Fe in nodules
Pyrolusite	MnO_2, occurs in Fe–Mn nodules
Carbonates—primarily formed by marine calcareous organisms:	
Calcite	$CaCO_3$, forms foraminiferal shells, coccoliths, mollusk shells
Aragonite	More soluble form of $CaCO_3$, in pteropod shells, corals
Phosphates—rare in sediments, common in bones and teeth:	
Apatite	$Ca_5(PO_4)_3F$
Sulfates:	
Barite	$BaSO_4$, found in certain primitive organisms and sediments rich in organic matter
Celestite	$SrSO_4$, found in certain radiolarians shells, rarely in sediment

*Biogenous minerals are in boldface. Hydrogenous minerals are in italics. All others are lithogenous in origin.

A sediment consisting of single-sized particles is considered *well sorted*. A *well-sorted sediment* is one that has a limited size range of particles; the other-sized particles have been removed, usually by currents. Beach sand is a typical well-sorted sediment in which particles have been segregated by size by wave action or currents. A *poorly sorted sediment* contains different-sized particles, usually indicating that little mechanical energy has acted to sort the particles. Glacial marine sediments, derived from mechanical wearing down of rocks by glaciers, are poorly sorted. Generally they contain particles ranging in size from boulders to fine silt (see Table 4-2).

Sands occur primarily along coastlines, where waves and currents remove the finer particles. Sands are rare in the deep ocean, although, as we shall see, they do occur there. *Mud*—a mixture of silt- and clay-sized particles—is the most common type of sediment in the deep ocean, as indicated in Table 4-3.

BIOGENOUS SEDIMENTS

More than half the deep-ocean bottom is covered by biogenous sediments, either calcarous (Fig. 4-2) or siliceous (Fig. 4-3, 4-4) as indicated in Table 4-3. Formation and distribution of these deposits are

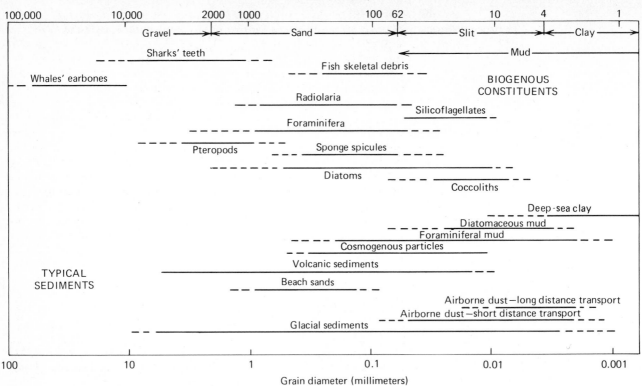

Figure 4-1

Grain-size distribution of common marine sediments and sediment sources.

TABLE 4-2

Classification of Sediment Particles By Size

SIZE FRACTION	PARTICLE DIAMETER (millimeters)
Boulders	> 256
Cobbles	64–256
Pebbles	4–64
Granules	2–4
Sand	0.062–2 (62–2000 μ*)
Silt	0.004–0.062 (4–62 μ)
Clay	< 0.004 (< 4 μ)

*μ is a *micron,* one millionth of a meter, or 39.4 millionths of an inch.

TABLE 4-3

*Abundance and Composition of Deep-Ocean Sediment**

CONSTITUENTS	CALCAREOUS MUD (percent)	Siliceous Muds DIATOMA-CEOUS (percent)	Siliceous Muds RADIO-LARIAN (percent)	BROWN MUD (percent)
Biogenous:				
Calcareous	65	7	4	8
Siliceous	2	70	54	1
Lithogenous and hydrogenous	33	23	42	91
Abundance in deep-ocean sediments	48	14		38

*Recalculated after R. R. Revelle, 1944. *Marine Bottom Sediment Collected in the Pacific Ocean by the* Carnegie *on its Seventh Cruise.* Carnegie Institution Publication 556, part 1.

88

governed by three processes: *productivity* of overlying surface waters, *destruction* of the shells before burial, and *dilution* by other kinds of sediment. Where nutrients (phosphate and nitrate) are resupplied to the surface layers and there is sufficient sunlight to support photosynthesis, marine organisms grow in abundance. The resistant shells of these organisms accumulate on the bottom and locally may form biogenous sediments.

Because of the vertical water movements bringing nutrients to the ocean surface in the high latitudes and along the equator, these waters are highly productive of marine life and produce distinctive deposits of biogenous sediments. Diatomaceous sediments (see Fig. 4-3) predominate in the high latitudes both in the North Pacific and surrounding Antarctica. Absence of predominantly diatomaceous sediment in the North Atlantic is presumably caused by mixing of diatom shells with abundant lithogenous sediment. Also, surface waters of the North Atlantic are relatively unproductive.

In the equatorial Pacific radiolarian-rich sediment (see Fig. 4-2

Figure 4-2

Some calcareous deep-sea sediments obtained by the *Challenger* expedition. (top) Foraminifera (Globigerina) mud; (bottom) pteropod mud. Foraminiferal shells are made of calcite, which is much less soluble than the aragonite forming the pteropod shells.

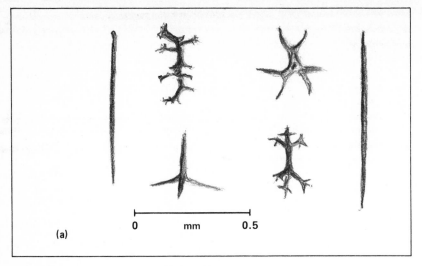

(a)

Figure 4-3A–C

Some common siliceous biogenous constituents in marine sediments: sponge spicules (a) radiolaria, (b) diatoms (c), and a typical diatomaceous sediment.

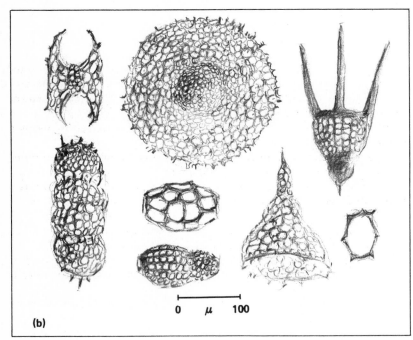

(b)

Figure 4-3D

Diatoms, siliceous constituents of deep-sea sediments, are typically about the size of a sand grain (0.05 to 0.1 millimeter). Broken or partially dissolved diatom shells are common in sediment deposits. (Photographs courtesy W. R. Reidel, Scripps Institution of Oceanography.)

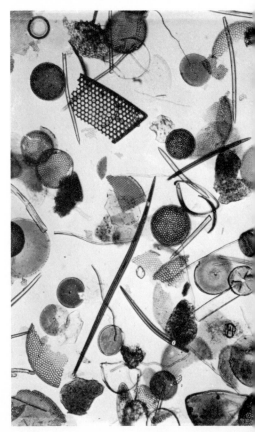

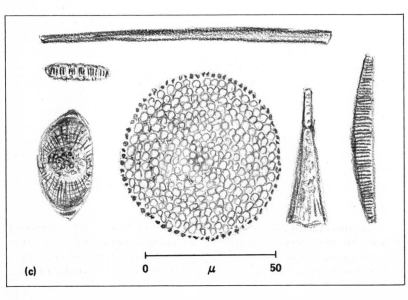

(c)

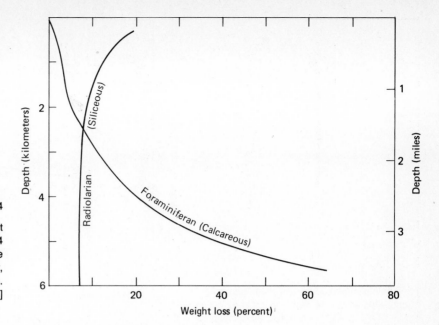

Figure 4-4

General relationship between depth and weight loss in particles due to dissolution after 4 months of submergence at various depths in the central Pacific Ocean. [After W. H. Berger, 1967. Foraminiferal ooze: solution at depths. *Science* 156(3773): 383–385.]

for typical organisms) accumulations are markedly thicker than elsewhere on the deep-ocean floor—that is, in excess of 600 meters and in some areas (120 to 170°W) over 1 kilometer thick. Accumulation rates of about 15 millimeters per thousand years are typical under the most productive parts of the equatorial Pacific, compared to about 1 millimeter per thousand years elsewhere.

A second factor controlling distribution of biogenous sediment is *destruction* of skeletal remains. Phosphatic and siliceous constituents dissolve when exposed to deep-ocean waters. Thus their survival depends in large measure on how sturdy the fragments were initially. Only the most resistant bones and most heavily silicified shells of radiolarians and diatoms are common sedimentary constituents. Thin or slightly silicified shell fragments are easily broken and readily dissolved; they do not survive long settling times and long exposures at the sediment surface. The longer the exposure, the greater the weight lost through dissolution.

Effects of particle destruction (see Fig. 4-4) can most readily be seen in the distribution of calcareous deep-ocean sediment. Surface ocean waters are usually saturated with calcium carbonate and so calcareous fragments there are not dissolved. At mid-depths lower temperatures and higher CO_2 contents of seawater cause dissolution of calcareous fragments, but rather slowly. Below about 4500 meters waters are rich in dissolved CO_2 and able to dissolve calcium carbonate much more readily. This deep relationship is shown in the decrease in calcium content of deep-ocean sediments at great depths in the ocean (see Fig. 4-5). Carbonate-rich sediments are common at relatively shallow depths in the ocean but almost completely absent below about 6000 meters, as Fig. 4-6 indicates. For instance, the relatively soluble aragonitic pteropod shells occur only on the shallow tops of submerged peaks in the Atlantic. Distribution of carbonate-rich foraminiferal and coccolith muds reflects this depth zonation. These sediments are most common on the shallower parts of the ocean bottom: on the Mid-Atlantic Ridge, the East Pacific Rise, and the Mid-Indian Ridge (see Fig. 4-12).

A third factor controlling distribution of biogenous sediment is *dilution* by lithogenous sediment. Sediments containing more than 30% biological constituents by volume are called *oozes*: a "diatom ooze" if diatoms were dominant, a "foraminiferal ooze" if shells of fora-

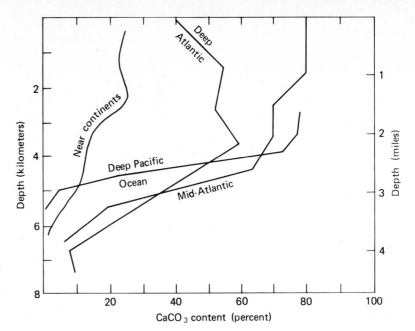

Figure 4-5

Distribution of calcium carbonate in sediment deposited at various depths in the deep ocean and near the continents.

Figure 4-6

Depth distribution of deep-ocean sediment. Minimum, maximum, and average depths are shown for each major sediment type. Frequency of occurrence of various depth zones is also shown.

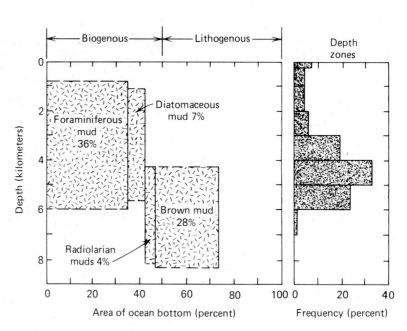

minifera were the most abundant carbonate constituent. For our purposes, we call such sediments *diatomaceous muds* if diatom shells constitute more than 30% by volume, or *radiolarian muds*, or *foraminiferal muds*, using the same standard. If none of these biogenous constituents dominates the sediment, then we shall refer to the deposit as a *deep-sea mud*, because of its red or red-brown color, especially in the Atlantic. Such sediments are sometimes called *red clays*.

Some biogenous remains are rarely recognized. Fish bones, for example, are rather fragile open structures made up of minute calcium phosphate crystals embedded in an organic matrix. When bones are broken and eaten, bacteria decompose the organic matrix, leaving only tiny phosphate crystals that commonly dissolve. Only the most resistant bones, such as sharks' teeth or whales' earbones, remain intact on the ocean bottom; they are recovered by dredging and are sometimes found in the centers of manganese nodules, as in Fig. 4-7.

92

Figure 4-7

Manganese nodules, 8 to 10 centimeters in diameter, on the ocean floor in certain parts of the South Pacific. (Photograph courtesy National Science Foundation.)

In estimating the rate at which biogenous sediment is formed, we assume a steady state—in other words, that conditions in the ocean are not changing through time. If the chemical composition of seawater is constant, then the amount of dissolved calcium, silica, and phosphate brought to the ocean each year must equal the amount removed during the year by organisms and incorporated into the sediment. Remember that only the amount actually reaching the sediment is included, not the amount removed by organisms from seawater, most of which is returned to the water by dissolution of particles as they sink to the bottom. This approach indicates that about 1.5 billion tons of calcium carbonate are deposited in sediments each year; most is probably removed by foraminifera and coccoliths. About 0.5 billion tons of siliceous sediment form each year. In total, it amounts to about 10% of the total sediment load brought to the ocean each year, ignoring the apparently small amount of phosphate added to sediments.

Hydrogenous constituents form through chemical-biological processes in seawater. Manganese nodules are the most conspicuous hydrogenous sediments. There is some indication of bacterial activity involved in formation of manganese nodules [also known as iron-manganese nodules (Fig. 4-7) because of their high iron content]. These elements, originally brought to the ocean by rivers and by discharges of hydrothermal vents on the ocean bottom, apparently precipitate in the presence of the abundant dissolved oxygen typically found in deep-ocean waters. These precipitates form tiny particles that are swept along by currents until they settle out or contact a surface.

The tiniest particles tend to adhere to solid surfaces; so they form a stain or coating around a solid nucleus, such as a fragment of volcanic rock or other debris. In the deep ocean, where sediment accumulates slowly, sharks' teeth and whales' earbones are common examples of such biological residues. Over time an iron-manganese coating builds up slowly around the original nucleus and eventually a nodule forms.

Nodules form by accretion and thus contain most particles generally found in seawater. Presumably they, too, have been incorporated in the nodules by striking the surface and adhering to it. Deep-sea nodules have high copper, cobalt, and nickel contents, which makes them attractive as a possible source of copper and other scarce metals.

Most manganese nodules form exceedingly slowly, only a few atomic layers per day—one of the slowest chemical reactions known. On the Blake Plateau of the continental rise off southeastern North America, the bottom is covered in many places with crusts or with such nodules, which apparently form at rates of about 1 millimeter per million years. In other areas, nodules seem to form 10 to 100 times faster. In some isolated or nearly isolated marine basins, such as the Baltic Sea, manganese nodules form at rates of 20 to 1000 millimeters per 1000 years. The speed record for nodule growth is the manganese coating that formed on a naval artillery shell fragment dredged off southern California; it formed at the rate of about 10 centimeters per 100 years.

Because of their slow growth, manganese nodules cannot reach any substantial size in areas receiving large amounts of other sediments; they would be buried too quickly. Consequently, nodules are rare in the Atlantic, which is well supplied with lithogenous sediment. In the central Pacific, where sediment accumulates slowly, nodules are quite common on the bottom. Manganese nodules are estimated to cover 20 to 50% of the deeper parts of the Pacific.

Near midocean ridge crests where hydrothermal vents are active, such as the East Pacific Rise, nearby sediment deposits are enriched in metals, precipitated from the vented waters. Apparently metal sulfides precipitate when the vented waters mix with overlying oxygen-rich bottom waters. Metal-rich particles formed from the chemical reactions are carried away by bottom currents, much like a smoke plume in the wind. The particles slowly settle out and increase the metal contents of sediments near the midocean ridges.

Particles from space also occur in marine sediments. About 30,000 tons per year of dust fall into the ocean, most apparently derived from meteorities. Cosmic spherules (Fig. 4-8), first recognized in deep-ocean sediment by John Murray, are about 200 to 300 microns in diameter. They are magnetic and largely composed of iron or iron-rich minerals.

Since the Industrial Revolution man has supplied substantial amounts of sediment to the ocean through discharges of municipal and industrial wastes. Because of the high level of industrialization

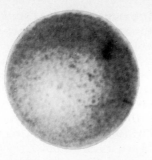

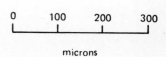

0   100   200   300

microns

Figure 4-8

Cosmic spherules from deep-ocean deposits. (*Challenger* Reports.)

along their coasts, the United States and northern Europe are large contributors of such man-made sediments. Most of these deposits appear to remain on continental shelves near cities.

LITHOGENOUS SEDIMENTS

Lithogenous constituents come from the mechanical and chemical breakdown of silicate rocks on land. Most rocks originally were formed at high temperatures and pressures in the absence of free oxygen and with little liquid water. These rocks break down at the earth's surface due to chemical and physical processes known as *weathering*. Soluble constituents are released and carried to the ocean, dissolved in river water. The remaining rock is often broken up by physical processes, such as freezing, into sand- and clay-sized grains, which are also carried suspended in running water.

Some minerals—for example, quartz and some feldspars—resist chemical alteration. Usually entering the ocean almost unaltered, they are common in sands and silts. The slightly altered remains of micalike minerals form the clay minerals, common in fine-grained deep-sea clays.

Rivers transport most of the lithogenous sediment to the ocean—about 20 billion tons each year, most of it coming from Asia, as shown in Table 4-4. The four largest rivers are in Asia: the Ganges and Brahmaputra of India and China's Hwang Ho (Yellow) and Yangtze. Together they supply about 25% of the total sediment discharge from continents (see Fig. 4-9 for their locations).

Except for the Ganges–Brahmaputra rivers and the Amazon and Congo rivers (which drain into the equatorial Atlantic), the world's major sediment-producing rivers drain into marginal seas. Consequently, most of the sediment discharged by rivers into the ocean cannot reach the deep-ocean floor directly.

TABLE 4-4

*River Discharge of Suspended Sediment from the Continents**

CONTINENT	AREA DRAINING TO OCEAN (millions of square kilometers)	ANNUAL SUSPENDED SEDIMENT DISCHARGE $\left(\dfrac{10^9 \text{ tons}}{\text{yr}}\right)$	$\left(\dfrac{\text{tons/km}^2}{\text{yr}}\right)$†
North America	20.4	2.0	96
South America	19.2	1.2	62
Africa	19.7	0.54	27
Australia	5.1	0.23	45
Europe	9.2	0.32	35
Asia	26.6	15.9	600
Total	100.2	20.2	

*From J. N. Holeman, 1968. The sediment yield of major rivers of the world. *Water Resources Research* **4**:737.

†Tons of sediment per square kilometer of drainage area per year.

Much lithogenous sediment comes from semiarid regions, usually mountainous areas, as indicated in Fig. 4-9. Normally rainfall in such areas is inadequate to support an erosion-resisting cover of vegetation but sufficient to erode soils. In Asia and India long-continued intensive agriculture has further increased sediment yield. Low-lying tropical areas usually supply relatively little sediment per unit of drainage basin because tropical vegetation inhibits particle movement. Also, intensive weathering of rocks in tropical climates tends to break them down completely, forming soluble components, rather than to leave discrete mineral particles. In polar regions low temperatures retard chemical decomposition of rocks because water, involved in most weathering processes, is frozen for much of the year. Furthermore, the glaciation of these high-latitude regions during the last Ice Age scoured out many lakes that still act as traps, inhibiting sediment transport to the ocean.

Figure 4-9

Major source areas of lithogenous sediment. The rivers shown discharge more than 100 million tons of suspended sediment each year. (Data from J. N. Holeman, 1968. The sediment yield of major rivers of the world. *Water Resources Research* 4: 737.)

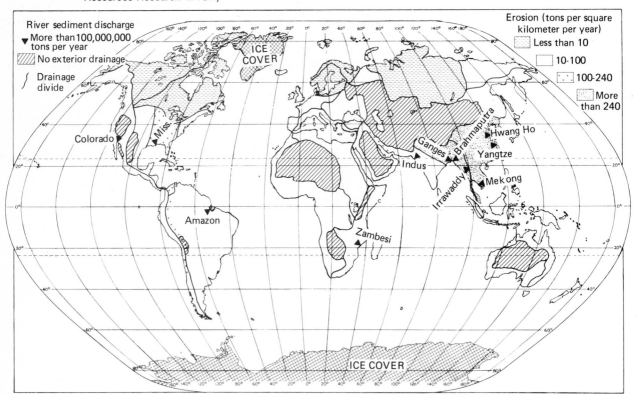

SEDIMENT TRANSPORT

Transport of sediment particles is determined primarily by physical characteristics. Particle size (see Fig. 4-1) and current speed (see Fig. 4-10) are most important. Vertical transport (sinking of particles through a column of water to the ocean floor) is easiest to visualize. The settling speed of a particle is controlled primarily by its diameter; large particles fall faster than small ones. Settling times for different sized grains are shown in tabular form.

PARTICLE	SETTLING VELOCITY (centimeters per second)	TIME TO SETTLE 4 KILOMETERS
Sand (100 μ)	2.5	1.8 days
Silt (10 μ)	0.025	185 days
Clay (1 μ)	0.00025	51 years

Obviously sands could settle out near the point where they enter the ocean, silts could be transported moderate distances, and clays could theoretically be transported throughout the world ocean before settling through the 4 kilometers of seawater to reach the ocean bottom.

But let us consider another aspect of sediment movement: lateral transport. Figure 4-10 indicates the current speeds necessary to suspend sediment particles and to transport them. Note that, for noncohesive sediments, the speed necessary to move sediment particles steadily decreases as particle size decreases so that even weak currents can erode and transport fine-grained sediment. If, however, particles are cohesive (stick together), the force required to erode the deposit first decreases as the grain size gets smaller and then increases so that erosion of cohesive silts and clays may require currents just as strong as those necessary to move sands and gravels.

Once eroded and set in motion, particles tend to settle out. This tendency is opposed by upward-directed but basically random water movements existing during turbulent flow. Even at current speed less than necessary to erode the sediment, particles may still be transported some distance before settling out of the water.

Surface and tidal currents (see Fig. 4-10) are often strong enough to erode and transport most common sediment types. Even deep-ocean currents can erode and transport noncohesive, fine-grained bottom deposits. This feature is especially important on the continental rises.

Sediment is initially transported by rivers either suspended in the water (usually about 90% of the river's sediment load) or rolling along the bottom as part of the so-called bed load. The bed load is commonly deposited near the head of the estuary, where the downstream flow of river water encounters upstream near-bottom currents caused by the estuarine circulation. At that point the downstream

Figure 4-10

Current speeds required for erosion of sediments and transport of grains as compared to settling speeds for sediment grains. (Note use of logarithmic scale.) (After B. C. Heezen and C. D. Hollister, 1964. Deep-sea current evidences from abyssal sediments. *Marine Geology* 1: 141.)

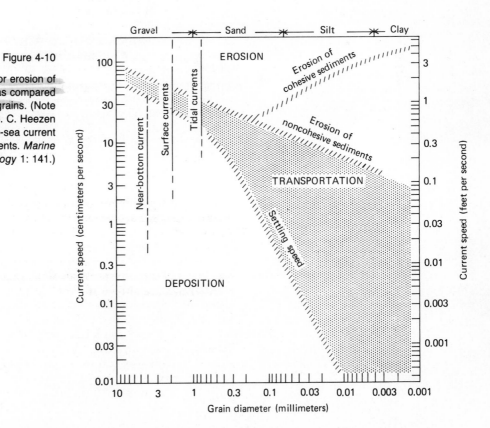

current is too weak to transport the relatively large grains that constitute the bed load and so they are deposited.

Organisms also cause sediment deposition. Many marine organisms obtain food by filtering water. Mineral particles removed along with their food are compacted into fecal pellets and then excreted. Oysters, mussels, clams, and many other attached and floating animals effectively remove particles from suspension and bind them into fecal pellets that accumulate nearby. Plants growing along the banks and on tidal flats also trap sediment by providing quiet sheltered areas where these small particles can settle out. Plants then prevent erosion of sediment deposits when tidal currents or waves scour the area.

As a result of these processes, most sediments are trapped and deposited near river mouths. Estuaries that have not been filled by sediment act as effective sediment traps; for example, rivers of the U.S. Atlantic Coast transport an estimated 20 million tons of sediment each year to their estuaries, but virtually none reaches the continental shelf. Thus the continental shelf of eastern North America is covered primarily by sediment laid down when sea level was lower. There are only small, isolated areas of modern sediment deposits.

When the sediment load of a river is large, it may fill its estuary; then the sediment load can be carried out into the ocean. The Mississippi River is a good example. Its large sediment load has not only filled the estuary but also created the Mississippi delta, a sediment deposit that extends into the Gulf of Mexico. Estimates for different rivers suggest that from 10 to 95% of a river's sediment load is deposited in its delta.

Perhaps as much as 100 million tons of lithogenous sediment per year are transported by winds to the ocean, mainly from deserts and high mountains having sparse vegetation and strong winds. (The distribution of deep-ocean sediments is shown in Fig. 4-11) Wind transport of sediment to the deep ocean seems to be especially important in middle latitudes, where there are few rivers owing to the arid climate. It seems to be less important in the Southern Hemisphere, where there is less land.

Particles less than 20 microns in diameter may be carried great distances by winds. Volcanic fragments smaller than about 10 microns may be carried around the world if an eruption is powerful enough to inject fine ash into the stratosphere. In 1956, for instance, a large volcanic eruption on the Kamchatka Peninsula, northeast of Japan, was later detected in the atmosphere over England at heights of 18,000 meters.

Particles eroded by winds from soils in arid or mountainous regions generally remain in the lower atmosphere, forming a "plume" of such material downwind. This situation is particularly noticeable off Africa, where Sahara sand is often blown hundreds of kilometers to sea. Precipitation quickly removes particles from the lower atmosphere; they serve as condensation nuclei for raindrops and snowflakes.

ATMOSPHERIC TRANSPORT Immense volumes of volcanic ash from large volcanic eruptions fall into the ocean each year. Very large events are infrequent; so our history of them is poorly documented. An eruption of the Indonesian volcano, Tambora, in 1815 released nearly 80 cubic kilometers (20 cubic miles) of ash and left an easily recognizable halo of volcanic ash in the sediments on the ocean bottom around the volcano. On August 26, 1883, Krakatoa, on an island between Java and Sumatra in Indonesia, erupted in one of the largest events ever studied. About 16 cubic kilometers (4 cubic miles) of ash was discharged to the atmos-

Figure 4-11
Distribution of deep-ocean sediments.

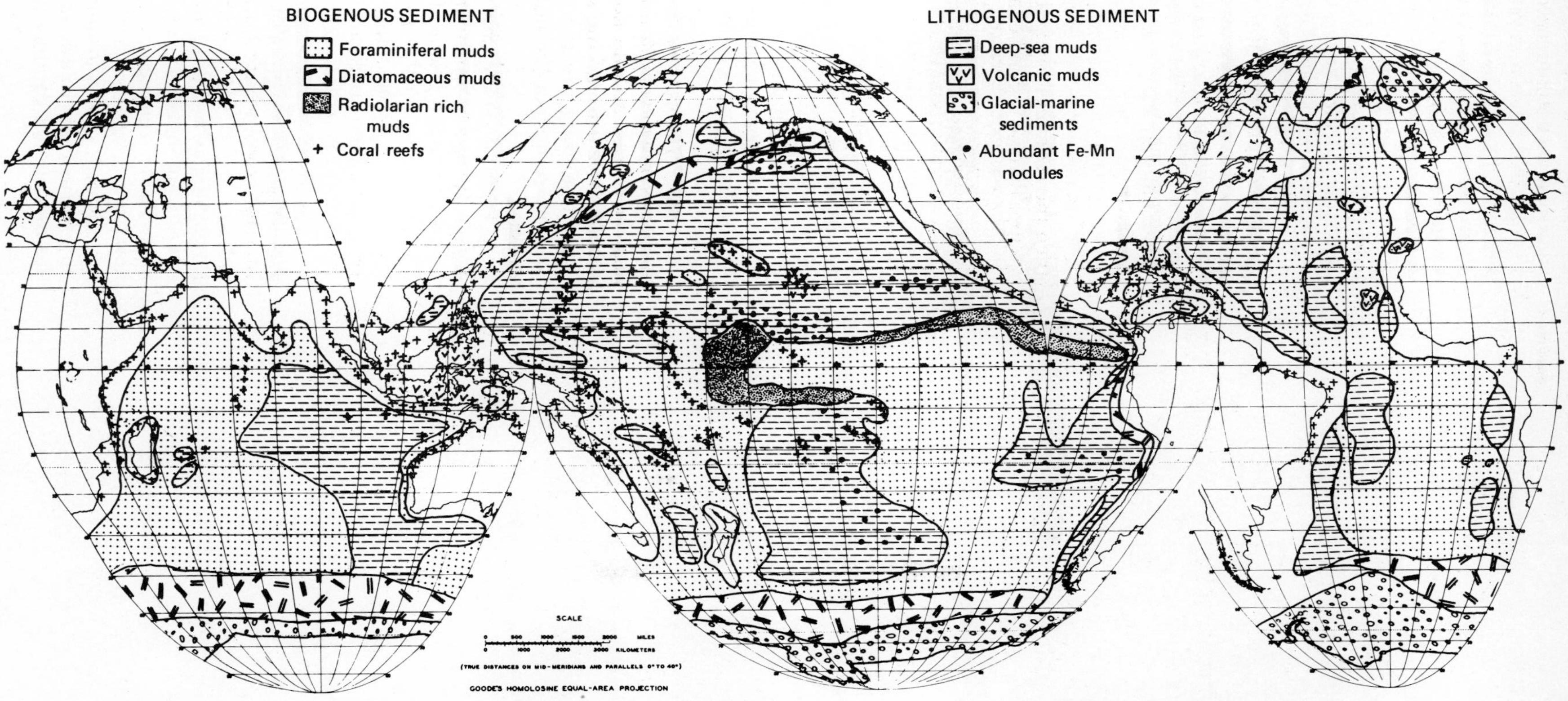
BIOGENOUS SEDIMENT
Foraminiferal muds
Diatomaceous muds
Radiolarian rich muds
+ Coral reefs
LITHOGENOUS SEDIMENT
Deep-sea muds
Volcanic muds
Glacial-marine sediments
• Abundant Fe-Mn nodules
SCALE
MILES
KILOMETERS
(TRUE DISTANCES ON MID-MERIDIANS AND PARALLELS 0° TO 40°)
GOODE'S HOMOLOSINE EQUAL-AREA PROJECTION

phere, much of it going into the stratosphere. An area of nearly 4 million square kilometers (about 1.5 million square miles) had ash deposits and large amounts were deposited on the sea floor. The May 18, 1980, eruption of Mt. St. Helens in Washington injected more than 4 cubic kilometers (about 1 cubic mile) of ash into the atmosphere. The plume of ash from the volcano extended hundreds of kilometers down wind (east and southeast) from the mountain.

Because of the enormous releases of energy during these volcanic eruptions, large amounts of ash are injected into the stratosphere to be transported around the world. Ash from Krakatoa was observed for several years over Europe, causing spectacular sunsets.

This wide dispersal of volcanic and wind-blown dust from deserts and high mountains apparently is a major source of sediment to the deep ocean, especially in the midlatitudes. Because most of the river-borne sediment is trapped in estuaries and on the continental shelves, the winds carrying up to 75% of the fine-grained lithogenous sediment deposited in open ocean areas beyond the influence of turbidity currents.

TURBIDITY CURRENTS

A turbidity current is a dense mixture of water and sediment that flows along the bottom, transporting sediment from continental margin to continental rise and even onto the deep-ocean floor. Such flows have never been directly observed in the ocean, although their occurrence in lakes is well documented.

An earthquake may set one off, as happened in November 1929, when a turbidity current triggered by the Grand Banks earthquake hurtled down the continental slope off Newfoundland, breaking telegraph cables along its way. Such a flow moves like a gigantic submarine avalanche at speeds of up to 100 kilometers per hour (60 miles per hour), finally depositing a thin layer of sediment over a large area of ocean bottom. A sudden large discharge of river-borne sediment may trigger a turbidity current, apparently strong enough to erode rock.

When the current first passes, it is moving at high speeds and carrying sand-sized particles. Later sections of the flow move progressively more slowly, transporting finer particles that are deposited on top of the heavy ones. The result is a rapidly formed sediment deposit showing distinct gradation in particle size, changing from coarsest grained at the bottom to the finest at the top, called *graded bedding*.

Turbidity currents also carry shells of shallow-water organisms and fragments of land plants onto the deep-ocean floor. Plant fragments—some still green—have been dredged from the ocean bottom following cable breaks. Turbidite layers containing remains of shallow-water organisms often alternate with layers of sediment that accumulated slowly particle by particle and contain organisms or exhibit other features typical of deep-ocean sediments.

Turidity currents are most active on narrow continental shelves, and least active on wide continental shelves. In the Atlantic, turbidity current deposits have buried most of the older topography so that currents flow from the edge of the continents almost to the midocean ridge. Similarly, sediment brought into the northern Indian Ocean can flow unimpeded over great distances of deep-ocean bottom. But in the Pacific trenches and ridges along the margins of the ocean basin now prevent movement of turbidity currents to the deep-ocean floor.

Besides carrying sediment and burying ocean bottom topography, turbidity currents are also thought to cut the submarine canyons that indent most continental margins. Whether turbidity currents are the sole agent responsible for these large features has been debated extensively, but there is good reason to suppose that in many areas they serve to remove the large volumes of sediment that would otherwise fill these canyons completely. Evidence for this statement comes from frequent breakage of submarine cables off the mouths of large rivers, such as the Congo.

CLAY MINERALS

Deep-sea muds cover the deeper parts of the ocean basins. The very fine grained deep-ocean sediment contains a distinctive suite of layered-silicate minerals (clays) derived from different climatic zones on land (see Table 4-5). The clay mineral fraction constitutes 60% or more of the nonbiogenous fraction of these sediments, as indicated in Fig. 4-12.

TABLE 4-5

Common Clay Minerals and Their Distribution In Marine Sediment*

MINERAL	ORIGIN	AREA OF ABUNDANCE IN OCEAN
Chlorite	Abrasion of metamorphic and sedimentary rocks, especially by glaciers in polar regions	Polar and subpolar regions
Montmorillonite group of minerals	Chemical alteration of volcanic ash; may form in place after deposition in seawater	South Pacific and Indian oceans near midocean ridges
Illite (hydromica group of minerals)	Chemical and mechanical alteration of mica minerals in rocks on continents	Near continents, especially common in North Atlantic
Kaolinite	Forms by decomposition of silicate minerals under intensive weathering in the formation of tropical soils	Equatorial Atlantic and Indian Ocean near Australia

Data from J. J. Griffin, H. Windom, and E. D. Goldberg, 1968. The distribution of clay minerals in the world ocean. Deep Sea Research **15: 433–459.*

Figure 4-12

Grain-size distribution in typical deep-ocean sediments. (After Sverdrup et al., 1942.)

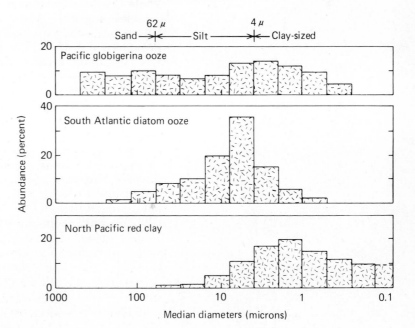

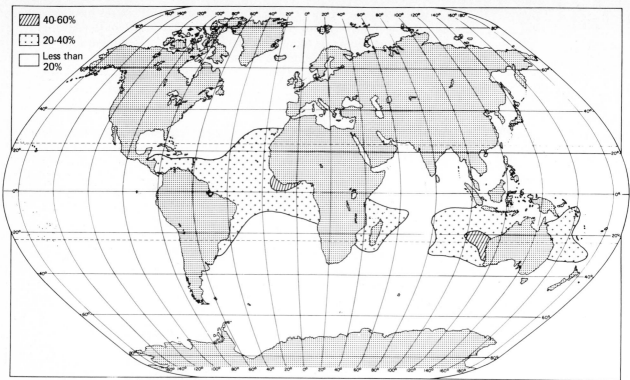

	40-60%
	20-40%
	Less than 20%

Figure 4-13

Distribution of kaolinite in the fine-grained portion (less than 2 microns) of deep-ocean sediment. (After J. J. Griffin, H. Windom, and E. D. Goldberg, 1968. Distribution of clay minerals in the world ocean. *Deep Sea Research* 15: 433–459.)

Clay minerals are small (see Fig. 4-1), less than 4 microns, and are transported by currents, winds, rivers, and glaciers. The distinctive composition of clay minerals permits us to determine the relative importance of these transport mechanisms in various parts of the ocean.

Tropical rivers contribute distinctive clay minerals *(kaolinite, gibbsite)* from highly weathered tropical soils. These clays are widely distributed (see Fig. 4-13) across the tropical ocean.

Illite, a group of minerals derived from slightly altered micas that generally occur in many different kinds of rocks, is also a common constituent of deep-ocean sediment. Illite is aboundant in the Atlantic Ocean basin (Fig. 4-14). In the central Pacific, where river-transported sediment does not occur, illite is also a common constituent. It is presumably carried there by winds, such as jet streams.

Another clay mineral, *montmorillonite,* derives partly from certain soils but is also formed in the ocean from in-place alteration of glass from volcanic eruptions. In the central Pacific, where submarine volcanism is common and other sediment sources far removed, this mineral occurs in great abundance, often to the exclusion of other clay minerals. The relatively high concentrations of montmorillonite near the Mid-Indian Ridge may also be due to the high incidence of volcanism in that region. Other common silicate minerals in deep-ocean sediments form from altered volcanic glass, such as zeolites.

At higher latitudes other clay minerals (primarily *chlorite*) are formed in Arctic soils. Because of the low temperatures, these clay minerals are much less altered than those derived from tropical soils. Ice transport of sediment is important in the polar and subpolar regions, usually in latitudes between 50 and 90° in both hemispheres. In midlatitudes jet streams are important transport routes for wind-blown material from the continents.

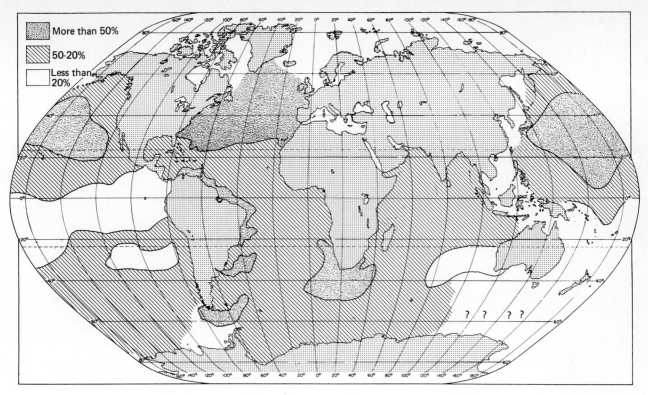

Figure 4-14

Distribution of illite in the fine-grained portion (less than 2 microns) of deep-ocean sediment. (After Griffin et al., 1968.)

DISTRIBUTION OF SEDIMENT DEPOSITS

Transport processes control the distribution of marine sediments. In general, river-transported sediment is restricted to continental margins. Most sediment remains on the shelf or rise except where turbidity currents are active and can carry material out into the deep ocean. Some deposits were formed by processes still active in the ocean; formed at some earlier time, they remain exposed on the ocean bottom but do not reflect present oceanic processes in areas where they occur.

Many ocean margins are covered by thick deposits of lithogenous sediment. Transport of this material from continents to the deep-ocean bottom is inhibited by the topography of most continental margins and the lack of modern sediment brought to the continental shelf by rivers, a result of the changes in sea level following the last retreat of the glaciers.

Deposits of sand and silt that were formed under conditions no longer existing in an area are known as *relict sediments*. They are recognized by distinctive features, such as shells of organisms that no longer grow there. Oyster shells found far out on the U.S. continental shelf, for example, indicate deposition at a time of greatly lowered sea level. Mastodon teeth and remains of extinct land animals indicate that the area was once dry and that its surficial deposits formed at that time. Other characteristics of relict sediments include iron stains or coatings on grains that could not have been formed under present marine conditions.

Where rivers bring sediment to the coastal ocean, relict deposits are buried. Alternatively, they may be reworked by the action of waves during periods of rising sea level so that the characteristic features of a relict deposit are destroyed or masked. About 70% of the world's continental shelves is now covered by relict sediments (see Fig. 4-15).

Sand deposited near the mouth of a river is moved along the

103

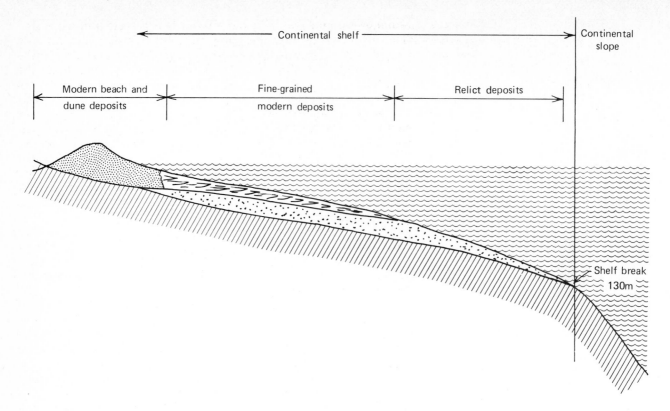

Continental shelf ⟶ Continental slope

Modern beach and dune deposits | Fine-grained modern deposits | Relict deposits

Shelf break
130m

Figure 4-15

Continental shelves in midlatitudes frequently have an active sandy beach–sand dune complex along the shore that merges with a finer-grained modern sediment deposit in a band parallel to the shore at mid-depths on the shelf. These presently active deposits overlie relict deposits that were deposited under conditions no longer prevailing in the region. These relict deposits are exposed at the outer edge of the shelf near the shelf break. All these unconsolidated sediments lie on the rocks of the submerged margin of the continental block.

coast by longshore currents. The association of sand beaches and river mouths is especially obvious on the Pacific Coast of the United States, where the largest beaches and dunes in Washington and Oregon are associated with the Columbia River mouth. Sands are common in areas of strong wave action because finer-grained materials are removed.

Slits and clays are carried farther seaward and deposited at depths where wave action is too weak to stir up sediment or erode the bottom. On many continental shelves this process occurs at depths of 50 to 150 meters.

Sands and silts deposited on continental shelves commonly accumulate at rates greater than 10 centimeters per 1000 years. This accumulation is relatively rapid and particles have insufficient time to react chemically with the seawater or with near-bottom waters as they lie on the ocean bottom before being buried. Coastal ocean sediments therefore retain many of the characteristics that they acquired during weathering. These rapidly accumulating deposits tend to be rather dark colored—gray, greenish, or sometimes brownish—and they contain abundant organic matter (1 to 3% is not uncommon).

In polar regions glacial marine sediments are the most common sediments on the ocean bottom. These deposits contain all sizes of particles—from boulders to silt (see Fig. 4-1) deposited by melting icebergs from continental glaciers. Sea ice normally contains sediment only if it has gone aground and incorporated material from the shallow continental shelf, which happens in the Arctic Sea. When such sea ice melts, its sediment load is deposited on the bottom, where it mixes with the remains of marine organisms. In the Antarctic icebergs carry sediment picked up from the bottom and also material trapped in the ice when it froze.

Reef-building corals can grow in areas where the temperature during the coldest month of the year is 18°C or warmer. Many continental shelves within this belt are covered by biogenous sediment that usually consists of shallow-water calcareous algae, shallow-water foraminifera, and coral. In the United States reefs and associated sediments occur around the southern tip of Florida. They are also common on the shallow Bahama Banks, Bermuda, and off the Yucatan Peninsula of Mexico, and are especially extensive around the northern coast of Australia, where the *Great Barrier Reef* is 150 kilometers wide and extends for about 2500 kilometers.

Near active volcanoes, the ocean also receives substantial amounts of volcanic rock fragments, usually the relatively large particles (greater than 50 microns in diameter) formed during eruptions. Sediments deposited in the Indonesian area, around the Aleutians off Alaska, and near western Pacific island arcs contain abundant volcanic debris. Volcanic sediments are not restricted to the Pacific Ocean; ash occurs in all ocean basins, transported by winds from major volcanic eruptions.

Sediments that accumulate fairly rapidly near continents (see Table 4-6), transported either by running water, turbidity currents, or ice, cover about 25% of the ocean bottom. Accumulations are thickest in marginal ocean basins; although accounting for only about 2% of the ocean area, they contain about one-sixth of all oceanic sediment.

Much of the deep-ocean floor is covered by *pelagic sediments* that accumulate slowly, particle by particle, at rates between 1 and 10 millimeters per thousand years. Deposits formed in this way differ substantially from those formed by turbidity currents. They blanket the original ocean bottom topography, faintly preserving its outlines. A comparison has often been made with fallen snow on land. Sometimes, however, as with a heavy accumulation of snow, deposits on sides of hills slump down into nearby valleys. In this case; the effect is similar to a turbidity current, which first fills the valleys and eventually buries the hills, leaving a gently sloping smooth surface.

Clays and very small particles that escape the river mouth can be carried great distances. Even relatively large clay particles (1 micron) require about 50 years to settle through 4 kilometers of water. Smaller particles (less than 0.5 micron), which are more typical of the particles in the deep ocean, theoretically require up to 1000 years to sink to the bottom in still water. Most particles, however, are removed from suspension fairly rapidly, probably due to filtration of seawater by planktonic organisms that ingest particles and incorporate them in fecal pellets. These relatively heavy pellets sink at rates of 10 meters or more per day.

TABLE 4-6

Typical Sediment Accumulation Rates

AREA	AVERAGE (RANGE) ACCUMULATION RATE (centimeters per thousand years)
Continental margin	
Continental shelf	30 (15–40)
Continental slope	20
Fjord (Saanich Inlet, British Columbia)	400
Fraser River delta (British Columbia)	700,000
Marginal ocean basins	
Black Sea	30
Gulf of California	100
Gulf of Mexico	10
Clyde Sea	500
Deep-ocean sediments	
Coccolith muds	1 (0.2–3)
Deep-sea muds	0.1 (0.03–0.8)

Deep-ocean sediments accumulate slowly, generally at rates of less than 1 centimeter per thousand years. Thus the particles not only spend years suspended in ocean water but also remain exposed on the ocean bottom for centuries before they are finally buried and sealed off from contact with near-bottom waters. Burrowing organisms further increase particle exposure by stirring up the deposits. As a result, there is usually ample time for particles in deep-ocean sediments to react with seawater. Abundant dissolved oxygen in deep-ocean waters causes iron to be converted to the ferric state (iron rust), giving red clays their typical color. The presence of dissolved oxygen also favors utilization of organic matter by organisms or its destruction by inorganic processes; consequently, deep-ocean sediments normally contain less than 1% organic matter.

Over most of the deep ocean the sediment cover is relatively thin—usually between 0.5 and 1 kilometer, averaging about 600 meters. Thin sediment deposits generally occur on midoceanic ridges or rises, where large areas may be completely swept clear of sediment, apparently by strong currents. Sediments accumulate there in certain protected "pockets." Sediment thickness increases away from the ridges and is generally greatest over the oldest crust near the continents along the ocean basin margins (shown in Fig. 4-16). In general, these sediments are flat lying and in many areas apparently consist of alternating layers of turbidites (deposits formed by turbidity currents) and particle-by-particle accumulations. Such stratified sediments are especially common in the Atlantic and Indian oceans and in parts of the Pacific. There is little evidence for disturbance of these sediments since their deposition many millions of years ago.

Manganese nodules are common in the central portions of the Pacific, far from major sediment sources. In the Atlantic manganese nodules are less common; an exception is the Blake Plateau off the southeastern United States, where the Gulf Stream keeps the ocean bottom free of sediment.

Some general relationships between sediment-transport process and sediment distribution are summarized diagrammatically in Fig. 4-17. Note in Fig. 4-17 that recent ice ages and the present polar ice cover have controlled the distribution of glacial marine sediments, both on the shallow and deep-ocean bottom. A less obvious effect of the present climatic regime is the formation of cold, dense bottom waters in the Antarctic that dissolve carbonate sediments in the deepest part of the ocean.

Effects of high productivity can be seen on the deep-ocean bottom and on the continental shelf. On the deep-ocean bottom (Fig.

Figure 4-16

Sediments are thickest near ocean basin margins because of heavy discharge from continents, probably greatest during times of lowered sea level. Also, ocean basins are youngest near midoceanic ridges.

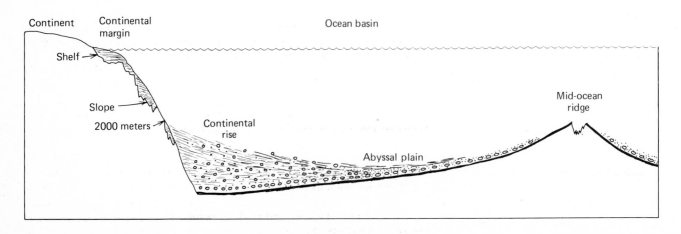

Iron-manganese sediment   Carbonate ooze   Turbidite deposits   Pelagic deposits

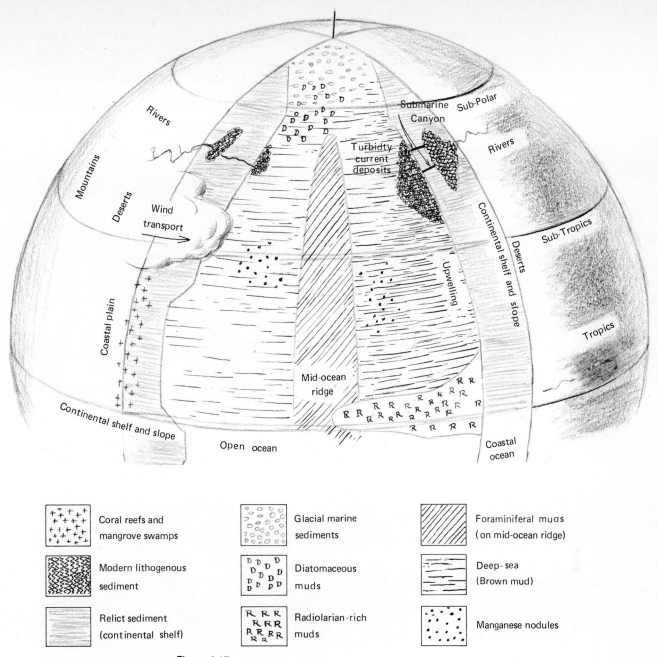

Figure 4-17

Coral reefs and mangrove swamps

Modern lithogenous sediment

Relict sediment (continental shelf)

Glacial marine sediments

Diatomaceous muds

Radiolarian-rich muds

Foraminiferal muds (on mid-ocean ridge)

Deep-sea (Brown mud)

Manganese nodules

Schematic representation of the general distribution of sediment in the deep and coastal ocean. [Modified in part after K. O. Emery, 1968. Relict sediments of continental shelves of world. *American Association of Petroleum Geologists Bulletin* 52(3): 445–464.]

4-17) bands of diatomaceous muds in high latitudes and radiolarian muds in equatorial regions directly reflect the high biological productivity of surface waters in these regions. On the continental shelf the abundance of recent carbonate sediment in tropical waters is again a result of the locally high carbonate productivity. Figure 4-17 shows that turbidity currents leave thick sediment deposits on and near continental margins. Note also that sediment cover is thinnest at mid-ocean ridges and thicker near continents on the older ocean floor.

AGE DETERMINATION OF SEDIMENT DEPOSITS

The most common techniques used to date sedimentary deposits employ fossils, such as those in Figs. 4-18 and 4-19. Fossils show changes through time that paleontologists use to assign ages to their assemblages. Time zones are defined in terms of the presence or absence of several different fossils. Microorganisms—foraminifera, radiolaria, and coccoliths (Fig. 4-18)—are commonly used in marine sediments, for a large number can be recovered from a small volume of sediment.

Figure 4-18

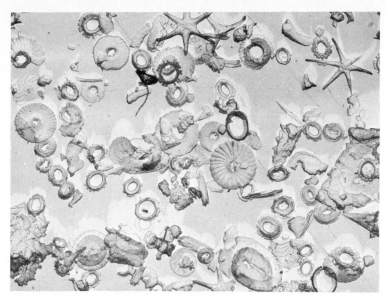

Fossil coccoliths and discoasters (calcareous nannoplankton) from sediment cores. In this electron photomicrograph, the fossils are magnified several thousand times; they range from 2 to 25 microns in diameter, equivalent to fine silt particles. The star-shaped fossils in the upper portion of the photograph are discoasters, the remains of nearly extinct organisms thought to have been related to coccoliths. Disks and rings are coccoliths, an abundant constituent of nannoplankton now living in the ocean. (Photograph courtesy Deep-Sea Drilling Project, Scripps Institution of Oceanography, under contract to the National Science Foundation.)

Figure 4-19

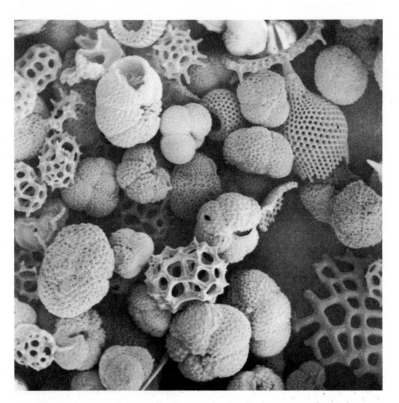

Skeletons of foraminifera and radiolaria taken from sediment in the western Pacific Ocean in waters 2660 meters deep. The core penetrated 131 meters into the ocean bottom. The fossils have been magnified about 160 times by an electron-scanning microscope. (Photograph courtesy Deep-Sea Drilling Project, Scripps Institution of Oceanography, under contract to the National Science Foundation.)

Using studies of fossil assemblages, it is possible to establish time zones and correlate widely separated deposits. Eventually the history of an entire ocean basin over the past 200 million years may be worked out from fossils buried in its sediments.

Where absolute ages in years are needed, radioactive substances are used as clocks. Radionuclides decay at known, unvarying rates; the time elapsed since the rock or shell formed is reflected in the amount of decay product and the amount of the original radionuclide present in a deposit. Depending on the age of the sediment, different radionuclides, depicted in Fig. 4-20, are used.

For recently deposited sediment, carbon-14 is a useful radion-

Figure 4-20

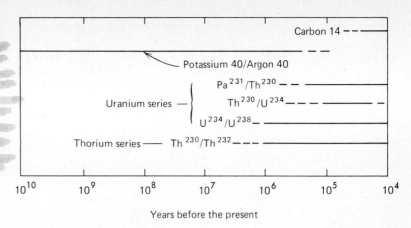

Years before the present

uclide. Cosmic rays bombard the earth, forming carbon-14 in the upper atmosphere, and all living organisms have carbon-14 atoms in their tissues. When an organism dies, no further carbon is taken up or exchanged with atmosphere or ocean water. Instead the amount of carbon-14 begins to decrease at a steady rate. One-half is gone in 5600 years (the *half-life* of this radionuclide), and after 11,200 years only one-quarter of the original amount remains. Thus by comparing the amount of carbon-14 per unit amount of carbon in a fossil with that in organisms presently living under similar conditions, it is possible to determine the time elapsed since that organism was alive. Assuming that the organism died and was incorporated in the layer when it formed, we can assign an age to the deposit. Carbon-14 is useful in dating sediments younger than about 30,000 years.

Dating older sediments requires using radionuclides with longer half-lives. Potassium-40 is useful for this purpose. Most rock-forming minerals contain potassium, which includes 1.2% potassium-40 having a half-life of 1.3 billion years. When potassium-40 decays, 11% of the daughter product is argon-40, an element not present in the mineral when it forms. Thus a comparison of the amount of argon-40 in a mineral with its potassium content provides a measure of the time elapsed since that mineral formed. The more argon-40 in a mineral, the greater its age. In using this radionuclide to date sediment deposits, it is necessary to date minerals that formed when the deposit was laid down. Volcanic ash and glauconite (a micalike mineral) are often used in dating deep-ocean deposits. The age of a mineral must be further interpreted to calculate sediment age. Because of its long half-life, potassium-40 is useful only for sediments older than 100,000 years.

Other evidence has been used for dating different parts of the ocean floor. We have seen that magnetic reversals leave magnetic patterns in oceanic crustal rocks (Fig. 3-11). It turns out that this same record is preserved vertically in sediment cores. Particles in a sediment layer reflect the earth's magnetic orientation at the time that layer was deposited. This fact was first documented in the mid-1960s, when a record of a million-year-old Jarmilla event (Fig. 3-6) was found in a core.

A record of worldwide climatic changes is also preserved in oceanic sediment. Shells of organisms known to thrive where surface waters are cold, for instance, may be found in a sediment layer deposited many millions of years ago, taken from a part of the world that now has a tropical climate. Movements of continents, changed ocean current patterns, and major climatic changes have been documented and studied by using fossils from deep-ocean deposits. The record of the last Ice Age climate, preserved in marine fossils, is discussed in Chapter 6.

109

1. Describe the three major categories (by composition) of particles in oceanic sediment deposits. Where in the deep-ocean basins is each type most likely to dominate? What is the primary source of each?

2. On a outline map of the world, indicate where the five largest sediment-transporting rivers discharge their sediment. Also indicate where you think the sediment from each river is deposited.

3. Describe how a turbidity current forms, how it transports sediment, and the type of sediment deposit it forms. Where are turbidity currents most common?

4. Explain why calcareous biogenous deposits are rare or absent on the deepest parts of the ocean floor.

5. Define relict sediment. Explain why relict sediments are common on many continental shelves.

6. Draw a cross section on the ocean floor, showing the distribution of various types of sediments in relation to the continental margin and the midocean ridge.

7. List and briefly discuss the principal methods of determining the age of sediment deposits.

SUMMARY OUTLINE

Origin and classification of sediments
Lithogenous particles—derived from silicate rocks, usually from the continents
Biogenous particles—skeletons of plants and animals

Biogenous sediments
Cover more than half of ocean basin
Distribution controlled by production, destruction, dilution
Most rapid accumulations in regions of high productivity, especially in equatorial region
Destruction of fragile or soluble remains—leaves only sturdy and less soluble shells and skeletons
Dilution by lithogenous constituents occurs near the continents

Hydrogenous sediments
Formed by precipitation from seawater, such as Fe–Mn nodules
Minor sources—cosmogenous debris from space; man's contribution

Sources and transport of lithogenous sediment
Rivers—supply about 20 billion tons of lithogenous sediment per year
Wind erosion and transport—important near desert and high mountain regions
Biogenous sediments—about 2 billion tons formed annually

Sediment transport
Settling velocity—controlled by particle size
Most sediment trapped near river mouths or in estuaries

Relict sediment—covers about 70% of continental shelves
Wind transport—important for small particles in midlatitudes; removal by precipitation

Turbidity currents
Dense sediment-laden waters move along the bottom
Form characteristic deposits—graded bedding, displaced shallow-water plant and animal remains
Important means of sediment transport across continental shelves, especially narrow shelves
Bury ocean bottom topography, forming smooth, gently sloping plains.

Clay minerals—layered silicates indicative of climatic zone where they formed in soils
Transported by winds and rivers

Distribution of sediment deposits
Reflects effects of various transport processes
Thickest deposits near continents, cover about 25% of ocean bottom; accumulate at rates up to 10 centimeters per 1000 years
Manganese nodules—common in areas of slow sediment accumulation, such as North Pacific

Correlations and age determinations
Detailed studies of fossil assemblages and isotopic composition give relative ages
Radiometric techniques give "absolute" ages—carbon-14, potassium-40
Magnetic reversals

SELECTED REFERENCES

GARRELS, R. M., AND F. T. MACKENZIE. 1971. *Evolution of Sedimentary Rock.* Norton, New York, 397 pp. Intermediate-level textbook on sedimentary processes and their long-term geochemical implications.

SHEPARD, F. P. 1974. *Submarine Geology,* 3rd ed. Harper & Row, New York. 517 pp. General treatment of sediments in ocean, emphasizing continental shelf and adjacent environments; technical, descriptive.

TUREKIAN, K. K. 1976. *Oceans,* 2nd ed. Prentice-Hall, Englewood Cliffs, N.J. 149 pp. Emphasizes chemical aspects of sedimentary process; elementary.

5 SEAWATER

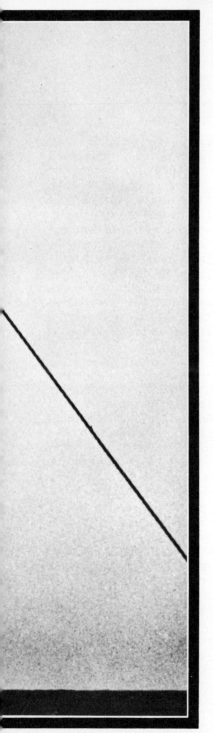

Oceanographers throughout the world have used Nansen bottles to collect samples of seawater at specified depths. Attached in a series to a hydrographic wire connected to a research vessel, the bottles reverse the fill when actuated by sliding weights called "messengers". Thermometers on the side of each bottle record water temperature *in situ*. (Photograph courtesy of Scripps Institution of Oceanography, University of California, San Diego.)

Seawater is a complex mixture. On the average, it is about 96.5% water containing 3.5% salt, a few parts per million of living things, and perhaps an equal amount of dust. Water itself has many unusual properties that are little affected by the presence of sea salts. An example is water's capacity to absorb and give up large quantities of energy as heat with little change in temperature. On the other hand, the ocean's suitability for life is strongly influenced by the dissolved materials. Minute differences in chemical composition can drastically alter its biological properties. Living organisms, in turn, affect the abundance of dissolved gases and salts in seawater. Finally, some important properties of ocean water, such as the way it transmits sound, are controlled partly by the structure of water and the behavior of its dissolved salts. But let us begin by considering the unique properties of water itself before considering the effects of adding sea salts.

MOLECULAR STRUCTURE OF WATER

Many of the important properties of water are a result of its unusual molecular structure. Two hydrogen atoms and one oxygen atom combine to form one water molecule. Each oxygen atom has two unshared electron pairs and two electron pairs shared with the hydrogen atoms. The unusual properties of water result from the structure of this molecule, specifically its associated electron cloud. This cloud has a definite shape, forming the "third and fourth corner" of the molecule, which resembles a child's four-pronged jack. The oxygen atom, being relatively large, accounts for most of the molecule's volume. The two hydrogen atoms are much smaller and are partially buried in the electron cloud of the oxygen atom, forming two of the four "prongs" of the molecule (see Fig. 5-1).

The other two "prongs" are formed by the unshared electrons of the oxygen atom. The hydrogen atoms are on one side of the molecule, the unshared electrons on the other side. So viewed on an atomic scale, a water molecule is essentially negative on one side (the side with the unshared electrons) and positive on the other (the side with the two hydrogen atoms). Such a molecule, whose electronic charges are separated, is called a *polar molecule*. It responds to nearby electrical charges. The unshared electrons on one molecule tend to form weak bonds, called *hydrogen bonds,* with hydrogen atoms of the adjacent molecule. This tendency for hydrogen bonding between water molecules accounts for many of water's unusual properties, including its large capacity for absorbing heat, its abnormally high

114

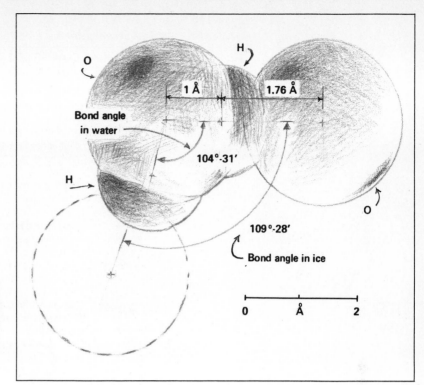

Figure 5-1

Schematic representation of a water molecule showing the bonding between the hydrogen atoms on the molecule and adjacent oxygen atoms in water molecules on the right and lower left. Note the separation between the side of the molecule having no hydrogen atoms and the side of the molecule having no hydrogen atom. (An angstrom, Å, is one-billionth of a meter.)

boiling point and freezing point, and its capacity to dissolve substances that are held together by ionic bonds; some of these properties are listed in Table 5-1.

Crystals and molecules are attracted to one another and held together by various forces. The weakest result from fleeting electronic interactions between molecules. These so-called *van der Waals bonds* form, break, and reform easily; in water they are greatly overshadowed in importance by the hydrogen bonds.

Ionic bonds form relatively strong bonds between adjacent molecules (or atoms) that have lost or gained electrons, thus acquiring either a positive charge, where an electron has been lost, or a negative charge, where an electron has been gained. One example of ionic bonding is the association between sodium (Na^+) and chlorine (Cl^-) to form table salt, sodium chloride. The bond is the result of strong attraction between closely spaced ions of opposite charge. To get some idea of the strength of these bonds, consider the amount of energy (expressed in kilocalories, a measure of heat energy) required to break each type of bond in 1 *mole* of a substance—for instance, water. One mole of water weighs 18 grams or about ½ ounce—that is its formula weight in grams—oxygen having an atomic weight of 16 and each hydrogen atom having an atomic weight of 1:

van der Waals bonds: 0.6 kilocalorie per mole
hydrogen bonds: 4.5 kilocalories per mole
ionic bonds: 10s of kilocalories per mole

At the relatively low temperatures and pressures of the earth's surface, water is one of the few substances that we generally see in all three *states of matter—solid, liquid,* and *gas. Water vapor*—the gas—is probably the easiest to understand. In water vapor each molecule exists separately and is little affected by other molecules. So each molecule is free to move with little restriction except for the walls of

115

TABLE 5-1

*Certain Anomalous Physical Properties of Water and Their Effect on Seawater**

PROPERTY	COMPARISON WITH OTHER SUBSTANCES	IMPORTANCE IN OCEAN
Heat capacity	Highest of all solids and liquids except liquid ammonia	Prevents extreme ranges in ocean temperature Heat transfer by currents is large
Latent heat of fusion	Highest except ammonia	Thermostatic effect at freezing point owing to uptake or release of latent heat
Latent heat of evaporation	Highest of all substances	Extremely important in heat and water transfer of atmosphere
Thermal expansion	Temperature of maximum density decreases with increasing salinity. For pure water, it is at 4°C.	Freshwater and dilute seawater reach maximum density at temperatures above freezing point
Surface tension	Highest of all liquids	Controls certain surface phenomena and drop formation and behavior
Dissolving power	Dissolves more substances and in greater quantities than any other liquid	
Dielectric consant	Pure water has the highest of all liquids	Of utmost importance in behavior of inorganic dissolved substances because of resulting high dissociation
Electrolytic dissociation	Very small	A neutral substance; yet contains both H^+ and OH^- ions
Transparency	Relatively great	Absorption of radiant energy is large in infrared and ultraviolet. In visible portion of energy spectrum there is relatively little selective absorption—hence is "colorless."

*Modified after Sverdrup et al., 1942.

the container in which we enclose it. This freedom to move accounts for the characteristics of a gas: it has neither size nor shape but expands readily to fill any container in which it is placed.

Water molecules striking the sides of a container exert *pressure*. Pressure can be increased in two ways: we can put more material in the container or we can shrink it, thereby increasing the rate of molecules striking the sides of the container; alternatively, we can make the molecules move faster and strike the container walls more frequently. We do the first by placing more material (even another gas) in the container, the second by increasing the temperature, which makes molecules move more rapidly.

At the opposite extreme in internal order is the solid—in this case, *ice*. Solids have a definite size and shape. Most are also crystalline—that is, they have a fixed internal structure. Usually solids break, or bend, when enough force is applied. This resistance to flow, or deformation, results from the orderly internal structure. Each atom, or molecule, has a position that it occupies for long periods of time. These atoms or molecules cannot move readily from position to position nor rotate in a given position. But at room temperatures there is always some vibration of atoms in their fixed position. Some movement is possible if there is a hole nearby into which a molecule can move. Just as with water vapor, movements of the atoms or molecules increase with increasing temperature so that movement between positions also increases with rising temperature.

Water molecules in ice are held together by hydrogen bonds, as shown in Fig. 5-2. The large oxygen atoms have definite positions and the hydrogen atoms in each molecule are also oriented in a regular manner. The oxygen atoms form six-sided puckered rings arranged in layers, each layer a mirror image of the adjacent layer. The result is a fairly open network of atoms that gives ice a somewhat lower density (approximately 0.92 gram per cubic centimeter) than that of liquid water (approximately 1 gram per cubic centimeter) because the molecules are not as tightly packed together.

Despite the openness of the ice structure such impurities as sea salts are not readily incorporated into the structure. Consequently, salt is excluded from sea ice. Excluded sea salts remain in pockets of unfrozen liquid (brine).

Liquid water has anomalous properties (see Table 5-1) intermediate between those of solids and gases. It apparently consists of two different typs of molecular aggregates (Fig. 5-3) in an equilibrium determined by temperature and pressure. The first component consists of clusters of hydrogen-bonded water molecules. These clusters form and reform rapidly—10 to 100 times during one-millionth of a microsecond (10^{-10} to 10^{-11} second). Although the lifetime of any individual one is extremely short, clusters persist long enough to influence the physical behavior of water. These structured portions of water are less dense than the unstructured portion. In some respects, this lowered density of the structured portion occurs for the same reasons that make ice less dense than liquid water. It seems likely,

Figure 5-2

Crystal structure of ordinary ice showing the fixed positions of water molecules (a). Note the six-sided rings formed by 24 water molecules. In the same volume of liquid water, 27 molecules would be present (b). The ice lattice is shown in (c).

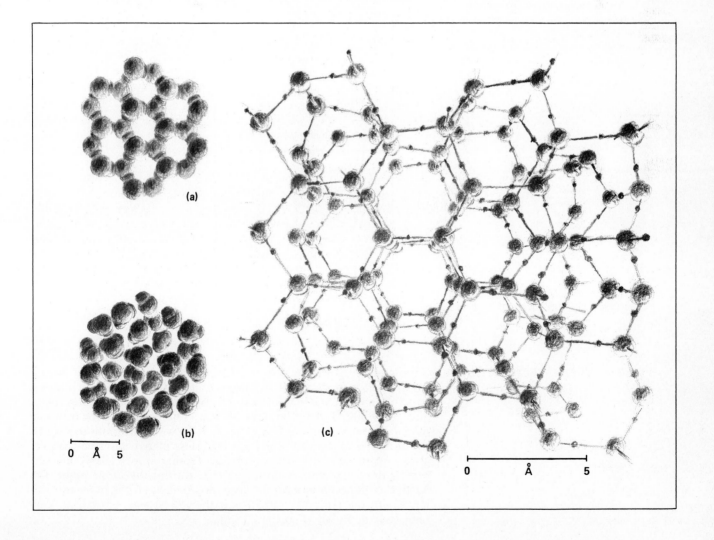

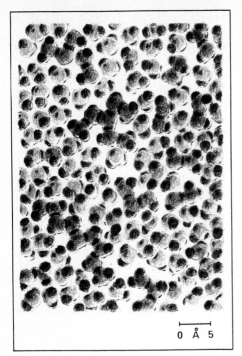

Figure 5-3

Schematic representation of liquid water at atmospheric pressure. Note the two types of water structures: areas of bonded molecules (ordered regions) and free water molecules between them.

however, that the ice structure is not exactly the same as that of the flickering clusters of molecules.

Although forming and reforming rapidly, clusters persist in liquid water until broken up by external forces. They apparently disappear when the pressure exceeds about 1000 atmospheres (a pressure reached only in the few deepest trenches of the ocean floor) or at temperatures exceeding 100°C (at atmospheric pressure), where liquid structures break up and individual molecules escape as water vapor.

The other constituent of liquid water is unstructured, consisting of "free" water molecules that surround the structured portions. These molecules move and rotate without restriction. Interactions with nearby molecules are weaker than in the structured portion. The "free" water portion of liquid water is denser than the structured portion, for molecules fit more closely together. Relative proportions of the structured and unstructured portions of liquid water vary with changes in temperature, pressure, and salt content and composition.

Because of its unusual molecular structure, liquid water is strikingly different from hydrogen compounds of elements chemically similar to oxygen, as indicated in Fig. 5-4. Water's melting point and boiling point occur at temperatures 90 and 170°C higher, respectively, than might be predicted from the chemical behavior of these other compounds. If water were a "normal" compound, it would occur only as a gas at Earth surface temperatures and pressures.

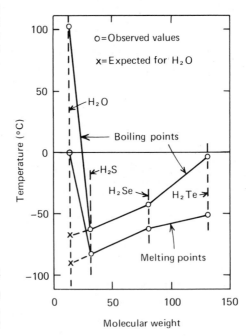

Figure 5-4

Melting and boiling points of water and chemically similar compounds. (After Horne, 1969.)

TEMPERATURE EFFECTS ON WATER

Temperature affects the internal structure of water and its properties. Much of the heat absorbed by water is used up in changing the internal structure so that water temperature rises less than other substances, after absorbing a given amount of heat. The large amount of water on Earth acts as a climatic buffer preventing wide variations in surface temperature, as experienced by a waterless planet. Surface temperatures on the moon, for example, go from about +135°C at noon during the lunar day to about −155°C during the lunar night. On Earth the highest temperature ever recorded was 57°C, at Death Valley, California (July 10, 1913), and the lowest was −68°C, at Verkhoyansk, in Russia (February 1892).

118

Earth's limited temperature range is controlled primarily by the abundance of water on its surface, for much of the incoming solar radiation goes into evaporating water and melting ice. In the ocean water temperatures rarely exceed 30°C or go below −1°C. Let us see what happens to water during these changes and how it affects the earth's heat balance.

First, we define a measure of heat, the *calorie,* as the amount of heat (a form of energy) required to raise the temperature of 1 gram of liquid water by 1 degree Celsius (1°C). This means that we can change the temperature of 1 gram of water by 50°C by supplying 50 calories of heat. Alternatively, we would change the temperature of 50 grams of water by 1°C with the same amount of heat.

Breaking bonds in ice and liquid water requires energy, usually heat. Conversely, heat is released when they reform. Consider what happens when 1 gram of ice just below the freezing point is heated slowly; let us keep track of the amount of heat added.

As we add heat to ice, its temperature increases about 2°C for each calorie of heat we add (see Fig. 5-5). When the ice reaches its melting point, 0°C, the temperature remains constant and instead the ice begins to melt. As long as ice and liquid water exist together in a container, the temperature of our system remains fixed at 0°C. After we have added about 80 calories per gram, the last bit of ice melts, leaving only liquid water.

As we continue to apply heat, water temperatures rise but now at a lower rate: 1°C per calorie of added heat. This rate of temperature change holds nearly constant between 0 and 100°C. At 100°C, the boiling point, the temperature rise again ceases and gas (water vapor) forms. At the boiling point, the same situation occurs that we experienced at the melting point. The temperature remains fixed as long as liquid and gas are present. After adding about 540 calories per gram, the last of the water evaporates. If we capture the vapor (steam) and continue heating it, we find that the temperature rises much more rapidly: about 2°C per calorie of heat added—approximately the same as with ice.

Heat added to the water is taken up in two forms. One is *sensible heat,* heat that we detect either through our sense of touch or measure with thermometers. This change in temperature results from the increased vibration of molecules or their motion in the gas state.

The other form is called *latent heat,* which is the energy necessary to break bonds in the water structures. As we have seen, at both the melting point and the boiling point of water there was no change in temperature as long as two states of matter existed together. The added energy went into breaking the bonds in the disappearing phase. This is called latent heat because we get back exactly the same amount of heat when the process is reversed. Condensing water vapor to form liquid water releases 540 calories per gram at 100°C; freezing water at 0°C releases 80 calories per gram. The difference between the latent heats of melting and evaporation arises from the fact that only a small fraction of the hydrogen bonds is broken when ice melts. All hydrogen bonds are broken when water evaporates.

Although ice freezes at 0°C and water boils at 100°C, it is possible for molecules to go from vapor to solid or liquid at other temperatures. The processes involved are similar to those described earlier, but the amount of heat involved is different. For example, the latent heat of evaporation changes as follows.

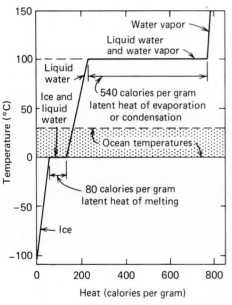

Figure 5-5

Temperature changes when heat is added or removed from ice, liquid water, or water vapor. Note that the temperature does not change when mixtures of ice and liquid water or liquid water and water vapor are present. (Redrawn from Gross, 1980.)

TEMPERATURE (°C)	LATENT HEAT OF EVAPORATION (calories per gram)
0	595
20	585
100	539

It takes more energy to remove a water molecule from liquid water at 0 or 20°C than it does at 100°C. This factor is important in the ocean, for most water evaporates from the ocean surface at temperatures around 18 to 20°C. If water had to reach 100°C before evaporating, we would have no water vapor in the atmosphere.

Both cluster size and number of molecules per cluster decrease with increasing temperature, as Fig. 5-6 illustrates. It is interesting to note in this figure that a large number of hydrogen-bonded water molecules remain bound together even after ice melts at 0°C.

Figure 5-6

Effect of changes in temperature on the relative abundance of unbroken hydrogen bonds, cluster size, and the number of molecules per cluster. (Data from G. Nemethy and H. A. Scheraga, 1962. Structure of water and hydrophobic bonding in protein. *Journal of Chemical Physics* 36: 3382–3400.)

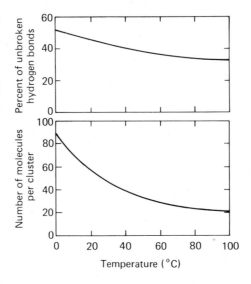

DENSITY Whether a substance sinks or floats in a liquid is determined primarily by its *density* (mass per unit volume, expressed in grams per cubic centimeter). A substance less dense than its surroundings tends to float upward. If its density is greater than its surroundings, it displaces less mass than its own mass, settles out of the liquid, and sinks. The sinking rate is determined by its relative density and the resistance it experiences while moving through the water. A small object generally experiences more drag per unit mass than a large object and so a small particle takes much longer to sink than a large object that experiences relatively less drag.

A density-stratified system of fluids is one in which lighter fluids float on heavier ones. A mass of fluid may sink through materials less dense than itself until it reaches a zone where the fluid below is more dense and the fluid above less dense. At this point, there is no force acting on the fluid and it tends to remain at that level. Furthermore, vertical displacement of the fluid will be countered by forces tending to keep it at the same relative position. (An example is the poussecafé, an after-dinner drink in which colorful liqueurs of varying densities are layered in a glass, each floating on a heavier one below.) In the ocean the density of fluids and solids determines, in most instances, whether a body of fluid or a solid remains in the upper layers or sinks to the bottom. Density differences drive currents in the ocean and atmosphere, as we shall see in later chapters.

Temperature, salinity, and pressure control seawater density. Warming increases the vibrations and movements of atoms and molecules so that they vibrate or move more, effectively occupying more volume and thus reducing density. For a constant mass, the density decreases as the volume increases. Remember that density is a ratio; it decreases because of either a decrease in the numerator or an increase in the denominator.

Ice, water vapor, and seawater behave like most materials, becoming less dense with increasing temperature. In any stable structure of liquid water or seawater, the least dense material will generaly be the warmest—if salinity is constant— and will occur at the top.

Pure liquid is unusual in having a density maximum at 4°C. It is an important factor in freshwater lakes of cold regions but not in the ocean. The anomalous density maximum does not occur in seawater of salinity greater than 24.7 parts per thousand (‰).

Ice at 0°C has a density of about 0.92 gram per cubic centimeter. Like most substances, ice becomes less dense as its temperature increases (Fig. 5-7). Since ice is about 8% less dense than liquid water, it floats on water. When liquid water at 0°C is warmed, its density increases slightly until it reaches a density maximum at 3.98°C. Once that temperature is reached, the density decreases with further temperature increases. In a freshwater lake, water cooled to 4°C sinks to the bottom and is protected there against further cooling. This process keeps the bottom of the lake ice free and supplies oxygen, dissolved in the water when it was in contact with the atmosphere, to bottom-dwelling animals. After the basin is filled with 4°C water, further cooling forms a less dense water layer on the lake surface, which eventually freezes if the weather remains cold long enough. If water behaved like most fluids and its density steadily increased with decreasing temperature, the coldest water would be found in the bottom of the lake and ice would also sink. This mode of freezing, with ice sinking to the bottom, would be far more efficient than freezing from the top.

Figure 5-7

Effect of temperature on density, expressed in grams per cubic centimeter, and specific volume, expressed in cubic centimeters per gram, for both ice and pure liquid water.

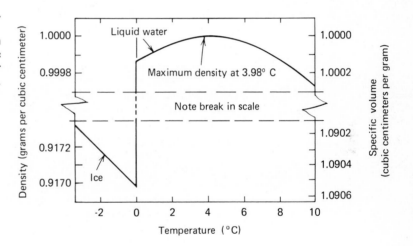

As noted, temperature has a profound effect on the relative proportions of the structured and "free" portions of liquid water. Because these two constituents have different densities, the temperature of water affects its density. One explanation for the anomalous density maximum in pure liqud water is that there is a relatively rapid increase in the amount of structured portion of water below 4°C. Consequently, water becomes less dense, for the denser "free" portion of the mix is diminished.

Adding salt to water (as in Fig. 5-8) or increasing the pressure causes a lowering of the temperature of maximum density. The amount of salt in seawater is sufficient to eliminate this anomalous density maximum. Furthermore, adding salt to water increases its density. Salinity of seawater is commonly the dominant factor controlling the density nearshore; temperature dominates in the open ocean.

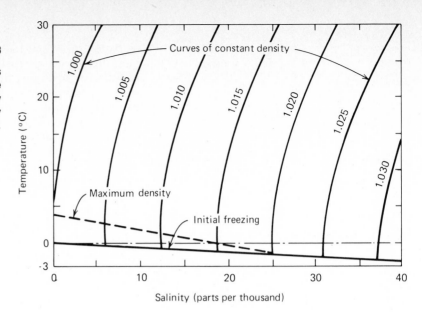

Figure 5-8

Variation in seawater density (in g/cm) as affected by temperature and salinity. Note the changes in temperature of maximum density and the point of initial freezing caused by changed salinity.

COMPOSITION OF SEA SALT

Almost anyone having seawater can determine salinity. The simplest way, although not the most precise, is to place a known amount of seawater in a pan and let it evaporate. The result is a gritty, bitter-tasting salt that never completely dries. The weight of the salt in a known volume of seawater provides a crude measure of salinity.

Six constituents—chloride (Cl^-), sodium (Na^+), sulfate (SO_4^{2-}), magnesium (Mg^{2+}), calcium (Ca^{2+}), and potassium (K^+)—constitute 99% of sea salts. The other elements present in sea salt add up to only 1%. As Fig. 5-9 shows, Na^+ and Cl^- alone make up 86% of sea salt.

Traces of nearly all the naturally occurring elements have been detected in sea salts, as shown in Fig. 5-10, and the job is still not

Figure 5-9

Major and minor constituents of seawater. The numbers indicate the amount of each constituent in grams, contained in a kilogram of seawater ($S = 34.7‰$).

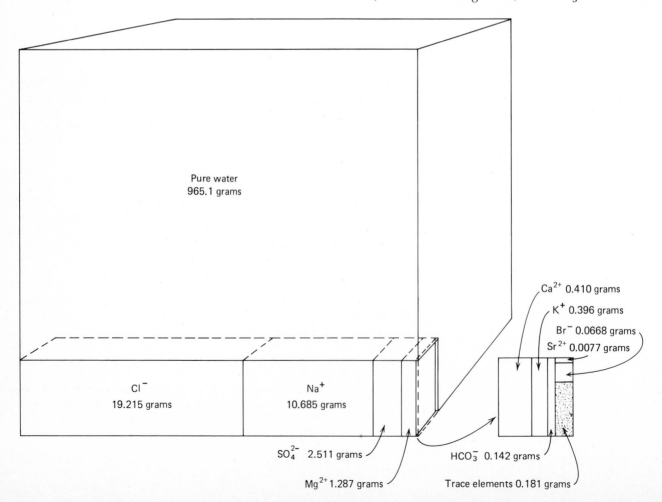

Pure water
965.1 grams

Ca^{2+} 0.410 grams

K^+ 0.396 grams

Br^- 0.0668 grams

Sr^{2+} 0.0077 grams

Cl^-
19.215 grams

Na^+
10.685 grams

SO_4^{2-} 2.511 grams

Mg^{2+} 1.287 grams

HCO_3^- 0.142 grams

Trace elements 0.181 grams

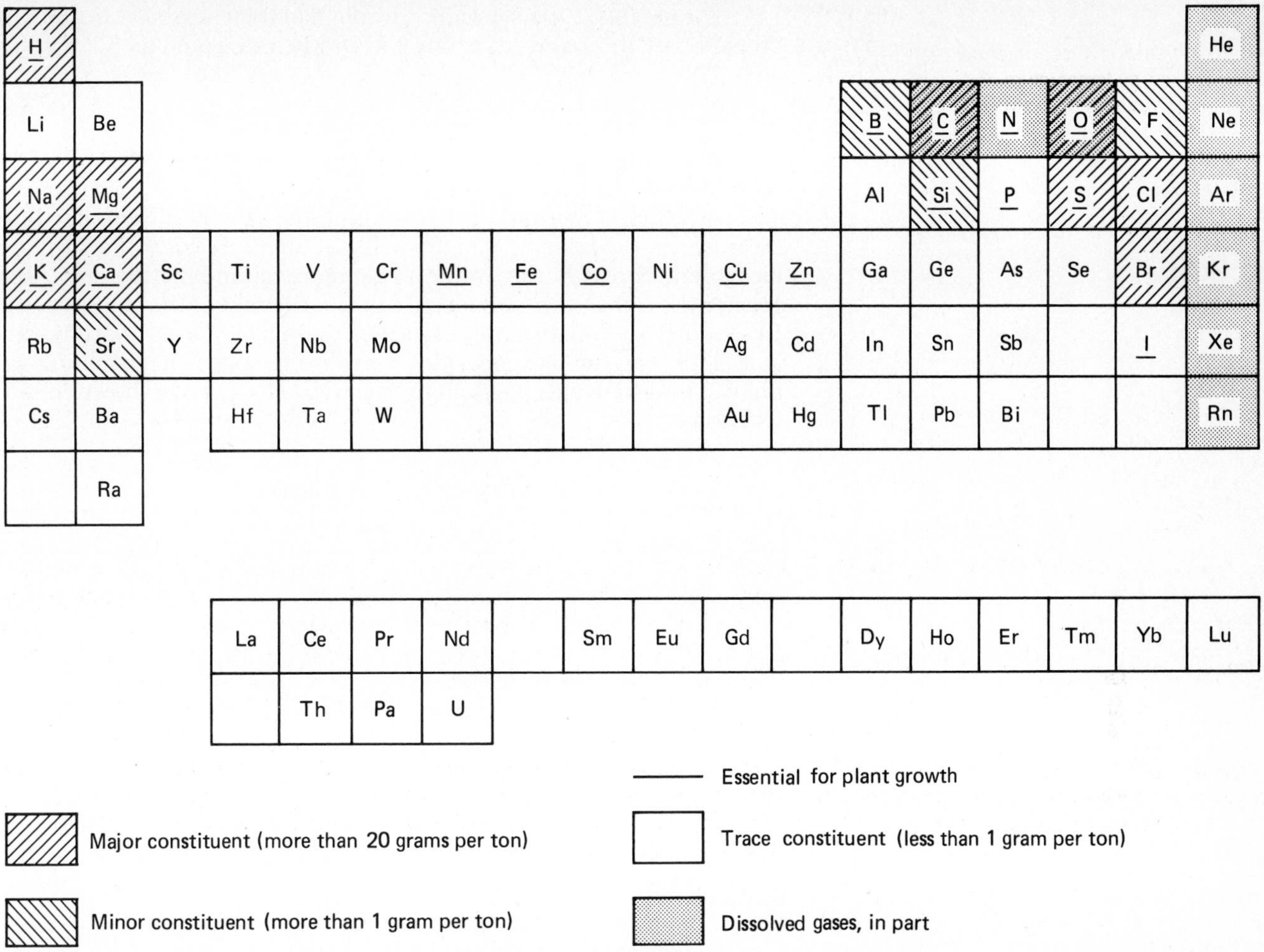

Figure 5-10

Elements detected in seawater, arranged as in the periodic table. Most nonconservative elements are involved in biological processes (underlined).

finished. Most of the analyses have been made on surface waters collected near the coast. Such waters are most likely to exhibit variability in the chemical composition of their dissolved salts. Furthermore, the development of each new analytical instrument opens new horizons for additional, more detailed studies of ocean chemistry.

To express the amount of dissolved salt present in seawater, oceanographers generally use two concepts—chlorinity and salinity, both based on chemical determinations.

Because ocean waters are well mixed, sea salts have a nearly constant composition. Therefore we can use the most abundant constituent, chloride (Cl^-), as the index of the amount of salt present in a given amount of seawater. *Chlorinity (Cl)* is defined as the amount of chlorine, in grams, in 1 kilogram of seawater (bromine and iodine are replaced by chlorine). It is expressed as parts per thousand, or per mil, written ‰.

Chlorinity indicates the amount of chlorine. Oceanographers commonly convert this to salinity, a measure of the total amount of dissolved salt. *Salinity (S)* is defined as the *total amount of solid material, in grams, dissolved in 1 kilogram of seawater* (iodine and bromine are replaced by chlorine and all organic matter destroyed). Salinity is also commonly expressed as parts per thousand.

Salinity *(S)* and chlorinity *(Cl)* are both measures of the saltiness of seawater. This relationship can be expressed mathematically by

$$S(‰) = 1.80655 \times \text{chlorinity}$$

For example, if a seawater sample had a measured chlorinity of 20 parts per thousand, its salinity would be calculated as follows:

$$S = (20 \times 1.80655) = 36.13‰$$

Oceanographers determine chlorinity to within 0.01 part per thousand.

A common method of determining the salinity of seawater is to measure its *conductivity*. Conductivity of water is controlled by ion movement through the water. The more abundant the ions, the greater the transmission of electricity and the higher the conductivity. In seawater the relative abundance of the major ions is nearly invariant. Thus conductivity provides a precise means of determining salinity. The major ions conducting electrical charges are shown in tabular form.

CONTRIBUTION TO CONDUCTIVITY

Ion	(percent)
Cl^-	64
Na^+	29
Mg^{2+}	3
SO_4^{2-}	2
Other	2

Although relative proportions of major elements in sea salts change little in the ocean, physical processes can change the amount of water in seawater. Therefore concentrations of the so-called *conservative properties*, such as sodium and magnesium and salinity, which are not involved in biological processes, are changed by the following processes.

1. Evaporation and precipitation (rainfall, snow).
2. Formation of insoluble precipitates whereby formerly dissolved elements or compounds settle out of seawater (salt formation by evaporation).
3. Mixing of water masses having different salinities.
4. Diffusion of dissolved materials from one water mass to another.
5. Movement of water masses within the ocean.
6. Freezing and thawing.

Salinity of seawater is most variable near the air–sea interface, at boundaries of ocean currents, and in coastal areas. Even though the composition of seawater changes little, those slight deviations that can be detected are extremely useful in tracing movements of water masses from their source to areas where they mix or otherwise lose their identity.

Many of the minor elements present in seawater exhibit pronounced changes in relative abundance because of biological and chemical processes. These we call *nonconservative properties*. Some nonconservative elements are incorporated in surface-ocean dwelling organisms and settle to the bottom when the organism dies, thus depleting that element in the surface layer. When organic matter is decomposed and the element released to seawater, it goes into near-bottom waters and is then involved in the slow circulation of these waters before eventually returning to the surface.

Other nonconservative constituents are associated with inorganic particles and may be removed near their point of entry into the ocean. Aluminum, for example, is usually associated with rock or soil particles, most of which settle out near the mouth of the river that brought them to the ocean. If the particles are wind transported and enter the ocean through its surface, the elements they contain are apt to be more widely dispersed than if brought in by rivers.

SEA SALT COMPOSITION AND RESIDENCE TIME

The composition of sea salt is controlled by chemical reactions between crustal rocks and seawater. The input rate of each element from all sources is matched by the removal rate of that element through reactions with weathered sediment particles from the continents, removed by marine organisms, and by reactions of seawater circulating through the hot oceanic crust as it cools. Thus the salinity of ocean water remains constant over millions of years.

The relative importance of contributions at the midocean ridges and rivers to seawater is not yet completely clear. Waters from the hydrothermal vents have not been studied enough to understand their role fully. It is clear, however, that hydrothermal circulation is a major removal process for magnesiums and sulfate from seawater and a primary source of lithium and rubidium. Other major elements are undoubtedly affected as well.

Each year rivers bring about 4 billion tons (4×10^{15} grams) of dissolved salts to the ocean. Nearly all the sodium chloride in river water is recycled sea salt that fell on land in rain, coming from sea-salt particles derived originally from the sea surface. The remaining salts dissolved in river water come from the chemical breakdown (weathering) of rocks on land. Although this amount of salt is only about one-thousandth of the total amount of salt in the ocean, it would seem reasonable to find that seawater is getting saltier through time.

All our data, however, indicate that the salinity of seawater has changed little. Direct measurements of salinity go back only a few hundred years and indicate no appreciable change. But this short period of observation is inadequate to decipher the ocean's billions of years of history.

Marine fossils preserved in rocks provide the best long-term information available to us. Fossils in these rocks—once marine sediment deposits—are usually found to be similar or closely related to organisms that now live in open ocean waters. This bit of evidence, combined with some chemical data, suggests that the salinity of the open ocean has changed little, if at all, during the past half-billion years of ocean history. Salinity of seawater before the era when animals developed preservable skeletons remains essentially unknown.

Because rivers are delivering salt to the ocean while seawater salinity remains unchanged, salt must be removed at about the same rate at which it is supplied, an example of the *steady-state condition* of the ocean, which is changing little—if at all—through time. Some of these salt-removing processes apparently involve complex chemical reactions with rocks or particles suspended in the ocean. In other cases, seawater is evaporated in isolated basins to form salt deposits in arid regions.

A useful concept for characterizing substances in seawater is *residence time*, the time required to replace the amount of a given substance in the ocean completely. This concept can work in two ways, using either the rate of addition or the rate of removal of elements incorporated in sediments depositing on the ocean bottom. Using the second option, we can define residence time *(T)* as

$$T = \frac{M}{R}$$

where M = total amount (grams) of the substance in the ocean
R = rate of removal by sediment (grams per year)

For example, we can calculate the residence time for sodium in seawater as follows:

$$T_{Na} = \frac{M_{Na}}{R_{Na}} = \frac{1.5 \times 10^{22} g}{5.7 \times 10^{13} \text{ g/yr}} = 2.6 \times 10^8 \text{ yr}$$

Sodium's calculated residence time in the ocean of 260 million years is one of the longest residence times for an element. Other elements, such as aluminum, have residence times as short as 100 years.

Residence time of an element in the sea is apparently related to its chemical behavior. Elements like sodium, which are little affected by sedimentary or biological processes, generally have residence times of many millions of years. Elements used by organisms or readily incorporated in sediments, such as aluminum or iron, tend to have much shorter residence times, ranging from a few hundred to a few thousand years.

We can calculate residence times for water too. There is a net removal, due to evaporation, of a layer of water about 10 centimeters thick from the ocean surface each year. This water falls on the land and returns to the ocean through river runoff. Recall that the ocean has an average depth of about 4000 meters. From this we obtain an approximate residence of 40,000 years for water.

SALT IN WATER   Natural waters usually contain some dissolved salt. River waters contain, on the average, about 0.01% dissolved salts and average seawater contains about 3.5% (35‰) of various salts (Table 5-2). Even rainwater usually contains small amounts of salts and gases that it has dissolved while falling through the atmosphere. These impurities are primarily a result of the remarkable dissolving powers of water.

Sodium chloride crystals consist of sodium ions (Na^+) and chloride ions (Cl^-) tightly bound together by ionic bonds. These bonds result from strong attraction of unlike charges on nearby ions. The charges arise because ions have lost or gained an electron. Sodium readily loses an electron, forming a sodium ion. Chlorine, on the other hand, readily accepts an extra electron, forming a chloride ion, Cl^-. Water molecules disrupt these bonds by orienting themselves around ions and effectively shielding adjacent ions from the influence of each other. As a result, the crystal disappears; we say that it has dissolved.

Dissolving salt affects the behavior of water in several ways. We have already discussed the effect on temperature of maximum density. Other properties affected by salts are as follows.

1. *Temperature of initial freezing*—lowered by increased salinity.
2. *Vapor pressure*—decreased by increased salinity.
3. *Osmotic pressure*—increased by increased salinity (water molecules move through a semipermeable membrane from a less saline solution to a more saline solution; the opposing pressure necessary to stop this movement is called the osmotic pressure).

Consider the temperature of initial freezing. Pure water freezes at 0°C and the temperature of the water–ice mixture remains fixed until there is only ice. In seawater the temperature of initial freezing is lowered by increased salinity. Furthermore, salts are excluded from the ice and remain in the liquid, causing the brine to become still more salty. So the temperature of freezing for the remaining liquid is still lower and the temperature of the system must drop before additional ice forms. Consequently, seawater has a lower temperature of initial freezing and no fixed freezing point, as observed in pure water.

TABLE 5-2
Elements Detected In Seawater[*]

ELEMENT		CHEMICAL FORM	CONCENTRATION (ppm)	ELEMENT		CHEMICAL FORM	CONCENTRATION (ppm)
silver	Ag	$AgCl_2^-$	0.0003	magnesium	Mg	Mg^{2+}	1350
aluminum	Al		0.01	manganese	Mn	Mn^{2+}	0.002
argon	Ar	Ar (gas)	0.6	molybdenum	Mo	MoO_4^{2-}	0.01
arsenic	As	$AsO_4H_2^-$	0.003	nitrogen	N	organic N, NO_3^-, NH_4^+ gas	0.5
gold	Au	$AuCl_4^-$	0.000011	sodium	Na	Na^+	10,500
boron	B	$B(OH)_3$	4.6	neodymium	Nd		0.00001
barium	Ba	Ba^{++}	0.03	neon	Ne	Ne (gas)	0.00014
beryllium	Be		0.0000006	nickel	Ni	Ni^{2+}	0.0054
bismuth	Bi		0.000017	oxygen	O	OH_2, O_2, SO_4^{2-}	857,000
bromine	Br	Br^-	65	phosphorus	P	PO_4H^{2-}	0.07
carbon	C	CO_3H^-, organic C	28	protoactinium	Pa		2×10^{-9}
calcium	Ca	Ca^{2+}	400	lead	Pb	Pb^{2+}	0.00003
cadmium	Cd	Cd^{2+}	0.00011	radium	Ra		6×10^{-11}
cerium	Ce		0.0004	rubidium	Rb	Rb^+	0.12
chlorine	Cl	Cl^-	19,000	radon	Rn	Rn (gas)	6×10^{-16}
cobalt	Co	Co^{2+}	0.00027	sulfur	S	SO_4^{2-}	885
chromium	Cr		0.00005	antimony	Sb		0.00033
cesium	Cs	Cs^+	0.0005	scandium	Sc		< 0.000004
copper	Cu	Cu^{2+}	0.003	selenium	Se		0.00009
fluorine	F	F^-	1.3	silicon	Si	$Si(OH)_4$	3
iron	Fe	$Fe(OH)_3$	0.01	tin	Sn		0.003
gallium	Ga		0.00003	strontium	Sr	Sr^{2+}	8.1
germanium	Ge	$Ge(OH)_4$	0.00007	tantalum	Ta		< 0.0000025
hydrogen	H	H_2O	108,000	thorium	Th		0.00005
helium	He	He (gas)	0.0000069	titanium	Ti		0.001
hafnium	Hf		< 0.000008	thallium	Tl	Tl^+	< 0.00001
mercury	Hg	$HgCl_4^{2-}$	0.00003	uranium	U	$UO_2(CO_3)_3^{4-}$	0.003
iodine	I	I^-, IO_3^-?	0.06	vanadium	V	$VO_5H_3^{2-}$	0.002
indium	In		< 0.02	tungsten	W	WO_4^{2-}	0.0001
potassium	K	K^+	380	xenon	Xe	Xe (gas)	0.000052
krypton	Kr	Kr (gas)	0.0025	yttrium	Y		0.0003
lanthanum	La		0.000012	zinc	Zn	Zn^{2+}	0.01
lithium	Li	Li^+	0.18	zirconium	Zr		0.000022

[*] After R. A. Horne, 1969. *Marine Chemistry: The Structure of Water and the Chemistry of the Hydrosphere.* Wiley-Interscience, New York. p. 153.

Sea salts affect the internal structure of water. Some dissolved ions, such as Na^+ and K^+, cause a shift in the equilibrium toward water's unstructured phase whereas other ions, such as Mg^{2+}, "favor" the structured portions. The effect of changing the relative proportions of these constituents can be seen in changes of such properties as *viscosity*, the internal resistance of a liquid to flow (see Fig. 5-11).

After a crystal has dissolved, water molecules remain associated with ions, forming an envelope or cloud that actually moves with the ion. It is postulated that about four water molecules are associated with each sodium ion, about two molecules with each chloride ion. Ions with higher charges, such as Mg^{2+} or SO_4^{2-}, also have these so-called *hydration atmospheres* (or *sheaths*), usually containing more water molecules. Like other structured parts of liquid water, the hydration atmospheres are affected by temperature and pressure. They are apparently destroyed by pressures exceeding 2000 atmospheres but are not completely destroyed by temperatures about 100°C.

Hydration sheaths affect the water around them. Some, such as NaCl and especially $MgSO_4$, "favor" the structured portion of water and actually increase viscosity, a property of water that responds sensitively to the relative abundance of the structured portion. Salts like potassium chloride (KCl) seem to "favor" the unstructured portion of the water and might be called *structure breakers*. They cause an initial reduction in the relative viscosity of water. Dissolved gases and insoluble or partially soluble substances also affect the relative abundance of structured and "free" water nearby.

Materials that dissolve completely, forming separate ions in water, are called *strong electrolytes*. Sodium chloride is an example. After a crystal dissolves, the ions form hydration sheaths that move more or less independently so that ions formed by strong electrolytes behave like separate ions. In general, strong electrolytes have little effect on water structure, as shown by their slight influence on water viscosity (see Fig. 5-11).

Not all materials separate completely into individual ions when they dissolve; these are called *weak electrolytes*. Magnesium sulfate ($MgSO_4$) is an example of a weak electrolyte. Only a fraction of the material dissociates into individual ions, as shown in Table 5-3. About 11% of the magnesium ions remains associated with sulfate ions. Sulfate ions are partially associated with Mg^{2+}, partially with Na^+.

Figure 5-11

Effect of various salts on viscosity of water 25°C. (After Horne, 1969.)

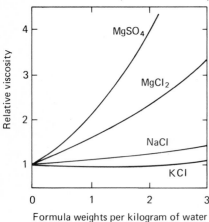

TABLE 5-3

Chemical Species of Some Major Constituents in Surface Seawater*

CATIONS	FREE ION (percent)	COMBINED WITH		
		SULFATE (percent)	BICARBONATE (percent)	CARBONATE (percent)
Na^+	99	1.2	0.01	–
Mg^{2+}	87	11	1	0.3
Ca^{2+}	91	8	1	0.2
K^+	99	1	–	–

ANIONS	FREE ION (percent)	COMBINED WITH			
		Na (percent)	Mg (percent)	Ca (percent)	K (percent)
SO_4^{2-}	54	21	21.5	3	0.5
HCO_3^-	69	8	19	4	–
CO_3^{2-}	9	17	67	7	–

*From R. M. Garrels and M. E. Thomson, 1962. A chemical model for seawater at 25°C and one atmosphere pressure. *American Journal of Science* 260: 57–66.

Still another mode of association is possible in seawater—the formation of *ion pairs*, associations of ions that remain together much of the time even though they retain their individual hydration sheaths. In human terms, we could think of it as a liaison rather than marriage. Ion pairs can have charges where the *valences* (amount of positive or negative charge) of individual members of the pair do not completely cancel. *Complex ions* are still more strongly attracted to each other, so much so that their hydration sheaths merge to form a single sheath, as represented in Fig. 5-12.

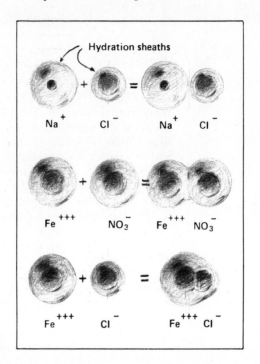

Figure 5-12

Schematic representation of hydration sheaths of ions formed by strong electrolytes (NaCl), ion pairs (Fe^3 + NO$_3^-$). (After Horne, 1969.)

DISSOLVED GASES

Gases are soluble in water, passing into a dissolved state at the air–water interface. Conversely, molecules of gas also pass through the interface in the opposite direction—back into the atmosphere. Water can only dissolve a limited amount of any substance at a given temperature and pressure; when that limiting value is reached, the amount of gas going into solution is the same as that going out. At this point, the water is *saturated* with the gas, which is then said to be present in *equilibrium concentration*.

Temperature, salinity, and pressure all affect the saturation concentration for a gas. In the normal range of oceanic salinity, temperature is the dominant factor, as indicated in Fig. 5-13. In general, such gases as nitrogen and oxygen or rare gases (helium, neon) that do not react chemically with water become less soluble in seawater as temperature or salinity increases. Seawater at all depths is saturated with most atmospheric gases. Exceptions are dissolved oxygen (O_2) and carbon dioxide (CO_2), which are involved in life processes.

Nitrogen (N_2), the most abundant gas in the atmosphere (Fig. 5-14), is also a common gas dissolved in seawater (Fig. 5-15). It is not involved in biological processes (with a few minor exceptions) and thus remains very near saturation throughout the ocean. Sometimes, in deep water, more nitrogen is present than can be accounted for by the temperature, salinity, and pressure of the water. We then say that the water is *supersaturated* with nitrogen, a condition that may be explained as follows.

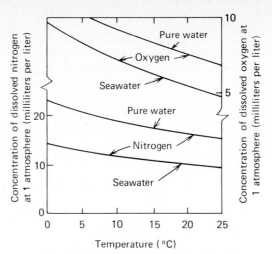

Figure 5-13

Solubility of oxygen and nitrogen in pure water and seawater. Note that both gases are more soluble in pure water.

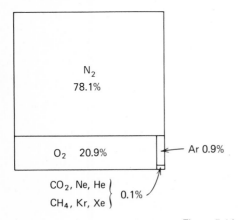

Figure 5-14

Relative abundance (by volume) of gases in dry atmosphere.

Figure 5-15

Relative abundance (by weight) of gases dissolved in surface seawater, $S = 36$ ‰, $T = 20°C$. Note the abundance of HCO_3^- and rare gases compared with their abundance in the atmosphere (see Fig. 3-14).

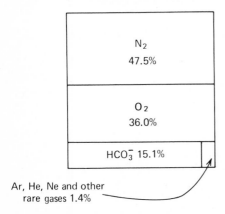

The solubility of a gas in seawater is determined primarily by temperature, pressure, and salinity. Since exchange of dissolved gas occurs at the sea surface, conditions at the surface control the amount of gas dissolved in water. As a water mass moves away from the surface where it last exchanged atmospheric gases, its temperature and salinity can change as a result of mixing with other water masses. Also, depth changes are accompanied by pressure changes. The observed slight supersaturations of nitrogen and rare gases in the deep ocean can thus be explained by considering these changes in physical conditions of the water mass since it was last at the surface.

A small amount of helium (He), a rare gas, is produced by radioactive decay in rocks and sediments. Otherwise the rare gases—argon (Ar), krypton (Kr), and xenon (Xe)—behave identically to nitrogen, allowing for their slightly different solubilities.

Oxygen is one of the most variable of the dissolved gases in seawater. Like other gases, it enters the ocean primarily through the surface. Oxygen is also produced by photosynthesis of plants in the sunlit layer of the ocean, usually restricted to a zone a few tens of meters deep just below the ocean surface. In the deep ocean dissolved oxygen is supplied by the sinking of cold, oxygen-rich waters in the Antarctic and Arctic regions. Oxygen is used by plants and animals at all depths in respiration.

In those parts of the ocean where rates of oxygen consumption by marine life exceed the resupply of dissolved oxygen, certain bacteria are able to derive their oxygen by breaking down compounds dissolved in seawater. First, nitrate ions are broken down, releasing nitrogen. Because there is little nitrate in most ocean waters, this process does not last long. Next, different organisms break down SO_4^{2-} to obtain oxygen and release hydrogen sulfide, H_2S, as a metabolic byproduct, causing the familiar rotten-egg smell often noticeable in salt marshes at low tide.

There is also some production of nitrogen by bacteria living in oxygen-deficient basin waters. These bacteria break down nitrate ions (NO_3^{2-}) from seawater and release nitrogen gas (N_2). In the absence of dissolved oxygen, most biological processes proceed slowly and the volume of ocean area deficient in oxygen is small; so this process is a minor factor in the dissolved-nitrogen picture. In addition, some formation of nitrate from dissolved nitrogen by bacteria occurs, but it appears to be minor.

CARBON DIOXIDE AND CARBONATE CYCLES

The carbon dioxide cycle in seawater is the most complicated of all the dissolved gases. Carbon dioxide occurs as a dissolved gas, as carbonic acid (H_2CO_3), and as the dissolved ions of carbonate (CO_3^{2-}) and bicarbonate (HCO_3^-). Its abundance is controlled by physical properties of seawater, by biological processes, such as photosynthesis and respiration, and by the formation and destruction of carbonate shells of marine plants and animals (see Fig. 5-16). If carbonate is used in plant production, the equilibrium shifts in response (in Fig. 5-16) and causes more bicarbonate to dissociate. Carbonate precipitation as calcium carbonate ($CaCO_3$) in plant and animal skeletons represents a net CO_2 loss to the cycle.

Another biologically significant aspect of CO_2 in the ocean is that the weak acid (H_2CO_3) acts as a *buffer*. Addition of acid (H^+) causes the equilibrium in Fig. 5-16 to shift, thereby, creating more bicarbonate. Bicarbonate ionizes very little so that the abundance of hydrogen ions (pH), which controls the relative alkalinity or acidity of a solution, remains fairly constant. Thus respiration and decomposition processes producing CO_2 hardly affect the pH of seawater, nor does the removal of CO_2 in photosynthesis. Hydrogen ions in seawater are freely accepted and given up by both carbonate and bicarbonate ions, creating a buffer against sudden, sharp changes in acidity.

Carbon dioxide is highly soluble in seawater and enters the ocean through the sea surface. It is also produced at nearly all depths by the respiration of plants and animals.

Calcium carbonate shells formed mostly in surface waters dissolve in intermediate and deep waters after the organism dies and sinks below the surface. On the ocean bottom less than about 4 kilometers deep (the depth varies in different ocean basins), the rate of supply of carbonate shells is greater than the rate of dissolution. The rate of resupply is much too slow, however, to prevent the dissolution of carbonate in the cold, CO_2-rich waters deeper than the 4-kilometer level. There the sediments are nearly devoid of carbonate because of its dissolution; this is the red mud area of the deep-ocean basins.

Over most of the ocean, removal of calcium carbonate is accomplished by biological processes. Inorganic precipitation seems to be a rare occurrence, restricted to shallow, warm, highly saline waters. The Bahama Banks, east of Florida, and areas around the Red Sea are among the few places where this process occurs. Seawater is warmed as it flows slowly over shallow areas, causing two effects: the solubility of $CaCO_3$ in water is diminished and the solubility of carbon dioxide decreases so that some escapes to the atmosphere. While in solution, each Ca^+ ion is balanced by the presence of two bicarbonate ions. If bicarbonate is lost by dissociation and escape of CO_2, then $CaCO_3$ (lime) precipitates as a solid coating on rounded grains (called oolites) or as tiny needles of the mineral aragonite.

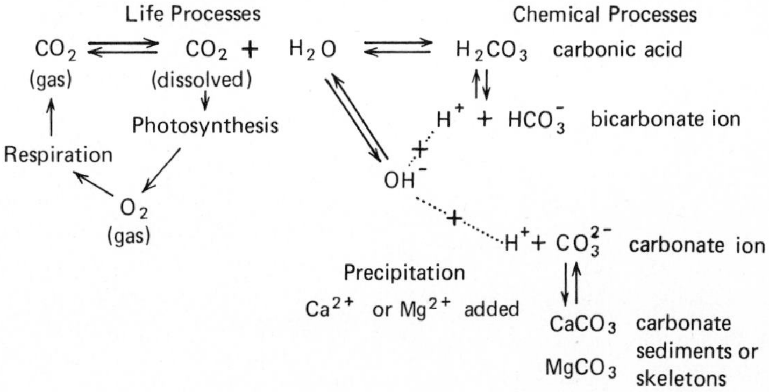

Figure 5-16

Carbonate–carbon dioxide cycle in the ocean. (After Horne, 1969.)

PHYSICAL PROPERTIES OF SEAWATER

Pressure in the ocean increases 1 atmosphere for each 10 meters of depth. Thus pressures in the deepest trenches are around 1100 atmospheres. Changes in pressure cause changes in physical properties of water.

Viscosity—resistance of a fluid to flow—in "normal" liquids increases as pressure increases. In a simple way, we can imagine that this process results from the increased resistance of atoms or molecules sliding past each other as they are pressed more closely together.

Liquid water responds in exactly the opposite way from a "normal" liquid; its viscosity decreases with increasing pressure. It reaches a minimum viscosity at pressures of about 500 to 1000 atmospheres, corresponding to depths of 5000 to 10,000 meters.

Ease of molecular movement in water (and therefore its viscosity) is related to the degree of binding of molecules and the amount of structuring of the water. The simplest explanation of the pressure effect on water is that increased pressure in the depth range to 5000 meters or more favors the destruction of the structured portions, thereby increasing the ease of movement of molecules. The result is a decrease in viscosity. At pressures exceeding 1000 atmospheres, the structured portion of the water has been completely destroyed and water then acts like a "normal" liquid. Its viscosity increases with further pressure increase.

Air trapped in seawater near the surface forms bubbles. These bubbles range in size from a few millimeters to a few microns in diameter and are formed by waves breaking and raindrop impacts. Bubbles provide an important route for gases in seawater to pass in and out of solution and for producing sea-salt particles that serve as condensation nuclei for raindrops and ice particles in the atmosphere.

Development and bursting of bubbles have been studied in detail via high-speed photography, as shown in Fig. 5-17. First a gas bubble forms (a) and rises toward the surface at rates of about 10 centimeters per second. Reaching the surface (b), the bubble has a cap or thin film of water that bursts (c), forming a thin spray of droplets (d) whose diameters are about 1 to 20 microns. Shortly thereafter a large droplet (e), about 100 microns in diameter, is formed by a jet rising rapidly from the bottom of the bubble and is ejected into the atmosphere at speeds of 10 meters per second. This droplet rises 10 centimeters or more above the sea surface. The smaller droplets are caught by wind and transported long distances in the atmosphere. The larger droplets rapidly fall back to the sea surface.

Airborne droplets evaporate, leaving tiny salt particles that are carried by winds. The residence time of most such particles in the atmosphere is probably short. Some, however, are carried high into the atmosphere, where they eventually act as nuclei around which raindrops or snowflakes form. In this manner, escaped salt is carried back to the earth's surface. The total amount of salt moved in this way is about 1 billion tons per year.

Composition of the salt particles closely resembles that of normal sea salt, although there is some enrichment in the more volatile materials that evaporate from the sea surface, such as boron and iodine.

Surface tension is the force necessary to break the filmlike surface layer of a liquid. Water has an unusually high surface tension—a clean needle, for instance, can float on a water surface. Surface tension causes an undeformed water drop to form a sphere. This situation can be seen by shaking a bottle of oil and vinegar—the vinegar forms droplets in the oil before the suspension separates.

High surface tension in water seems to result from strengthened local water structure near any interface, including the air–water boundary. Near an interface there is more hydrogen bonding of the water molecules, which tend to orient themselves with their oxygens pointing out of the liquid (Fig. 5-18). This structured zone extends about 10 molecules deep into the liquid. Because of the highly structured nature of interface water, certain ions tend to be excluded so

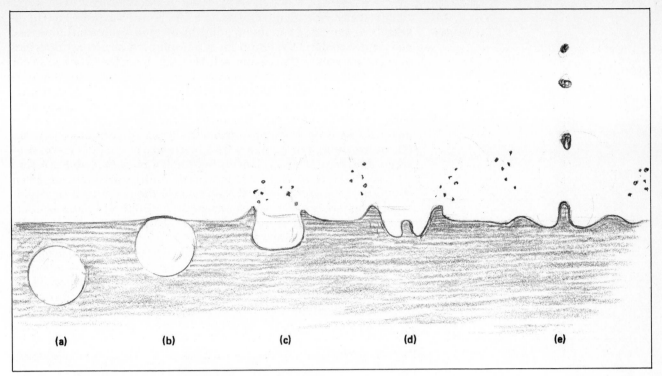

Figure 5-17

(a)   (b)   (c)   (d)   (e)

Bubble bursting at the sea surface, forming water droplets in the atmosphere. Many of these droplets evaporate, forming salt particles that serve as condensation nuclei for raindrops or snowflakes.

that the chemical composition of this zone may differ appreciably from that of bulk seawater. This selective exclusion probably affects the chemical composition of sea salts that pass into the atmosphere from breaking bubbles and other processes. In fact, several elements are more abundant in airborne salt particles than they are in normal seawater.

Surface films are sensitive to the presence of impurities. Many substances that are insoluble in water tend to collect at the interface. Of these substances, many are molecules whose various components have different chemical behavior. For example, surface-active

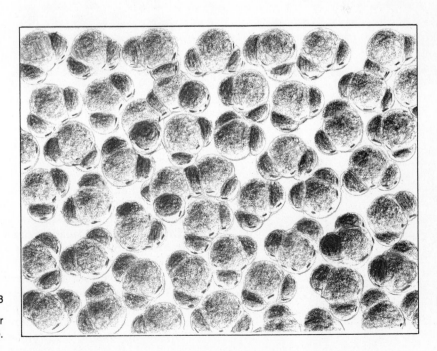

Figure 5-18

Changes in water structure near the air−water interface at the top of figure.

133

agents—substances that concentrate at or change the properties of an interface—have one end that readily dissolves in water while the other end of the molecule (often a long, complicated structure) does not dissolve in water. The water-soluble end is called the *hydrophilic* ("water-loving") end, the insoluble end the *hydrophobic* ("water-hating") end; this relationship is illustrated in Fig. 5-19 if you look closely.

Surface-active agents change the surface tension of seawater, generally lowering it. Surface "slicks" or films, where surface-active materials have been swept together by wind or current action, are especially common in coastal waters, particularly where currents converge. These areas have relatively few small waves at the water surface and their relative smoothness compared to adjacent waters has given rise to the descriptive term *slick*. Normally they contain a variety of organic materials of an oily or otherwise hydrophobic nature.

Detergents and soaps are familiar surface-active agents. Their long molecules surround dirt or grease, substances easily attached to the hydrophobic end of the surface-active molecule. The hydrophilic end is directed toward the surrounding water and the whole aggregate moves as if it were soluble so that the otherwise insoluble dirt or grease can "wash away" in the water. Comparable surface-active materials occur naturally in the ocean, apparently formed by marine organisms.

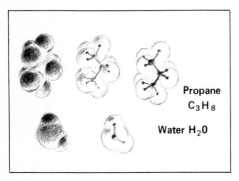

Propane
C_3H_8

Water H_2O

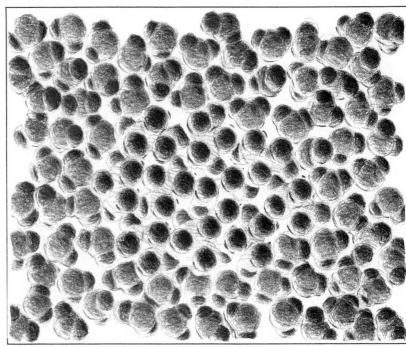

Figure 5-19

Changes in water structure in the vicinity of a nonpolar, hydrophobic material (right). Note that the polar water molecules exclude the nonpolar propane molecule from the water structure (below).

PARTICLES IN SEAWATER The behavior of particles dispersed in ocean water influences the chemical and biological makeup of both seawater and sediment. The total amount of particles in the ocean is about 10^{16} grams, roughly half the amount of sediment brought to the ocean by rivers each year. But particles are relatively rare in deep-ocean waters, about 10 to 20 parts per billion, equivalent to 1 gram of particles in 100 tons of seawater.

Principal sources of particles are rivers, airborne dust, biological processes, and resuspension from the bottom. Only the smallest river-derived particles reach the open ocean, those having settling velocities less than 5×10^{-3} centimeter per second (equivalent to particles less than 5 microns). Biological particles produced in the ocean are larger, ranging in size from 1 micron to 1 millimeter. Particles in near-bottom waters are usually larger; for they have been resuspended from the bottom by currents. The particles in near-bottom waters rarely penetrate more than a few hundred meters above the bottom. Their concentrations are often high in areas of strong deep-ocean boundary currents.

As particles sink, they are removed, or reduced in size by grazing zooplankton, or dissolved, in the case of calcareous or siliceous particles. In dissolving, particles release various constituents (nutrients, silica, certain metals) to the deep subsurface waters and thus markedly alter the chemical composition of deep-ocean waters. Particles play a dominant role in the transfer of constituents from surface waters to the deeper zones.

Biogenous particles—shell fragments, fecal pellets—are most abundant; they constitute between 30 and 70% of the total particulate matter in the ocean. Particulate carbon contents are about 25%. The biogenous particles are associated with upwelling areas and are transported by currents. Lithogenous particles constitute about 2% of the total particulate matter; high values occur near glaciers and in the trade wind zones.

Particle concentrations are low near the equator, apparently due to the high productivity leading to effective filtering by zooplankton. Removal of particles by filter-feeding zooplankton is a principal mechanism for removing particles from ocean waters.

Large particles, such as fecal pellets, are rapidly removed from the water by settling (about 100 meters per day) and are little altered by biological or chemical processes. Small particles sink more slowly; the smallest particles may totally dissolve before reaching the bottom. Biological particles are destroyed in two stages. First, they are broken up, either mechanically by grazing or by chemical dissolution. Then the smaller fragments are chemically dissolved. Dissolution is especially significant for particles smaller than 10 microns.

Addition or removal of constituents from particles is the primary mechanism controlling residence times of several trace elements in seawater: thorium, plutonium, lead, iron, and copper. These processes are important in the removal of pollutants from seawater, including many of the radioactive materials that fall into the ocean from atmospheric testing of nuclear devices.

SOUND IN SEAWATER

Behavior of sound in seawater is controlled by temperature, pressure, and salinity. Sound is a form of mechanical energy, consisting of the regular alternation of pressure or stress in an elastic substance. Of all the various forms of radiated energy, sound penetrates ocean water best and so sounds made by organisms or physical processes can be detected at relatively great distances. Sound is also used to determine depth to the bottom, to detect other vessels or objects (including fish), and for emergency signaling using the technique known as *sonar,* an acronym for "sound navigation and ranging"—an underwater equivalent of radar.

The speed of sound in water is increased by increasing pressure (which increases with depth), temperature, and salinity. In most of the ocean the effect of salinity is less important than the effect of pressure. Near the ocean surface, temperature effects dominate. At greater depths pressure effects are appreciable, as indicated in Fig. 5-20(a). Sound in water can be slowed by bubbles or high concentrations of dissolved gases in some bays and harbors.

Intensity of a sound wave traveling through water diminishes as energy is absorbed. Absorption of sound in seawater is far greater than in pure water at frequencies less than 100 kilocycles per second (kilohertz) but its essentially identical to pure water at higher frequencies, as Fig. 5-20(b) demonstrates. The presence of $MgSO_4$ is the cause of this anomaly. As noted, $MgSO_4$ in seawater is typically surrounded by a large, bulky hydrated aggregate. This complex aggregate absorbs acoustical energy over a large frequency, which accounts for seawater's unusual sound absorption.

Sound passing through water can be considered as traveling along paths called *rays.* Some rays are direct; in other words, they pass directly from source to receiver. Others are indirect and may involve reflection from boundaries, such as in the air–water interface or the water–sediment interface. This reflection of sound is the basis for the echo-sounding technique for determining water depths.

As sound passes through water of varying sound speed (caused by variations in temperature, salinity, and pressure), the rays are bent toward regions of lower sound speed. Thus sounds originating in the surface ocean tend to be bent or refracted downward into the depth zone of lowered sound speed. Sound originating below such a layer

Figure 5-20

Acoustical properties of seawater: (a) typical profiles showing variation in speed of sound with depth in various latitudes; (b) absorption coefficients in seawater and distilled water. (After R. J. Ureck, 1969. *Principles of Underwater Sound for Engineers.* McGraw-Hill, New York.)

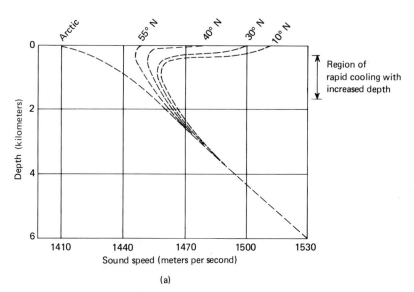

(a)

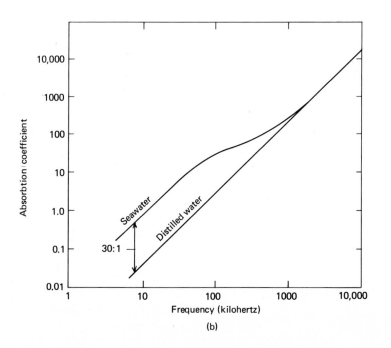

(b)

is similarly bent upward. Consequently, the sound ray tends to be trapped as if it were in a channel. This *sound channel* is a typical feature of the open ocean at depths of around 1000 meters at midlatitudes to near the surface in polar regions and it greatly extends the transmission range of underwater sound. This sound channel has been used for long-range underwater signaling, called *sofar* (sound fixing and ranging). Emergency locating of downed aircraft crews is just one example of its utility. Other sound channels occur near the surface and in shallow waters less than 200 meters deep, owing to temperature changes or physical boundaries.

Marine organisms interfere with sound-ranging operations in several ways. Systems that passively listen to detect characteristic noises of ships passing or of other activities are especially troubled by the noises made by fishes—especially drumfish, grunts, snappers, and croakers, whose names suggest some of the sounds they make. These sounds have been confused with ships' engines, propellers, and other equipment. The utility of such noise to the fish is not always apparent. Mammals, including whales, dolphins, and seals, also produce noises, apparently for communication and for their own form of sound ranging.

More active sound-ranging systems are also affected by the mere presence of organisms. Many marine organisms have minute swim bladders that are highly efficient reflectors of sound energy. Consequently, they register as much larger targets on a detection device and can disrupt the operation of the equipment.

REVIEW QUESTIONS

1. Draw a diagram of a water molecule. Explain why water is a polar molecule.

2. Describe and contrast the molecular structures of ice, liquid water, and water vapor. Discuss the significance of the different structures on the amount of energy required (or given off) by changing from one structure to another.

3. Discuss the reasons why the presence of the ocean acts as a thermostat, preventing extremes of temperatures in coastal regions.

4. What factors control the density of seawater?

5. Describe a stable density distribution; an unstable density distribution.

6. List the six major constituents of seawater and indicate the percentage of each constituent in seawater ($S = 34.7‰$).

7. Calculate the salinity of seawater with chlorinity of 19‰; 21‰.

8. Define residence time. Discuss its significance and give examples of elements having long residence times; short residence times.

9. Discuss the difference between conservative and nonconservative behavior of constituents in seawater. Give examples of each type of behavior.

10. Compare the composition of gases in the atmosphere with those dissolved in seawater. Explain the differences.

11. Explain the effects of high pressure on the molecular structure of liquid water and the effects on water viscosity.

12. Explain the carbon dioxide–carbonate cycle in seawater without living organisms or carbonate solids.

13. Draw a diagram showing the changes in sound velocity with depth in midlatitudes. Explain what causes this "sound channel."

Seawater

a living soup; 96.5% water, 3.5% salts, little organic matter, few particles

Molecular structure of water

Water—a polar molecule

Many unusual properties result from hydrogen bonds

Ionic bonds readily broken in water; weak van der Waals bonds also present

Molecules move freely in water vapor; exert pressure on walls of container

Ice, held together by hydrogen bonds, has regular crystal structure; salt is excluded from structure; less dense than water

Liquid water—an anomalous substance made of structured and unstructured portions in a dynamic equilibrium

Temperature effects on water

Heat energy absorbed in changes of stage with addition of heat

Latent heat utilized in breaking bonds; heat is released in condensation of water vapor and freezing of water

Cluster size and number of molecules per cluster decrease with increasing temperature

Density

Sinking (or floating) is determined by density relationships with surrounding water

Seawater density dependent on temperature, salinity, and pressure

Density of seawater decreases with increasing temperature

Freshwater has a density maximum at 3.98°C

Composition of sea salt

Six ions constitute 99% of sea salts—Cl^-, Na^+, SO_4^{2-}, Mg^{2+}, Ca^{2+}, K^+

Chlorinity (grams per kilogram of parts per thousand)—amount of chlorine in seawater, a measure of salinity

Salinity—total amount of dissolved salts in parts per thousand (‰) in seawater

Conservative elements are those whose relative proportions in seawater are usually constant; their concentrations change with addition or removal of water

Nonconservative elements—variable, generally lower concentrations that depend on biological or chemical processes.

Salt composition and residence times

Salts dissolved from rocks enter seawater; they are eventually removed by incorporation in sediments

Residence time of an element in seawater is related to chemical behavior in organisms and with sediments

Salt in water

Water shields ions of unlike charge from one another; thereby breaking ionic bonds

Temperature of initial freezing is reduced by increased salinity

Different ions affect structuring properties of liquid water, cause formation of hydration atmospheres around the ion

Strong electrolytes dissolve completely; weak electrolytes dissociate only in part

Other modes of association—complex ions and ion pairs

Dissolved gases

Equilibrium concentration depends on solubility of a gas

Saturation concentration affected by salinity, temperature, and pressure conditions at the ocean surface

Nitrogen—near saturation throughout the ocean

Oxygen—produced by photosynthesis; utilized by plants and animals at all depths

Carbon dioxide and carbonate cycles

Carbon dioxide occurs as a dissolved gas, carbonic acid, carbonate, and bicarbonate

H_2CO_3 acts as a buffer

Calcium carbonate removed from seawater, mostly by plants and animals

Some inorganic precipitation of carbonate in warm waters

Physical properties of seawater

Pressure and viscosity increase in direct proportion in "normal" liquids

At less than 100 atmospheres pressure, increased pressure reduces water viscosity because it "favors" the unstructured mode

Bubbles and sea salt

Bubbles promote solution, evaporation, and transfer of gases

Bursting bubbles release salts to the atmosphere

Surface tension

High surface tension results from strengthened local water structure near any interface

Surface-active agents change surface tension of water, cause hydrophobic substances to "dissolve" in water

Electrical conductivity can be used to measure salinity

Particles in seawater

Affect chemistry of seawater by dissolving and either taking or releasing materials from surface waters

Large particles sink rapidly, contributing to sediments

Sound in the sea—a form of mechanical energy
  Speed increased by increasing temperature, salinity, pressure
  Sound channel (approximately 1000 meters depth) used for long-range acoustic signaling
Many fishes and mammals use sound to communicate and to locate food

SELECTED REFERENCES

BROECKER, W. S. 1974. *Chemical Oceanography*. Harcourt Brace Jovanovich, New York. 214 pp. Intermediate-level discussion of processes affecting ocean chemistry.

DEMING, H. G. 1975. *Water: The Fountain of Opportunity*. Oxford University Press, New York. 342 pp. General treatment of water; Chapter 15 discusses desalination of seawater.

GROSS, M. G. 1980 *Oceanography*, 4th ed. Charles E. Merrill, Columbus, Ohio. 145 pp. Elementary discussion of seawater.

HORNE, R. A. 1969. *Marine Chemistry: The Structure of Water and the Chemistry of the Hydrosphere*. Wiley-Interscience, New York. 568 pp. Advanced treatment of the subject; also a reference handbook.

MACINTYRE, FERRAN. 1970. Why the sea is salt. *Scientific American* 223 (November 1970): 104–115.

RILEY, J. P., AND R. CHESTER. 1971. *Introduction to Marine Chemistry*. Academic Press, New York. 465 pp. Intermediate in difficulty.

TEMPERATURE, SALINITY, AND DENSITY

This drifting buoy will follow water movement of eddies in the Gulf Stream. It will be tracked by satellite with position and temperature reports available twice daily for up to a year. These buoys are considered expendable but may be recovered and repowered for another experiment. (Photograph by Peter Wiebe, Woods Hole Oceanographic Institution.)

Two different approaches are used to describe the behavior of ocean waters on a global scale. The first is the concept of the *budget*. Heat and water budgets are constructed to show the major sources and what happens to each in the ocean. Regarding the water budget, it is important to remember that the amount of water on the earth's surface is nearly fixed. New sources need not be considered, only redistribution of water on the earth (except over hundreds of millions of years). The second approach determines routes followed by ocean water. This procedure includes studying *distributions of water temperature and salinity*. Both approaches are used in this chapter.

Ocean and atmosphere are closely coupled and a complete description of either requires detailed pictures of both oceanic and atmospheric conditions and current patterns throughout the earth over the course of a year. Despite enormous advances in oceanography in the past century, such a complete descriptive series is still not available. By using ships, buoys, aircraft, and/or satellites simultaneously, however, enough information about temperature and salinity distributions is accumulated to portray ocean conditions as they exist over a season.

LIGHT IN THE OCEAN   Radiation from the sun striking the earth's surface supplies almost all the energy that heats the ocean surface and warms the lower portion of the atmosphere. Part of this incoming solar radiation is the visible part of the spectrum and it provides energy needed by plants for photosynthesis, on land and in the ocean. After passing into the surface of the ocean, most of this energy is converted into heat, either raising water temperatures or causing evaporation.

The spectrum of radiant energy from the sun is filtered once as it passes through the atmosphere (Fig. 6-1) and is further filtered in the surface ocean. Within the first 10 centimeters of even pure water, virtually all the infrared portion of the spectrum is absorbed and changed into heat. Within the first meter of seawater, about 60% of the entering radiation is absorbed and about 80% is absorbed in the first 10 meters. Only about 1% remains at 140 meters in the clearest subtropical ocean waters.

In coastal waters abundant floating marine organisms, suspended sediment particles, and various dissolved organic substances absorb light at even shallower depths. Near Cape Cod, Massachusetts, for instance, only 1% of the surface light commonly penetrates to a

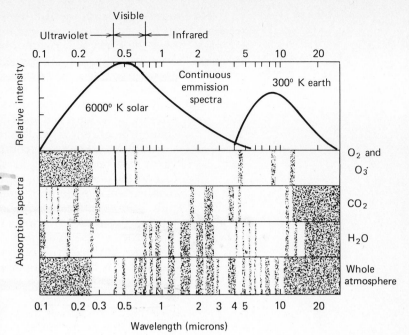

Figure 6-1

Emission spectrum of the sun and Earth as compared to the absorption spectrum for the atmosphere and several important constituents. (The dark areas are absorption bands where energy of that wavelength is absorbed.) The earth radiates to space as much energy as it received from the sun. The major difference is the shift from the relatively short-wave radiation of the sun, which we experience as visible light, to the longer-wave infrared radiation, which we perceive as heat. The amount of heat received from the sun is balanced by the amount of heat radiated back to space from the earth. (After Weyl, 1970.)

Figure 6-2

Spectrum of extinction for visible light in 1 meter of seawater in areas removed from sources of freshwater discharge. A is the curve for extremely pure water in the open ocean; B is for relatively turbid tropical-subtropical ocean water; C is for ocean water in midlatitudes; D is for clearest coastal water; and E, F, and G are for coastal waters of increasing turbidity. All wavelengths in the violet-to-yellow range are transmitted freely in open ocean water. Near coasts, the presence of dissolved pigments ("yellow substance" of organic origin) strongly inhibits transmission in the violet and blue ranges. (After Dietrich, 1963.)

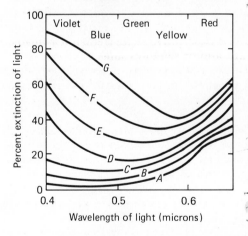

ATMOSPHERIC CIRCULATION AND HEAT BUDGET

depth of 16 meters. In such waters the maximum transparency shifts from the bluish region typical of clear oceanic waters to longer wavelengths, as shown in Fig. 6-2. In very turbid coastal waters the peak transparency occurs around 0.6 micron, in the yellow range. The Thames River water near New London, Connecticut, shows maximum transparency in the red, at wavelengths between 0.65 and 0.7 micron. In highly polluted waters absorption of all light takes place within a few centimeters of the water surface.

Far from the coast, ocean water often has a deep luminous blue color quite unlike the greenish or brownish colors common to coastal waters. The deep blue color indicates an absence of particles. In these areas, usually restricted to the central portions of the deep-ocean basins, the color of the water is thought to result from *scattering* of light rays within the water. A similar type of scattering is responsible for the blue color of the clean atmosphere.

The amount of light reflected from the ocean is controlled by the state of the sea surface and the angle at which the sun's rays strike the water. When the sun is directly overhead, only about 2% of incoming radiation is reflected; the remainder enters the water. When the sun is near the horizon, nearly all incoming radiation is reflected. Waves on the sea surface generally increase the amount of light reflected by as much as 50%, but when the sun is very near the horizon, waves decrease the amount of reflection.

One way to study average conditions on the earth as a whole is to construct a *budget* of energy or of a substance, such as water. Budgets indicate where materials or energy comes from (*sources*) and where they go (*sinks*), but often they cannot give detailed information about the rates at which these transfers occur or the routes they follow. Thus our generalized picture may differ substantially from conditions at any specific time or location.

First, we consider the earth's *heat budget*. The earth is essentially a sphere that is exposed to the sun's radiation from a distance of 149 million kilometers (about 90 million miles). The earth's atmosphere, directly beneath the sun, receives about 2 calories per square centimeter per minute ($cal\ cm^{-2}\ min^{-1}$), and radiant energy is distributed over the earth as it rotates about its axis every 24 hours. So, on the average, about 0.5 calorie per square centimeter per minute strikes any given point at the top of the atmosphere, as Fig. 6-3(a) indicates, for the earth's surface is dark for part of the 24-hour day.

There is also a small amount of heat coming from the earth's interior, as we learned in Chapter 2; it may be disregarded in this chapter because it is small compared to solar radiation.

Earth's atmosphere contains large amounts of nitrogen and oxygen, plus small and locally variable amounts of dust, water vapor, carbon dioxide, and ozone. Although present in minute quantities, dust, carbon dioxide, and water absorb some incoming solar radiation; thus only about half of the solar radiation coming to earth actually reaches the earth's surface, as indicated in Fig. 6-4. The remainder is absorbed by the atmosphere or scattered and reflected back to space. Estimates of the earth's reflectivity (albedo) range from 0.30 to 0.35. Figure 6-4 is based on an albedo of 0.35. Clouds, typically covering about 50% of the earth's surface, account for about two-thirds of the back-scattered radiation.

Figure 6-3

Incoming solar radiation above the earth's atmosphere. (a) shows the effect of latitude on the average solar radiation (compared to the equator) in the earth's surface in the Northern Hemisphere at the autumnal equinox, September 23; (b) shows the instantaneous maximum of incoming solar radiation during the summer solstice (June 21) when the sun is directly overhead at 23.5°N. Note the change between the equator and the pole.

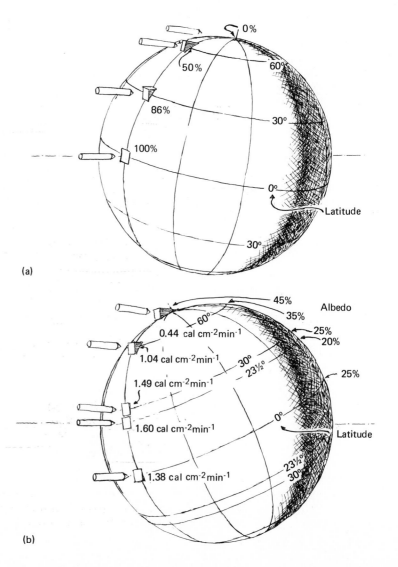

(a)

(b)

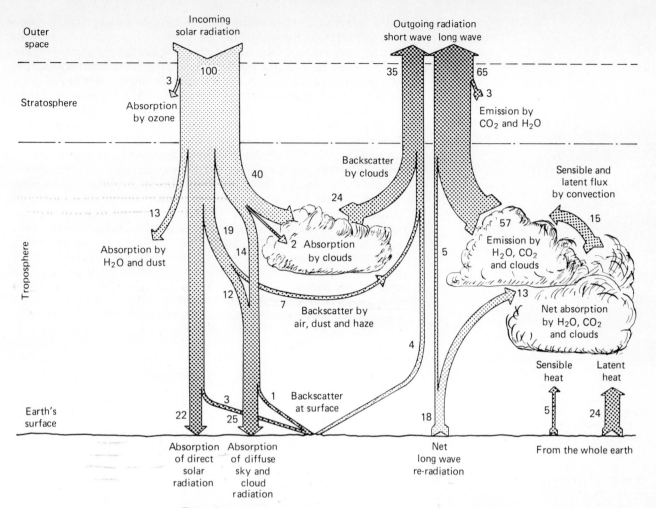

Outer space

Stratosphere

Troposphere

Earth's surface

Incoming solar radiation

100

3 ↘ Absorption by ozone

13 ↘ Absorption by H₂O and dust

40

19

14

12

7

Backscatter by clouds

24

2 Absorption by clouds

Backscatter by air, dust and haze

3

1 Backscatter at surface

22 25

Absorption of direct solar radiation

Absorption of diffuse sky and cloud radiation

Outgoing radiation
short wave long wave

35 65

↘ 3 Emission by CO₂ and H₂O

57

5 Emission by H₂O, CO₂ and clouds

13 Net absorption by H₂O, CO₂ and clouds

4

18 Net long wave re-radiation

Sensible and latent flux by convection

15

Sensible heat

Latent heat

5 24

From the whole earth

Figure 6-4

Average annual heat budget of the earth showing the major interactions with incoming solar radiation and loss of heat from Earth.

Despite the large amount of heat received by the earth from the sun each year, historical records indicate that the earth's surface temperatures over the past few thousand years have been so constant that any significant changes are extremely difficult to document. Therefore the earth radiates back to space as much energy as it receives from the sun. The sun's surface, being extremely hot, radiates energy, including the visible portion of the electromagnetic spectrum. Because the earth's surface is so much cooler, its radiation occurs in the infrared portion of the electromagnetic spectrum (see Fig. 6-1).

Of the energy lost to space, only about 5% is radiated back directly from the earth's surface (this is called *back-radiation*). About 60% is radiated back to space, primarily from clouds (see Fig. 6-4). The atmosphere is nearly opaque to the earth's radiations so that it acts much like a blanket in keeping the surface warmer than would be the case under a more transparent atmosphere. The so-called *greenhouse effect* is caused by strong absorption of infrared radiation by carbon dioxide and water-vapor molecules. Without these constituents, Earth surface temperatures might drop as low as −20°C, which is the temperature at the top of the cloud layer. Instead ocean surface temperatures average about 17.5°C and the land about 14°C.

Relatively little heat is transferred to the atmosphere by direct heating *(sensible heat)*. Most atmospheric heating takes place through release of latent heat during condensation of water vapor and almost all this water vapor comes from the ocean.

The earth's axis of rotation is inclined 23.5° to the plane of the earth's movement around the sun. This factor causes the amount of radiant energy at the top of the atmosphere to vary throughout the year for any point on earth, creating the change in seasons. Those points where the sun is directly overhead at noon receive the maximum amount of *insolation* (incident solar radiation). Points north and south receive less and no direct radiation at all is received where the sun's rays just graze the earth's surface [see Fig. 6-3(a)].

Averaged over a year, the maximum amount of incoming solar energy is received in low latitudes so that insolation is unevenly distributed over the earth's surface. The atmosphere, however, radiates energy back to space at a nearly constant rate, as shown in Fig. 6-5. As a result, the earth receives most of its heat in low latitudes (between about 40°N and 40°S whereas there is a net heat loss to space in the high latitudes between 40 and 90° in both hemispheres but especially pronounced in the hemisphere experiencing winter.

Large inputs of solar energy into tropical regions cause high ocean surface temperatures there, as seen in Figs. 6-6 and 6-7. Maximum open ocean temperatures, between 26 and 29°C, occur just north of the equator, at about 5°N (excluding isolated basins). Temperature distribution is strongly modified by ocean currents and by the unequal distribution of landmasses in the two hemispheres.

Note that the temperature difference between the warmest and coldest month of the year is small—less than 3°C—in the low latitudes (10°N to 10°S) and near the North and South poles. In the midlatitudes, around 30°N and 30°S, the annual variation is largest (see Fig. 6-8).

Heat transfer from low to high latitudes takes place in the atmosphere and the ocean. In this somewhat oversimplified picture, atmosphere and ocean function as a simple heat engine, transforming heat energy into motion. Atmospheric motions are experienced as winds (see Fig. 6-9), oceanic movements as currents. Although ocean currents move far more slowly than winds, the large heat capacity of water (compared to air) enables it to carry far more heat per volume than air does. Ocean currents carry more heat than the winds between about 30°N and 30°S. Nearer the poles, the atmosphere carries more heat than the ocean circulation (see Fig. 6-10).

Figure 6-5

About four times as much solar energy reaches the earth's surface at the equator as at the poles, but radiation of energy to space is nearly the same everywhere on the earth.

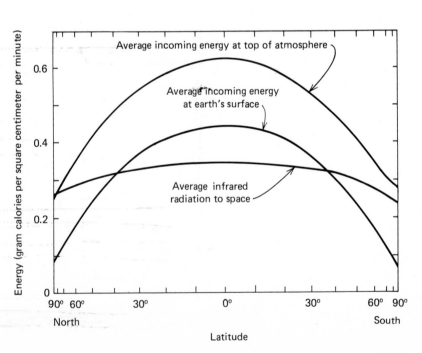

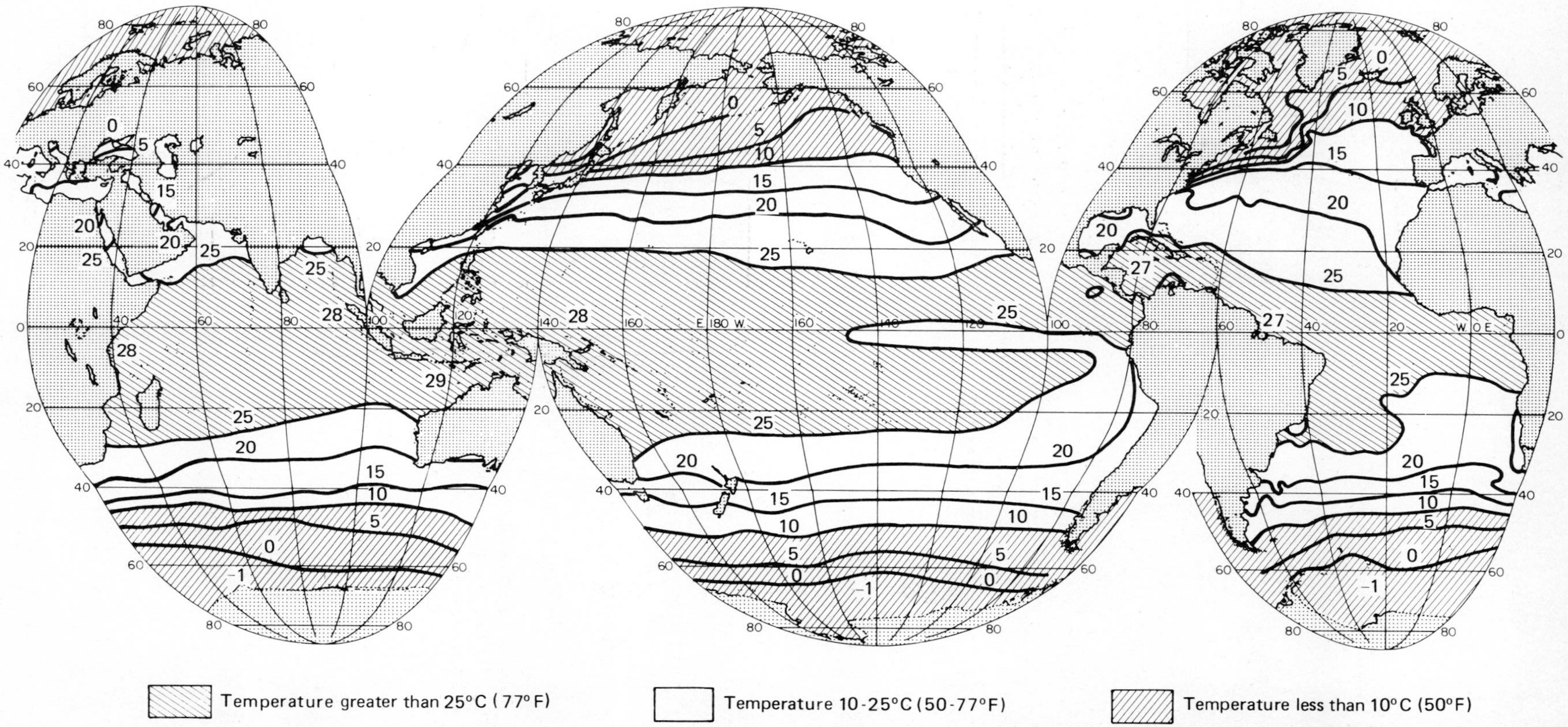

Figure 6-6

Ocean surface temperatures in February. Note that isotherms (lines connecting points with surface temperature) tend to parallel the equator. (After Sverdrup et al., 1942.)

Temperature less than 10°C (50°F)

Temperature 10-25°C (50-77°F)

Temperature greater than 25°C (77°F)

Figure 6-7

Ocean surface temperatures in August. Below about a hundred meters of the surface, temperatures drop sharply except at very high latitudes. (After Sverdrup et al., 1942.)

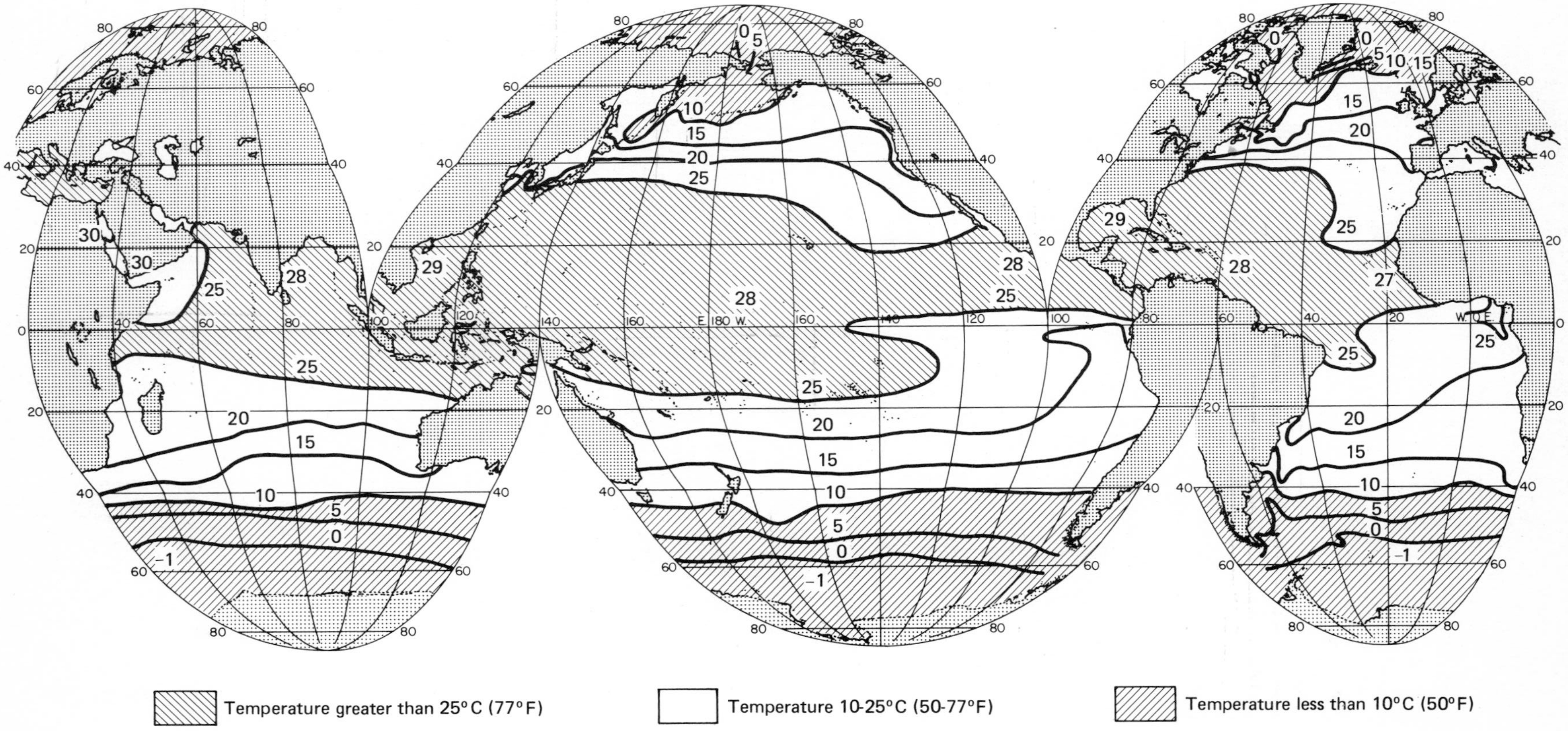

Temperature less than 10°C (50°F)

Temperature 10-25°C (50-77°F)

Temperature greater than 25°C (77°F)

148

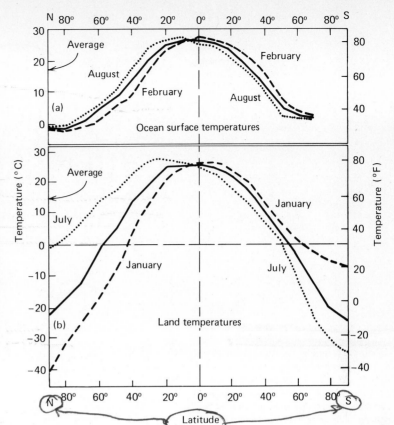

Figure 6-8

Average and temperature ranges at the earth's surface. (a) Average temperature ranges for the ocean surface. Note that the maximum variation occurs in midlatitudes, around 40°N and 40°S. (After G. Wüst, W. Brogmus, and E. Noodt, 1954. Die zonale verteilung von salzghehalt, niederschlag, verdunstung, temperatur und dichte an der oberflache der ozeane. *Kieler Meeresforschungen* 10:137–61.) (b) Average temperature ranges for land (or ice-covered) areas. Note that the extreme temperature variations occur in polar regions, in contrast to the ocean. (Data from *Smithsonian Physical Tables*, 1964.)

Figure 6-9

Schematic representation of the planetary circulation on the earth's atmosphere.

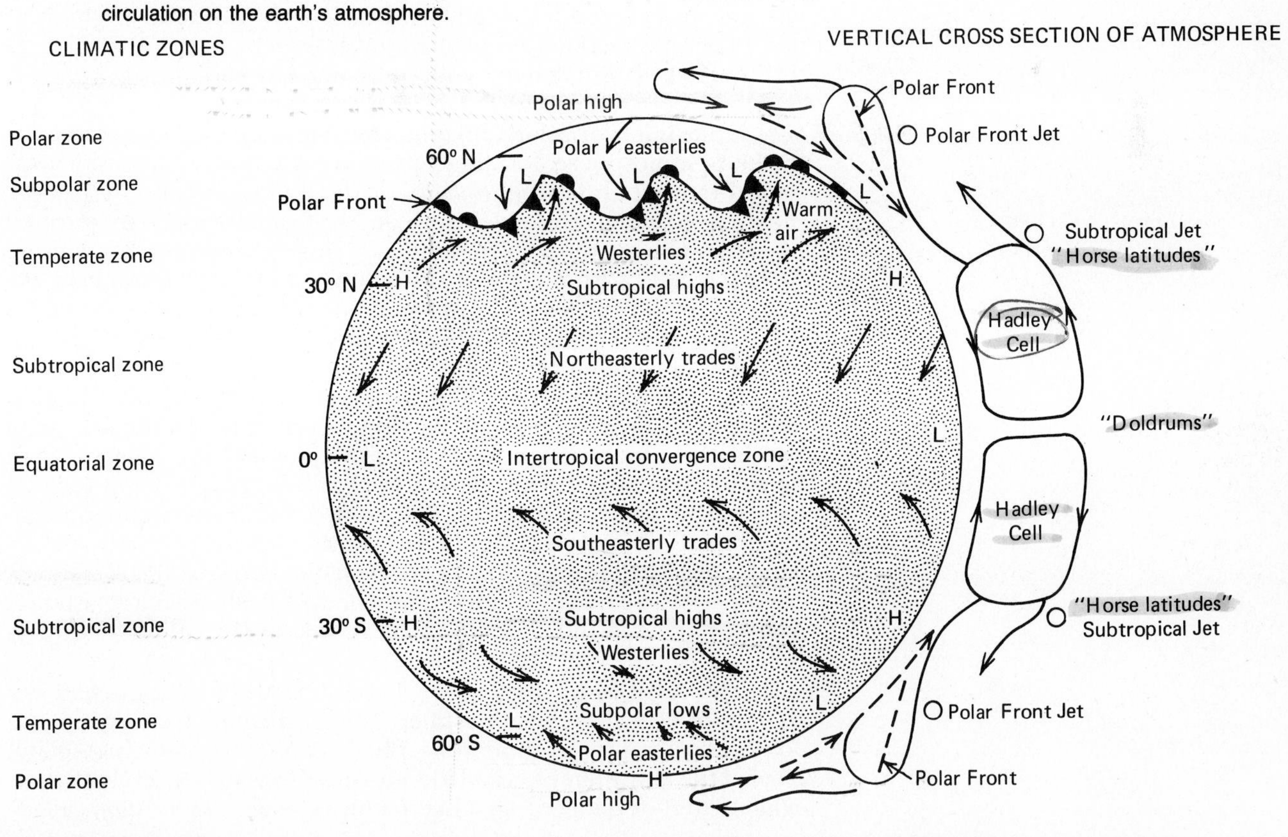

CLIMATIC ZONES

VERTICAL CROSS SECTION OF ATMOSPHERE

Polar zone
Subpolar zone
Temperate zone
Subtropical zone
Equatorial zone
Subtropical zone
Temperate zone
Polar zone

60° N
30° N — H
0° — L
30° S — H
60° S

Polar high
Polar Front
Polar easterlies
Westerlies
Warm air
Subtropical highs
Northeasterly trades
Intertropical convergence zone
Southeasterly trades
Subtropical highs
Westerlies
Subpolar lows
Polar easterlies
Polar high

Polar Front
Polar Front Jet
Subtropical Jet "Horse latitudes"
Hadley Cell
"Doldrums"
Hadley Cell
"Horse latitudes" Subtropical Jet
Polar Front Jet
Polar Front

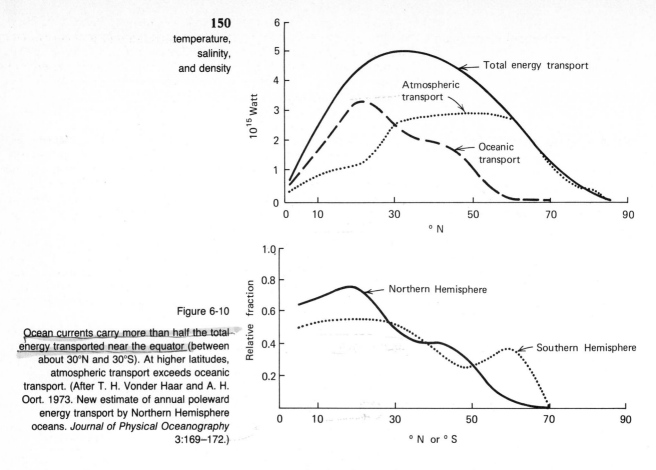

Figure 6-10

Ocean currents carry more than half the total
energy transported near the equator (between
about 30°N and 30°S). At higher latitudes,
atmospheric transport exceeds oceanic
transport. (After T. H. Vonder Haar and A. H.
Oort. 1973. New estimate of annual poleward
energy transport by Northern Hemisphere
oceans. *Journal of Physical Oceanography*
3:169–172.)

Atmospheric circulation is controlled primarily by minute amounts of water vapor, ozone, and carbon dioxide. Although nitrogen constitutes about 78% of the atmosphere by volume, it has little direct effect on circulation: carbon dioxide and water vapor together cause the greenhouse effect previously mentioned.

Absorption of solar radiation (see Fig. 6-5) causes warming of the lower atmosphere, forming a pronounced change in density *(tropopause)* that separates the *stratospheric* (high-altitude) and *tropospheric* (near-earth) circulation systems. In the stratosphere gases are warmed by absorption of ultraviolet radiation; thus temperature of ionized gases increases with increased altitude. Gases in the troposphere are warmed by the earth's surface; consequently, temperature decreases with increased altitude. Weather systems, with vigorous vertical convection and formation of clouds, all occur in the troposphere.

On a featureless, nonrotating water-covered earth, the simplest atmospheric circulation would be a single cell. Warm moist air would rise in the equatorial regions and move toward the poles in both hemispheres, carrying heat as latent heat (water vapor) and sensible heat (warm air). When the air was cooled and water vapor precipitated, it would sink and flow back toward the equator in a surface wind. Such a simple cell does occur near the equator: the *Hadley cell*, named for the British meteorologist who first postulated its existence. This cell includes the persistent trade winds of low latitudes (see Fig. 6-9).

The earth is, of course, not featureless and it rotates. And the amount of heat that the atmosphere must transport from equator to poles is too great for a simple atmospheric pattern to maintain stability. Instead the atmospheric circulation breaks down into several cells and exhibits well-developed, wavelike patterns, especially in the midlatitudes where the westerly winds prevail, as illustrated in Fig. 6-9. These waves give rise to the series of *fronts* (sharp boundaries between two air masses) that dominate weather patterns through much of North America and northern Europe. Such waves (or fronts) commonly separate cold, dry polar air masses from warmer, more humid air masses of midlatitudes.

OCEANIC WATER AND HEAT BUDGETS

The amount of liquid water at the earth's surface has remained essentially constant for the past 5000 years—since the retreat of continental glaciers from North America and Europe. The ocean is the major water reservoir; the small amounts in lakes, rivers, and atmosphere are essentially in transit back to their source. Groundwater also eventually finds it way back to the ocean, although taking much longer to make the trip.

A convenient way to visualize the earth's water budget is expressed in Table 6-1. About 97 centimeters is evaporated, on the average, from the ocean surface each year. Of this amount, 88 centimeters falls as rain on the ocean. The remainder falls on land (along with rain from local evaporation) and eventually makes its way into rivers to return to the ocean.

Yet another way of describing the water budget or the hydrological cycle is to express amounts in thousands of cubic kilometers, as in Fig. 6-11.

TABLE 6-1

*Water Budget of the Earth**

AREA	PRECIPITATION		EVAPORATION	
	(10³ km³/yr)	(cm/yr)	(10³ km³/yr)	(cm/yr)
Ocean	320	88	350	97
Continent	100	67	70	47
Entire earth	420	82	420	82

*After Albert Defant, 1961. *Physical Oceanography.* Pergamon Press, Elmsford, N.Y. Vol. I, p. 235.

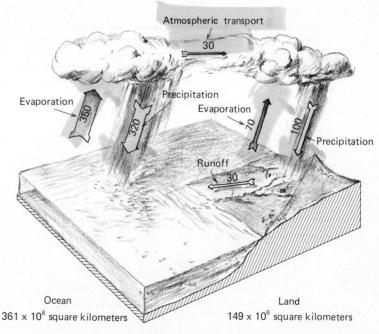

Figure 6-11

Schematic representation of total amounts of liquid water involved in the hydrological cycle each year, in thousands of cubic kilometers.

Evaporation from the ocean surface is most active in subtropical areas, where it is favored by clear skies and dry winds. The sides of ocean basins where dry winds blow from the continents are especially favorable for evaporation.

Neither is precipitation uniform over the entire ocean, as Fig. 6-12 shows. Equatorial and midlatitude regions experience heavy precipitation; in the tropics particularly, large amounts of rainfall result from an abundance of warm, moist air rising in the equatorial atmospheric circulation. As rising air cools, its ability to hold water vapor diminishes and precipitation results. Satellite photographs (see Fig. 6-13) commonly show a line of clouds just north of the equator because of this circulation; this is known as the intertropical convergence. The high rainfall in the equatorial region (shown in Fig. 6-12) causes lower surface salinities in the equatorial ocean, as seen in Fig. 6-14.

Because of evaporation and precipitation distributions, salinity varies over the ocean surface. Highest salinities occur in open-ocean subtropical areas, where evaporation is pronounced and no significant sources of freshwater exist. Near coasts, rivers discharge freshwater into the ocean and salinities are lowered. Surface waters of high-latitude oceans are commonly low in salinity because of relatively heavy precipitation and little evaporation.

Continents (especially those with mountain ranges) and ocean currents affect salinity distributions. The Atlantic Ocean, for example, has a higher salinity than the Pacific. Mountain ranges, especially the Andes Mountains of South America, cause winds from the Pacific to give up their moisture before blowing across to the Atlantic. High mountains deflect winds to higher altitudes where they are cooled and their water vapor condenses to fall as rain on the mountains. On the other hand, the Atlantic's water vapor is carried to the Pacific by winds blowing across Panama.

Figure 6-12

Relationship between oceanic evaporation and precipitation at various latitudes (a) and the salinity of surface waters (b). (After Wüst et al., 1954.)

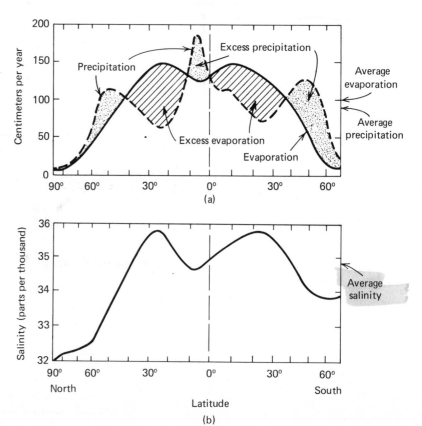

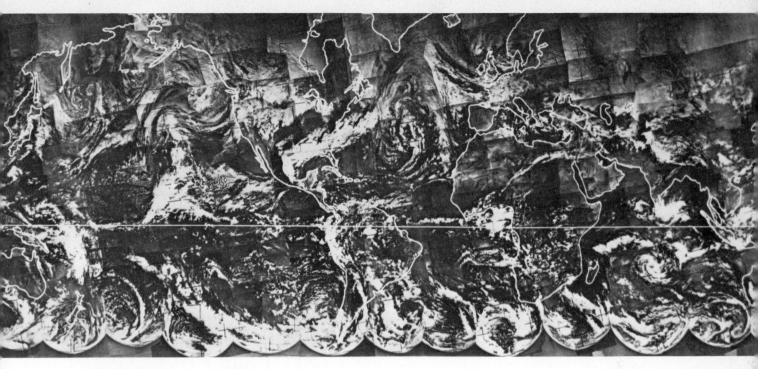

Figure 6-13

Composite photograph (photomosaic) of the earth surface prepared from 450 individual pictures taken by NASA satellite Tiros IX during the 24 hours of February 13, 1965—the first complete picture of the world's weather. Continents are outlined by white lines. Large areas of white are clouds or the permanent ice covers on Antarctica (bottom margin). Note the large spiral weather systems south of Alaska and Greenland. Also note the thin line of clouds along the equator, especially noticeable in the Atlantic. (Photograph courtesy NASA.)

A water budget can be constructed for a single region and expressed as Evaporation = Precipitation + Runoff + Contribution from currents

$$E = P + R + C$$

where E = the amount of water lost through evaporation

P = the amount of water gained through precipitation

R = the amount of water brought to the region by rivers

C = the relative amount of freshwater moved by currents

The last quantity (C) can be positive if water of lower salinity flows into the region or negative if more saline water flows into the region. Unless there is a pronounced salinity change, the amount of water gained by precipitation and river runoff will equal the amount lost through evaporation and carried out of the region.

To summarize, surface-water salinity in the open ocean is controlled primarily by the local balance between evaporation and precipitation. Where evaporation exceeds precipitation, surface waters are more saline than average ocean water ($S = 34.7‰$). Freezing and melting of sea ice affect salinity in polar regions.

Where precipitation and river runoff exceed evaporation, surface-water salinities are lower than for the average ocean. Low-salinity waters are typical of three regions: the tropics because of the large rainfall there; high latitudes because of relatively high precipitation and limited evaporation; and along continental margins because of local rainfall and river runoff. Consequently, high surface-water salinities in the open ocean tend to occur near the centers of the ocean basins, at midlatitudes (Fig. 6-14).

Highest surface-water salinities occur in partially isolated basins of tropical areas, such as the Red Sea, the Arabian Gulf, and the Mediterranean Sea. Here water losses due to evaporation are not

153

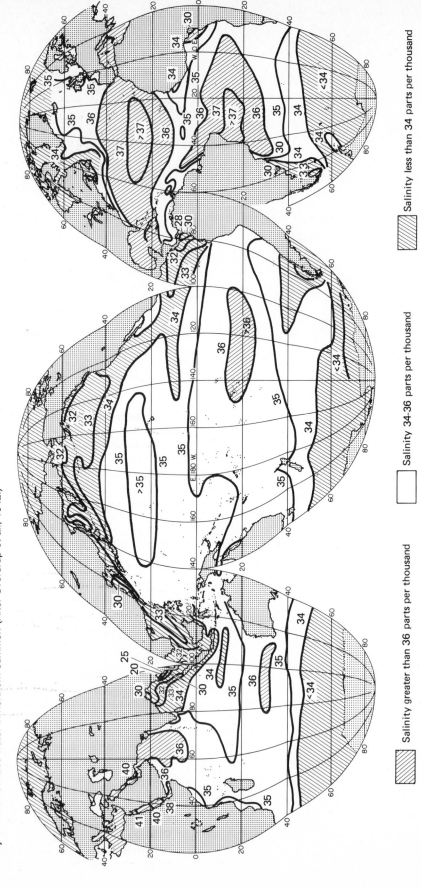

Figure 6-14
Salinity of ocean surface water in northern summer. (After Sverdrup et al., 1942.)

Salinity less than 34 parts per thousand

Salinity 34-36 parts per thousand

Salinity greater than 36 parts per thousand

154

replaced by river runoff or precipitation. The restricted communication with adjacent ocean waters prevents ready renewal of the surface waters. Consequently, the salinity of surface waters is 40‰ or more in such basins as the Arabian Gulf and the northern Red Sea. These basins occur at midlatitudes, around 30°N, where several of the world's great land deserts occur—such as the Sahara in Africa and Mojave in California. The arid Australian continent is an example of the low rainfall at comparable latitudes in the Southern Hemisphere.

Heat and water budgets are intimately related. Evaporation of water removes heat from the ocean surface whereas precipitation contributes large amounts of heat to the atmosphere.

A *regional heat budget* for a part of the ocean can be expressed as follows:

Heat gained = Heat lost

$$Q_s + Q_c = Q_e + Q_r + Q_h$$

where Q_s = energy gained from solar radiation

Q_c = energy gained (or lost) by ocean currents

Q_e = energy lost through evaporation

Q_r = energy lost by radiation

Q_h = energy lost by direct heating of the atmosphere (sensible heat)

Areas of relatively high surface salinity supply heat to the atmosphere in the form of water vapor, which is transported by winds to the midlatitudes. There water vapor condenses so that heat is given off, accompanied by precipitation. Thus the poleward transport of heat and water is reflected in the relatively low surface salinities at midlatitudes. Regional salinity differences, therefore, are another manifestation of global heat transport.

DENSITY OF SEAWATER   Density of seawater is determined primarily by temperature and salinity. (Pressure is important only in the deep ocean and will be ignored in this discussion.) Decreases in temperature and increases in salinity cause increases in density. The effects of temperature and salinity can be seen in Fig. 6-15: changing salinity from 19 to 26‰ (at 30°C) has the same effect as changing the temperature from 31 to 12°C at a constant salinity of 20‰.

In the ocean water masses generally form thin horizontal layers arranged according to water density; such waters are said to be *vertically stratified*. In the *stable-density* configuration, the densest water is on the bottom, the least dense on top. A stable situation can also be achieved by supplying energy to a stratified system and mixing the layers so that density stratification no longer exists. This would be a *neutrally stable system*. If one layer is more dense than the one beneath it, the system is unstable; the denser layer tends to sink to its appropriate density level thus creating a stable structure.

Because ocean water normally has stable stratification, the most dense water might be expected to occupy the deepest part of a given ocean basin. Where such is not the case, it usually indicates that some physical barrier—say, a shallow sill—is preventing lateral spreading of dense water. For example, dense waters formed in the Arctic Ocean cannot readily flow into the North Atlantic because of submarine ridges between Greenland and Scotland.

Figure 6-15

Variation in seawater density as affected by
temperature and salinity. Note the changes in
temperature of maximum density and the point
of initial freezing caused by changed salinity.

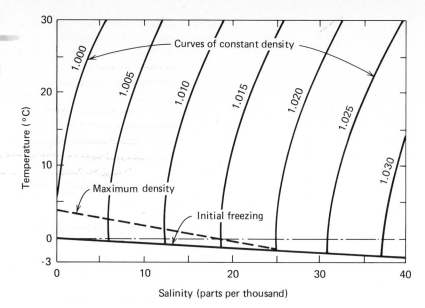

In the absence of such barriers, a water mass tends to seek its appropriate density level and to spread laterally, often as a thin layer. Consequently, lateral changes in water temperature and salinity are much smaller than vertical changes.

Energy must be supplied if a density-stratified system is to be mixed. Below the surface of the open ocean, vigorous mixing is rare. In coastal waters, on the other hand, mixing is common because of tidal currents and movement of waters across rough bottom topography. Near the ocean surface, energy sufficient to cause mixing is supplied by the wind, passage of waves, and surface heating and cooling. In high-latitude areas, the density structure is weak enough to permit widespread vertical circulation through the ocean depths.

This pattern is quite different from that of the lower atmosphere, where strong vertical circulation is a characteristic feature. The ocean is heated and cooled at its upper surface. Heating of the surface inhibits vertical circulation and powers a very inefficient heat engine. In contrast, the lower atmosphere is cooled at the top by radiation to space and heated at the bottom through contact with land and ocean. Consequently, a vertical circulation is quite well developed through the lower parts of the atmosphere.

Three general depth zones can be identified in the ocean; the surface, pycnocline, and deep zones, schematically represented in Fig. 6-16. The *surface zone* is in contact with the atmosphere and undergoes

Figure 6-16

Schematic representation of the horizontal layering of ocean waters.

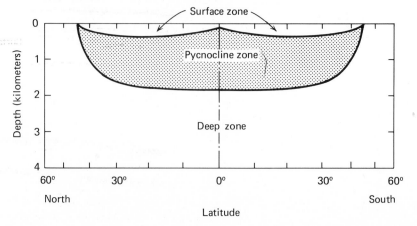

substantial changes because of seasonal variability in local water budgets (evaporation, precipitation) and heat budgets (incoming solar radiation, evaporation). The surface zone contains the least dense water masses in the ocean and is the ultimate source of all marine food production because there is sufficient light for photosynthesis by plants.

The surface zone averages about 100 meters in thickness and constitutes about 2% of the ocean's volume. Surface waters are generally quite warm—except at high latitudes, where extensive mixing with deeper layers may prevent development of a warm surface zone even during summer. Near the surface, waters have nearly neutral stability because of strong mixing and vertical water movements are quite common. Winds, waves, and strong cooling of surface waters cause mixing in the open ocean.

In the *pycnocline zone*, water density changes appreciably with increasing depth and thus forms a layer of much greater stability than is provided by the overlying surface waters. The formation of a pycnocline zone results from marked changes with depth in either salinity or temperature. Marked temperature changes with depth are common in the open ocean, as Fig. 6-17 indicates, and there the pycnocline coincides with the *thermocline*, or zone of sharp temperature change. Thermoclines are especially prominent in open ocean areas, where the surface layers are strongly heated during much of the year.

At higher latitudes there is less surface heating by incoming solar radiation. There the lowered surface salinity caused by abundant precipitation causes a *halocline*, a marked change in salinity with depth. The halocline is well developed over much of the North Pacific Ocean, for example, where it is an important factor in the stability of the pycnocline zone. The halocline is also important in most coastal ocean areas.

The pycnocline partially isolates the deep ocean from surface processes because low-density surface waters cannot readily move downward through it. Deep-ocean waters move slowly upward through the pycnocline as part of the worldwide deep-ocean circulation. This is a significant factor in oceanic circulation at midlatitudes and low latitudes. Seasonal changes in temperature and salinity are quite small in waters below the pycnocline.

The *deep zone* contains about 80% of the ocean's waters and lies at depths of about 2 kilometers or more—except at high latitudes, where deep waters are in contact with the atmosphere. The relationship between deep-ocean waters and high-latitude areas accounts for the low temperatures of the subsurface ocean. Dissolved oxygen is also supplied to the deep ocean through these high-latitude exposures of cold water and the movements of the water masses formed there.

Figure 6-17

Variations with depth in temperature cause the thermocline. Similar variations in salinity cause the halocline. The pycnocline can be caused by either independently or by both occurring together.

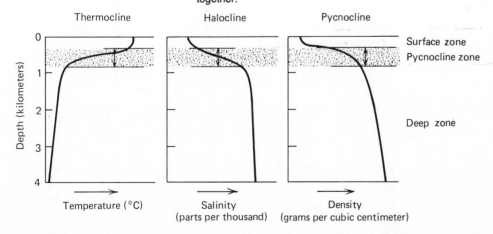

Figures 6-18 and 6-19 show vertical distribution of temperature and salinity in all three ocean basins. The profiles display temperature and salinity distributions along the western sides of the Atlantic and Indian oceans and the central portion of the Pacific Ocean. Note both the similarities and the differences between this and the generalized diagram (Fig. 6-16) we previously considered. Note also the complexities of temperature and salinity distribution evident on even such a simplified diagram.

The base of the surface zone corresponds to the 10°C-isotherm contour on the vertical section. The base of the pycnocline can be taken as the 4°C isotherm. Note that there are strong similarities between the distributions of pycnoclines and surface zones in the three ocean basins but that they are by no means identical. Also note that

Figure 6-18

Vertical distribution of temperature in the three ocean basins. Vertical exaggeration is approximately a thousandfold. (After Dietrich, 1963.)

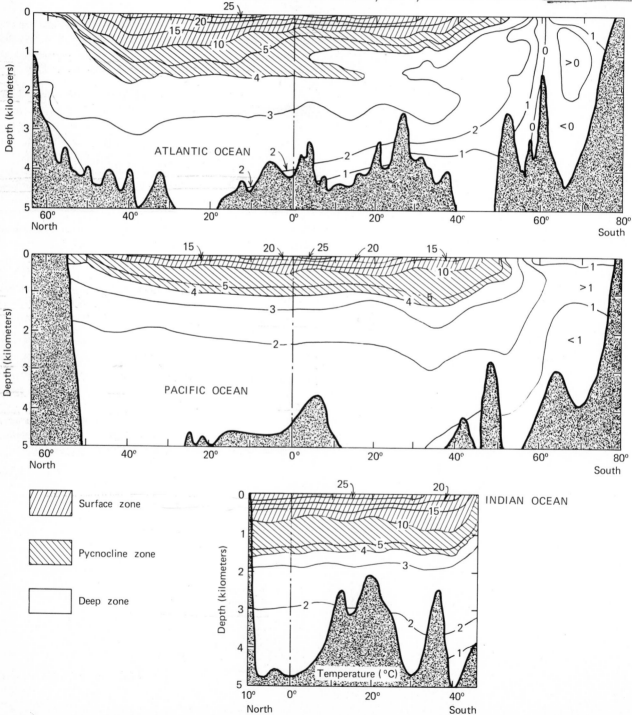

there are differences between hemispheres even within the same ocean basin.

The vertical temperature and salinity sections show that distinctly different strata or water masses exist at intermediate depths. For example, there is a relatively low-salinity water mass about 1 kilometer below the surface, extending from the Antarctic region into the South Atlantic Ocean. Through analysis of such sections oceanographers are able to determine where water masses form and how they move below the ocean surface. This technique has been one of the most rewarding means of mapping subsurface currents. Such studies have demonstrated that the dominant movement of near-bottom currents is generally north–south in contrast to the primarily east–west movement of currents at the surface.

Figure 6-19

Vertical distribution of salinity in the three basins. Vertical exaggeration is approximately a thousandfold. (After Dietrich, 1963.)

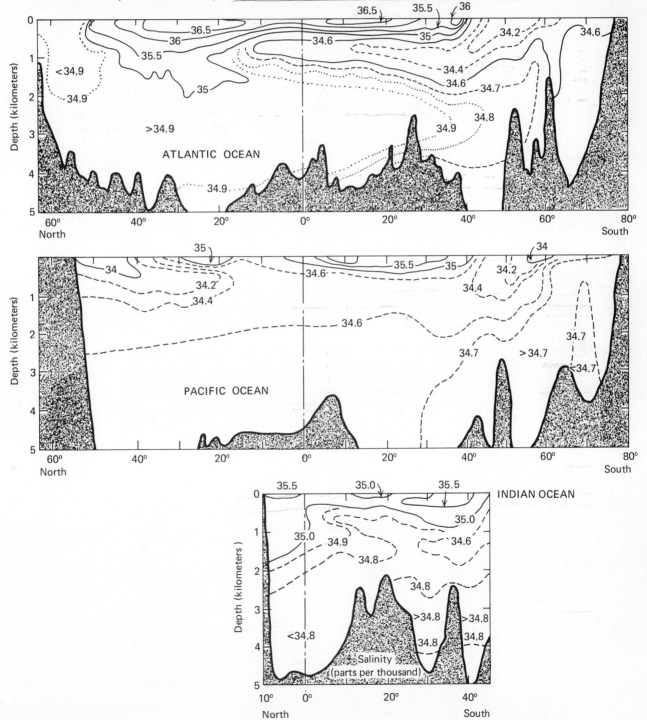

Another aspect of the ocean illustrated by the vertical section is that most of the ocean has much lower temperatures (see Fig. 6-20) than might have been predicted from examining the ocean surface alone. The average temperature of the entire ocean is 3.5°C; in contrast, the average ocean surface temperature is about 17.5°C. So it is the surface ocean that serves to buffer our climate against short-term changes. Most of the deep ocean is too isolated to have much effect on short-term climatic conditions, although it may influence long-term climatic changes over hundreds or thousands of years.

Water masses with characteristic temperatures and salinities are formed in certain ocean areas, usually partially isolated seas or on continental shelves, listed in Table 6-2. If a mass of surface water becomes more dense than the water below, it will sink below the surface. From that point, until it returns again to the surface, the

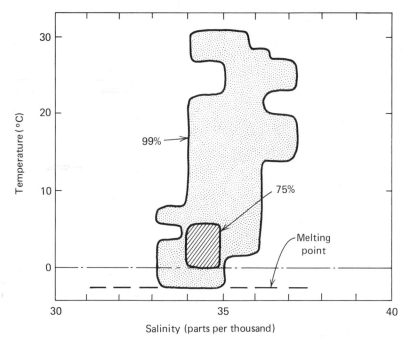

Figure 6-20

Temperature and salinity of 99% of ocean water are represented by points within the stippled area enclosed by the 99% contour. The shaded area represents the range of temperature and salinity of 75% of the water in the world ocean. (After R. B. Montgomery, 1958. Water characteristics of the Atlantic Ocean and the world ocean. *Deep Sea Research* 5:134–148.)

TABLE 6-2

*Some Major Water Masses of the Ocean**

WATER MASS	TEMPERATURE RANGE (°C)	SALINITY RANGE (‰)
Antarctic bottom water	−0.4	34.66
North Atlantic deep water	3–4	34.9–35.0
North Atlantic central water	4–17	35.1–36.2
Mediterranean water	6–10	35.3–36.4
North polar water	−1–+2	34.9
South Atlantic central water	5–16	34.3–35.6
Subantarctic water	3–9	33.8–34.5
Antarctic intermediate water	3–5	34.1–34.6
Indian equatorial water	4–16	34.8–35.2
Indian central water	6–15	34.5–35.4
Red Sea water	9	35.5
Pacific subarctic water	2–10	33.5–34.4
Western north Pacific water	7–16	34.1–34.6
Pacific equatorial water	6–16	34.5–35.2
Eastern south Pacific water	9–16	34.3–35.1
Antarctic circumpolar water	0.5–2.5	34.7–34.8

*After Albert Defant, 1961. *Physical Oceanography*. Pergamon Press, Elmsford, N.Y. Vol. I, p. 217.

temperature and salinity of the waters change primarily by mixing with other water masses. In the deep open ocean, where there is little energy to cause mixing, this process can be very slow. Temperature and salinity can therefore be used in the open ocean as "tags" to identify individual water masses as they spread laterally between other water masses or move along the ocean floor.

The so-called *core* technique has been used to study movement of well-defined subsurface water bodies (having characteristic temperature and salinity) as they spread out from a source region. It is a horizontal representation of the most likely paths of water particles. It allows flow across density surfaces, for the density of the core water mass usually varies. An example is the warm, saline water (6 to 10°C, 35.3 to 36.4‰) formed in the Mediterranean Sea that flows out through the Strait of Gibraltar near the bottom and spreads laterally at mid-depths in the North Atlantic Ocean, pictured in Fig. 6-21(a). Its distinctive temperature and salinity permit it to be clearly distinguished from water masses above and below. As the water moves away from its source [Fig. 6-21(b)], it mixes with adjacent waters and grad-

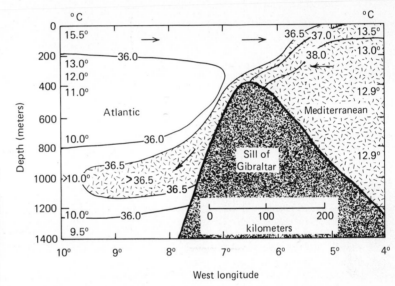

Figure 6-21(a)

Vertical distribution of salinity and temperature in a cross section (east–west) through the Strait of Gibraltar at the entrance to the Mediterranean. Arrows show the main direction of water movement. (After G. Schott, 1942. *Geographie des Atlantisches Ozeans.* Boysen, Hamburg.)

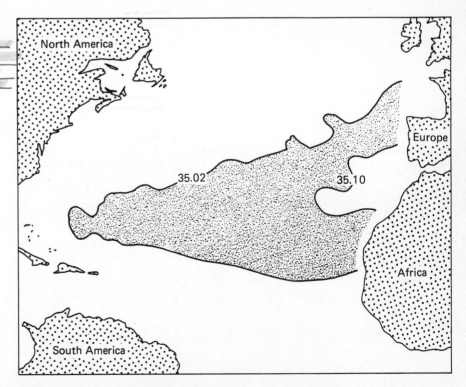

Figure 6-21(b)

The waters from the Mediterranean (salinity greater than 35‰) spread out as a layer a few hundred meters thick that can be recognized all the way across the North Atlantic at depths around 1000 meters. (After L. V. Washington and W. R. Wright, 1970. North Atlantic Ocean atlas. *Woods Hole Oceanographic Institution Atlas Series,* vol. 4.)

ually loses its distinctive characteristics. Comparison of the values observed at any given point where that mass is identifiable with the original values provides an estimate of the amount of mixing with other waters that has occurred since the water mass initially formed.

To study mixing, a *temperature–salinity (T-S) diagram* is constructed. Points are plotted to correspond to the measured temperatures and salinities of individual water samples. If conditions are constant and a water mass is homogeneous, the entire mass is represented by a single point on a *T-S* diagram. More commonly, there is some variation within the mass, as well as mixing with adjacent masses, resulting in a *T-S curve* that indicates a continuum of properties within the water column at each station (Fig. 6-22). This *T-S* curve remains the same (or within a small range of difference) for

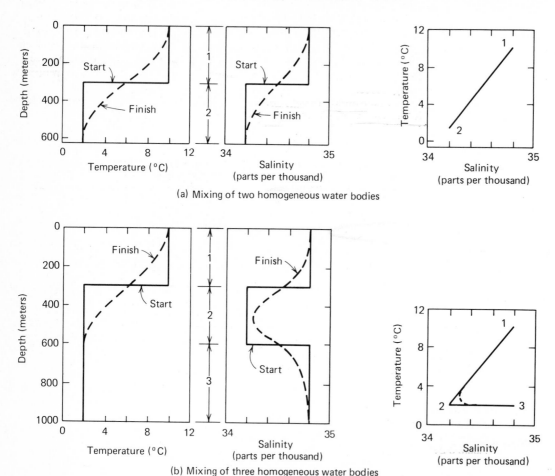

(a) Mixing of two homogeneous water bodies

(b) Mixing of three homogeneous water bodies

Figure 6-22

Temperature–salinity relationships resulting from mixing of water masses. (a) Water mass 1 (0 to 300 meters of depth; $T = 10°C$, $S = 34.8‰$) mixes with water mass 2 (300 to 600 meters depth; $T = 2°C$, $S = 34.2‰$). Initially the boundary between the two masses is sharp and represented by sudden, large changes in temperature and salinity with depth. After mixing, the boundary is more diffuse and the changes in temperature and salinity occur over a depth range of nearly 600 meters. On the *T-S* diagram, mixing of these two water masses is shown by the straight line connecting the *T-S* points characteristic of the two original water masses. (b) Mixing of three water masses can be represented in a similar manner. Water mass 3 (600 to 1500 meters depth; $T = 2°C$, $S = 34.8‰$) mixes with water mass 2 but not with water mass 1. The initial sharp boundaries between masses are gradually obliterated. On the *T-S* diagram, the mixing process is shown by two straight lines, indicating that water masses 1 and 2 mix and that water masses 2 and 3 mix.

each location at any point in time because the water masses present have formed in a constant-source area and have the same properties when they reach any given station.

Consider the two water masses in Fig. 6-22(a), each with its characteristic temperature and salinity. As they mix, the newly formed waters will have temperatures and salinities that plot on the line connecting the two original points. Furthermore, a position on this line indicates the relative proportions of the waters that have been mixed. For example, a mixture of equal amounts of two different waters will fall midway between them on the line connecting the two points on the T-S diagram.

Adding a third water mass results in a more complicated diagram, but the same general relationship holds. From analyses of T-S curves it is possible to determine the sources of the water masses present at that station. Figure 6-23 shows diagramatically how mixing of subsurface water layers might appear on a T-S diagram. A characteristic reversal point is associated with the core (central depth) of water mass B. It can be distinguished until the mass has completely lost its identity, although it becomes less and less obvious as mixing progresses.

Each major ocean area has a set of distinctive T-S curves. A set of curves is shown for the North Atlantic in Fig. 6-24.

Figure 6-23

Use of a T-S diagram to study subsurface water movements. Water mass B (intermediate density) flows beneath surface layer A and above the denser water mass C. A series of oceanographic stations taken at three locations show different but related T-S curves. At station 1, closest to the source of B, its characteristic temperature and salinity plot at point B on the T-S diagram. Farther from the source, the water mass mixes with those above and below so that it no longer has its original value, although the general shape of the T-S diagram is still quite discernible. At station 3, farthest from the source, water mass B is not present and masses A and C mix together, resulting in a continuum of T-S values intermediate between the two.

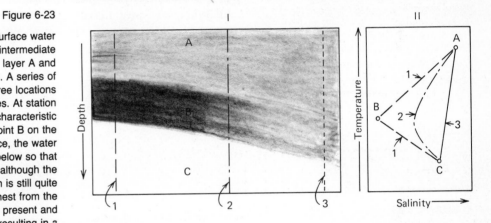

Figure 6-24

Temperature–salinity curves for major water masses in the Atlantic. (Data from Sverdrup et al., 1942.)

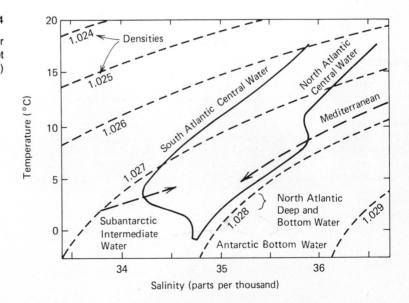

A few materials have been injected into the ocean in quantities large enough to permit tracing the movements of water masses. Such techniques are simplest when the substance introduced does not occur naturally in the ocean. Radioactive wastes from the manufacture of nuclear weapons and their large-scale testing in the atmosphere during the early 1960s provided tracers that also permit estimates of the rates of the mixing processes.

Tritium, a radioactive hydrogen isotope with a half-life of 12.3 years, is one example of such a tracer. Produced during hydrogen bomb tests, tritium immediately reacts with oxygen, forming radioactive water. This labeled water falls as rain or snow and eventually moves through the ocean like normal water. A small amount of tritium occurs naturally in the ocean because of cosmic ray bombardment of the upper atmosphere. But the bomb-produced tritium is far more abundant in ocean surface waters and thus easily detected.

Distributions of tritium in the deep ocean illustrate how oceanic processes control the fates of substances injected into the ocean surface. Tritium from the bomb tests is incorporated in river waters. When these slightly radioactive waters flow in the ocean, initially they remain in surface waters, above the pycnocline (see Fig. 6-25). In the high latitudes the newly formed water masses incorporate some of this radioactive water when they sink below the surface; thus they carry a distinctive "signature" of high-tritium contents. In the western Atlantic such water masses form in the Greenland and Norwegian seas, sink to the bottom, and flow southward to form the North Atlantic deep waters. The high-tritium values around 50°N (see Fig. 6-25) come from the outflow of Arctic Ocean surface waters, which

Figure 6-25

Distribution of tritium in the western Atlantic in 1972–1973. Note the high concentrations in the surface waters around 60°N and the penetration of tritium in the bottom waters and intermediate waters of the North Atlantic. No tritium had penetrated into the subsurface waters of the South Atlantic. (After Ostlund et al., 1976.)

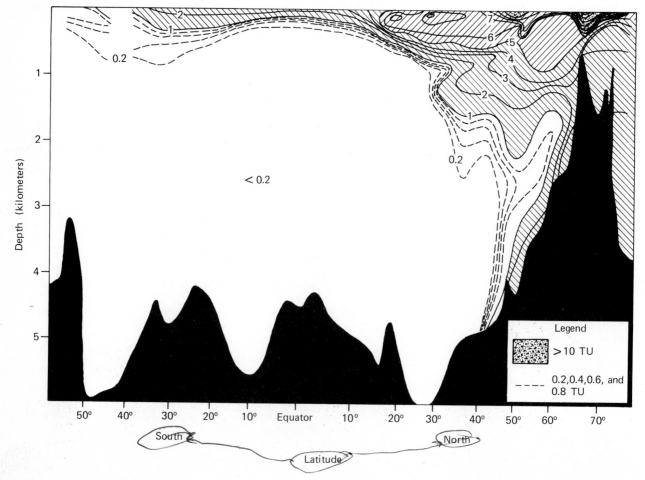

received large amounts of tritium during the 1960s from hydrogen bomb testing in the northern USSR.

Studies of the distributions of such transient traces provide needed information about the role of the ocean in taking up carbon dioxide from the burning of fossil fuels (coal, oil) since the Industrial Revolution. About half the carbon dioxide released to the atmosphere since the 1850s has apparently gone into the ocean. The remainder remains primarily in the atmosphere, where many scientists fear that it may cause a global warming of the earth's climate due to the greenhouse effect.

Carbon dioxide injected into deep waters would not return to the atmosphere for hundreds of years, providing long-term storage of the excess and thereby reducing the effects of carbon dioxide in the atmosphere. But any carbon dioxide that remains in the surface waters is able to exchange freely with the atmosphere and provides no long-term storage. So it is important to know where the carbon dioxide injected into the ocean is stored.

Because of the large amounts of carbon dioxide naturally present in seawater, it is nearly impossible to detect the slight increase caused by human activities. By using tritium, oceanographers will be able to determine how carbon dioxide behaves in the ocean and to improve our ability to predict its effects on world climate.

SEA ICE Sea-ice formation is a conspicuous and important ocean phenomenon (shown in Fig. 6-26) at high latitudes. Sea ice exists year round in the central Arctic and in some Antarctic bays, extending in winter across the entire Arctic and far out to sea around the Antarctic continent effectively doubling the size of the ice mass. During most winters it forms in bays as far south as Virginia on the Atlantic Coast of the United States. During glacial times it was even more widespread than at present, as shown in Fig. 6-27.

In addition to its local importance, sea-ice formation influences the ocean far beyond ice-covered areas. For instance, Antarctic bottom water forms through mixing of extremely cold surface waters and brines released by the freezing of sea ice.

When seawater reaches its temperature of initial freezing, clouds of tiny ice particles form, making the water slightly turbid. The ocean surface becomes dull and no longer reflects the sky. Initially disklike, the ice particles grow and form hexagonal *spicules* 1 to 2 centimeters long. Sea salt is excluded, for it does not fit into the ice-crystal structure.

The water remaining becomes slightly saltier so that its freezing point is lowered. As freezing continues, convection cells form at the underside of the ice layer. Water comes up from below to replace the increasingly saline brines, which tend to form finger-shaped parcels and stream downward.

Ice crystals form a layer of *slush,* which covers the ocean surface like a blanket. Some crystals grow downward and cross connections form between crystals. The result is a thin, plastic layer of ice in which small cells of seawater are trapped. One kilogram of newly formed sea ice contains about 800 grams of ice (salinity 0‰) and 200 grams of seawater (salinity 35‰). Therefore the average salinity of newly formed ice is about 7‰, depending on the temperature at which it was formed. At temperatures near the freezing point, ice forms slowly and so brines can escape. Relatively little seawater is enclosed in the cells and the salinity of the ice is low. At lower temperatures, ice forms more rapidly and much more seawater is trapped. In this case, ice salinity is higher, although always less than the seawater from which it formed.

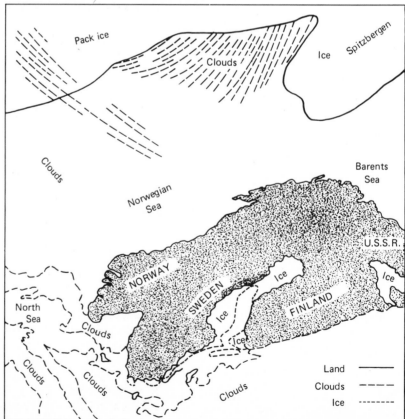

Figure 6-26

Ice pack in the Arctic Ocean north of Scandinavia, which itself is covered with snow, as seen by the Nimbus-4 weather satellite on April 13, 1970. The pack ice is the irregular white area near the top of the photograph. The Gulf of Bothnia in the Baltic Sea (lower center) is ice covered. The filmy streaks in the center of the photograph are clouds. (Photograph and explanatory diagram courtesy NASA.)

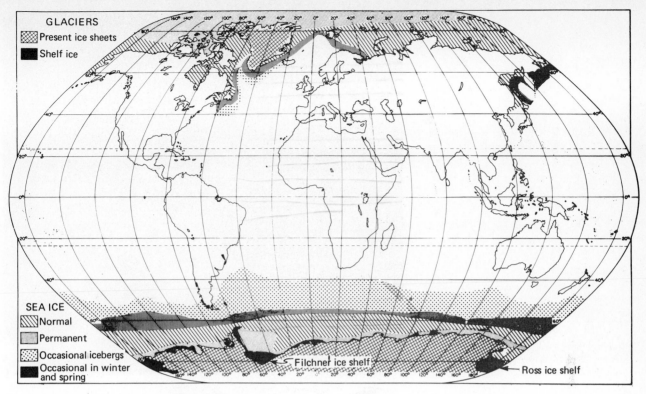

GLACIERS
☒ Present ice sheets
■ Shelf ice

SEA ICE
▨ Normal
▦ Permanent
⬚ Occasional icebergs
■ Occasional in winter and spring

Filchner ice shelf

Ross ice shelf

Figure 6-27

Distribution of sea ice and probable maximum extent of land ice during glacial Pleistocene times.

As the weather grows colder and heat is continually removed from the sea surface, new ice is formed. Water in the cells freezes and their interior walls become thicker. Concentrated brines remain in the ever-smaller cell cavities and eventually certain salts begin to crystallize out. Sodium sulfate crystals form at $-8.2°C$ whereas sodium chloride does not crystallize above $-23°C$. Salt crystals and ice crystals may intergrow in a completely solid ice structure if temperatures are low enough.

As sea ice ages, the brine cells migrate. Many flow into the seawater below. Through time the ice retains less and less salt; eventually it may be nearly pure freshwater. Pools of water from melting of such nearly pure ice have saved the lives of several polar explorers who ran short of drinking water.

Sheets of newly formed sea ice break into pieces under stress of wind and waves. These sheets are called *pancake ice* because that term so aptly describes their appearance in the early stages of formation, as shown in Fig. 6-28. As pancakes coalesce, they form young sea ice, which is usually flexible and retains traces of the pancakes from which it formed. Currents and winds pile up the ice units so that they form hummocks or pressure ridges. With the freezing of winter snows on the surface, growth occurs from the top as well as from the bottom. This thick, solid mass may break into large *floes* in response to winds and surface currents; they constantly move and shift, freeze together and break loose, buckle up and flatten out as the pack is dragged by currents flowing beneath.

During the course of a freezing season sea ice may freeze to a thickness of 2 meters at high latitudes. When more than 1 year old, it forms fields of coherent ice. Where ice never completely melts away but continues to grow during subsequent seasons, as in the central Arctic Ocean, adjacent units are piled on top and an average thickness of 3 to 4 meters is achieved. During summer the central Arctic ice pack melts to a thickness of about 2 meters.

167

Figure 6-28

Wave action on newly formed sea ice forms pancake ice. (Official U.S. Navy photograph.)

The behavior of oceanic pack ice should not be confused with the huge masses of land ice that cover Greenland and Antarctica throughout the year. The latter sometimes extends out over bays and sheltered waters as a permanent floating *ice shelf*. Pieces of glacial ice sometimes break off in the ocean, particularly when glaciers flow out of valleys, as in western Greenland. Irregular pieces are known as *icebergs,* as shown in Figs. 6-29 and 6-30; larger, more regular, and usually flat-topped pieces are called *tabular* icebergs, as in Fig. 6-31. Such icebergs may be tens to hundreds of kilometers long. Icebergs rarely project more than 35 meters above water; five times as much is usually below water.

Icebergs are carried by currents out of the polar regions at speeds of about 15 to 20 kilometers per day. In the North Atlantic

Figure 6-29

An iceberg in pack ice off the coast of Labrador in March 1962. The blocks of ice and snow on top of the berg are as big as houses. (Official U.S. Navy photograph.)

Figure 6-30

A tabular iceberg in Nelville Bay, Greenland. The icebreaker approaching the iceberg is approximately 90 meters long. [The iceberg extends 24 meters long.] The iceberg extends 24 meters above the water and [is about 1.1 kilometers long] and about 0.75 kilometer wide. The iceberg extends about 150 meters below the surface. (Official U.S. Coast Guard photograph.)

Figure 6-31

An iceberg along the Grand Banks of Newfoundland is tracked by a U.S. Coast Guard airplane for the International Ice Patrol. (Official U.S. Coast Guard photograph.)

they are carried southward by the Labrador Current until they reach the vicinity of the New York–Europe shipping lanes. Icebergs from western Greenland travel up to 2500 kilometers, reaching 45°N, which is the vicinity of the Grand Banks.

Currents around Antarctica keep icebergs relatively close to the continent. They are known to move up past the tip of South America to about 40°S in the Atlantic but rarely reach beyond 50°S in the Pacific Ocean. In 1894 a piece of ice was spotted at 26°30'S, only about 350 kilometers form the Tropic of Capricorn, which is commonly taken as the margin of the subtropical ocean. Icebergs typically have about a 4-year life span in the ocean.

CLIMATIC REGIONS

The open ocean can be divided into east–west trending climatic regions. These regions, indicated in Fig. 6-32, generally exhibit characteristic ranges in surface-water characteristics. In addition, they have relatively stable boundaries that change little through time. Both temperature and salinity of surface waters in such areas are related primarily to incoming solar radiation and the relative amounts of evaporation and precipitation.

The Arctic Ocean and the band around Antarctica are the major *polar-ocean* areas. Ice occurs at the surface most of the year and keeps surface temperatures at or near the freezing point despite intense insolation during local summer. In winter there is little direct sunlight. Relatively low surface salinities, combined with the ice cover, act to inhibit vertical mixing in surface waters.

Subpolar regions are affected by seasonal migration of sea ice. In winter intense cooling of surface waters and freezing of sea ice contribute to the formation of dense water masses that flow into deep subsurface ocean. Sea ice disappears during summer except for isolated blocks and icebergs. Enough heating exists to melt the ice and raise water surface temperatures to about 5°C. Abundant light and the stable layer formed by surface warming combine to provide good

Figure 6-32

Oceanic climatic zones and their relationships to terrestrial climatic zones: (1) tundra, (2) boreal forest, (3) temperate, (4) desert, (5) savannah, (6) steppe, (7) tropical rain forest, (8) icecap. (After D. V. Bogdanov, 1963. Map of the natural zones of the ocean. *Deep Sea Research* 10:520–523.)

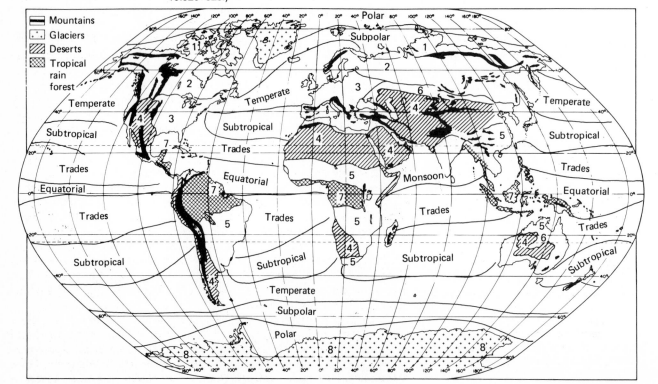

conditions for pulses of phytoplankton (marine algae) growth called blooms. This rich plant crop supports large populations of *zooplankters* (minute marine animals), birds, and whales. Diatom shells (frustules) sink to the bottom, forming diatomaceous sediments.

The *temperate* region corresponds to the band of westerly winds where there are strong storms, especially during winter, with gales and heavy precipitation. Extensive mixing resupplies nutrient elements from deep water to the surface layers, supporting a marine food chain that includes large populations of fish. Some of the world's major fisheries are in this region.

Subtropical regions coincide with the nearly stationary high-pressure cells of the midlatitudes. Winds are weak and so surface currents tend to be weak also. Clear skies, dry air, and abundant sunshine lead to extensive evaporation so that surface salinities are generally high. The pycnocline is well developed and prevents ready exchange with water beneath. Plants growing in surface waters deplete the supply of nutrient elements; and because there is little circulation of deep water to surface layers, plant productivity is limited in this region. Surface waters and floating materials from a vast area tend to converge at the center of ocean basins in subtropical latitudes. For instance, the Sargasso Sea lies in the Atlantic's subtropical current system and derives its name from the floating seaweed that accumulates there.

The *tropical* (trade wind) regions are characterized by persistent winds blowing from the northeast in the Northern Hemisphere and from the southeast in the Southern Hemisphere. These winds cause the equatorial currents. The trade winds also cause moderate seas in contrast to the relative calm of subtropical latitudes. Waters in tropical oceans originate in subtropical regions and are therefore more saline than average seawater. Near the equator, precipitation increases, causing decreases in surface-water salinity.

In *equatorial* regions surface waters remain warm throughout the year. Annual temperature variations are small. Warm, moist air generally rises near the equator, causing heavy precipitation and relatively low surface salinities. In the Atlantic and much of the Pacific winds tend to be weak. Sailors named parts of this region the *Doldrums* because their ships were often becalmed there.

In the Indian Ocean the monsoons complicate the current patterns. Vertical water movement along the equator supplies nutrients to the surface layers and causes the area to be highly productive of phytoplankton and the organisms that feed on them.

The ocean moderates climatic conditions on land. Coastal areas typically have a *maritime climate*, where the annual temperature range is generally much narrower than in regions far inland, as shown in Fig. 6-33. Far from the ocean, *continental climates* prevail, with their characteristically wide temperature ranges.

Oceans store large amounts of heat because of water's high heat capacity, permitting heat storage without an accompanying large temperature increase. Furthermore, the ocean (or any large body of water) stores heat through several meters of water due to mixing, and thus retards loss of heat by back-radiation at night. In contrast, rocks on land have a low heat capacity; so they become hot during the day. But because this heat is not readily transferred to deeper rocks beneath the surface, most of it is lost at night by back-radiation. This situation can be observed in any desert area, where intense daytime heat is followed by chilly nights.

During summer continents heat up and warm the atmosphere above them. The relatively less dense continental air mass rises and is replaced by moist air from the ocean. As the oceanic air mass rises, it cools and releases moisture, often as heavy rains. This is the *monsoon* circulation, especially well developed over southeast Asia and India

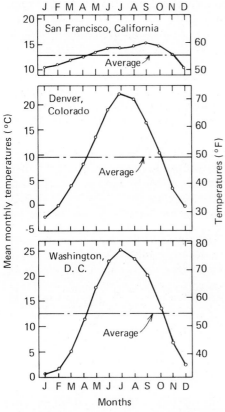

Figure 6-33

Temperature variation at three locations: San Francisco has a typical maritime climate, with cool summers and relatively warm winters; Denver has a typical continental climate—cold winters, warm summers; Washington, D.C. has a generally continental climate, although tempered somewhat by its proximity to Chesapeake Bay.

as a result of seasonal atmospheric circulation shifts caused by the warming and cooling of the Asian landmass. A similar air-circulation pattern prevails throughout much of the United States midcontinent during summer. Moist air from the Gulf of Mexico supplies large amounts of moisture during the summer, usually as thundershowers.

Mountain ranges block the flow of air from the ocean and thereby affect climatic conditions over large continental regions. North–south mountain ranges of the western United States block the flow of air from the Pacific, creating a rainshadow over interior valleys so that they receive little precipitation. Air cools as it rises to flow over the mountains. Rainfall is sufficient to support a temperate rainforest on the western slopes of the Coast Range of Washington whereas deserts or near-desert conditions occur east (downwind) of the mountains. The coastal portion of the western United States thus has a maritime climate, characterized by moderate temperatures, winter and summer, and abundant rainfall. Continental climates with wide temperature ranges are common in the interiors of the continent, away from the ocean's moderating influence (see Fig. 6-33).

On the Eurasian continent, by contrast, the Alps, Himalayas, and associated mountains generally run east–west. These ranges have little effect on the flow of maritime air over Europe and Asia and the maritime climate extends many hundreds of kilometers inland. These east–west ranges do, however, block north–south air movements and probably cause the climate of the Mediterranean region to be warmer and drier than it might be in their absence.

OCEAN INFLUENCE ON LAND CLIMATES

Large masses of unusually warm—or cold—surface waters in the ocean apparently influence weather on land. Such anomalously warm or cold waters have been studied in the North Pacific and their influence on North American weather documented. These masses are 1000 to 2000 kilometers (600 to 1200 miles) across, 200 to 300 meters (600 to 1000 feet) thick, and persist for many years as they are moved across the ocean by the major current systems. The vast size of these water bodies and the amount of heat energy they contain can apparently "steer" major weather systems and storms as they move across the ocean and onto the continent.

In the 1960s and 1970s a persistent cold surface water mass in the central Pacific caused a shift in the prevailing westerly winds blowing across the eastern United States (see Fig. 6-34). Cold, dry polar air from Canada displaced warm, moist air from the Gulf of Mexico and the Atlantic. As a result, much colder winter temperatures occurred in the southeastern United States. A comparable situation prevailed during the unusually cold winter of 1976, the coldest in 177 years in the U.S. plains states and the eastern seaboard. The relative positions of the warm and cold masses shifted during the winters of 1971 to 1975, causing warm winters along the Atlantic Coast of the United States.

Some scientists believe that it will be possible to predict whether the coming winter will be relatively mild or especially cold along the Pacific Coast and in the southeastern United States several months ahead of time.

ICE AGE CLIMATES

The ocean is also involved in longer-term changes in global climate. The most dramatic of these climatic changes was the last glacial stage of our present Ice Age, which ended about 10,000 years ago. The record of that glacial stage has been studied to determine what causes these shifts in the earth's climate and to improve our predictive ability.

Fossils preserved in marine sediments were used to reconstruct

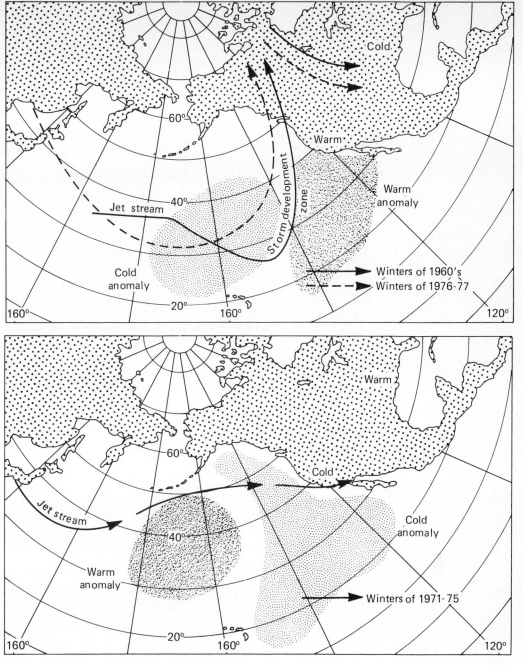

Figure 6-34

The positions of unusually cold or warm surface water masses affect the paths of the jet stream across the continent. This, in turn, controls movements of cold polar air from Canada or warm, moist air from the Atlantic or Gulf of Mexico into the United States. The positions of such pools are shown for the unusually cold winters of the 1960s and 1976–1977. The bottom shows the positions during the mild winters of 1971–1975. (Data courtesy National Science Foundation.)

distributions of ocean surface salinity and temperature 18,000 years ago, when the most recent glacial stage was at its maximum. Using sediment from carefully dated layers in sediment cores, scientists analyzed the chemical composition, abundance, and distribution of foraminifera, radiolarians, and other microfossils that float and are moved by currents. Knowing the conditions under which these planktonic organisms now grow, scientists were able to use statistical correlation techniques to determine the salinity and temperatures of the ocean surface waters in which these organisms grew.

The ocean surface during the last glacial stage was quite different from present conditions (see Fig. 6-35). For example, surface ocean waters were cooler by about 2.3°C on the average; the land was about 6.5°C cooler. But the central gyres of the major ocean basins were little changed in position or temperature. The Gulf Stream flowed directly eastward from the Carolinas to Spain. Cold polar waters advanced southward; thus temperature changes between subtropical

173

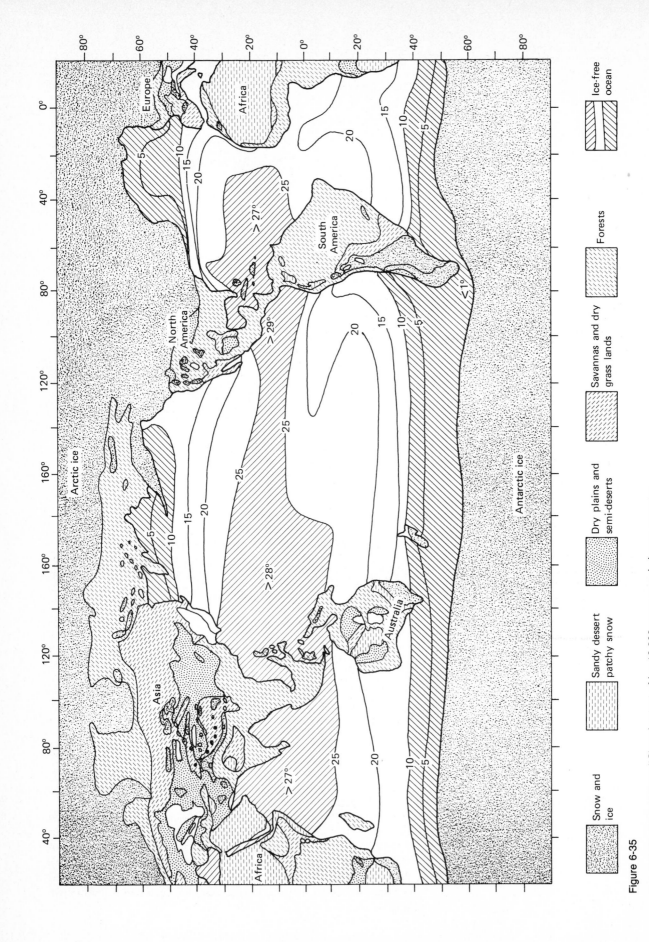

Figure 6-35

Sea surface temperatures (°C) and extent of ice 18,000 years ago during an average August. Continental outlines correspond to sea level 85 meters below its present position. Compare to sea surface temperatures shown in Fig. 6-7. (After CLIMAP.)

Legend:

- Snow and ice
- Sandy dessert patchy snow
- Dry plains and semi-deserts
- Savannas and dry grass lands
- Forests
- Ice-free ocean

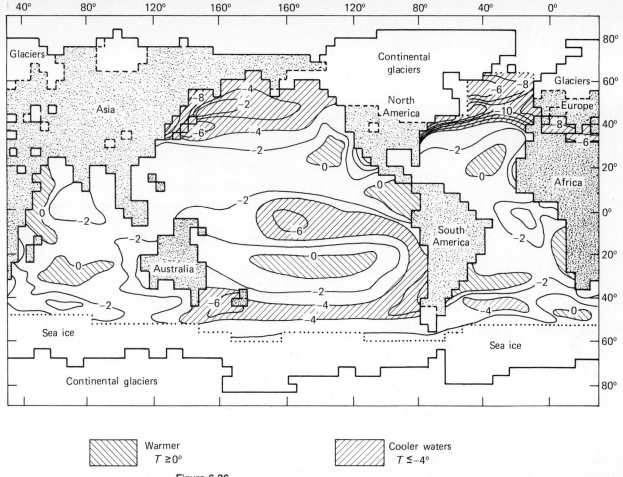

Glaciers

Asia

−8

−4

−2

−6

−4

−2

−2

0

0

−2

0

−2

−2

Australia

0

−6

0

−2

−2

−6

−4

−4

Sea ice

Continental glaciers

Continental glaciers

North America

−2

0

0

South America

−2

−2

−2

0

Sea ice

Glaciers

−6

−8

−10

Europe

8

−6

Africa

80°
60°
40°
20°
0°
20°
40°
60°
80°

Warmer
$T \geq 0°$

Cooler waters
$T \leq -4°$

Figure 6-36

Differences between sea surface temperatures 18,000 years ago and modern August temperatures. During the most recent glacial stage sea surface temperatures were generally lower except for small areas in the subtropics.

Temperatures were markedly lower in the western North Atlantic and North Pacific. Dotted lines indicate the limits of sea ice. Dashed lines indicate margins of glaciers on land. (After CLIMAP, 1976. The surface of the ice-age earth. *Science* 191:1131–1144.)

and subpolar waters were much more pronounced than at present. There was much more upwelling along the equator, causing the equatorial ocean to be much colder than it is today. Temperature changes were especially noticeable in the North Atlantic, which was markedly cooler (see Fig. 6-36).

Changes in the earth's orbit around the sun seem to be a major factor controlling the waxing and waning of the glacial stages. Because of these changes, the volume of ice on Earth reaches a maximum every 23,000, 42,000, and 100,000 years. At present, the earth's surface is warmer than it has been for 98% of the past half million years. Thus any natural change in climate will probably bring colder temperatures and cause problems with food production in high-latitude agricultural areas, such as the USSR and Canada.

REVIEW QUESTIONS

1. Describe how budgets are used to study ocean processes. Discuss the earth's heat budget.

2. Describe the "greenhouse effect" and what causes it.

3. Explain the causes of seasonal changes in temperature in the midlatitudes of the ocean basins.

4. What processes cause the minimum temperature of seawater to be about −1.8°C and the maximum temperature to be about 30°C?

5. Using a diagram, show the water budget of the earth. Explain how the water budget and the heat budget of the earth are related.

6. Contrast the warming and cooling of the ocean and atmosphere.

7. Define and illustrate with graphs (showing variations with depth) the thermocline, halocline, and pycnocline.

8. Discuss and illustrate with diagrams the principal depth zones of the open ocean.

9. What is a water mass? Where do the principal subsurface water masses form?

10. Describe how pack ice forms.

11. How do icebergs differ from pack ice?

12. Describe how the climate on land is influenced by ocean processes.

13. Describe how the Ice-age ocean differed from present ocean conditions.

SUMMARY OUTLINE

Ocean and atmosphere—closely coupled systems

Approaches used:
   Budgets and distribution of properties

Light in the ocean
   Solar radiation—source of energy for heating ocean and atmosphere and for circulation
   Infrared (heat) absorbed at surface
   Turbidity—increases absorption and scattering of solar energy; pure water is most transparent to blue green light
   Blue color caused by scattering in particle-free water

Atmospheric circulation and heat budget
   Inclination of axis of rotation causes seasons
   Heat received in low latitudes; lost in high altitudes, especially in winter hemisphere
      Heat radiates to space as infrared
      Latent heat of water dominates heat transport in atmosphere
   Atmospheric circulation typically forms cells and wavelike frontal patterns

Oceanic water and heat budget
   Evaporation highest in subtropical areas
   Precipitation heaviest in equatorial regions and in high latitudes
   Salinity controlled by balance between evaporation and precipitation
      High salinity common in central ocean areas
      Low salinity in coastal regions caused by river runoff and by high precipitation and little evaporation in high latitudes
      Poleward heat transport reflected in low salinities at high latitudes

Density of seawater
   Controlled primarily by temperature and sa-

linity; pressure effect noticeable only at great depths in ocean
   Usually stable density structure in ocean
      Vertical stratification inhibits vertical movements of water
      Surface, pycnocline, and deep zones
   Thermocline—sharp temperature gradient
   Halocline—sharp salinity gradient
   Pynocline—sharp density gradient

Water masses
   Temperature and salinity used to distinguish subsurface water masses and to study mixing

Sea ice
   Ice spicules form slush, later plastic layer with enclosed brine cells
   Salt crystallizes from bines; brine cells migrate
   Pancake ice forms; later coalesces in floes as season advances
   Icebergs derived from glaciers on land

Climatic regions
   Generally east–west trending
   Surface waters respond to local climate, especially winds

Ocean influence on climate
   Pools of unusually warm or cold waters deflect jet streams
   Influence storm tracks and air mass movements

Ice Age climates
   Sea surface temperatures studied by using fossils in ocean sediments
   Ice-age ocean about 2.3°C cooler than present ocean
   Colder in polar regions and along equator (upwelling)
   Cycles of ice ages controlled by changes in earth's orbit

SELECTED REFERENCES

IMBRIE, JOHN, AND KATHERINE PALMER IMBRIE. 1979. *Ice Ages: Solving the Mystery*. Enslow Publishers, Short Hills, N.J. 224 pp. Easily readible account of the research on the causes of the ice ages.

KNAUSS, JOHN A. 1978. *Introduction to Physical Oceanography*. Prentice-Hall, Inc., Englewood Cliffs, N.J., 338 pp. Intermediate-level treatment of basic principles of physical oceanography. Assumes knowledge of calculus.

PERRY, A. H., AND J. M. WALKER. 1977. *The Ocean-Atmosphere System*. Longman, London, 160 pp. Emphasizes interactions between ocean and atmosphere.

PICKARD, G. L. 1979. *Descriptive Physical Oceanography*, 3rd ed. Pergamon Press, New York. 233 pp. Elementary treatment of physical oceanography.

RIEHL, HERBERT. 1979. *Climate and Weather in the Tropics*. Academic Press, New York. 611 pp. Nonmathematical treatment of equatorial atmospheric-oceanic interactions. Intermediate level.

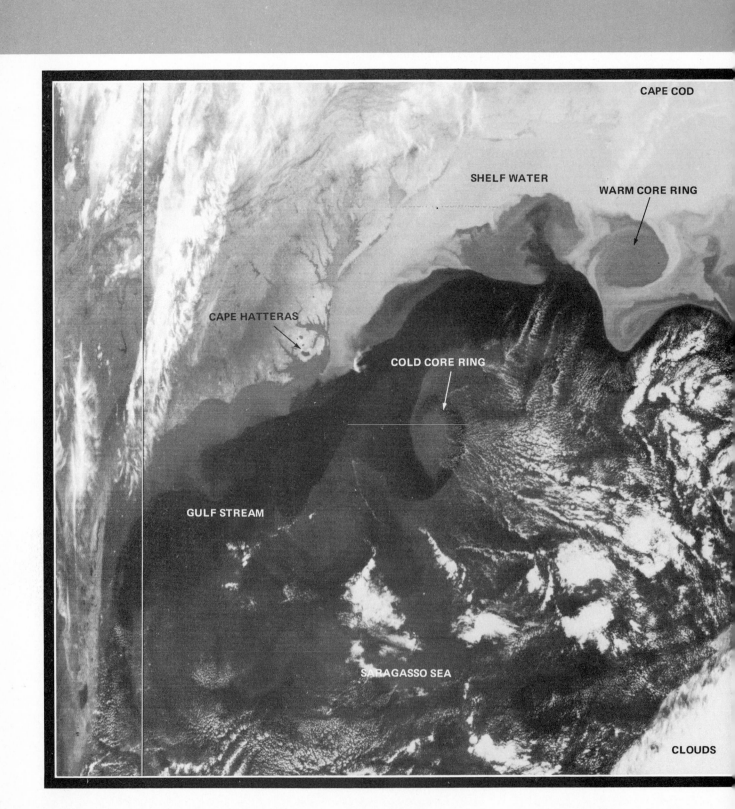

OCEAN CIRCULATION

The warm waters of the Gulf Stream (darkest tone) offshore from the United States (upper left) as deducted by a heat-sensing satellite on April 28, 1974. Two eddies are shown: a warm eddy in the colder (lighter tone) waters in the upper center, and a cold eddy (swirl of darker tone) in the lower left area. Clouds (white tone) partially obscure the Sargasso Sea in the lower right corner. Note that the warm core ring circulates clockwise; cold core rings circulate counterclockwise. (Photograph courtesy of Woods Hole Oceanographic Institution.)

*C*urrents—large-scale water movements—occur everywhere in the ocean. Major ocean currents are driven primarily by winds and by unequal heating and cooling of ocean waters. Both result from unequal heating of the earth's surface. Ocean currents contribute to the heat transport from tropics to poles, thereby partially equalizing Earth surface temperature. In this chapter we consider the major ocean circulation, what causes it, and how currents vary in space and time.

CURRENT MEASUREMENTS

Modern techniques for current observation involve automated buoys with meters that record the direction and speed of water movements and then relay the data to a ship that services the buoy. Other buoy systems transmit data back to shore stations—in some cases, using satellites.

The current measurements on which our ocean current maps are based were made primarily by navigators who observed how a ship's course was deflected by movements of water through which it had passed. A navigator, having predicted that his ship would be in one location, after steaming for a time found that his final position differed from the prediction. Assuming no errors in navigation and in the absence of winds, the displacement (called the *set*) is caused by currents. Measuring and recording such deflections of ships' courses over many years permitted Matthew Fontaine Maury to compile such observations in the files of the U.S. Navy and to draw early charts of ocean currents. Today we still rely on similar information sources.

Movements of floating objects also provide information about average surface currents. Wreckage of ships or glass floats used in Japanese fishing nets move across the Pacific, providing evidence of a current flowing eastward. The wreckage of the ship *Jeanett,* crushed in Arctic ice in 1881 north of the Siberian coast, was discovered near southwestern Greenland 3 years later. This fact provided compelling evidence of a current from Siberia toward Greenland and gave an indication of its average speed.

Observations averaged over many years provide a picture of average currents. When the effects of variable wind and short-term tidal currents cancel out, only the average current can be detected. Minor variations and even some seasonal variations are difficult or impossible to detect with such data.

Direct current observations are made by current meters, such as in Fig. 7-1. Propellers or rotors on the meter measure current

Figure 7-1

Ekman current meter was used to record current speed and direction. The meter was suspended below the surface on a wire hung from the ship. The propeller was turned by the current and the number of turns recorded by the dials. The vane oriented the meter in the current. A recording compass (not visible) recorded how many turns occurred in each orientation of the meter. Modern current meters perform the same functions and record results on magnetic tape for computer processing of the results. (Photograph courtesy Woods Hole Oceanographic Institution.)

speed. There are many ingenious methods for recording these data over long periods so that both short-term fluctuations and long-term average currents can be measured.

Using meters, current speeds and directions can be determined at the surface and at all depths in the ocean. Meters can be suspended from anchored ships, but the ship's own motion in the water makes it difficult to interpret the results. Mooring of meters on buoys or fixed platforms is more satisfactory although expensive in deep-ocean areas. Because of the difficulties of measuring currents in the deep ocean, relatively few direct measurements are available. Such measurements are primarilily in areas having strong currents like the Gulf Stream and the North Atlantic.

Distributions of materials recently injected into the ocean in large quantities have been used to study how subsurface water masses form and move through the deep ocean. The most useful have been radioactive materials from nuclear weapons tests. These tracers have been used to study the formation and movements of waters in the North Atlantic.

Other methods are used to study currents below the surface. Near-surface currents can be studied with *parachute drogues*—parachutes in the water rigged to be moved by currents (see Fig. 7-2). The drogue is attached to a surface float that can be tracked by ship. Continual tracking of the float provides a direct indication of current speed and direction at the depth at which the parachute was placed.

The *Swallow Float* (invented by the English oceanographer John Swallow) is a cylinder whose density is carefully adjusted so that it floats at a preselected depth, where it moves with the water. In the float is a powerful sound source called a *pinger*. A ship at the surface or a fixed-listening station receives the "pings" and by recording their location traces the movements of the float as it moves in the deep current. Just as in the case of the current meter, net currents can be calculated from long-term average movements of these floats.

All these deep-ocean current-measurement techniques are expensive; consequently, data on subsurface currents are limited. Lacking a source of subsurface-current information similar to that compiled for hundreds of years by surface ships, most of the data on deep-ocean currents are deduced from changes in ocean water properties, such as the temperature of deep waters or variations in dissolved-oxygen concentrations.

In a later section we shall see that it is also possible to determine current speed and direction indirectly. This so-called *geostrophic approach* uses precise temperature and salinity determinations to map the density variations that cause currents.

Figure 7-2

Parachute drogue rigged to permit tracking movement of subsurface water masses. (Redrawn after Von Arx, 1962.)

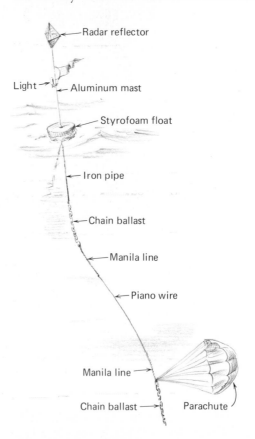

Radar reflector

Light — Aluminum mast

Styrofoam float

Iron pipe

Chain ballast

Manila line

Piano wire

Manila line

Chain ballast — Parachute

SURFACE CURRENTS IN THE OPEN OCEAN

Current patterns are more or less similar in all three major ocean basins despite their geographic differences (see Fig. 7-3). Much of the ocean surface in equatorial regions is dominated by the westerly flow of water in both the *North* and *South Equatorial currents*, caused primarily by the trade winds. These are separate equatorial currents in the Northern and Southern hemispheres, separated by a narrow cur-

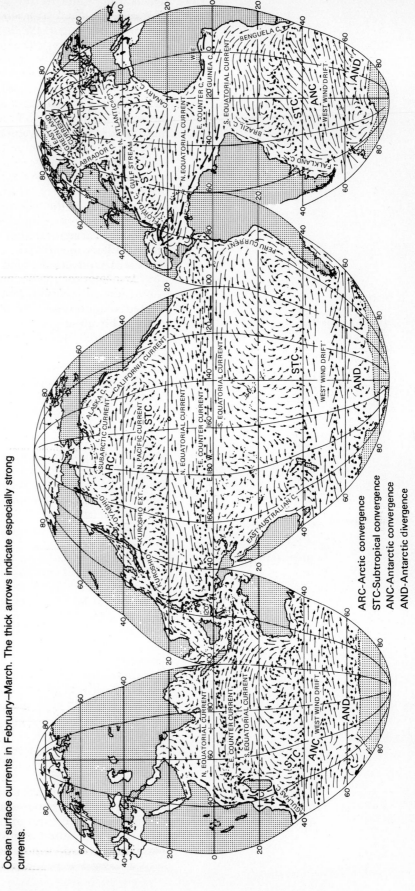

Figure 7-3

Ocean surface currents in February–March. The thick arrows indicate especially strong currents.

ARC- Arctic convergence
STC- Subtropical convergence
ANC- Antarctic convergence
AND- Antarctic divergence

rent flowing eastward, the Equatorial Countercurrent, in the region of the light and variable winds called the *Doldrums*. Associated with each of these equatorial currents is a current *gyre*—a nearly closed current system. The current gyres are elongated east–west and lie primarily in the subtropical regions, centered around the 30°N and 30°S latitudes.

In addition to an equatorial current, each of these current gyres includes a major east–west current flowing in a direction opposite to the equatorial currents. This is the *West Wind Drift,* the largest and most important current in the Southern Hemisphere. The *North Atlantic Current* is the continuation of the *Gulf Stream system* in the Atlantic and occupies a position similar to the West Wind Drift in the other ocean basins. The *North Pacific Current* is an extension of the *Kuroshio,* the Pacific analog to the Gulf Stream. The Indian Ocean has an equatorial current in the Southern Hemisphere. Near Asia, seasonally variable winds caused by the heating and cooling of the land cause a seasonally variable current system—the *monsoon currents.* The *Somali Current* follows northward along the east African coast in northern summer. Despite its seasonal nature, the Somali Current is one of the world's strongest currents and causes extensive upwelling.

In the subpolar and polar regions of all ocean basins there are smaller current gyres. These high-latitude gyres circulate "opposite" to the subtropical gyres. Because of the position of the continents, subpolar gyres are well developed in the Northern Hemisphere, where their movement is counterclockwise (see Fig. 7-3). (Note the clockwise direction of the North Atlantic Current–Gulf Stream system–Northern Equatorial Current systems, which form the North Atlantic subtropical gyre.)

Subpolar gyres occur in the Southern Hemisphere as well but are not as well developed. Because Antarctica occupies a central position in the Southern Hemisphere circulation, the West Wind Drift essentially flows around it rather than being broken up and deflected toward the South Pole by continental land barriers. Small, clockwise gyres can, however, be seen in the vicinity of Antarctica. (Note the counterclockwise direction of the West Wind Drift–Peru Current–Southern Equatorial Current system.)

BOUNDARY CURRENTS

Ocean surface circulation is dominated by east–west currents, such as the equatorial currents or the West Wind Drift. These currents involve east–west movements of large volumes of water generally remaining in the same climatic zone. The waters have ample opportunity to adjust to the temperature of the climatic regime in which the current is located. These currents are deflected when they encounter the continents and the water flows are split into *boundary currents* that flow more or less north–south.

Boundary currents are especially important to us as land-dwellers. They include the strongest currents in the ocean (the *western boundary currents*); they also play a major role in transporting heat from tropics to polar regions. Furthermore, the currents form boundaries to the circulation of coastal waters. Where boundary currents are strong, the break between coastal and open ocean circulation is sharp, which is the case on the western side of ocean basins, particularly in the Northern Hemisphere. Where boundary currents are relatively weak—as on the eastern side of ocean basins—the separation between coastal and open ocean circulation is more diffuse.

The characteristics of boundary currents on different sides of the ocean differ substantially, as Table 7-1 indicates. Because of the rotation of the earth on its axis, there is a tendency for current gyres to be displaced toward the west. This results in deep, narrow, fast-

TABLE 7-1

Comparison of Boundary Current Systems in the Northern Hemisphere

TYPE OF CURRENT (example)	GENERAL FEATURES	SPEED	TRANSPORT (millions of cubic meters per second)	SPECIAL FEATURES
Eastern boundary currents California Current Canaries Current	Broad, ≈ 1000 km Shallow, ≤ 500 m	Slow, tens of kilometers per day	Small, typically 10–15	Diffuse boundaries separating from coastal currents Coastal upwelling common Waters derived from midlatitudes
Western boundary currents Gulf Stream Kuroshio Somali (seasonal, caused by Monsoons)	Narrow, ≤ 100 km Deep—substantial transport to depths of 2 km	Swift, hundreds of kilometers per day	Large, usually 50 or greater	Sharp boundary with coastal circulation system Little or no coastal upwelling; waters tend to be depleted in nutrients, unproductive Waters derived from trade wind belts

moving currents on the western boundaries of ocean basins (Gulf Stream, Kuroshio) in contrast to relatively shallow, broad, and slow currents along the eastern boundaries (Canary Current, California Current). This tendency is well developed and clearly discernible in the Northern Hemisphere but not so obvious in the Southern Hemisphere.

Surface water displaced to the west tends to pile up there, deepening the pycnocline. Phytoplankton grow in this accumulation of surface water and deplete the nutrient supply, which is not replaced by nutrient-rich waters from below the pycnocline. So productivity of marine organisms tend to be limited on the western side of ocean basins except where vertical mixing is especially strong. In regions dominated by eastern boundary currents, prevailing winds blow surface waters away from ocean basin margins. Subsurface waters are drawn upward to replace them, in a process known as *upwelling*, which is discussed in Chapter 10. For this reason, productivity of marine plants is highest along eastern basin margins.

The Gulf Stream system is a prime example of a western boundary current. It includes several currents: the Florida Current (see Fig. 7-3), extending from the tip of Florida to Cape Hatteras, North Carolina; the Gulf Stream, extending from Cape Hatteras to the tip of the Grand Banks, Newfoundland; and its eastern extension, which forms the North Atlantic Current with its several branches and many eddies.

Water moving in the Gulf Stream system comes from the equatorial currents, especially the Northern Equatorial Current. In the Atlantic substantial amounts of surface waters cross the equator. Part of the Southern Equatorial Current is deflected into the Northern Hemisphere by the eastern projection of South America. These currents feed the Gulf Stream system.

The Gulf Stream system forms a partial barrier between adjacent surface water masses (see Figure 7-4). Over the continental shelf and slope, the coastal water masses exhibit seasonally variable salinity and temperature whereas, in contrast, waters in the midst of the Gulf Stream are fairly constant, with temperatures typically about 20°C or higher and salinities of around 36‰. There is some movement of surface waters toward the center of the North Atlantic subtropical gyre. This area, known as the Sargasso Sea, has indistinct boundaries

185

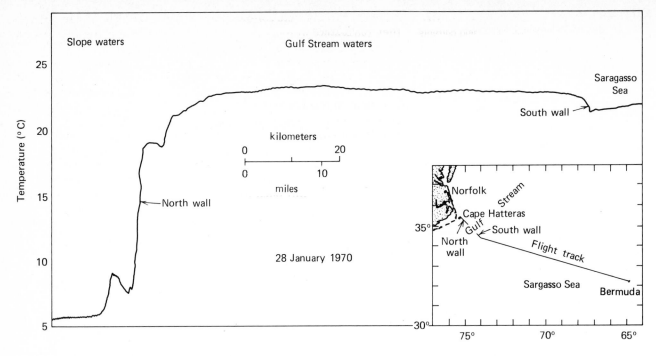

Figure 7-4

Sea surface temperatures recorded by an infrared radiometer on an aircraft crossing in the Gulf Stream, between Cape Hatteras and Bermuda on January 28, 1970. Note the sharp temperature change between slope waters and the Gulf Stream (the so-called North Wall) and the much less pronounced South Wall, the boundary between the Gulf Stream and the Sargasso Sea. [Data from Naval Oceanographic Office, 1970. *The Gulf Stream* 5(1):3.]

formed by current systems. Average surface temperatures are 20°C or more and salinities are around 36.6‰. The relatively high salinity is due to evaporation in excess of precipitation.

In the core of the Florida Current, speeds of 100 to 300 centimeters per second have been measured (corresponding to 2 to 6 knots). Current speeds are highest at the surface, in a band about 50 to 75 kilometers wide. Below the surface, at depths of 1500 to 2000 meters, current speeds are measured at 1 to 10 centimeters per second. This is truly a "deep-draft" current and so does not encroach on the continental shelf at this point. It flows along the edge of the North American continent until reaching Cape Hatteras, North Carolina, where it turns eastward into the Atlantic.

After passing Cape Hatteras, the Gulf Stream forms large curves or *meanders*, which develop, detach, and reform in complicated ways. The cause of these meanders is still not known but may be related to the ocean bottom topography over which the current moves; if so, this would be one case in which ocean bottom topography influences ocean surface currents. (In general, surface currents involve only water above the pycnocline—perhaps 100 meters thick—and are therefore unaffected by ocean bottom topography.) After passing the Grand Banks, off Newfoundland, the flow forms the rather diffuse North Atlantic Current, a relatively shallow surface current.

Separating the Gulf Stream from adjacent, slower-moving water masses are several *oceanic fronts*—sudden changes in water color, temperature (see Figure 7-4), and salinity. They are often spaced a few kilometers apart but sometimes nearly overlap. Even more obvious to an observer aboard ship is the change in color from greenish gray or bluish coastal waters to the intense cobalt blue of the Gulf Stream waters. Often the front is also marked by choppy waters. Such fronts mark *convergences*—areas toward which surface waters flow from different or opposing directions, tending to sink at the line where they meet.

Eastern boundary currents are much less spectacular. The waters move much more slowly and the boundaries between the boundary current and coastal currents are more diffuse. Eastern boundary currents can flow over the continental margin. Like all boundary currents, they participate in the global heat transport by moving cooler waters toward the equator, as in the case of the California Current.

Boundary currents are more changeable and far more complicated than the major currents shown in Fig. 7-3. Irregularities of continental coastlines cause local eddies in coastal ocean currents and seasonal wind shifts cause major changes in currents, especially in the coastal ocean. The northward-setting Davidson Current of the Washington–Oregon coast, for instance, is well developed in winter, when strong winds from the southwest hold freshwater along the coast and drive the surface waters northward. During summer, when north winds tend to drive water offshore, the Davidson Current disappears until the winds shift back to the southwest. Along the Atlantic Coast of the United States, the extension of the Labrador Current brings cold northern waters as far south as Virginia in winter. During summer cold surface waters from the Labrador Current extend no farther south than Cape Cod, Massachusetts.

FORCES CAUSING CURRENTS

Surface ocean currents are primarily wind driven. The prevailing winds supply much of the energy that drives surface water movements, as is clear by comparing the generalized surface winds in February, illustrated in Fig. 7-5(a), with the ocean surface currents in February and March shown in Fig. 7-3. We have already discussed how seasonal wind changes cause surface current shifts in the monsoon areas of the Indian Ocean and along some coastal areas. (In the next section we discuss how the wind causes water movements and how they are altered by the earth's rotation.)

In addition to the *drift currents* caused by the wind, other currents—called *geostrophic currents*—result from the distribution of water density, which is, in turn, controlled by water temperature and salinity. For our purposes, it is convenient to discuss these two driving forces separately. In the ocean their effects combine to produce the currents previously described. Generally currents below the depth to which wind effects penetrate are geostrophic currents.

Both drift and geostrophic currents move water horizontally. Other density-controlled currents are responsible for vertical water movements that supply deep and bottom water masses to all ocean basins. Because both heat and salinity are involved, this circulation is referred to as *thermohaline circulation*. It controls the vertical distribution of temperature and salinity in the ocean.

The earth's rotation about its axis, which causes an apparent deflection in the trajectory of artillery shells and rockets, also acts as a modifying influence on winds and ocean currents. Except near the equator, winds and ocean currents follow curved paths instead of the straight ones that a simple theory would predict for their movement if the earth were not rotating. To account for this discrepancy, the *Coriolis effect* (sometimes called the Coriolis force) is included in predictions of long-range wind or water movements. This effect does not set winds or waters in motion; it deflects them after they are in motion.

To understand the Coriolis effect, consider what happens to a rocket fired toward the North Pole from the equator. As it sits poised on the launching pad, the rocket is already moving eastward at a speed of approximately 1670 kilometers per hour because of the earth's rotation (see Fig. 7-6). After launching, the rocket still moves eastward at this speed, plus the speed imparted to it during the launch. If the rocket were aimed *along* the equator, it would move in a straight line with no deflection—that is, it would not demonstrate the Coriolis effect—just as it would on a stationary earth.

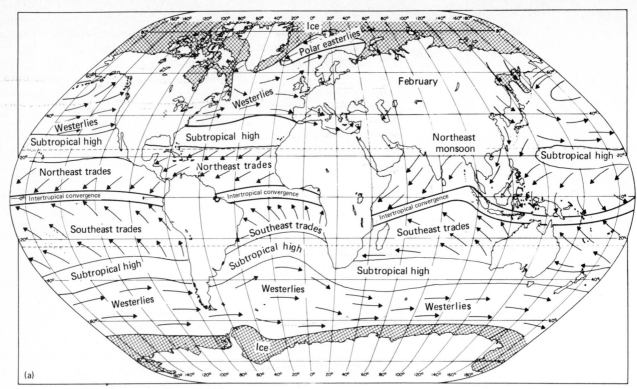

Figure 7-5(a)

Generalized surface winds over the ocean in
February.

Figure 7-5(b)

Generalized surface winds over the ocean in
August.

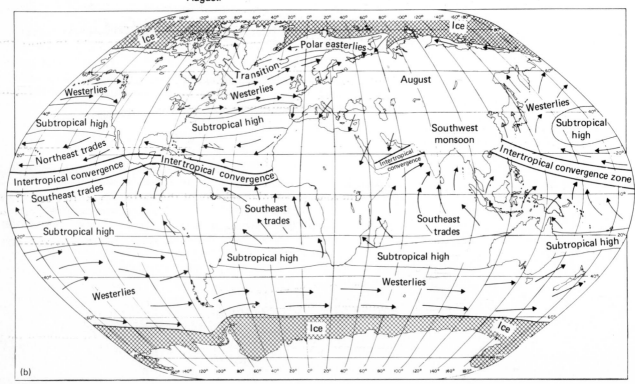

The rocket moving northward from the equator moves over portions of the earth where the surface is moving eastward at progressively slower speeds than that of the equator. At the pole, the surface has no lateral eastward movement (just as there is no lateral movement at the center of a phonograph record whereas the outer edge is moving rapidly) relative to the center. At New Orleans, for example (approximate latitude 30°N), the earth's surface moves eastward with a speed of about 1500 kilometers per hour. But the rocket when it passes overhead still has a net eastward speed of 1700 kilometers per hour and is therefore moving eastward faster than the earth beneath it. Consequently, the rocket seems to veer to the right when an earthbound observer looks along the flight path. An observer on the moon, however, would see that the rocket was traveling a straight line while the earth turned beneath it. The veering to the right observed by an observer on earth would increase if the launching position were moved northward. This apparent increase is simply a result of more rapid changes in the earth's eastward movement as one moves nearer the pole (see Fig. 7-6).

A similar argument can be developed for the Southern Hemisphere (see Fig. 7-6). The deflection is still eastward but appears, to an observer on earth looking along the path of the rocket, like a deflection to the left.

So far we have described the effect of the Coriolis force on particles or water masses moving north or south. But particles moving east or west are also deflected unless moving along the equator. To understand this point, we must consider a water-covered rotating earth. The water is attracted to the surface by gravity but is slightly deformed by the centrifugal force caused by the earth's rotation. Thus the water envelope is slightly flattened at the poles and widened at the equator in equilibrium with the speed of the earth's rotation.

The earth rotates toward the east so that a particle that starts moving eastward (say at midlatitudes) is moving slightly faster than the earth, a factor that increases the centrifugal force acting on it. This result, in turn, causes it to be deflected toward the equator or toward the right.

Conversely, a particle moving westward is moving opposite to the earth's rotation and thus the centrifugal force is slightly less. It is deflected away from the equator or again to the right. In short, the Coriolis effect always acts to deflect particles to the right of their original path in the Northern Hemisphere (to the left in the Southern Hemisphere). It arises from the earth's rotation and the movement of the particle. A particle at rest does not experience this effect nor does a particle moving along the equator.

The Coriolis effect (f) can be calculated as follows:

$$f = 2\,(\omega \sin \phi)$$

where ω = angular velocity of the earth, 7.3×10^{-5} sec^{-1}, corresponding to the complete rotation of the earth in 1 day

ϕ = geographic latitude; $\sin \phi = 0$ at the equator, $\sin \phi = 1$ at the poles

The Coriolis effect acts as a force operating in a direction perpendicular to the current. It deflects the current as noted. Theoretically it could cause water parcels to follow spiral paths as they move in the general direction of the current (see Fig. 7-7), although they are not recognized in the ocean because other forces, including friction, affect the moving water.

Wind blowing across water drags the surface along, which sets

Figure 7-6

Note the change in speed of the earth's surface moving in an eastward direction going from the equator to either pole. A rocket moving from the equator to the North Pole would apparently be deflected to the right—the Coriolis effect.

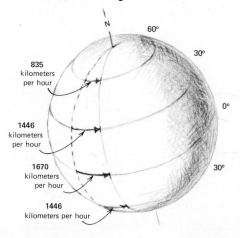

N

60°

30°

835 kilometers per hour

1446 kilometers per hour

1670 kilometers per hour

1446 kilometers per hour

0°

30°

a thin layer in motion. This layer, in turn, drags on the one beneath, setting it in motion. The process continues downward, involving successively deeper layers. Transfer of momentum between layers is inefficient and energy is lost. As a result, current speed decreases with depth below the surface. In an infinite ocean on a nonrotating earth, the water would always move in the same direction as the wind that set it in motion. A similar effect is seen in *storm surges*—large amounts of water piled up on a coastline by storm winds blowing the water ahead of them. Because the distances and times are relatively short, complications resulting from the earth's rotation are minimal.

Because the earth does rotate about its axis, however, movements of surface waters are deflected to the right of the wind in the Northern Hemisphere, as shown in Fig. 7-8. This tendency was observed by, among others, Fridtjof Nansen, who while studying the drift of polar ice when his ship *Fram* was frozen in polar ice during his 1893–1896 expedition, found that ice moved 20 to 40° to the right of the wind.

Using Nansen's observations, the physicist Walfrid Ekman showed mathematically that such effects can be explained by using a simple uniform ocean with no boundaries as a model. Each layer, Ekman demonstrated, sets in motion the layer beneath, with the result that the latter moves somewhat more slowly than the former and its motion is deflected to the right of the layer above. If this effect is represented by arrows *(vectors)* whose direction indicates current direction and whose length indicates speed, the change in current direction and speed with increasing depth forms a spiral when viewed from above (see Fig. 7-8), now called the *Ekman spiral*. Figure 7-8 shows the Ekman spiral for the Northern Hemisphere. A similar spiral drawn for the Southern Hemisphere would exhibit the opposite sense of deflection, but current speeds would decrease at the same rate.

Wind effects penetrate to considerable depths below the ocean surface, determined in part by the stability of the water column. The usual limit for wind effects (such as drift currents) is taken to be the depth at which the subsurface current is exactly opposite to the surface current. At that depth the current speed is about 4% of the surface current. Presence of a pycnocline may limit the depth of the drift current. Under strong winds, wind-drift currents have been observed to extend to 100 meters below the surface.

Energy is transferred to this depth by *turbulence* in the surface

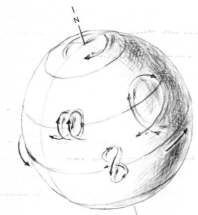

Figure 7-7

Some possible paths of particles moving freely over the earth's surface, showing the influence of the Coriolis effect.

Figure 7-8

Schematic representation of the Ekman spiral formed by a wind-driven current in deep water. Note the change in direction and decrease in speed with increased depth below the surface.

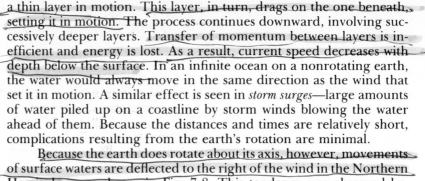
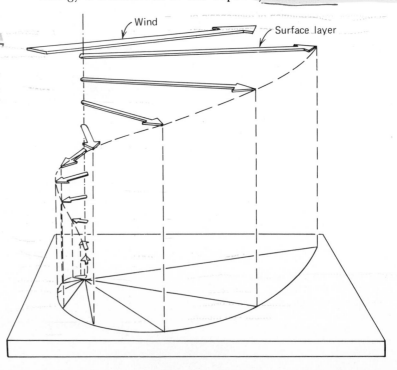

layer. Turbulence is the disorderly state of motion. Transfer of energy by turbulence allows energy to penetrate at least 100 times deeper into the ocean than could be accounted for by movements of water molecules alone.

Speed of surface currents set up by winds is about 2% of the speed of the wind that caused them. For instance, a wind blowing at 10 meters per second would cause a surface current of about 20 centimeters per second.

In shallow waters wind-generated currents are not deflected as much as predicted theoretically for an infinitely deep ocean. The ocean is not completely uniform. The pycnocline, for instance, inhibits downward transfer of momentum and materials from the surface into subsurface layers. Furthermore, the wind does not always blow long enough—probably a few days—from a single direction to establish a fully developed Ekman spiral. Under these conditions the deflection is less than the 45° predicted by the simple case shown in Fig. 7-8. Nansen's original ice observations, for instance, showed that the ideal Ekman spiral is not always encountered in the ocean.

UPWELLING

So far we have considered movements of individual layers of water set in motion by the wind. The net motion of the entire mass of moving water—predicted, using the Ekman spiral, as moving at right angles to the direction of the wind that set it in motion—is called the Ekman transport (Fig. 7-9).

In coastal regions Ekman transport of surface waters can produce upwelling. Where the wind blows somewhat parallel to a coast, it causes surface waters to move offshore. These surface waters are replaced by waters moving upward, typically from depths of 100 to 200 meters, as seen in Fig. 7-10. Upwelled subsurface waters are

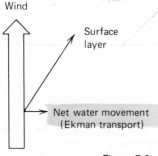

Figure 7-9

Schematic representation of relationship of wind, surface current, and net transport (also called Ekman transport) in a drift current in the Northern Hemisphere.

Figure 7-10

Schematic representation of water properties in an area of wind-induced upwelling. Note that the upwelled water comes from a depth of about 200 meters.

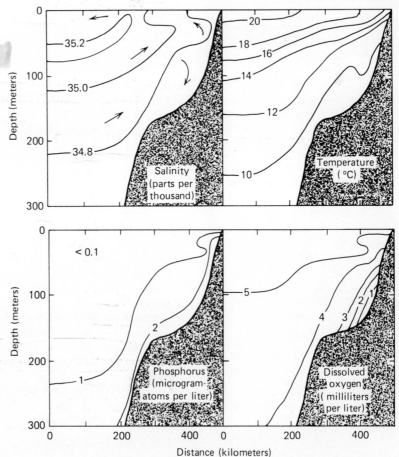

usually colder and contain less dissolved oxygen than surface waters and can thus be readily recognized. In addition, these upwelled waters are usually rich in nutrients (phosphates and nitrates) necessary to support extensive growths of phytoplankton, which feed other marine organisms. Consequently, upwelling areas are highly productive of fishes and other marine organisms. Half the world's fish production comes from upwelling areas.

Ekman transport also forces surface waters to move toward a coastline, which leads to sinking or *downwelling*, the opposite of upwelling. The surface waters moving toward the coast tend to accumulate there—depressing the pycnocline—and tend to hold river discharge near the coast. Winter conditions along the Washington–Oregon coast, causing the Davidson current, is one example of downwelling.

GEOSTROPHIC CURRENTS

Slight differences in water density cause water movements. Most common are the geostrophic currents in which the water flows in response to density-related forces and is deflected by the Coriolis effect.

If you examine the charts of prevailing winds (see Fig. 7-5) and remember that the net transport of surface waters is to the right of the wind in the Northern Hemisphere and to the left in the Southern, you can see that there is a distinct tendency for winds to move surface waters toward the subtropical regions. As Fig. 7-3 shows, these mid-latitude areas are zones of convergence for surface waters. Two things happen in an area of convergence (see Fig. 7-11): first, the water piles up, forming a hill; and secondly, it depresses the local pycnocline.

In the reverse case—where surface water is blown away from an area—a *divergence* occurs, illustrated in Fig. 7-12. A prominent divergence appears around Antarctica (see Fig. 7-3). In a divergence, subsurface waters move upward to replace the water moving away from the region, thereby causing the pycnocline to move upward.

As a result of the wind-driven water movements, the sea surface has a subtle topography. The net difference between the hills of water formed in the midlatitude convergences on the western side of the ocean basins and the divergence surrounding Antarctica is about 2 meters. The distances involved are about half the earth's circumference or at least 20,000 kilometers.

Despite this relatively low relief, water responds to these oceanic hills just as it would on land—by tending to run downhill. Let us consider a simple case of such a hill in the Northern Hemisphere and follow a water parcel along it (see Fig. 7-13) in order to see how the earth's rotation changes that path.

Initially the water parcel moves downslope, just as it does on land. The Coriolis effect, however, changes the path of the water parcel by deflecting it to the right. This process continues until the water follows a path that allows it to flow downhill just enough to keep moving but with most of its motion parallel to the side of the hill. If our hill were contoured to show lines of equal sea surface elevation, the path of the water parcel would nearly parallel these contour lines. On a frictionless ocean, water movement would exactly parallel the side of the hill. This balance—between the gravitational force that pulls the water downhill and the Coriolis effect that deflects it—gives rise to geostrophic currents. The currents are strongest on the steepest hills (where the lines of sea surface elevation are closest together) and weakest where the slope of the sea surface is most gentle.

Where a western boundary current passes near land, it is now possible to measure such sea surface slopes. Detailed surveys have shown that the Florida Current (part of the Gulf Stream system be-

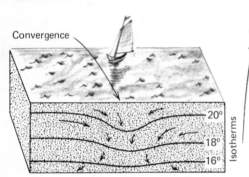

Figure 7-11

Schematic representation of a convergence in the open ocean. A slight hill of water forms and the surface layer thickens because of water accumulation.

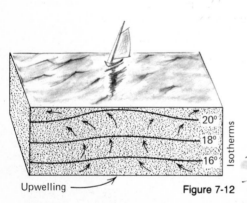

Figure 7-12

Schematic representation of a divergence in the open ocean. Note that the surface layer is thinned.

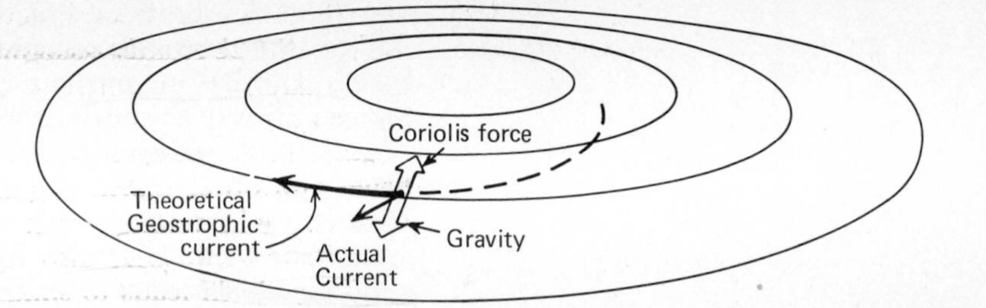

Figure 7-13

Schematic representation of water flow and balance of forces in a geostrophic current in the Northern Hemisphere.

tween Cuba and the Bahama Banks) has sea surface slopes of about 19 centimeters over a distance of approximately 200 kilometers. At this location, current speeds have been measured at 150 centimeters per second (nearly 3 knots) or more. This rapid current has a much steeper slope than currents in which waters move only a few kilometers per day.

It may soon be possible to survey the elevation of the ocean surface by satellites. Then oceanographers can map the major ocean currents by using the topography, making allowance for the slight deviations arising from the frictional effects in the ocean. Direct-measurement techniques are not yet available to survey the entire ocean surface; therefore surface topography must be determined by indirect methods. Usually a reference depth is chosen where no currents are active. It is called *depth of no horizontal motion* and is designated as the base level for subsequent calculations. Assuming that the weight of a water column of fixed dimensions overlying this chosen surface is everywhere, it is possible to calculate the height of the sea surface and the *dynamic topography*—the topography to which a water parcel responds.

Let us see how the procedure works. Assume that the weight of each water column of fixed base area is equal—that is, that there is an equal mass of water and salt above each unit of our fixed surface. Now calculate the height of the column of water above any given location for which there are precise measurements of temperature and salinity; knowing the density, we can then calculate relative topography.

Remember that the mass of water in the column is fixed. A column of less dense water occupies more volume and thus stands higher above the reference surface than a column of more dense water. By this indirect method, it is possible to calculate and map dynamic topography for various parts of the ocean. Furthermore, it is possible to prepare such maps for parts of the ocean well below the surface. We do so by selecting the reference surface and the depth interval used in the calculation. The resulting map of dynamic topography is interpreted in the same way that a surveyed map of sea surface topography would be. The steeper the dynamic topography, the stronger the currents. Current charts can be prepared for subsurface circulation systems, using these techniques.

THERMOHALINE CIRCULATION   So far we primarily discussed horizontal water movements, but massive vertical water movements also control temperature and salinity distributions in most of the ocean's depths. The *thermohaline circulation* is responsible for the movement of waters from polar regions—specifically, from the North Atlantic and Antarctic—through the ocean basins and their slow return to the surface.

Thermohaline circulation is driven by density differences that occur by warming ocean waters in equatorial regions and cooling them in polar regions, as well as evaporation and freezing of seawater. When the density of waters at the surface equals or exceeds water density at depth because of cooling, the water column becomes unstable. The denser water mass sinks, thereby displacing less dense waters beneath. As our previous discussion of density and stability showed, sinking continues until the water mass reaches its appropriate density level. In a stable water column the waters beneath will be slightly denser and the waters above will be slightly less dense; at this level the newly implaced mass of water tends to spread laterally, forming a thin layer. Such processes give rise to the intricately layered structures discovered by precise temperature and salinity measurements.

In semi-isolated ocean areas of polar regions, lowered temperature and increased salinity, resulting from the freezing of sea ice, cause dense water masses to form. Being denser than other waters, they sink to the bottom. Low water temperatures result from intensive cooling in polar regions, where the surface waters reach the freezing point of seawater. Such a combination of conditions is found in the Norwegian Sea, in the Labrador Sea near Greenland, and near Antarctica in the Weddell Sea, as indicated in Fig. 7-14.

The waters of the North Atlantic are the saltiest of the major ocean basins. This salty water is carried into high latitudes by the Gulf Stream. Near Greenland, it is intensively cooled to less than 0°C. When surface waters reach a critical density, they sink suddenly and flow as a mass along the bottom of the North Atlantic basin, especially along the western side of the basin. Such water masses apparently form intermittently during winter.

Submarine ridges between Greenland and Scotland, the region forming the entrance to the Arctic basin, prevent any bottom waters formed in the Arctic Sea from entering the main part of the Atlantic. The Bering sill effectively isolates the Arctic from the Pacific Ocean and so the deep, cold water masses enter the Pacific only from Antarctica (see Fig. 7-14).

Large quantities of Antarctic Bottom Water form in the Weddell Sea, a partially isolated embayment in Antarctica. There surface waters are chilled to temperatures of -1.9°C. At this temperature and a salinity of 34.62‰, the water sinks to the bottom of the adjacent deep-ocean basin where it forms the densest water mass in the open ocean. In the process of sinking, it mixes with other waters and is warmed to -0.9°C. After circulating around Antarctica and mixing with other water masses, cold dense Antarctic Bottom Waters move northward into the deeper parts of all three major ocean basins. Using temperature as a tracer for water masses, we can follow Antarctic Bottom Waters (characteristics given in Table 6-2) as far north as the edges of the Grand Banks (45°N) in the North Atlantic. In the Pacific mixtures of these waters reach the Aleutian Islands (50°N).

Deep-water movement through ocean basins is controlled by ocean bottom topography. The Romanche Trench, for example, provides a path for Antarctic waters to flow into the deep basins of the eastern portion of the South Atlantic; direct entry of this water mass into the eastern side of the South Atlantic is blocked, however, by the Walvis Ridge between Tristan da Cunha and South Africa.

Near-bottom currents commonly move much slower than surface currents. Speeds of 1 to 2 centimeters per second are typical—except along the western basin margins, where speeds of 10 centimeters per second have been calculated. This is another manifestation of the strong boundary currents along the western side of ocean basins.

Bottom waters are formed in large quantities. To compensate for the production of bottom water, there must be a gradual upward movement of waters toward the surface in all ocean basins. This upward movement opposes the downward movement of heat and dissolved gases from the surface.

Figure 7-14

Variation in temperature of bottom waters at depths greater than 4 kilometers and the inferred water movements. Blank areas are less than 4 kilometers deep. [After G. Wüst, 1935. Die stratosphäre. Deutsche Atlantische Exped. Meteor, 1925–27. *Wiss Erg.* 6(1):288 pp.]

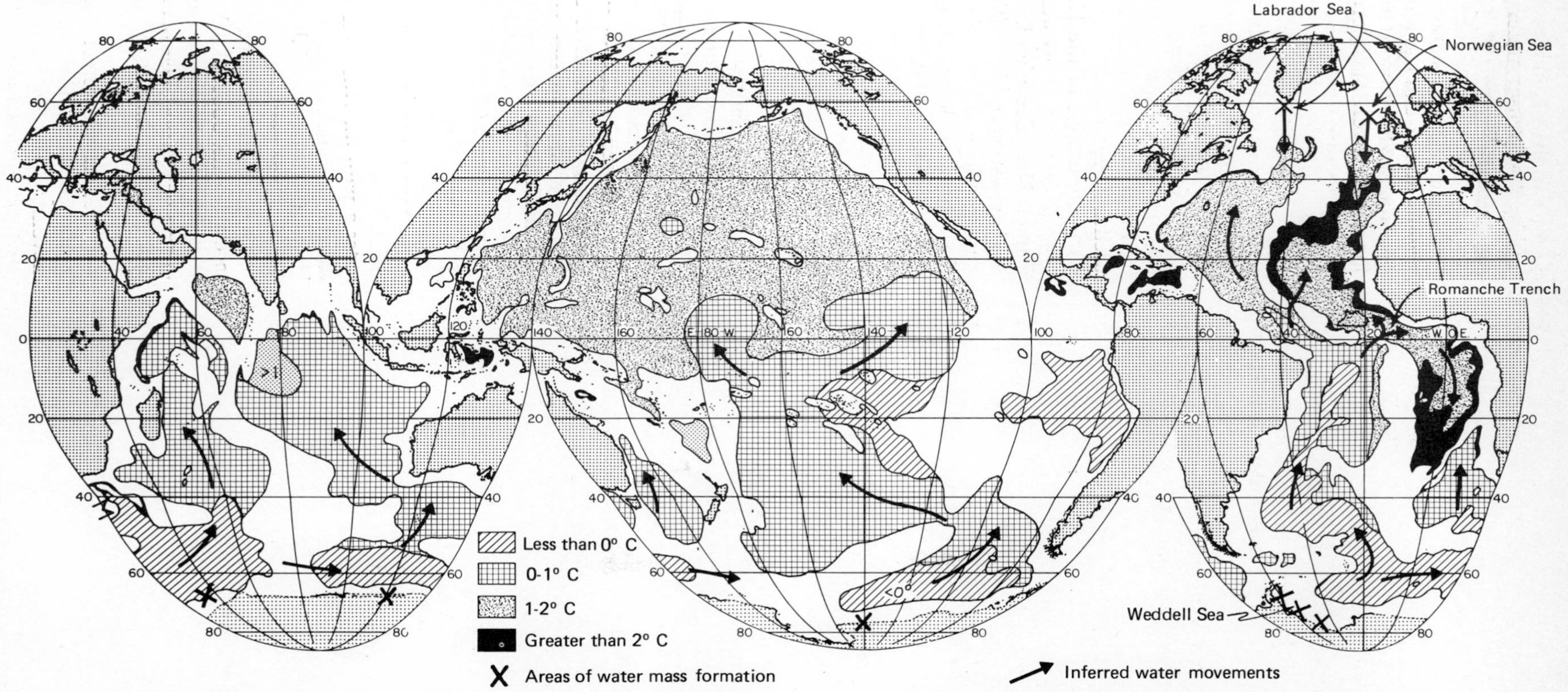

Less than 0° C

0–1° C

1–2° C

Greater than 2° C

X Areas of water mass formation

Inferred water movements

Norwegian Sea

Romanche Trench

Labrador Sea

Weddell Sea

Various techniques are used to estimate deep-water *residence times* (a measure of the time necessary to replace bottom waters by newly formed water masses). Radioactive carbon-14 has been used to determine the time elapsed since the dissolved-carbon compounds in the water were last at the surface. The resultant data suggest residence times ranging from 500 to 1000 years. If we take 1000 years to be a characteristic residence time for deep-ocean waters, it amounts to an upward movement of about 4 meters per year.

ATLANTIC OCEAN CIRCULATION—A THREE-DIMENSIONAL VIEW

Distributions and movements of water masses have been mapped by using temperature and salinity distributions as previously explained. Plotting the distribution of these values as in Fig. 7-15 graphically demonstrates where different water masses form in the ocean. Using the geostrophic approximation discussed earlier, we can estimate their probable movement rates. Because the Atlantic Ocean is especially well known, we shall use it to illustrate the complexities of circulation throughout an entire ocean basin.

Over most of the ocean, circulation of surface waters is almost completely unconnected with movements of subsurface waters. Figure 6-19, in which salinity is plotted, shows that the relatively high-salinity surface water (greater than 35‰ in the Atlantic) extends down to a few hundred meters at most. There is an exception in the North Atlantic (about 30°N), where the warm saline water from the Mediterranean Sea occurs at a depth of about 1 kilometer. If water temperature were plotted instead of salinity, we would have a similar picture, as seen in Fig. 6-18 (top).

In the West Wind Drift around Antarctica, there is little separation between surface and subsurface circulation. The small temperature or salinity change with depth does not hinder vertical circulation. This largest of all currents is the prime communication link for surface and subsurface waters of the three ocean basins. Note that the waters in the West Wind Drift are among the lowest salinity waters found in the Atlantic. (Waters of salinity less than 34.8‰ are shown by diagonal patterns in Fig. 7-15).

Recall that surface currents are dominantly east–west with north–south movements confined to the boundary currents along the continents. In contrast, the subsurface circulation is primarily north–south. As in the surface currents, the strongest near-bottom currents are associated with the western boundaries of the ocean basins. In these boundary currents, speeds of about 10 centimeters per second are common. Throughout most of the deep-water masses the waters have slow net movements, about 1 centimeter per second, although they may move more rapidly for short periods because of large-scale disturbances, much like deep-ocean "storms."

Earlier we discussed the water masses that flow along the ocean bottom because of their relatively high density. Cold bottom water from the Antarctic (Antarctic Bottom Water) is a conspicuous feature of the deep circulation up to about 45°N. Deep waters from the North Atlantic (North Atlantic Deep Water) are also conspicuous along the ocean bottom down to the point where they flow out over the Antarctic Bottom Water. North Atlantic deep water flows southward at depths between 2 and 4 kilometers and mixes extensively with the waters circulating in the West Wind Drift before flowing into the deep Pacific basin.

A somewhat smaller intermediate water mass forms in the region of the Antarctic convergence around 50°S. Because of the low salinities in that region, the water density is not high enough for it to sink to the bottom. Instead the Antarctic intermediate water sinks about 1 kilometer and spreads northward, crosses the equator, and is recognizable to about 20°N.

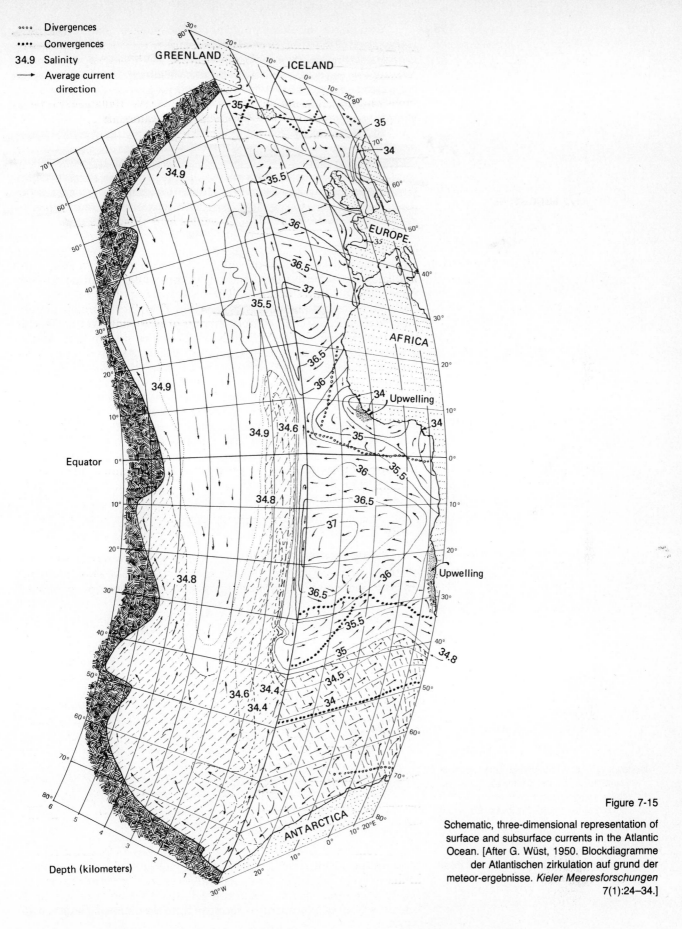

Figure 7-15

Schematic, three-dimensional representation of surface and subsurface currents in the Atlantic Ocean. [After G. Wüst, 1950. Blockdiagramme der Atlantischen zirkulation auf grund der meteor-ergebnisse. *Kieler Meeresforschungen* 7(1):24–34.]

Two points should be noted about the deep circulation. The first is that cold water flowing toward the equator (which eventually rises to the surface) is a form of heat transport. We previously mentioned warm-water transport to high latitudes by surface currents. The return of cold water to low latitudes is the counterflow and it takes place in both surface and subsurface waters. It corresponds to the cold-air or cold-water return in a household heating system.

The second point is that deep-ocean currents transport water across the equator. The circulation of surface waters of the Northern and Southern hemispheres is almost completely separate. Except for some South Atlantic surface water transported into the Northern Hemisphere where the equatorial current is deflected by the South American continent and some areas in the western Pacific, there is almost no movement of surface waters across the equator.

MONSOON CURRENTS

In addition to the wind-driven major ocean currents, there are smaller-scale currents that change directions or even disappear as the regional winds shift directions. The most striking of seasonal current changes is the monsoon circulation of the northern Indian Ocean. This seasonally variable current system is intimately associated with the monsoon winds. During summer (May to September) Asia is greatly warmed relative to the adjacent ocean. As the warmed continental air rises, it draws air from the Indian Ocean toward the land, as shown in Fig. 7-16(a). In this case, the Southwest Monsoon Current replaces the Northern Equatorial Current, as shown in Fig. 7-16(b), in the Indian Ocean. In winter (November to March) the reverse takes place. Air over Asia is much colder than air over the ocean and so the flow is from the land toward the ocean, as shown in Fig. 7-16(c); the Northern Equatorial Current reappears, as shown in Fig. 7-16(d), and the Indian Ocean circulation resembles that of the other ocean basins.

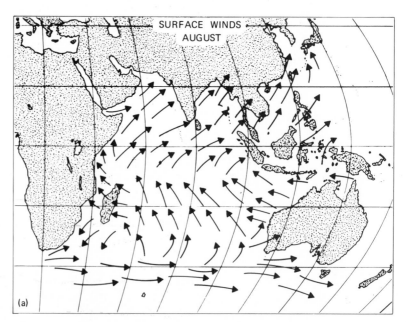

Figure 7-16

Surface winds (a) and monsoon currents (b) in summer during the Southwest Monsoon. Surface winds (c) and monsoon currents (d) in winter during the Northeast Monsoon.

CURRENT MEANDERS AND RINGS

Up until now only average currents and conditions were described. Observations taken decades apart and separated by hundreds of kilometers have been combined in order to study currents, especially in open ocean areas. Thus until recently little was known about short-

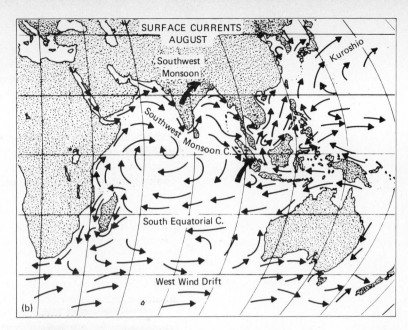

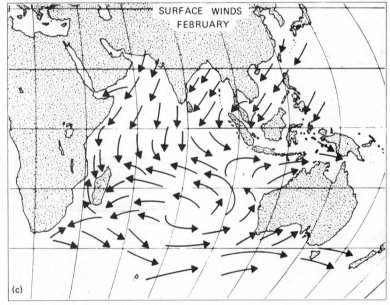

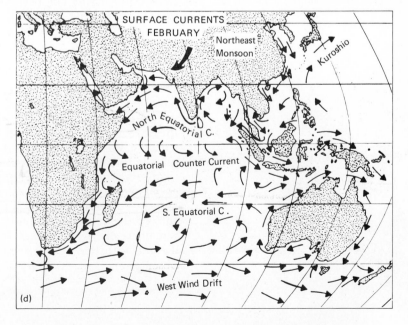

Figure 7-16 (continued)

term processes affecting open ocean currents. The broader views of the ocean surface available from satellites have shown events, previously unknown, as they happen. One example is the development of current meanders and subsequent formation of rings in the Gulf Stream and its extension, the North Atlantic Current.

A pronounced temperature change—the so-called North Wall—separates the cooler coastal waters from the warmer Gulf Stream waters (see Fig. 7-4). On the other side, the Gulf Stream forms the northern boundary of the Sargasso Sea, where the waters are slightly cooler than in the Gulf Stream. Such strong contrasts in surface-water temperatures can be detected by infrared sensors on low-flying aircraft or satellites.

Gulf Stream meanders, while present most of the time, are especially well developed after passage of a storm. Such meanders have been extensively mapped by using several ships operating together with aircraft or satellites with infrared sensors. The meanders move slowly northeastward with the Gulf Stream at speeds of 8 to 25 centimeters per second (7 to 22 kilometers per day).

If a meander becomes too large, it forms a ring and detaches itself from the Gulf Stream to move with the waters flowing southwestward on either side of the Gulf Stream. A ring is 100 to 300 kilometers (60 to 200 miles) across and is bounded by a nearly circular system of swift currents (90 centimeters per second or 78 kilometers per day) that keeps the ring together and contains the waters in it. The ring moves with the waters around it, usually southwestward, at speeds of 5 to 10 kilometers per day. Rings and their associated ring currents extend to depths of 2 kilometers; so normally they do not go up on the shelf, where the waters are only 200 meters deep (see insert Fig. 7-17).

Rings form on both sides of the Gulf Stream. Those forming on the north side consist of a core of warm water surrounded by colder slope water (shown in Fig. 7-17). They are fairly easily detected by aircraft or satellites. Cold rings (also called cyclonic rings) form on the south side of the Gulf Stream and inject cooler water into the Sargasso Sea (see Fig. 7-18). As the rings move southwestward, the surface waters warm up. And as their temperatures reach those of the surrounding waters, they become more difficult to detect.

To see how rings behave in the ocean, we can cite the history of a warm core ring that probably formed in late August 1970 north of the Gulf Stream and south of Cape Cod (see Fig. 7-17). When first sighted in early September, it was apparently moving northward, but after encountering the continental slope, the ring was deflected. Subsequent ship and aircraft surveys showed it moving southwestward at speeds between 5 to 10 kilometers per day. By late December it had reached Cape Hatteras, where it was deflected into the Gulf Stream by the continental slope. In early January the ring was resorbed, apparently by a meander of the Gulf Stream. But other rings have been tracked for periods up to 3 years.

A cold core ring was tracked for 7 months as it moved through the Sargasso Sea from south of Cape Cod to near Cape Hatteras, where it was resorbed into the Gulf Stream (see Fig. 7-18). When initially formed, the waters in the ring were 15°C, nearly 10°C cooler than surrounding waters. The surface waters were greenish, smelling of seaweed—in other words, resembling coastal waters rather than the surrounding Sargasso Sea waters. Satellite infrared photographs showed the details of formation [illustrated in Fig. 7-18(a)] as the ring remained nearly stationary for a month. Later the ring briefly remerged with the Gulf Stream and moved northeastward for nearly

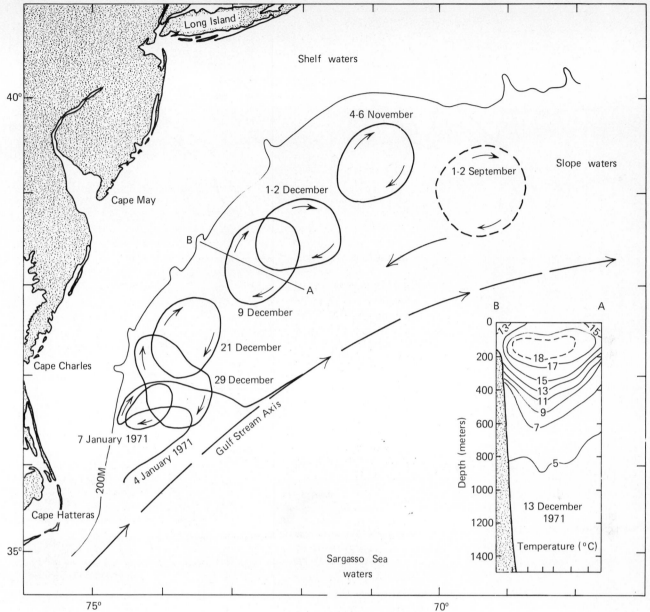

Figure 7-17

Movement of a large, warm core ring as shown by the movement of the 15°C isotherm at 200-meter depth that surrounds the mass of warm water. The ring apparently formed in late August 1970 and was resorbed into the Gulf Stream in early January 1971, after moving southwestward at speeds of 5 to 7 centimeters per second for nearly 5 months. Current speed around the ring reached 90 centimeters per second. The insert shows a temperature profile through the ring, along line A–B. (After Naval Oceanographic Office, 1971. *The Gulf Stream* 6(1):Washington, D.C.)

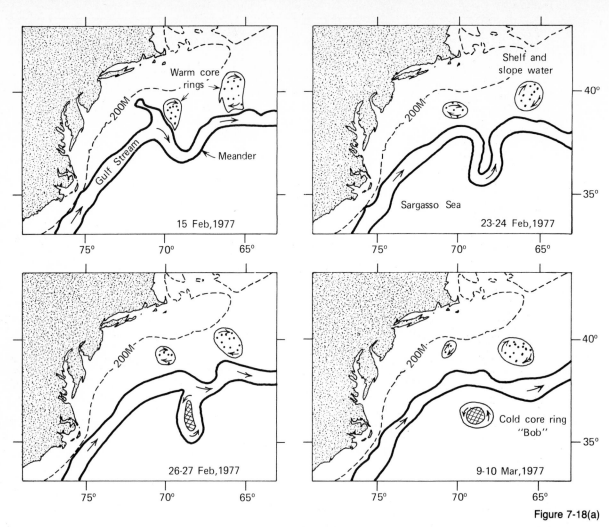

Figure 7-18(a)

A cold core ring, called "Bob," formed from a Gulf Stream meander in February 1977 and began to move southwestward through Sargasso Sea waters. Note the two warm core rings north of the Gulf Stream. Data from NOAA-5 satellite. (After P. L. Richardson, 1980. Gulf Stream ring trajectories. *Journal of Physical Oceanography* 10:90–104.)

a month when it again separated and resumed its southwesterly drift. Finally in September, nearly 7 months later, the ring was again resorbed into the Gulf Stream off Cape Hatteras, North Carolina [see Fig. 7-18(c)].

Although only recently discovered, rings are clearly significant oceanic phenomena. They transport heat, momentum, dissolved constituents, and weakly swimming organisms. It is interesting to note that eddies apparently do not cause a net transport of water because of their complex history of formation, movement in an opposite direction to the Gulf Stream, and then resorption into the stream. Rings apparently can have a complex history of several periods of formation, separate existence, and then later resorption.

The rings are closely associated with the strong western boundary currents, such as the Gulf Stream or the Kuroshio in the Pacific. Other comparable but weaker features called *eddies* are especially abundant in the western portions of the Atlantic and Pacific oceans.

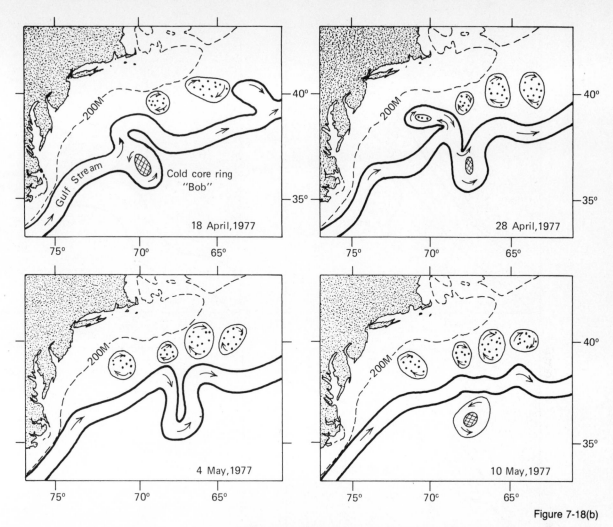

Figure 7-18(b)

Cold core ring "Bob" briefly reattached itself to the Gulf Stream again, forming a meander on 18 April 1977 that traveled nearly 300 kilometers northeastward before detaching to reform as a ring about 20 days later. Note that two new warm core rings formed during this period. Data from satellites (infrared), ship surveys, and tracking buoys released in the ring. (After P. L. Richardson, 1980.)

These eddies (also called *mesoscale eddies*) extend from the sea surface to the ocean bottom, are 200 to 400 kilometers across (see Fig. 7-19), and take several months to pass a location. Currents associated with eddies are weaker than in rings, a few tens of centimeters per second. Still, currents in these eddies are a hundred times more energetic than the average deep-ocean currents. Eddies are abundant in some areas (shown in Fig. 7-20) and are the deep ocean's equivalent of atmospheric storms and weather systems. The role of eddies in oceanic processes has not been worked out.

LANGMUIR CIRCULATION   Momentum imparted to surface ocean waters by winds blowing across the ocean also causes small-scale vertical movements in near-surface waters. Heat, bubbles, dissolved gases, and other substances are quickly mixed through surface waters to depths of a few meters to a few tens of meters. Part of this mixing results from wind-generated turbulence—random motion of water parcels of many different sizes.

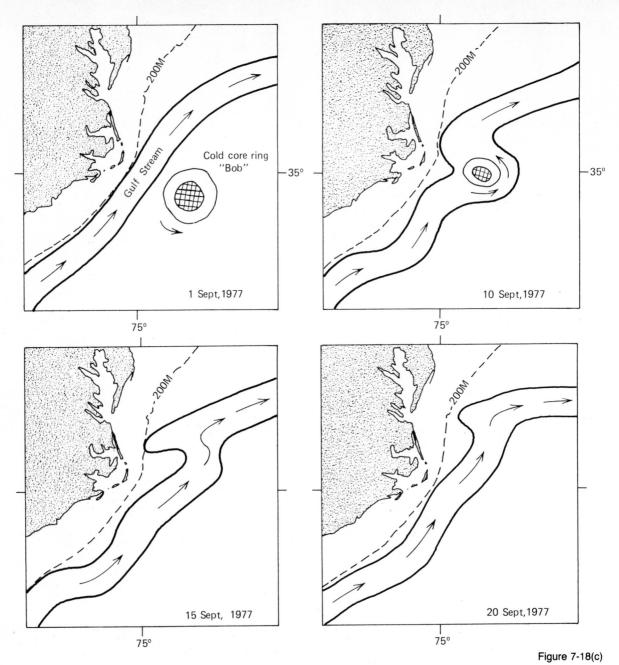

Figure 7-18(c)

Cold core ring "Bob" coalesced with the Gulf Stream near Cape Hatteras, N.C., about September 10, 1977, forming a meander that moved northeastward with the Gulf Stream. (After P. L. Richardson, 1980.)

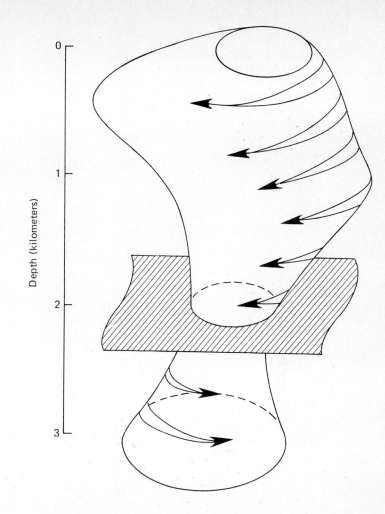

Figure 7-19

Schematic representation of an eddy observed in the North Atlantic Ocean. The diameter of the eddy is about 100 kilometers. Note that the upper portion of the eddy rotates clockwise; the lower portion rotates counterclockwise.

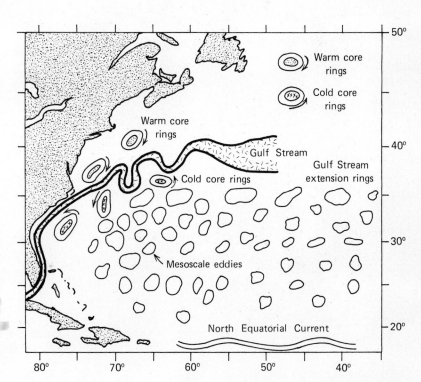

Figure 7-20

Relationship of warm core and cold core rings to the Gulf Stream and eddies. Rings have stronger currents and are closely associated with the Gulf Stream. The eddies have weaker currents and are not obviously associated with the Gulf Stream. Eddies cover 15 to 30% of the Sargasso Sea, southeast of the Gulf Stream. (Data courtesy National Science Foundation.)

Figure 7-21

(a) Oil surface films and debris form windrows along lines of convergence in the Langmuir circulation set up by a steady wind. (b) Foam lines caused by Langmuir circulation near a floating offshore petroleum drilling platform in the North Sea. (Photograph courtesy Mobil Oil Corporation.)

(a)

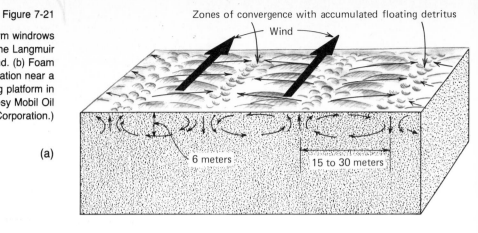

(b)

In addition to the Ekman spiral (already discussed), winds cause cellular circulation patterns to be set up; these cells are known as *Langmuir cells* after their discoverer, Irving Langmuir. In Langmuir cells, water moves with screwlike motions, in helical vortices alternately right and left handed (see Fig. 7-21). The long axes of these cells generally parallel the wind direction. The cells tend to be regularly spaced and are often arranged in staggered parallel rows. Because of the counterrotation of the cells, alternate convergences and divergences are formed at the surface. Floating debris collects in the convergences, such as the conspicuous floating sargassum weed whose distinctive pattern on the water surface can be seen by observers from high-flying aircraft. Between the lines of convergence are lines of divergences where water moves upward and along the surface of each cell toward the next convergence.

Langmuir cells form when wind speeds exceed a few kilometers per hour. The higher the wind speed, the more vigorous the circulation. When evaporation and cooling of surface waters tend to increase water density and favor convective movements, Langmuir cells can form at relatively low wind speeds. Increased stability resulting from surface warming or lowered salinity tends to inhibit cell formation so that stronger winds are required before it can be set up.

Jet streams of water move downward under the convergences. In large lakes, where this type of circulation has been extensively studied, downwelling streams have been observed to extend 7 meters below the surface. Their speeds have been measured at about 4 centimeters per second. In the adjacent divergences, upwelling waters moved at speeds of about 1.5 centimeters per second.

The vertical dimension of these cells is dependent on wind speed and the vertical density structure of the surface layers. If the mixed layer is deep, there is relatively little hindrance to the vertical extent of the cells. If the mixed layer is shallow, the cells may not be able to penetrate the pycnocline. Such vertical water circulation may partially control the depth of the pycnocline. This is also an important mechanism for transporting heat, momentum, and substances from the surface to subsurface layers.

REVIEW QUESTIONS

1. What causes the principal open-ocean surface currents?
2. On an outline map of the world, indicate the location of major surface currents and label them.
3. Describe the monsoon circulation of the northern Indian Ocean and the atmospheric processes that cause it.
4. Contrast eastern and western boundary currents.
5. How are warm core (and cold core) rings formed by the Gulf Stream? Using a diagram, show their relative positions and movements.
6. Describe Langmuir circulation and explain the processes that cause it. How can Langmuir circulation be detected at the ocean surface?
7. Describe ocean surface currents and their relation to the prevailing wind systems.
8. Draw a diagram of an Ekman spiral and describe how it forms.
9. Explain Ekman transport and its role in upwelling.
10. Describe geostrophic circulation.
11. Explain the thermohaline circulation of the deep ocean.

Currents—large-scale water movements

Current measurements
> Mapping of average current set from displaced ship's course and floating objects
> Direct current observations—current meters, buoys

Surface currents in the open ocean
> Gyre—nearly, closed set of currents, usually elongated east–west
> East–west currents, such as equatorial currents, West Wind Drift
> Boundary currents—eastern and western
> Seasonally variable currents—coastal, monsoon—controlled primarily by wind

Boundary currents
> Western boundary currents—Gulf Stream system—strong, narrow, deep
> Eastern boundary currents—California Current,
> Alaska Current—broad, shallow, sluggish
> Oceanic fronts—sharp discontinuities in water properties mark convergences

Forces causing surface currents
> Prevailing wind systems—primary driving force
>> Drift currents—caused directly by wind, primarily in surface layers
>> Geostrophic currents—caused indirectly by wind and its effect on density distribution
> Coriolis effect—deflecting force arising from Earth's rotation; deflects currents to right in Northern Hemisphere, to the left in Southern Hemisphere

Ekman spiral—systematic decrease in current speed and change in direction with increasing distance below surface affected by wind
> Surface current flows 45° to right of wind in infinite, homogeneous Northern Hemisphere ocean, to the left in the Southern Hemisphere
> Surface current about 2% of speed of the wind causing it
> Net movement of surface layer is perpendicular to wind—known as Ekman transport, to the right in Northern Hemisphere, left in Southern Hemisphere
> Movement of surface waters offshore causing upwelling, vertical movements of subsurface waters

Sinking or downwelling is caused by surface waters blown landward

Geostrophic currents
> Caused by small density differences resulting from variations in temperature and salinity
> Involves balance between Coriolis effect and gravitational attraction
> Dynamic topography computed from density distribution (controlled by temperature and salinity)
> Strongest currents where dynamic topography is steepest, weakest where slopes are gentle

Thermohaline circulation
> Vertical water circulation driven by density differences
> Bottom waters formed in North Atlantic and in Weddell Sea (Antarctica)
> Primarily a north–south circulation
> Gradually rises to surface through pycnocline over entire ocean

Atlantic circulation
> Intermediate waters form at high latitudes and flow toward equator
> Antarctic bottom waters flow generally northward

Monsoon currents
> Controlled by winds
> Seasonally variable
> Well developed in northern Indian Ocean

Current meanders and rings—Gulf Stream
> Meander—wavelike current pattern
> Rings—nearly closed current systems
>> 100–300 kilometers across; up to 2 kilometers deep; move 5–10 kilometers per day; last as long as 3 years
>> Form on both sides of Gulf Stream
>> Move southwestward, counter to Gulf Stream
>> Recombine with Gulf Stream
>> Langmuir circulation—organized set of helical vortices in surface waters, caused by winds
>> Transport heat and momentum downward
>> Mix surface waters, often to depths below wave-affected zone

SELECTED REFERENCES

DIETRICH, GUNTER, KURT KALLE, WOLFGANG KRAUSS, AND GEROLD SIEDLER. 1980. *General Oceanography: An Introduction*, 2nd ed. Interscience, London. 626 pp. Technical-level treatment of physical oceanography.

KNAUSS, JOHN A. 1978. *Introduction to Physical Oceanography*. Prentice-Hall, Inc., Englewood Cliffs, N.J. 338 pp. Intermediate-level treatment.

STOMMEL, HENRY. 1965. *The Gulf Stream: A Physical and Dynamical Description*, 2nd ed. University of California Press, Berkeley. 248 pp. Describes present understanding of Gulf Stream system; intermediate-to-technical level.

VON ARX, W. S. 1962. *An Introduction to Physical Oceanography*. Addison-Wesley, Reading, MA. 422 pp. Physical and geophysical aspects of ocean systems are emphasized; little descriptive material; intermediate level.

8 WAVES

Giant Breakers. More than 40-foot high waves offer a real challenge to even the expert surfers at Makaha beach a short distance from Honolulu. The International Surfing Championship meet is held at this spot each January. (Photograph courtesy of Hawaii Visitors Bureau.)

Waves—disturbances of the water surface—can be seen at any beach. And seafarers have observed waves for thousands of years. Yet despite an abundance of observations, an understanding of sea waves has come slowly. The ancients knew that waves were somehow generated by wind, but not until the nineteenth century were the first mathematical descriptions of waves developed. In this chapter we study the features of waves, how they are formed, and some of the ways that they affect the ocean.

Waves are important to us in many ways. They make and remake beaches each year and in the process entertain millions of surfers and swimmers. Waves from a single storm can kill hundreds of people and can result in millions of dollars in damage to low-lying coastal regions through flooding and beach erosion. And waves must be considered in the design and construction of docks, breakwaters, and jetties along the coast because all too often they are responsible for the failure or even destruction of these structures. Nor can waves be neglected when designing or operating the largest ships.

The ocean surface displays a complex and continually changing pattern—a pattern that never exactly repeats itself no matter how long we watch. Ocean waves come in many sizes and shapes, ranging from tiny ripples formed by a light breeze, through enormous storm waves, tens of meters high, to the tides (which are also waves, as we see in Chapter 9).

Because of their complexity, ocean waves usually do not lend themselves to accurate description or complete explanation in simple terms. Nevertheless, we commonly work with simplified explanations and descriptions that help us understand wave phenomena; moreover, most advances in the study of waves have come through the use of appropriate simplifications. Initially our discussions involve such simplified or *ideal waves*.

IDEAL PROGRESSIVE WAVES

To start, let us consider simple *progressive waves* and their parts as they pass a fixed point—say, a piling. We can make such waves in a laboratory wave tank or by steadily bobbing the end of a pencil in a basin of water or a still pond surface. Then we see a series of waves, each wave consisting of a *crest*—the highest point of the wave—and a *trough*—the lowest part of the wave. The vertical distance between any crest and the succeeding trough is the *wave height H* and the horizontal distance between successive crests or successive troughs is the *wave-*

length L; these wave constituents are illustrated in Fig. 8-1. The time (usually measured in seconds) that it takes for successive crests or troughs to pass our fixed point is the *wave period T,* which is rather easily measured with a stopwatch. We can express the same information by counting the number of waves that pass our fixed point in a given length of time. Doing so gives the *frequency (1/T),* which is expressed in hertz, or cycles per second. For individual progressive waves, the speed (*C,* in meters per second) can be calculated from the simple relationship

$$C = \frac{L}{T}$$

where L = the wavelength in meters

T = the wave period in seconds

Where the wave height is low, crests and troughs tend to be rounded and may be approximated mathematically by a *sine wave* (a mathematical expression for a smooth, regular oscillation). As wave height increases, sea waves normally have crests more sharply pointed than simple sine waves and can be approximated by more complicated mathematical curves, such as the sharp-pointed, rounded-trough mathematical curves known as *trochoids* (see Fig. 8-1).

Wave steepness (H/L), the ratio of wave height to wavelength, is a measure of wave stability. It is also the factor that determines whether a small boat glides smoothly over low waves or pitches through steep, choppy waves. When wave steepness exceeds 1/7, waves become unstable and begin to break (see Fig. 8-1) by raveling of the oversteepened crests, forming spilling breakers. The angle at the crest must be 120° or greater for the wave to remain stable; we examine this factor in detail when we discuss breakers and surf.

So far we have considered only movements of the water surface—the crests and troughs moving together that make up a wave train. But what happens to the water itself as waves pass? How is the motion of the water related to the motion of the wave-form? These questions have been studied by using wave tanks with bits of material floating on the surface or dyed bits of water or oil droplets below the surface.

When small waves move through deep water, individual bits of water move in circular orbits that are vertical and nearly closed. The water moves forward as the crest passes, then vertically, and finally backward as the trough passes; this series of movements is diagrammed in Fig. 8-2. The orbit is retraced as each subsequent wave passes; and after each wave has passed, the water parcel is found

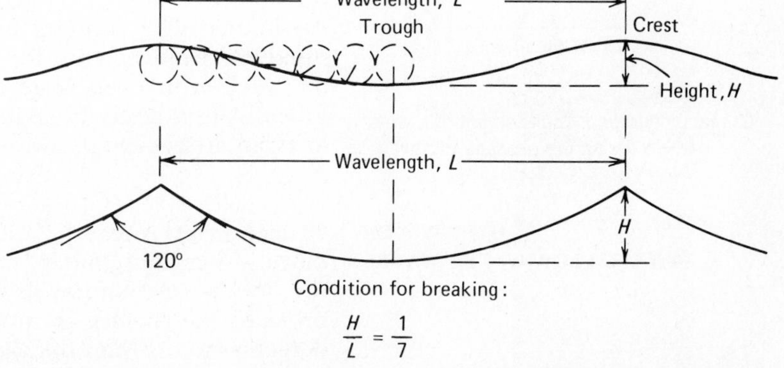

Figure 8-1

Two idealized cases for simple waves, indicating their various parts. Relatively small waves can be described as simple sine waves. Larger waves tend to be more sharp pointed than a simple sine curve. There are limits on the size to which a wave can grow. Waves commonly break when the angle at the crest is less than 120° or the ratio of wave height to wavelength is $H/L = 1/7$.

Condition for breaking:

$$\frac{H}{L} = \frac{1}{7}$$

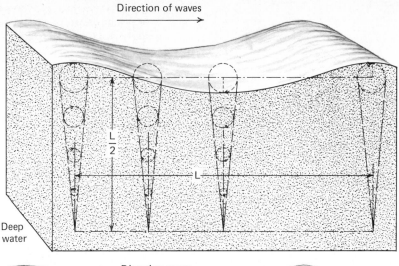

Direction of waves

L/2

L

Deep
water

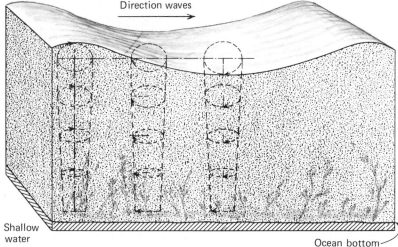

Direction waves

Shallow
water

Ocean bottom

Figure 8-2

Movements of water particles caused by the passage of waves in deep water (a) and shallow water (b). Note that the particles tend to move at the surface in circular orbits that become smaller with depth. Near the bottom, orbits are flattened. Little movement occurs at depths greater than $L/2$.

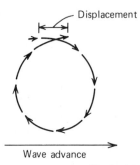

Displacement

Wave advance

Figure 8-3

Orbital motion and displacement of a water particle during the passage of a wave.

nearly in its original position. But there is some slight net movement of the water because the water moves forward slightly faster as the crest passes than it moves backward under the trough. The result is a slight forward displacement of the water in the direction of wave motion and perpendicular to the wave crests, as shown in Fig. 8-3.

Note that if the water moved forward with the waveform, ships would never have been successful, for no ship ever built could withstand the forces exerted by water movements on such a scale. You may have experienced this situation yourself. When one floats in the ocean near the beach but beyond the breakers, there is only a gentle rocking motion as waves pass because there is little net movement of the water. If, however, one tries to stand where waves are breaking, the immediate pounding by the breakers demonstrates large-scale rapid water movements, for the water in the breakers does move with the waveform. Even large ships are damaged when hit by tons of water from a large, breaking wave. Consequently, ships are routed to avoid areas of high waves.

**DEEP-WATER
AND SHALLOW-WATER WAVES**

In deep water, where the water depth is greater than $L/2$, water parcels move in nearly stationary circular orbits; such waves, unaffected by the bottom, are known as *deep-water waves*. The diameter of these orbits at the surface is approximately equal to the wave height. It decreases to one-half the wave height at a depth of $L/9$ and is nearly zero at a depth of $L/2$ (see Fig. 8-2). At depths greater than $L/2$, the water is moved little by wave passage; thus a submarine is essentially undisturbed by waves when it is submerged to depths greater than $L/2$.

In deep water the *wave speed C,* in meters per second, may be calculated by the equations

$$C = \frac{gT}{2\pi} = 1.56T \qquad C = \sqrt{\frac{gL}{2\pi}} = 1.25 \sqrt{L}$$

where T = the wave period in seconds

L = the wavelength in meters

g = the acceleration of gravity, 9.8 meters

per second per second

$\sqrt{}$ = indicates the square root

Of the two equations, the first is the more useful because wave period is easily measured whereas wavelength is difficult to determine.

Where water depths are less than $L/20$, the motion of the water parcels is strongly affected by the presence of the bottom (see Fig. 8-2); these waves are called *shallow-water waves.* Orbits of water parcels at the surface may be only slightly deformed, usually forming an ellipse (a flattened circle with its long axis parallel to the bottom). Near the bottom, wave action may be felt as the water particles move back and forth; vertical water movements are prevented by the proximity of the bottom. Sometimes we observe movements of water parcels in shallow waters as waves pass over them—for example, where bottom-attached plants are moved with the water, as seen in Fig. 8-2. The speed of shallow-water waves, C, in meters per second, can be calculated by

$$C = \sqrt{gd} = 3.1 \sqrt{d}$$

where g = the acceleration of gravity, 9.8 meters

per second per second

d = the water depth in meters

So far we have discussed the behavior of simple (or ideal) waves identical with others in a *wave train.* Although simple uniform waves are rare at sea, the concept is useful in analyzing the more complicated real waves. Even the most complicated waves may be separated by mathematical techniques into groups of simple waves that, when added together, reproduce the complex original ocean waves. An observed profile of a sea is shown in Fig. 8-4(a). This group of waves has been analyzed to determine which wave frequencies occur in the wave spectrum; the various components (each a simple sine wave) are shown in Fig. 8-4(b). Combining these waves will result in a wave essentially identical to the one observed.

Now that we have considered individual ocean waves and calculated their speeds, we can place them in context: in the ocean waves usually occur as wave trains or as a system of waves of many wavelengths, each wave moving at a speed corresponding to its own wavelength.

Suppose that we produce a wave train in a laboratory wave tank and then carefully observe the results. If we follow a single wave in the resulting wave train, we find that it advances through the group. Indeed, the individual waves move at a speed twice that of the group. As each wave approaches the front, it gradually loses height. Finally,

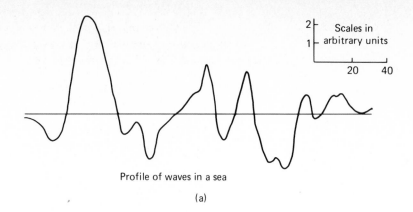

Profile of waves in a sea

(a)

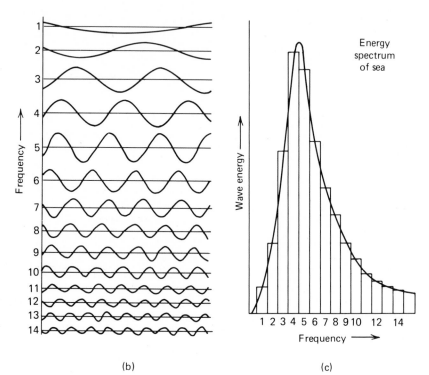

Energy spectrum of sea

Figure 8-4

An observed profile of waves in a sea (a). Such a complicated wave pattern can be described as consisting of many different sets of sine waves (b), all superposed. The lower-frequency waves contain more energy than the higher-frequency ones (c). The energy in a wave is proportional to its height.

(b)

(c)

it disappears at the front of the wave train, to be followed by another, later-formed wave that has also moved forward from the rear. New waves continually form at the back of the wave train while others disappear at the front.

In deep-water waves, wave energy travels at one-half the speed of individual waves. In shallow water individual waves are slowed down until the individual wave speed equals the *group speed*.

FORCES CAUSING WAVES   Formation and behavior of waves involve two types of forces: those that initially disturb the water and those that act to restore the equilibrium or still-water condition. The disturbing forces are familiar to us. We have all made small waves by tossing a pebble into water. If the water surface was initially still, we observed a group of waves changing continuously as they moved away from the disturbance. Sudden impulses, such as explosions or earthquakes, cause some of the longest waves in the ocean. If the disturbance, such as an explosion, affects only a small area, the waves will move away from that

216

point, much as the waves moved away from our pebble. But if the disturbance affects a large area, as a great earthquake can, the resulting *seismic sea waves* (also called *tsunamis*) behave as if they had been generated along a line.

Winds are the most common disturbing force acting on the ocean surface and, consequently, cause most ocean waves (note the large amount of energy associated with wind waves as illustrated in Fig. 8-5). Winds are highly variable and so wind waves vary greatly in different ocean regions and with the seasons (We have more to say about wind waves in the next section.)

The attraction of the sun and moon on ocean water causes the longest waves of all—the tides. Because the attractive forces of the sun and the moon act continuously on the ocean water, the tides are not free to move independently as a seismic sea wave does; such waves, where the disturbing force is continuously applied, are known as *forced waves*—in contrast to the *free waves*, which move independently of the disturbance that caused them. An explosion-generated wave is an example of a free wave. Wind waves have characteristics of both free and forced waves.

Once a wave has formed, restoring forces act to damp out the wave and to restore the initial or equilibrium state. Depending on the size of the wave, different forces may be involved. For the smallest waves (wavelength less than 1.7 centimeters, period less than 0.1 second), the dominant restoring force is *surface tension*, in which the water surface tends to act like a drum head, smoothing out the waves. Such waves, called *capillary waves*, are round crested with V-shaped troughs. For waves with periods between 1 second and about 5 minutes, gravity is the dominant restoring force; this range includes most of the waves we see. Because of the influence of gravity, such waves are known as *gravity waves*.

The largest waves—tides and seismic sea waves—are essentially shallow-water waves and involve substantial water movements. In addition to gravity effects, the Coriolis effect is important for waves whose periods exceed 5 minutes.

Waves transmit energy gained from the disturbance that formed them. This energy is in two forms. One-half is potential energy, de-

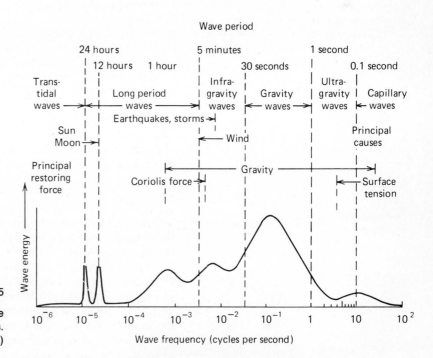

Figure 8-5

Schematic representation of the relative amounts of energy in waves of different periods. (After Kinsman, 1965.)

pending on the position of the water above or below the still-water level; the potential energy advances with the group speed of the individual waves. The rest of the wave energy—known as kinetic energy—is possessed by the water moving as the wave passes. There is a continual transformation of potential energy to kinetic energy and back to potential energy.

The total energy in a wave is proportional to the square of the wave height; in other words, doubling the wave height increases its energy by a factor of 4. An enormous amount of energy is contained in each wave. A swell with a wave height of 2 meters, for example, has energy equivalent to 1200 calories per square meter of ocean surface; a 4-meter swell has 4800 calories per square meter. Nearly all the wave energy is dissipated as heat when the wave strikes a coastline. Sensitive seismographs record the pounding of surf on beaches many hundreds of kilometers away as faint earth tremors. The energy in ocean waves may some day be harnessed on a large scale for human purposes.

FORMATION OF SEA AND SWELL

Waves are disturbances of the ocean surface that pass rapidly across the water but with little accompanying water movement. Stated another way, waves are manifestations of energy moving across the ocean surface. Now we shall see how the energy of the wind is supplied to the ocean surface and how wind waves are formed.

Wave formation by the wind is easily observed. In fact, wind speed at sea can be estimated from wave conditions, as indicated in Table 8-1. Even a gentle breeze results in the immediate formation of ripples or capillary waves, which form more or less regular arcs of long radius, often on top of earlier-formed waves. Ripples play an important role in wind-wave formation by providing the surface roughness necessary for the wind to pull or push the water: in short, they provide the "grip" for the wind.

Ripples are short lived. If the wind dies, they disappear almost immediately; but if the wind continues to blow, ripples grow and are gradually transformed into larger waves, usually short and choppy ones. These latter waves continue to grow as long as they continue to receive more energy than is lost through such processes as wave breaking. Energy is gained through the pushing-and-dragging effect of the wind. The amount of energy gained by the waves depends on such factors as sea roughness, the specific waveform, and the relative speed of the wind and waves. Choppy, newly formed *seas* provide a much better grip for the wind than smooth-crested *swells*.

The largest wind waves (see Fig. 8-6) are formed by storms—often a series of storms—at sea. The size of the waves formed depends on the amount of energy supplied by the wind. The relevant factors operating here are the wind speed, the length of time that the wind blows in a constant direction, and the *fetch*—the maximum distance that the wind flows in a constant direction. The process is illustrated in Fig. 8-7. Usually waves are present on the ocean at the time new waves begin to form. Either the older waves will be destroyed by the storm or newly formed waves will be generated on top of the old ones. There is continuous interaction between waves. Wave crests coincide, forming momentarily new and higher waves. Seconds later the wave crests may no longer coincide but instead cancel each other; then the wave crests disappear.

As the winds continue to blow, waves grow in size, as illustrated in Fig. 8-7, until they reach a maximum size—defined as the point at which the energy supplied by the wind is equaled by the energy lost by breaking waves, called *whitecaps*. When this condition is reached, we refer to it as a *fully developed sea*.

TABLE 8-1

Appearance of the Sea at Various Wind Speeds

BEAU-FORT NUM-BER*	WIND SPEED (kilometers per hour)	SEAMAN'S TERM	EFFECTS OBSERVED AT SEA
0	under 1	Calm	Sea like a mirror
1	1–5	Light air	Ripples with appearance of scales; no foam crests
2	6–11	Light breeze	Small wavelets; crests of glassy appearance, not breaking
3	12–19	Gentle breeze	Large wavelets; crests begin to break; scattered whitecaps
4	20–28	Moderate breeze	Small waves, becoming longer; numerous whitecaps
5	29–38	Fresh breeze	Moderate waves, taking longer form; many whitecaps; some spray
6	39–49	Strong breeze	Larger waves forming; whitecaps everywhere; more spray
7	50–61	Moderate gale	Sea heaps up; white foam from breaking waves begins to be blown in streaks
8	62–74	Fresh gale	Moderately high waves of greater length; edges of crests begin to break into spindrift; foam is blown in well-marked streaks
9	75–88	Strong gale	High waves; sea begins to roll; dense streaks of foam; spray may reduce visibility
10	89–102	Whole gale	Very high waves with overhanging crests; sea takes white appearance as foam is blown in very dense streaks; rolling is heavy and visibility reduced
11	103–117	Storm	Exceptionally high waves; sea covered with white-foam patches; visibility still more reduced
12	118–133		
13	134–149		
14	150–166	Hurricane	Air filled with foam; sea completely white with driving spray; visibility greatly reduced
15	167–183		
16	184–201		
17	202–220		

Beaufort numbers, still used to indicate approximate wind speed, were devised in 1806 by the English admiral Sir Francis Beaufort, based on the amount of sail a fully rigged warship of his day could carry in a wind of a given strength. Modified from U.S. Naval Oceanographic Office, 1958. *American Practical Navigator* (Bowditch), rev. ed., H.O. Publ. No. 9, Washington, D.C., p. 1069.

Given wind speed, duration, and fetch, it is possible to predict the size of waves generated by a given storm. Waves of many different sizes and periods are present in a fully developed sea, but waves with a relatively limited range of periods will predominate for a steady wind with a fixed speed (see Fig. 8-8). Such predictions are complicated because winds almost never blow at a constant speed; winds are just as likely to be gusty at sea as on land and just as likely to change direction as not.

Initially the waves in a sea are steep, chaotic, and sharp crested, often reaching the theoretical limit of stability ($H/L = 1/7$, when the waves either break or have their crests blown off by the wind; the factors affecting wave steepness are illustrated in Fig. 8-9. As waves continue to develop, their speed approaches, then equals, and finally exceeds the wind speed; as this happens wave steepness decreases. As

Figure 8-6

A chaotic sea surface is caused by combined waves of all sizes in an area where waves are formed. (Photograph courtesy Woods Hole Oceanographic Institution.)

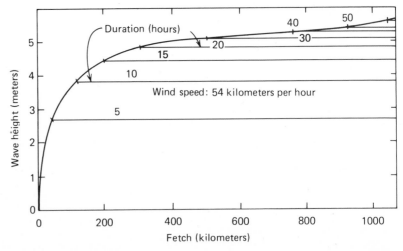

Figure 8-7

Growth of wave height under a constant wind of increasing duration acting over different length and fetch. Note the rapid change in wave height during the first 10 hours as compared with the change between 40 and 50 hours. (After Sverdrup et al., 1942.)

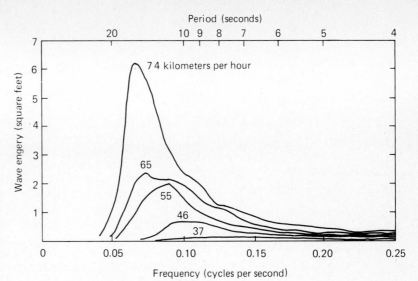

Figure 8-8

In a fully developed sea, most of the wave energy occurs in a relatively restricted range of wave periods (or wave frequencies). Note that changes in wind speeds cause marked changes in wave energy and wave period. (After G. Neumann and W. J. Pierson, 1966. *Principles of Physical Oceanography*. Prentice-Hall, Englewood Cliffs, N.J.)

waves travel out of the generating area, or if the wind dies, the sharp-crested, mountainous, and unpredictable sea is gradually transformed into smoother, long-crested, longer-period waves—*swell*. These waves can travel far because they lose little energy due to viscous forces.

Let us examine some of the processes that cause waves to change from sea to swell as they move through calmer ocean areas. One of these processes is the spreading of waves due to variations in the direction of the winds that formed them. Unless destroyed or influenced in some way by ocean boundaries, waves continue to travel for long distances in the direction that the wind was blowing when they formed. Because winds are rarely constant for long in terms of either speed or direction, waves formed in a storm will move away from the area and "fan out" as they move (angular dispersion). In this case, the wave energy is spread over a larger area, causing a reduction in wave height.

At the same time that the waves are fanning out, they are also separating by wavelength, a process known as *dispersion*. Remember that the speed of deep-water waves C, in meters per second, can be calculated by

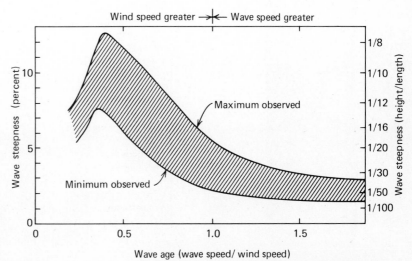

Figure 8-9

Variation in wave steepness with wave age (wave speed/wind speed). The upper and lower curves are probable maximum and minimum values of wave steepness for a given wave age. (After H. U. Sverdrup and W. H. Munk, 1947. *Wind, Sea and Swell: Theory or Relations for Forecasting*. U.S. Naval Oceanographic Office, H. O. Publ. 601. Washington, D.C.)

221

$$C = \frac{L}{T} = \sqrt{\frac{gL}{2\pi}} = 1.25 \sqrt{L}$$

where L = the wavelength in meters

T = the wave period in seconds

Thus longer waves travel faster than shorter waves. As a result, complex waves of varying wavelengths formed in the generating area are sorted as they move away from the storm area, the long waves preceding the shorter waves. Consequently, the first waves to reach a particular coast from a distant large storm will be those having the longest periods. Island-dwellers, sensitive to the normal wave period on their coasts, may be warned of approaching hurricanes by the arrival of such abnormally long waves, which travel faster than the storm.

Swell travels great distances, crossing entire oceans before encountering a coastline. Storms in the North Atlantic, for example, form waves that end up as surf on the coast of Morocco, about 3000 kilometers from their point of origin. On extremely calm summer days, very long period swell strikes the southern coast of England after traveling about 10,000 kilometers (roughly one-quarter of the way around the earth) from storm areas in the South Atlantic. Similarly, in the Pacific, waves from Antarctic storms have been detected on the Alaskan coast, more than 10,000 kilometers away.

WAVE HEIGHT   Despite an abundance of data, observations of wave height leave much to be desired. An observer on a moving ship with no fixed reference points for use in making estimates does not always provide the most reliable information. Still, more than 40,000 observations made from sailing ships, as classified in Table 8-2, indicate that about one-half the waves in the ocean are 2 meters or less in height. Only about 10 to 15% of the ocean waves exceeds 6 meters in height, even in such notoriously stormy areas as the North Atlantic or in the Roaring Forties of the southern oceans.

TABLE 8-2

Relative Frequency, in Percent, of Wave Heights in Various Ocean Areas*

OCEAN	WAVE HEIGHT (meters)					
	< 1	1–1.5	1.5–2	2–4	4–6.1	> 6.1
North Atlantic (Newfoundland to England)	20	20	20	15	10	15
North Pacific (Latitude of Oregon and south of Alaska Peninsula)	25	20	20	15	10	10
South Pacific (West Wind belt latitude of southern Chile)	5	20	20	20	15	15
Southern Indian Ocean (Madagascar and northern Australia)	35	25	20	15	5	5
Whole ocean	20	25	20	15	10	10

*After H. B. Bigelow and W. T. Edmondson. 1947. *Wind Waves at Sea, Breakers and Surf.* U.S. Naval Oceanographic Office, H.O. Publ. No. 602, Washington, D.C., 177 pp.

How do we report wave height? When we look out over a stretch of water, there are waves of many different heights. It turns out, however, that the scene is not totally random and can be described statistically. Detailed studies of ocean waves show a nearly constant relationship between waves of various heights. One useful index is that based on the height of the *significant waves*—the average of the highest one-third of the waves present—as shown in Table 8-3. Setting the height of the significant waves at 1, we find that the most frequent waves are about one-half as high and the average waves are about 0.61. The highest 10% will be about 1.29 times higher than the significant waves. Thus given the wave height for part of the wave spectrum, we can predict the other parts of the spectrum.

The largest waves occur in the open ocean, where they are formed by strong winds blowing over large bodies of water. Such waves occur most frequently at stormy latitudes, where the storms tend to come in groups traveling in the same direction, with only short periods separating them. Thus the waves of one storm often have no chance to decay or travel out of the area before the next storm arrives to add still more energy to the waves, causing them to grow still larger. Typhoons and hurricanes (as the one in Fig. 8-10) do not form exceptionally large waves because their winds, although very strong, do not blow long enough from one direction.

There are reliable reports of waves 13 to 15 meters high in the North and South Atlantic and the southern Indian Ocean. It appears that, for several reasons, these ocean regions rarely produce waves much higher. First, winds rarely blow from one direction long enough to produce waves that are significantly higher. When the wind changes direction, waves produced under previous wind systems are destroyed or greatly modified. Also, the stormy areas of all oceans experience

TABLE 8-3

Wave-Height Characteristics*

WAVES	RELATIVE HEIGHT
Most frequent waves	0.50
Average waves	0.61
Significant (highest one-third)	1.00
Highest 10%	1.29

*U.S. Naval Oceanographic Office, 1958. p. 730.

Figure 8-10

Hurricane driving waves against the North Bayshore retaining wall at Biscayne Bay, Miami, Florida, September 21, 1964.) (Photograph courtesy NOAA.)

equally severe storms at one time or another. Thus the major difference between ocean areas is the maximum fetch over which the wind can act. In the North Atlantic the maximum effective fetch is about 1000 kilometers. With such a fetch, a wind blowing about 70 kilometers per hour can produce waves about 11 meters high; with an unlimited fetch, the same wind could produce waves about 15 meters high.

As might be expected when considering ocean basin dimensions, the Pacific holds the records for giant waves. The largest deep-water wave for which we have reliable data was measured in the North Pacific on February 7, 1933. The Navy tanker *U.S.S. Ramapo* encountered a prolonged weather disturbance that had an unobstructed fetch of many thousands of kilometers. The ship, steaming in the direction of wave travel, was relatively stable and the ship's officers were able to measure wave height (Fig. 8-11) that showed that one wave was at least 34 meters high. The wave period was clocked at 14.8 seconds and the wave speed at 102 kilometers per hour, somewhat faster than the theoretically predicted wave speed.

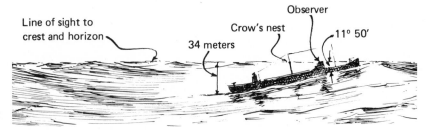

Figure 8-11

Measurement of a wave 34 meters high by the *U.S.S. Ramapo* in the Pacific Ocean, February 7, 1933. This is the largest wave ever measured in a reliable manner.

WAVES IN SHALLOW WATER

An impressive amount of energy is dissipated by breaking waves in the surf. A single wave 1.2 meters high, with a 10-second period, striking the entire West Coast of the United States is estimated to release 50 million horsepower. Most of this energy is released as heat, but it is not detectable because water has a high heat capacity and, perhaps more important, there is extensive mixing in the *surf zone*—where waves break, forming surf—so that the heat is mixed through a large volume of water.

Some waves are destroyed when they encounter opposing winds, others interact and some cancel each other, but most end up as breakers when they encounter the bottom at a coastline. Except for the very longest waves, such as seismic sea waves or the tides, most waves move through the deep ocean without experiencing any effects of the bottom. As waves approach the coast, they are increasingly affected by the presence of the bottom, changing gradually from deep-water to shallow-water waves. The wavelength and speed continually decrease whereas the wave period remains constant. At the same time, the wave height first decreases slightly and then increases rapidly as water depths decrease to one-tenth the wavelength and the wave crests crowd closer together. (This series of events is illustrated in Fig. 8-12.)

Furthermore, the direction of wave approach changes on entering shallower water so that we commonly see breakers nearly parallel to the coastline when they reach the beach even though they may have approached the coast from many different directions. This process, known as *wave refraction,* occurs because the part of the wave still in deeper water moves faster than the part that has entered the shallower water. The result is to rotate the crest more so that it is more parallel to the bottom depth contours in the shallow water, as shown in Fig. 8-13.

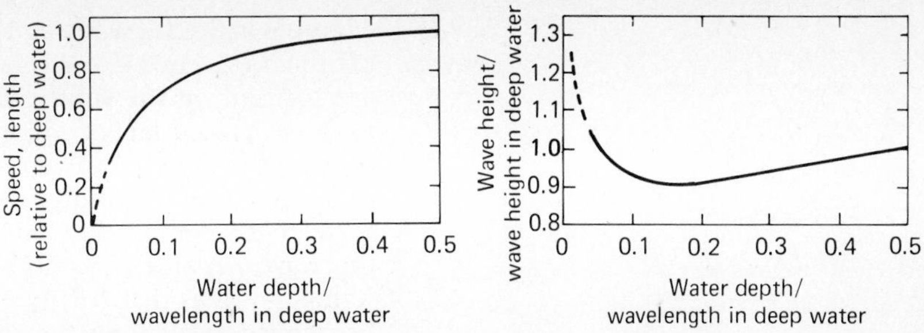

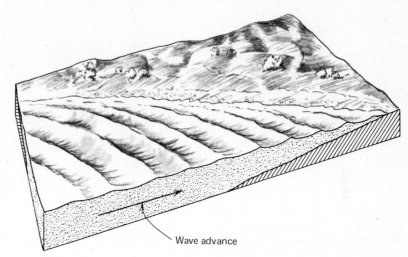

Figure 8-12

Waves change as they enter shallow water. Speed and wavelength decrease as the water becomes more shallow; wave height decreases and then increases.

Left graph: y-axis: Speed, length (relative to deep water), 0 to 1.0; x-axis: Water depth/wavelength in deep water, 0 to 0.5

Right graph: y-axis: Wave height/wave height in deep water, 0.8 to 1.3; x-axis: Water depth/wavelength in deep water, 0 to 0.5

Figure 8-13

Refraction of a uniform wave train advancing at an angle to a straight coastline over a gently sloping, uniform bottom. Note the bend in the crests as the waves approach the beach. Such waves would cause a longshore current moving to the right near the beach.

Wave advance

In the simple case just discussed, the ocean bottom was sloping uniformly away from the beach. Obviously such is not always the case and ocean bottom irregularities cause pronounced wave refraction. Submarine ridges and canyons, for example, cause wave refraction such that the wave energy is concentrated on the headlands and spread out over the bays, as shown in Fig. 8-14. More rapid erosion of the headlands occurs; the eroded material is usually deposited in adjacent bays, eventually creating a simpler, less rugged coastline. An

Figure 8-14

Wave refraction causes equal amounts of energy between orthogonals at 1 and 1' to be concentrated on the headland and to be spread out over the adjacent bay. This increases erosion of the headland and causes sand deposition in the quiet waters of the bay.

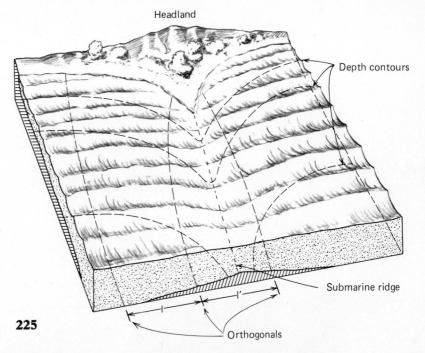

Headland

Depth contours

Submarine ridge

Orthogonals

225

example is the refraction of waves by the Hudson submarine canyon off the New York–New Jersey shore, where waves of certain periods are focused on the southern shore of Long Island and the entrance to New York Harbor (illustrated in Fig. 8-15); much less energy reaches the New Jersey coast.

As waves encounter shallow water, wave height increases and wavelength decreases. Consequently wave steepness (H/L) increases; the wave becomes unstable when the wave height is about 0.8 the water depth and it forms a breaker. The belt of nearly continuous breaking waves along the shore or over a submerged bank or bar is known as *surf*. These breaking waves are distinctly different from the breaking of oversteepened waves in deeper water, where the tops are blown off by the wind.

Several types of breakers are shown in Fig. 8-16 and classified in Table 8-4. The *spilling* type and the *plunging* type behave differently and form under different circumstances. The spilling breaker (see Fig. 8-17) is easily visualized as an oversteepened wave where the unstable top spills over the front of the wave as it travels toward the beach. In a spilling breaker the waveform advances, but wave height (i.e., wave energy) is gradually lost.

Figure 8-15

Submarine topography, especially Hudson Channel (a), causes complicated wave-refraction patterns near the entrance to New York Harbor (b). (After W. J. Pierson, G. Neumann, and R. W. James, 1955. *Practical Methods for Observing and Forecasting Ocean Waves by Means of Wave Spectra and Statistics.* U.S. Naval Oceanographic Office, H. O. Publ. 603. Washington, D.C.)

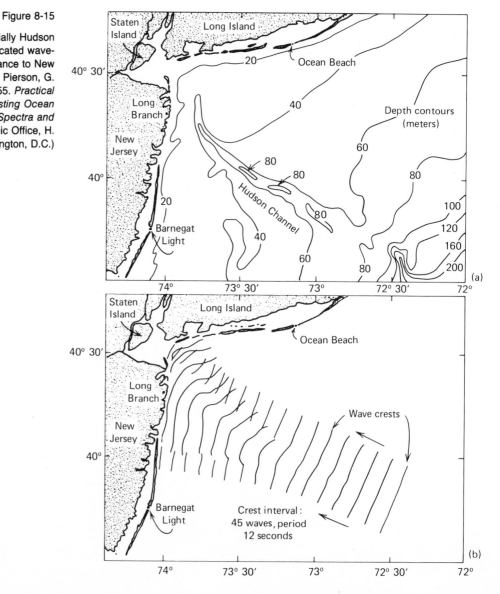

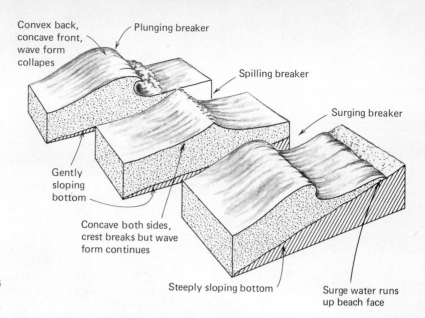

Convex back, concave front, wave form collapes

Plunging breaker

Spilling breaker

Surging breaker

Gently sloping bottom

Concave both sides, crest breaks but wave form continues

Steeply sloping bottom

Surge water runs up beach face

Figure 8-16

Types of breakers.

TABLE 8-4

Types of Breakers and Beach Characteristics Associated with Each*

BREAKER TYPE	DESCRIPTION	RELATIVE BEACH SLOPE	RATIO OF WATER DEPTH TO WAVE HEIGHT
Spilling	Turbulent water and bubbles spill down front of wave; most common type	Flat	1.2
Plunging	Crest curls over large air pocket; smooth splashup usually follows	Moderately steep	0.9
Collapsing	Breaking occurs over lower half of wave; minimal air pocket and usually no splashup; bubbles and foam present	Steep	0.8
Surging	Wave slides up and down beach with little or no bubble production	Steep	Near 0

*After Cyril J. Galvin, 1968. Breaker type classification on three laboratory beaches. *Journal of Geophysical Research* **73** (12):3655.

Plunging breakers are more spectacular (see Fig. 8-18). The wave crest typically curls over, forming a large air pocket. When the wave breaks, a large splash of water and foam is usually thrown into the air. These waves are excellent for surfing. Plunging breakers tend to form from long, gentle swells ($H/L = 0.005$) over a gently sloping bottom with rocky irregularities. On even steeper bottoms they may break over the lower half of the wave with little upward splash. This is known as a *collapsing* type of breaker, shown in Fig. 8-19.

Generally surf is a mixture of various types of breakers, the result of different type waves coming into the beach and the complex and uneven bottom topography offshore, which changes because of tidal action.

If a wave strikes a barrier, such as a vertical wall, it may be reflected, its energy being transferred to another wave, which travels in a different direction. *Wave reflection* may be seen when small waves in a bathtub are reflected from the sides. In other cases, a wave striking a steeply dipping barrier may form a surging breaker or a turbulent

Figure 8-17

Spilling breakers on a rocky beach, Stradbroke
Island, Queensland, Australia. (Photograph
courtesy Australian Tourist Commission.)

Figure 8-18

A plunging breaker on the south shore of Long
Island, New York. The spray is blown seaward
by strong winds.

Figure 8-19

Changes in the water surface through time as a breaker advances toward the beach. The numbers above the profiles indicate the number of 0.06-second intervals elapsed since the profile shown on the front of the block. Thus 10 indicates a profile of the water surface 0.6 second after the first profile. (After C. J. Galvin, 1968. Breaker-type classification on the three laboratory beaches. *Journal of Geophysical Research* 73: 3651–3659.)

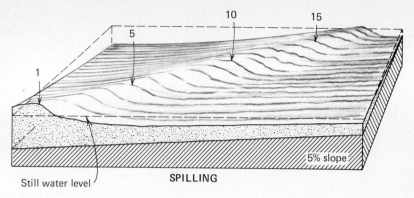

Still water level

SPILLING

5% slope

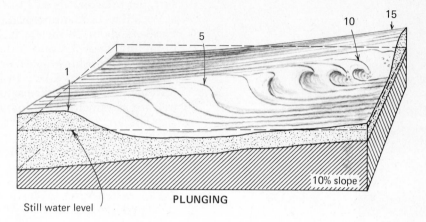

Still water level

PLUNGING

10% slope

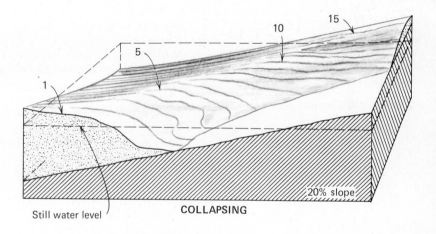

Still water level

COLLAPSING

20% slope

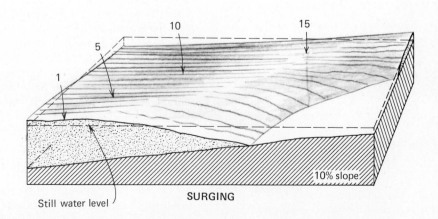

Still water level

SURGING

10% slope

229

wall of water, which then moves up the barrier as the wave advances and runs back down when the wave retreats.

Surf height depends on the height and steepness of the waves offshore and, to a certain extent, on the offshore bottom topography. Thus the breakers and surf may be only a few centimeters high on a lake or a protected ocean beach or many meters high on an open beach. Reports of spectacular surf come from lighthouses built in exposed positions. Minot's lighthouse, 30 meters tall, on a ledge on the south side of Massachusetts Bay, for instance, is often engulfed by spray from breakers and the glass in the lighthouse at Tillamook Rock, Oregon, 49 meters above the sea, has often been struck by waves.

We have no record of observations of the waves that cause such surf, but breakers about 14 meters high twice damaged a breakwater at Wick Bay, Scotland, moving blocks weighing as much as 2600 tons. Breakers about 20 meters high have been reported at the entrance to San Francisco Bay and at the Columbia River estuary on the Pacific Coast when onshore gales were blowing. Waves and breakers at a river or harbor are likely to be especially high when the incoming waves encounter a current setting in the opposite direction. Ships may wait for days before finding a time when incoming waves and tidal currents are right, thus permitting them to enter the harbor with safety.

Large *tsunamis,* or *seismic sea waves* having very long periods, are apparently caused by sudden movements of the ocean bottom resulting from earthquakes or volcanic eruptions. They behave like shallow-water waves even when passing through the deep ocean. An earthquake in the Aleutian Islands on April 1, 1946, for example, caused a tsunami with a 15-minute period and a wavelength of 150 kilometers. Even being in the Pacific Ocean, where the average depth is about 4300 meters, the wave speed was controlled by the bottom (L/d = 150/4.3); yet it still traveled about 800 kilometers per hour. In deep water such wave crests were estimated to be about a half meter high, which would be undetectable to ships, especially considering the extremely long wavelength.

Like wind waves, tsunamis eventually encounter the coast (Fig. 8-20), often with catastrophic results because of their great speed and height. As the waves from the Aleutian earthquake hit the Hawaiian Islands, they were driven ashore in a few places as a rapidly moving wall of water up to 6 meters high. These waves also formed enormous breakers that towered up to 16 meters above sea level, where the water was funneled in a valley. More than 150 people were killed in Hawaii and property damage was extensive. Apparently the wave was highest in the Aleutian Islands, where a reinforced concrete lighthouse and radio tower 33 meters above sea level were destroyed at Scotch Cap, Alaska. Japan has been hit by about 150 tsunamis. The Great Hoei Tokaido–Nankaido tsunami of 1707 killed 30,000 persons and destroyed 8000 homes.

Because of their frequent occurrence in areas bordering the Pacific (see Fig. 8-20), an international warning net operates seismographs to detect large earthquakes likely to cause tsunamis and to warn areas that may be hit. As a result, the 1957 tsunami killed no one in Hawaii even though water levels were locally higher than in 1946. Still, hundreds are killed each year. In 1979 tsunamis killed 540 in Indonesia, 100 in New Guinea, and 500 (estimated) in Columbia, South America.

INTERNAL WAVES

So far we have spoken only about progressive surface waves. But other types of waves in the ocean are not as easily observed and hence not as well known. *Internal waves,* as in Fig. 8-21, occur within the

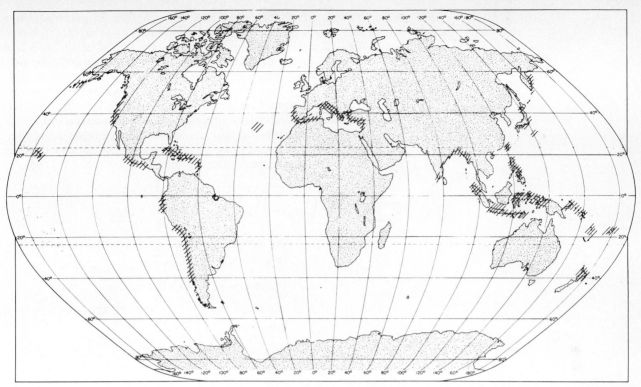

Figure 8-20

Diagonally hatched areas are affected by seismic sea waves generated at nearby tectonically active areas by earthquakes and volcanic eruptions.

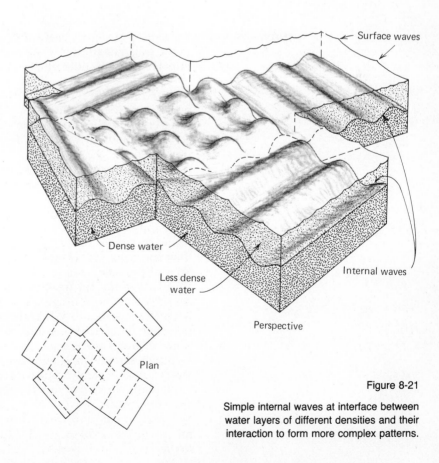

Surface waves

Dense water

Less dense water

Internal waves

Perspective

Plan

Figure 8-21

Simple internal waves at interface between water layers of different densities and their interaction to form more complex patterns.

ocean rather than at the ocean surface, although, in principle, they are similar to surface waves. Surface waves occur at an interface between air and water and internal waves are found at an interface between water layers of different densities—for example, the pycnocline. Because the pycnocline is associated either with the halocline or the thermocline, internal waves can be indirectly observed by studying changes in temperature or salinity at a given depth and fixed location. Internal waves cause mixing below the ocean surface.

Internal waves, it is thought, move as smoothly undulating shallow-water waves. Neglecting the earth's rotation, the speed C of an internal wave in meters per second is given by

$$C = \sqrt{g \left(\frac{\rho - \rho'}{\rho} \right) \left(\frac{dd'}{d + d'} \right)}$$

where ρ, ρ' = the densities of the lower and upper layers, respectively

d = the depth in meters below the interface

d' = the height of the free surface above the interface

g = the acceleration of gravity

Surfaces where internal waves form involve only small density differences between two water layers rather than the much larger density difference between air and water. As a result, internal wave heights can be much greater than surface waves and they generally move much slower.

STANDING WAVES

Standing (or *stationary*) *waves* are yet another type of wave phenomenon in the ocean; standing waves are also important in lakes and play an important role in tidal phenomena. Standing waves can be generated experimentally by first tilting a round-bottomed dish of water and then setting it on a table; the water's surface will appear to tilt first one way and then the other. This type of movement is distinctly different from the progressive waves described where the waves move across the body of water, as if we had dropped a pebble in the dish of water.

It is noticeable that in the standing wave part of the water surface does not move vertically but acts as a sort of hinge about which the rest of the water surface tilts. This stationary line or point is known as the *node* and the parts of the water surface having the greatest vertical movement—known as the *antinodes*—are situated at the walls of the container. More complicated stationary waves may have more than one node and several antinodes in addition to those at the boundaries of the container.

In a stationary wave (as shown in Fig. 8-22) the maximum horizontal water movement occurs at the nodes when the water surface is exactly horizontal. When the water surface is tilted most, there is no water motion. In contrast to the continual orbital motion of the water in progressive waves, in a stationary wave the water flows for a distinct period, stops, and then reverses the flow direction. Also, the waveform alternately appears and disappears and the water does not move in the circular or nearly circular and continuous orbits associated with progressive waves. Standing waves, also known as *seiches* (pronounced "saysh"), are characteristic of steep-sided basins and are well known in many lakes and in the tidal phenomena of nearly closed basins, such as the Red Sea.

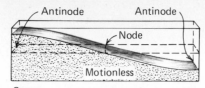

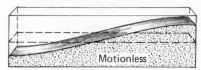

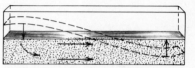

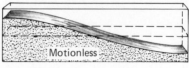

Figure 8-22

Simple standing wave, with one node, shown at quarter-period intervals.

Start

Quarter period later

Half period later

Three quarter period later

One period later

One and one quarter periods later

Like progressive waves, standing waves are modified by their surroundings. They are reflected by vertical boundaries and partially absorbed or obliterated by gently sloping bottoms. Standing waves are refracted by moving into depths substantially less than one-half their wavelength. Standing waves in large basins, such as the Great Lakes, are influenced by the Coriolis effect. The resulting wave, instead of simply sloshing back and forth, has a rotary motion around the edges of the basin.

STORM SURGES

In a storm surge strong winds pile up water along a coast, causing sea level to rise. When a northwest gale blows across the North Sea, for example, with a fetch of 900 kilometers from Scotland to the Netherlands, sea level can rise more than 3 meters. On January 31–February 1, 1953, a storm surge, combined with high tides and strong waves, broke through dikes and dunes, causing disastrous flooding of the low-lying Dutch and English coasts (Fig. 8-23).

The approximate height of a storm surge can be predicted, based on the wind speed and direction, fetch, water depth, and shape of the ocean basin. Other factors, such as currents, astronomical tides, and seiches set up by storms, complicate the calculations, making accurate predictions difficult.

A second type of storm surge resembles a large wave that moves with the storm or hurricane that caused it. First comes a gradual change in water level, the *forerunner,* a few hours ahead of the storm's arrival. It is apparently caused by the regional wind system and may cause sea level to fall slightly along a wide stretch of coastline.

When the hurricane center passes, it causes a sharp rise in water level called the *surge.* This surge usually lasts about 2½ to 5 hours; rises in sea level of 3 to 4 meters have been observed—usually localized but slightly offset from the storm's center. Combined with extremely high waves generated by the storm, hurricane surges can be extremely destructive.

Following the storm, sea level continues to rise and fall as oscillations set up by the storm pass. They are more or less free, wavelike motions of the water surface and have been termed the *wake* of the

233

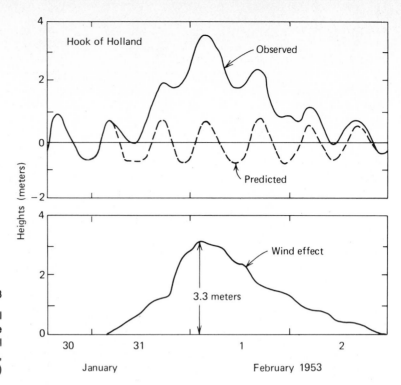

Figure 8-23

Storm surge of 1953 in the North Sea. Sea-level changes resulting from the surge were estimated by subtracting the predicted level from the observed sea levels. (After Groen, 1967.)

storm, like the wake left by the passage of a ship through the water. These resurgences can be quite dangerous, particularly because they are often not expected once the storm itself has subsided.

Tropical storms and hurricanes frequently cause storm surges along the Gulf Coast of the United States. In 1900 Galveston, Texas was destroyed and about 2000 people were killed by a storm surge resulting from a hurricane. In 1969 the second strongest recorded storm to hit the Gulf Coast caused millions of dollars of damage; even with advance warning, it killed several hundred people. A disastrous storm surge occurred in 1876 on the Bay of Bengal, when 100,000 people were killed. In 1970 a storm surge hit the same area, killing an estimated half million people.

REVIEW QUESTIONS

1. Draw an ideal wave and label the parts.
2. Diagrammatically show the changes in orbits of water particles at various depths below the water surface and near the bottom for deep-water and shallow-water waves. Explain the differences between the two types of waves.
3. List the forces (or causes) that produce waves in the oceans.
4. Explain the differences between capillary and gravity waves.
5. Define sea and swell. Describe how each forms.
6. What factors limit the maximum size of wind waves in the ocean?
7. Describe seismic sea waves (tsunamis). How and where are they formed? What areas are mostly likely to experience tsunamis?
8. Draw a diagram showing a simple standing wave in a basin. How is a standing wave different from a progressive wave?
9. Describe a storm surge. What causes storm surges?

Ideal waves—disturbances of the water surface

　　Crest—highest part; trough—lowest part of wave·form

　　Wave height—vertical distance from bottom of wave trough to top of wave crest

　　Wavelength—horizontal distance between successive wave crests (or troughs)

　　Wave period—in progressive waves, time for successive wave crests or troughs to pass a fixed point; in standing waves, time for water surface to assume initial position

　　Sine wave—smooth-crested, smooth-troughed ideal waveform used to describe low ocean waves (a mathematical expression)

　　Trochoid—sharp-crested, flattened trough; ideal waveform useful to describe larger waves (another mathematical expression)

　　Wave steepness—ratio of wave height to wave length

Deep- and shallow-water waves

　　Deep-water wave—wave unaffected by ocean bottom, water deeper than $L/2$

　　Shallow-water wave—wave affected by ocean bottom, in water less than $L/20$ deep; wave speed $C = \sqrt{gd}$

Forces causing waves

　　Seismic sea waves—formed by sudden movement of ocean bottom due to earthquake or sediment slump

　　Forced waves—formed by disturbance continuously applied

　　Free waves—move independently of disturbance

　　Capillary waves—ripples, round crested, V troughed; wavelength less than 1.7 centimeters; wave period 0.1 second; surface tension is dominant restoring force

　　Gravity waves—most common waves; gravity is dominant restoring force; period 1 second to 5 minutes

Formation of sea and swell

　　Sea—waves under influence of the wind that formed them; short, sharp crests, chaotic, unpredictable

　　Swell—waves outside generating area; long, smooth crests

　　Energy for most waves supplied by wind

　　　　Ripples form first, grow into larger waves and eventually into fully developed seas

　　Swells form by gradual transformations

　　　　Angular dispersion occurs because of difference in direction

　　　　Waves separate according to wavelength, longer waves travel faster than short ones

Wave height

　　About 90% of wave less than 6 meters high

　　Highest waves: Pacific—34 meters; other oceans—15 meters

Waves in shallow water

　　Refraction—change in wave advance direction on entering shallow water; energy focused on headlands

　　Reflection—energy reflected, transferred to newly formed wave moving in direction opposite to original wave

　　Breakers—unstable waves losing energy; classified as plunging, spilling, collapsing, or surging

　　Surf—band of breakers parallel to coast

Internal waves

　　Internal waves—occur at boundaries between water layers

Standing waves (seiches)—occur in nearly closed basins; entire surface tilts

　　Nodes—water level constant; maximum currents

　　Antinodes—maximum vertical water movements and changes in water surface level

Storm surges—*strong winds move waters toward coast*

　　Creates standing wave in some basins—for instance, North Sea

　　Surge—preceded by forerunner and followed by pronounced oscillations of water level

SELECTED REFERENCES

Bascom, Willard. 1980. *Waves and Beaches: The Dynamics of the Ocean Survey*, rev. ed. Doubleday Anchor Books, Garden City, N.Y. 366 pp. Elementary.

Clancy, E. P. 1968. *The Tides*. Doubleday, Garden City, N.Y. 228 pp. Nontechnical.

Kinsman, Blair. 1965. *Wind Waves: Their Generation and Propagation on the Ocean Surface*. Prentice-Hall, Englewood Cliffs, N.J. 676 pp. Advanced mathematical treatment; good descriptions.

Redfield, A. C. 1980. *The Tides of the Waters of New England and New York*. Woods Hole Oceanographic Institution, Woods Hole, MA. 108 pp. Elementary treatment of tides in the coastal ocean, emphasizing New England and New York.

Russell, R. C. H., and D. M. MacMillan. 1954. *Waves and Tides*. Hutchinson, London, 348 pp. Elementary.

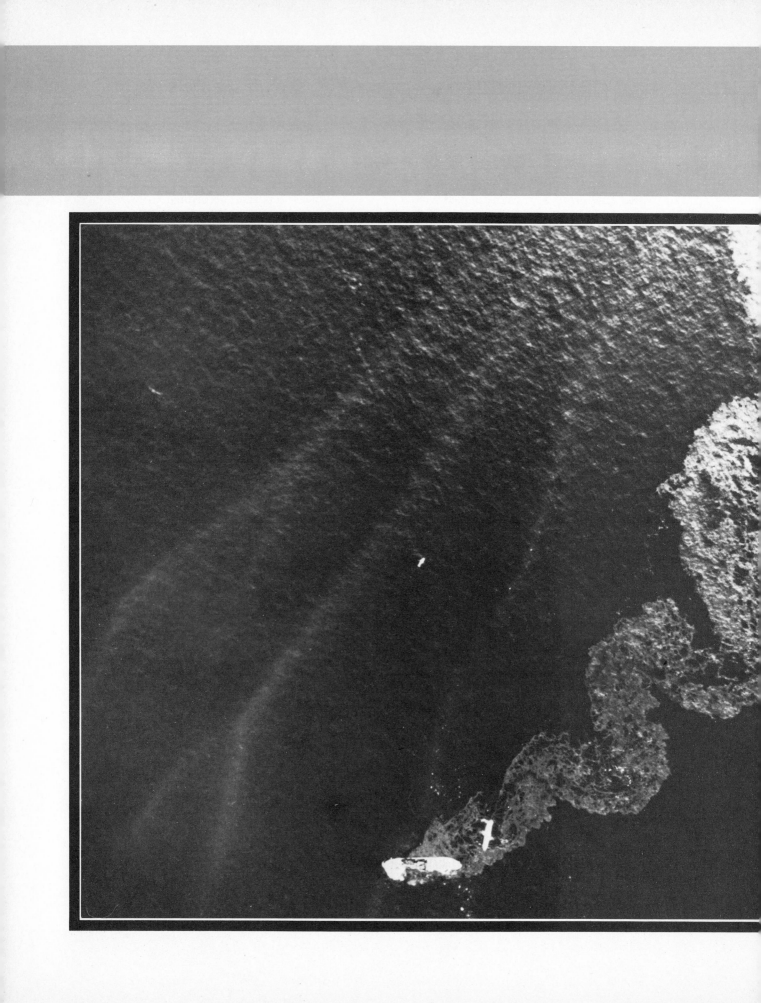

9 TIDES AND TIDAL CURRENTS

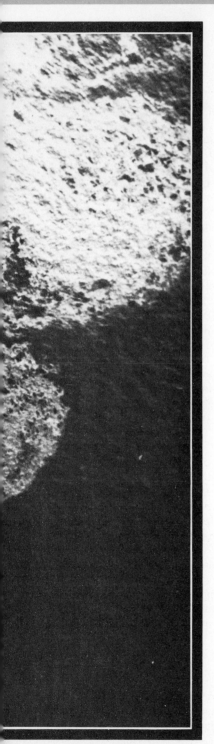

This aerial photograph shows an oil slick from a wrecked tanker 54 kilometers (28 miles) off the coast of Nantucket Island, Massachusetts being moved by winds and currents. (Photograph courtesy of National Aeronautics and Space Administration.)

Tides are the pulse of the ocean. Their effects are felt most keenly in coastal areas, where the periodic rise and fall of the ocean surface, with alternate submersion and exposure of the intertidal zone, modulates plant and animal behavior. Nor are humans oblivious to the tides. Extremely low tides are occasions for digging clams; extremely high tides combined with storms cause flooding of low-lying areas and often extensive damage to coastal installations. Tidal currents, accompanying the rise and fall of the tide, are by far the strongest currents in the coastal ocean; even large, modern ships prefer to depart a port with an outgoing tide and to enter on an incoming tide just as sailors did in antiquity. Tidal currents disperse pollutants in the coastal ocean.

Like waves, tides are easily observed and have been studied since ancient times. Pliny the Elder (A.D. 23–79) currently attributed tides to the effect of the sun and the moon. The small tides of the Mediterranean Sea could easily be ignored and most ancient writers did so. Two famous military men got into trouble by ignoring the tides when their conquests took them into unfamiliar ocean areas. Julius Caesar lost part of his fleet and sustained damage to most of his ships during a night high tide on the coast of England. Alexander the Great got into similar difficulties at the mouth of the Indus River in the northern Indian Ocean.

Tides, however, were well known to those who lived around the North Atlantic. The Venerable Bede (A.D. 673–735) wrote of the extensive tidal observations made by medieval British priests and exhibited a clear understanding of the relationship of tidal phenomena to the moon. He pointed out that separate tidal predictions must be made for each area and that 19 years of observations (corresponding to a complete lunar cycle) were required in order to compile accurate tide tables.

Among the oldest records of oceanographic observations is the "Flood at London Bridge," a set of tide tables for London Bridge dating from the late twelfth or early thirteenth century. Probably other seafaring people also had some means of predicting tides, but any records have been lost so that we can only speculate about them. Tables of tides and tidal currents represent one of our most successful efforts at predicting events in the ocean.

TIDAL CURVES Compared with most oceanographic observations, measurements of the tide can be quite simple, requiring only a measuring pole firmly attached to a piling or stuck in the bottom; such a pole is shown in

Fig. 9-1. At intervals we record the height of the still-water surface on the pole—for instance, every hour. When we plot the height of the water surface at each interval of time, the result is a *tidal curve*.

More elaborate installations are needed for the continuous tidal observations made in most major ports. A simplified diagram of such a *tidal station* is shown in Fig. 9-2. An enclosed basin is constructed with a narrow connection to the ocean so that the water level in the basin is always equal to the undisturbed sea level outside but is not itself disturbed by wind waves. Some of the uncertainty in tidal ob-

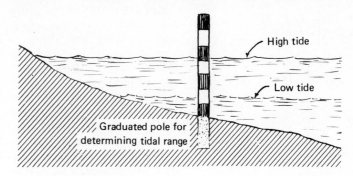

Figure 9-1

Graduated pole for determining tidal range.

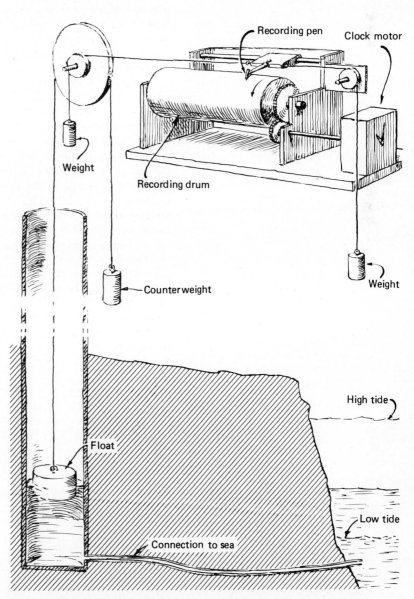

Figure 9-2

Simplified diagram of automatic tide gage.

servations caused by waves or other disturbances of the water surface is avoided. A float on the water surface in the basin is connected to a marker, often a pencil, which draws the tidal curve on a clock-driven, paper-covered drum, indicating the changing sea level. The mechanism operates continuously and requires a minimum of maintenance. Modern tide gages work automatically and the observations are recorded on magnetic tape for later computer processing.

TYPES OF TIDES

Examination of tidal curves shows that each port has a different tide. Tides are grouped into three types, based on the number of highs and lows per day, the relationship between the heights of successive highs or lows, and the time between corresponding high (or low) stands of sea level. Most tidal curves show two high tides and two low tides per *tidal day*—about 24 hours, 50 minutes. This period corresponds to the time between successive passes of the moon over any point on the earth. The time, either 12 hours and 25 minutes or 24 hours, 50 minutes, between successive high (or low) tides is known as the *tidal period*.

A few ocean areas, such as parts of the Gulf of Mexico, have only one high tide and one low tide each day (see Fig. 9-3)—called *daily tides* or *diurnal tides*. Most North Atlantic ports have two high and two low tides that are approximately equal; these are *semidaily* or *semidiurnal* tides. They are relatively easy to predict because high tides tend to occur at a regular time after the moon has crossed the meridian for that port. Tidal predictions for ports with semidaily tides were made for centuries, based primarily on the lunar cycle.

Tidal curves from U.S. Pacific ports also show two high tides and two low tides per tidal day, but the highs are usually quite different in height and the low tides differ as well. These *mixed tides* are shown in Fig. 9-4. The higher of the two high tides is called *higher high water* (abbreviated HHW); the other is called *lower high water* (LHW). There is a similar nomenclature for the low tides—*lower low water* (LLW) and *higher low water* (HLW).

Mixed tides are not as easy to predict as semidaily tides because the timing of the high and low tide *stands* (when there is no appreciable change in the height of the tide) does not bear a simple relationship to the passage of the moon over the meridian of the station.

From a record of only a few days' length, we can classify the type of tide characteristic of any harbor (see Fig. 9-5). Typical tidal curves are shown in Fig. 9-6. We can, for example, measure the *tidal range* (the difference between the highest and lowest tide levels) and the

Figure 9-3

Typical tide curves for (a) semidaily tide, New York Harbor, April 1920 and (b) daily tide, Pensacola, Florida. (After H. A. Marmer, 1930. *The Sea.* D. Appleton, New York.)

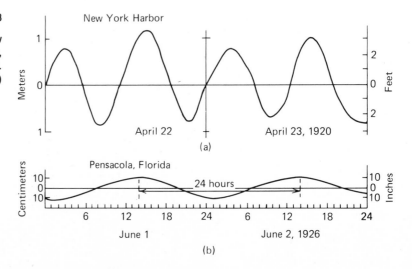

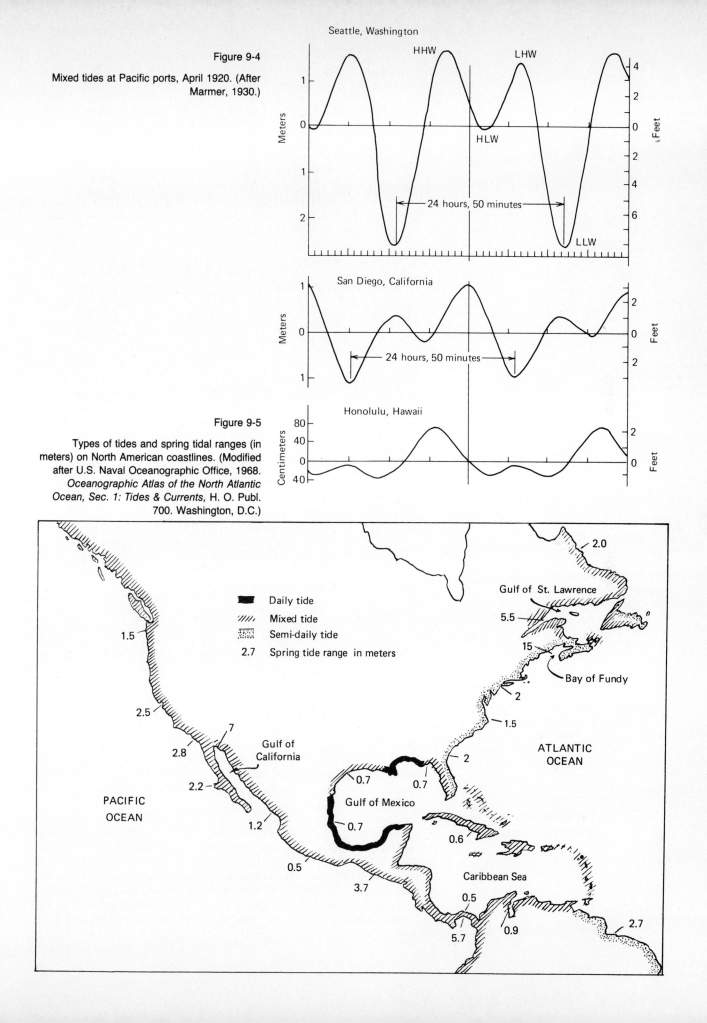

Figure 9-4

Mixed tides at Pacific ports, April 1920. (After Marmer, 1930.)

Seattle, Washington

HHW LHW

HLW

24 hours, 50 minutes

LLW

San Diego, California

24 hours, 50 minutes

Honolulu, Hawaii

Figure 9-5

Types of tides and spring tidal ranges (in meters) on North American coastlines. (Modified after U.S. Naval Oceanographic Office, 1968. *Oceanographic Atlas of the North Atlantic Ocean, Sec. 1: Tides & Currents,* H. O. Publ. 700. Washington, D.C.)

Daily tide
Mixed tide
Semi-daily tide
2.7 Spring tide range in meters

PACIFIC OCEAN

ATLANTIC OCEAN

Gulf of St. Lawrence

Bay of Fundy

Gulf of California

Gulf of Mexico

Caribbean Sea

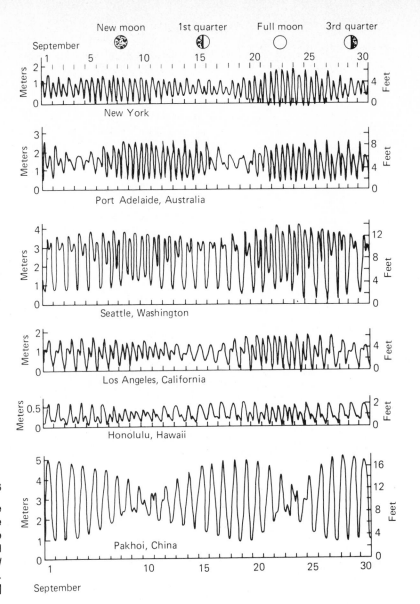

Figure 9-6

Tidal variations at certain ports during the course of a month. Note the variations in the timing of the spring and neap tides relative to the new and full moon. [After U.S. Naval Oceanographic Office, 1958. *American Practical Navigator*, rev. ed. (Bowditch). H. M. Publ. No. 9. Washington, D.C.]

daily inequality (the difference between the heights of successive high or low tides). But the tide also changes from week to week. With a record of several weeks' duration, we see a pattern in the changes of tidal range. *Spring tides* occur near the times of full and new moons and the spring tidal range is larger than the *mean tidal range* (the difference between mean high and mean low tides) or the *mean daily range*. During the first and third quarters the tidal range is least; these are the *neap tides*. As Fig. 9-6 shows, there is substantial variation in the tides at the same place during the month. Other, less striking variations occur over periods of several years.

TIDE-GENERATING FORCES AND THE EQUILIBRIUM TIDE

Although it had long been known that the tides were closely related to the movements of the sun and moon, it remained for Sir Isaac Newton (1642–1727) to lay the foundation for understanding the mechanics of the tides. He began by making several simplifying assumptions; his equilibrium theory of the tides assumes a static ocean completely covering a nonrotating Earth with no continents.

Gravitational attraction pulls the earth and moon toward each

243
tide-generating forces
and the equilibrium tide

other while centrifugal forces, acting in the opposite direction, keep them apart, as illustrated in Fig. 9-7. The earth and moon thus act like twin planets revolving about a common center, which, in turn, moves around the sun. If the earth and moon were the same size, the center of revolution of the system would be located midway between them. The moon, however, is only about 1/82 the mass of the earth; consequently, the center of revolution of the earth–moon system is located nearer the earth, about 4700 kilometers from the earth's center. This situation is analogous to an adult and a small child on a seesaw: the adult must sit closer to the pivot to achieve a balance.

There are small unbalanced forces in the system. Consider the moon's gravitational attraction on the earth's surface. Remember that the force of gravity is inversely proportional to the square of the distance separating the two objects; so doubling the separation reduces the force of gravity to one-fourth. A parcel of water located on the earth's surface is only 59 earth radii away at a point nearest the moon but 61 earth radii away when it is on the opposite side of the earth (see Fig. 9-7). Therefore the moon's gravitational attraction is greatest on the side of the earth nearest the moon and least on the opposite side of the earth. Centrifugal forces, however, are equal over the earth's surface. On the side nearest the moon, the attraction of the moon exceeds the centrifugal force so that the water is attracted toward the moon. On the opposite side of the earth, the centrifugal forces overbalance the attraction of the moon so that there, too, a force acts on the water, effectively dragging it away from the earth. These unbalanced forces on the earth's surface, shown in Fig. 9-8, are the tide-generating forces associated with the moon.

To see how the tides are created, let us look at these forces in more detail. Such forces can be represented by vectors (shown in Fig. 9-9 as arrows) that point in the direction in which the forces act. The length of the arrow (the vector) corresponds to the relative strength

Figure 9-7

Schematic representation of the earth–moon system revolving around their common center, M. Note that the ocean nearest the moon is about 59 earth radii (r) from the moon compared to 61 earth radii for the side farthest from the moon. At the center of the earth, gravitational attraction of the moon is balanced by the centrifugal forces.

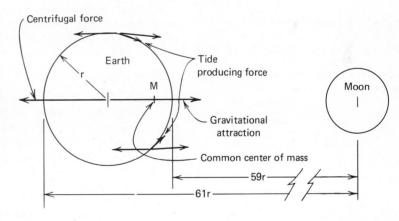

Figure 9-8

Horizontal component of the tide-producing forces acting on the ocean surface when the moon is in the plane of the earth's equator (left) and when the moon is above the equatorial plane (right). Note that the tide-producing forces shift their orientation as the moon's position changes. (After U.S. Naval Oceanographic Office, 1958.)

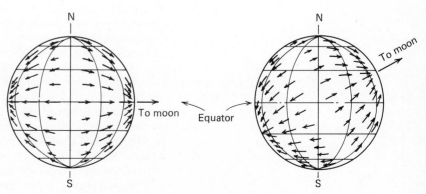

Figure 9-9

Equilibrium tide (shown by the "fences") when
the moon is above the plane of the earth's
equator. The height of the tidal bulge is shown
by the height of the radial "fences." Note
variations in the tide on any parallel of latitude
as the earth rotates. (After W. S. von Arx, 1962.
An Introduction to Physical Oceanography.
Addison-Wesley, Reading, MA)

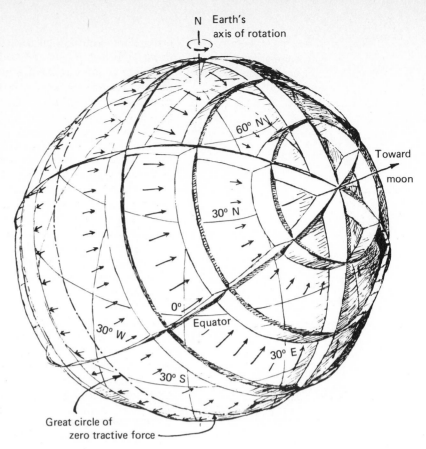

of the force. Each vector can be resolved into two components, acting
at right angles to each other. One component acts in a vertical direction, perpendicular to the earth's surface; the other acts in a horizontal
direction, parallel to the earth's surface at that point. The vectors
point toward the moon on one side of the earth (and away from the
moon on the opposite side). The relative strength of the force acting
in a horizontal sense (parallel to the earth's surface) varies over the
earth (see Fig. 9-9).

Directly beneath the moon and on the opposite side of the earth,
tide-generating forces act solely in a vertical direction. But these vertical components have little effect, for they are counteracted by gravity, which is about 9 million times stronger. Horizontal components
of the tide-generating force are also weak, but they are comparable
in strength to other forces acting on the ocean's surface.

These horizontal or tractive forces cause the *equilibrium tide,*
which consists of a tidal bulge on the side of the earth nearest the
moon and on the side opposite the moon. As a result of these horizontal forces, the water covering the earth forms an egg-shaped water
envelope on our imaginary, nonrotating, water-covered earth. The
solid earth itself also responds to these tide-generating forces and
deforms slightly but much less than ocean waters.

If we permit the earth to rotate beneath its deformed, watery
covering, we can see how the equilibrium tide theory explains successive high and low tides. If the moon is in the plane of the earth's
equator, the tidal bulges of the equilibrium tide will also be centered
on the equator. A tide gage at any point of the equator would register
high tide when that point is directly under the moon. After the earth
rotates 90°, the tide gage would register low tide, which is located
midway between the two tidal bulges. After rotating another 90°, the
tide gage is directly opposite the moon and registers high tide.

244

This simplified model explains semidaily tides with two equal high tides and two equal low tides per tidal day. Remember that the earth rotates beneath its slightly deformed water cover. The more or less egg-shaped deformed water surface, corresponding to the equilibrium tide, remains fixed in space, its location determined by the location of the moon in our simplified case.

The moon, however, does not maintain a fixed position relative to the earth. It moves from a position 28.5° north of the equator to 28.5° south of the equator. When the moon changes its position, so does the orientation of the tide-generating forces and the position of the equilibrium tide.

In order to demonstrate the effect of the earth's rotation on the height of the tides, imagine a purely theoretical case of tide-generating forces on a water-covered earth. We shall identify points on the earth's surface by their real names, but the conditions described do not necessarily apply on the real earth at those points because the example illustrates conditions as they would be without continental barriers to water movements.

If the moon were, for instance, over the site of Miami, there would be a tidal bulge there and also one near the west coast of Australia. These bulges would be particularly well developed because of being in line with the tide-generating forces. After the earth rotated 180°, these points would again experience a high tide—but a lower one because Miami and Australia would no longer be in line with the moon. Tide-generating forces would now be greatest off northern Chile and in the Bay of Bengal–South China Sea area. (Use of a globe will be helpful in demonstrating this point.) This example shows that the equilibrium theory of the tides not only explains the semidaily tides but the daily inequalities as well.

A similar analysis could be made for the sun–earth system to determine the sun's influence on the tides. There are, however, several important differences. First, the greater distance of the sun from the earth (approximately 23,000 earth radii) is only partially compensated for by its greater mass (330,000 earth masses) so that the solar tide-generating forces are only about 47% as powerful as those of the moon. Secondly, solar tides have a period of about 12 hours rather than the 12 hour, 25 minute period of the lunar tide. Finally, the sun's position relative to the earth's equator also changes—from 23.5°N to 23.5°S of the equator—but it requires a full year to make the complete cycle, in contrast to the monthly changes of the moon's position from 28.5°N to 28.5°S.

The equilibrium theory also explains variations in tidal range between the spring and neap tides. Spring tides occur every 2 weeks, usually within a few days of the new and full moons (see Fig. 9-6). Considering their positions in space (shown in Fig. 9-10), we see that during the time of the full and new moons the solar and lunar tide-generating forces act together, causing large tidal bulges and hence greater tidal ranges. During the first and third quarters of the moon tide-generating forces partially counteract each other by acting in different directions, resulting in the lowest tidal range—the neap tides.

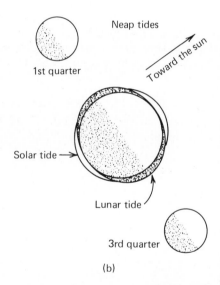

Figure 9-10

Relative positions of sun, moon, and Earth during spring and neap tides.

DYNAMICAL THEORY OF THE TIDE

Since Newton's time mathematicians and physicists have investigated tides by considering the ocean's response to the tide-generating forces. This *dynamical approach* involves a dynamic rather than a static ocean. Like the equilibrium theory, it also requires some simplifying assumptions in order to deal with the complicated interactions. Among them is the neglect of vertical forces, which is not too serious because they are quite small compared with the earth's gravitational attraction. Perhaps the most important difference between the dynamical treatment and the equilibrium theory is that the former considers the effect of the basin's shape and depth on the tide. It also recognizes that ocean tides do not occur in a static ocean but involve substantial water movements and so we must also consider the Coriolis effect.

The dynamical theory treats tides as a type of wave phenomenon. As we have seen, tides have many characteristics in common with both progressive and standing waves. Tides can be resolved mathematically into several components and each treated separately. The tide-generating forces can also be resolved into *tidal constituents,* the most important being those due to the moon and the sun; some important constituents are shown in Table 9-1. Because of the changing positions of these bodies relative to the earth, as many as 62 tidal constituents are used to make tidal predictions, although the four principal ones usually account for about 70% of the tidal range.

The response of a bay or harbor—or an entire ocean basin—to each of the tidal constituents can be considered separately as a *partial tide.* The tide for any location thus consists of the combination of these partial tides (illustrated in Fig. 9-11), just as the complicated waves in a sea can be reconstructed by combining several simple wave trains.

Study of tidal curves indicates how the partial tides must be combined for a given port. Then tidal predictions are made by combining partial tides in the appropriate way, normally by using computers. Some of the earliest tide-predicting machines were simple computers. Development of large, high-speed computers permits more sophisticated tidal models to be studied in greater detail with fewer simplifying assumptions than was previously possible.

Tides can be considered long-period waves. Neglecting the earth's curvature, the two water bulges are the crests and the intervening low areas are the troughs of a simple wave. Because of their immense size relative to the ocean basins, such waves should behave as shallow-water waves. Their wavelength, one-half of the earth's circumference, is about 20,000 kilometers whereas the ocean basins have an average depth of only about 4 kilometers and so $d/L = 4/20,000$. This is much smaller than the limit of 1/20 for shallow-water waves. If tides were free waves, they would move at a speed of $\sqrt{gd}$ or $\sqrt{9.8 \times 4000}$ meters per second, or about 200 meters per second

TABLE 9-1

*Some Important Constituents of Tide-Generating Forces**

CONSTITUENT	SYMBOL	PERIOD (hours)	RELATIVE AMPLITUDE	DESCRIPTION
Semidaily tides				
	M_2	12.4	100.00	Main lunar constituent
	S_2	12.0	46.6	Main solar constituent
	N_2	12.7	19.2	Lunar constituent due to changing distance between earth and moon
Daily tides				
	K_1	23.9	58.4	Solilunar constituent
	O_1	25.8	41.5	Main lunar constituent
	P_1	24.1	19.3	Main solar constituent
Long-period tides (spring and neap tides)				
	M_1	327.9	17.2	Main fortnightly constituent

*After A. Defant, 1958. *Ebb and Flow.* University of Michigan Press, Ann Arbor.

Figure 9-11

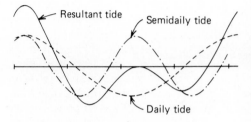

A daily and a semidaily tide when combined produce a mixed tide. (After H. A. Marmer, 1926. *The Tide.* Appleton-Century-Crofts, New York.)

(720 kilometers per hour). But in order to "keep up with the moon," the tides need to move around the earth in 24 hours, 50 minutes, which means that they must move 1600 kilometers per hour at the equator.

In order for the tide waves to move fast enough at the equator to keep up with the moon, the ocean would need to be about 22 kilometers deep. Because it is much shallower, the tidal bulges move as forced waves whose speed is determined by the movements of the moon. The position of the tidal bulges relative to the moon is determined by a balance between the attraction of the sun and moon and frictional effects of the ocean bottom.

Because the world ocean is cut by the north–south-trending continents into basins, it is not possible for the tide to move east–west across the earth as a forced wave except around Antarctica, where the absence of continents permits the tide to sweep unimpeded through the ocean. Let us see how the tide behaves in the Atlantic Ocean.

After the tide enters the South Atlantic, it moves northward. In the South Atlantic its course is relatively simple, although doubtlessly influenced by the irregular ocean boundaries and probably by the irregular bottom topography as well. As it progresses northward, the wave from the Antarctic tide also interacts with the independent tide of the Atlantic.

Even if the southern end of the Atlantic Ocean were closed off, it would still have tides. The tide in a basin like the Atlantic behaves to some degree like a standing wave, besides having some characteristics of a progressive wave. Every basin has its own natural period for standing waves. If that natural period is approximately 12 hours, the standing wave component of the tide will be well developed. If the period of the basin is substantially greater than or less than 12 hours, the resemblance of the tide to a standing wave is lessened.

Even if a basin lacks a standing wave tide, it often has a tide that is distinctly different from that in the adjacent ocean. Tides in the North Atlantic are altered by reflection of the tide wave from the complicated coastline and a standing wave is set up, but it is not the simple standing wave discussed previously. Since substantial water movements are involved, it is modified by the Coriolis effect. The resulting standing wave is a swirling motion such as might result if we swirled water in a round-bottomed cup. Let us see why such a wave forms and how it behaves.

We shall choose a channel in the Northern Hemisphere and set the water moving northward associated with a standing wave. Because of the Coriolis effect, it will be deflected toward the right, causing a tilted water surface, as shown in Fig. 9-12. When the water flows back again, its surface will be tilted in the opposite direction. The tilted wave surface rotates around the basin, once for each wave period. Near the center of such a system, called an *amphidrome system,* is a point where the water level does not change, the *amphidromic point.*

Figure 9-13 shows the movement of the high water associated with the principal lunar partial tide in the Atlantic. There we see a large amphidrome system in the North Atlantic. Other, smaller amphidrome systems not shown on the map occur in the English Channel and the North Sea. Points located near an amphidromic point for one partial tide will not be affected by that partial tide but will probably be affected by another. For example, an area near the amphidromic point for a semidaily tide may well have a daily-type tide.

Because of their dimensions and hence their natural periods, each ocean basin responds more readily to certain constituents of the tide-generating forces than to others. The Gulf of Mexico appears to have a natural period of about 24 hours; consequently, it responds

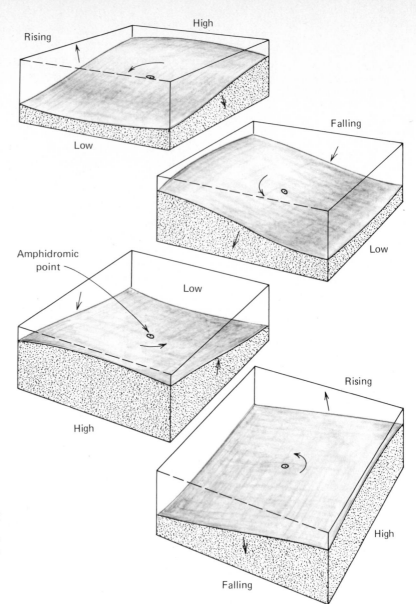

Figure 9-12

Amphidromic motion of a standing wave tide in a Northern Hemisphere embayment. (After von Arx, 1962.)

more to the daily tidal constituents than to the semidaily constituents and much of the Gulf has a daily tide (see Fig. 9-5). The Atlantic Ocean, on the other hand, responds more readily to the semidaily constituent of the tide-generating forces and so tends to have a semidaily type of tide. Because of their size and complex shapes, the Caribbean Sea and the Pacific and Indian oceans respond to both the daily and semidaily tidal forces and thus have mixed types of tides.

Each sea, or bay, is affected by the tide in much the same way the Atlantic is affected by the wave from the Antarctic. A tide wave advancing through the bay is reflected by the coast. If the basin has a natural resonance of the appropriate period, a standing wave may be excited as well. The tide in a bay, harbor, or sea is greatly influenced not only by the magnitude of the ocean tide at its mouth but also by the natural period of its basin and by the cross section of the opening through which the tide wave must pass. The small tide range (less than 0.6 meter) of the Mediterranean Sea, for example, is a result of the small opening through the narrow Strait of Gibraltar into the Atlantic that inhibits exchange of water during each tidal cycle. The small tidal range in several marginal seas of the Pacific can be similarly explained.

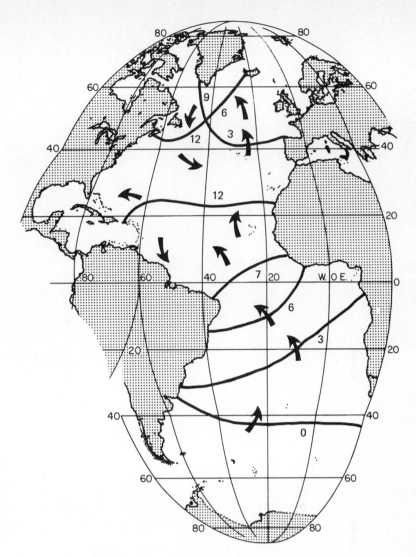

Figure 9-13

Locations of the high water of the principal lunar partial tide—the major tidal component caused by the moon. The numbers indicate the number of hours since the moon crossed the Greenwich meridian. (After Defant, 1958.)

Where the natural period of a basin is near the tidal period, it is possible to set up large standing waves, giving rise to exceptional tidal ranges. One example is the Bay of Fundy, whose natural period is apparently about 12 hours. Spring tides in the inner part of the bay have ranges of 15 meters (see Fig. 9-5) because of an especially favorable situation. A standing wave is combined with a narrowing valley, which funnels the tide and increases its range. The resulting tidal bore is illustrated in Fig. 9-14. Large tidal ranges (about 13 meters) also occur on the Normandy coast of France and at the head of the Gulf of California (about 7 meters).

The most successful techniques for studying and predicting these complicated interactions involve harmonic techniques in which the response of the basin to each of the various tide-generating forces is more or less isolated and then combined with all the other responses to other forces in order to construct a prediction of the tide for a given location at a given time. With even more sophisticated computers, it should eventually be possible to make such predictions based solely on physical principles and using the known characteristics of the oceans and of the particular area for which a prediction is being calculated. But at present tidal predictions still involve the recombination of information gained from many years of observations for each port.

TIDAL CURRENTS

Tidal currents are horizontal water movements associated with the tidal rise and fall of the sea surface. The relationship between these two types of water movement is not necessarily a direct one. Not every seacoast has tidal currents and a few areas have tidal currents but no tides.

First, let us look at the currents we would expect to find in a tide consisting only of a simple progressive wave. As we saw earlier, tides must be considered shallow-water waves because of their extreme length and the relative shallowness of the ocean basins. In such a tide, the crest of the wave is high tide and the trough is low tide. Orbital motions of water caused by the tides are ellipses, greatly flattened circles with their long axes parallel to the ocean bottom. In other words, most of the water motions associated with the tides consist of horizontal motions, with little vertical motion involved.

Near coasts are the familiar reversing tidal currents in which water periodically flows in one direction for awhile and then reverses to flow in the opposite direction. Such tidal currents can be compared to water movement associated with the passage of a progressive wave (see Fig. 9-15). As the wave crest moves toward the coast, the water also moves toward the coast; this corresponds to the *flood current.* As the wave trough moves toward the coast, the water moves away from the coast, corresponding to the *ebb current.* Each time that the current changes directions a period of no current, known as *slack water,* intervenes.

In fact, tides and tidal currents are rarely that simple. Progressive waves are reflected by the shore so that the tide observed in most coastal areas consists of several progressive waves moving in different directions, as well as the standing-wave component that we discussed previously. Other complicating factors include friction effects on the waves and nontidal currents, such as river currents. As a result, no universally applicable generalization can be made concerning high and low tides and their relation to the times of slack water or maximum currents. Just like the tides, tidal current predictions depend on the analysis of observations of tidal currents taken over a long period of time.

In contrast to simple observations of tides, tidal current studies are difficult and expensive, for they involve measurements of currents. For example, a study of the tidal currents in Long Island Sound

Figure 9-14

(a) High tide and (b) low tide at Market Slip, Saint John, New Brunswick, on the Bay of Fundy. At the northern end of the bay on the Petitcodiac River, the profile of the incoming tide surges up the river as a steep wave, known as a tidal bore (c), photographed here at Moncton, New Brunswick. (Photographs courtesy New Brunswick Travel Bureau.)

Figure 9-15

Relationship of tide and tidal currents in an idealized tide consisting of a simple progressive wave.

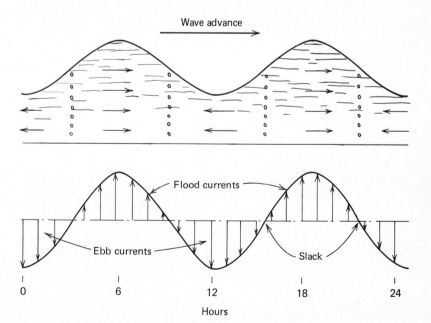

in the 1960s required 3 years to make the measurements and involved 160 separate buoy stations, each with one to three current meters operating for periods of 4 to 15 or 29 days, depending on the station location. Because of the expense involved, our knowledge of tidal currents is usually restricted to coastal areas with much shipping (for example, the North Sea) or harbors and bays.

Tidal currents through a tidal period in New York Harbor are illustrated in Fig. 9-16. Beginning with slack water before high tide, the flood current increases until it reaches a maximum and then decreases again until it is slack water about an hour after high tide. After this slack, the current ebbs as the tide falls. The current again reaches a maximum and then decreases until about 2 hours after low tide, when it is slack water again. Elsewhere in New York Harbor, slightly different tidal current patterns may be observed. Tidal currents nearshore, for example, usually turn sooner than in the middle of the channel because of friction along the channel walls. Tugboat captains often take advantage of these nearshore tidal currents to avoid bucking strong midchannel currents.

Tidal currents are altered by winds or river runoff. River runoff, for example, can prolong and strengthen ebb currents because more water must move out of the harbor on the ebb than comes in on the flood. Also, different tides will have different tidal currents. In areas with large daily inequalities between successive high (or low) tides, there may be days with continuous ebb (or flood) currents that change in strength during the day.

The strength of a tidal current depends on the volume of water that must flow through an opening and the size of the opening. Thus it is not possible to predict the strength of the tidal current, given only the tidal range. The large tidal range in the Gulf of Maine, for instance, is accompanied by weak tidal currents. Conversely, Nantucket Sound has strong tidal currents but a small tidal range. At a given location, however, the relative strength of the tidal currents is generally proportional to the relative tidal range for that day. A spring tide, for instance, is usually accompanied by stronger tidal currents than a neap tide.

In general, tidal currents are the strongest currents in coastal regions. In the English Channel and the Strait of Dover the tidal currents locally have speeds of 2.5 to 3 meters per second (about 9 to 10 kilometers per hour). Tidal currents of 1.2 meters per second (2.4 knots) occur at the Narrows, the entrance to New York Harbor. At Admiralty Inlet, the entrance to Puget Sound, tidal currents have velocities of 2.4 meters per second (4.7 knots).

In the open ocean, tidal currents are not restricted by the coastline and they exhibit rotary patterns quite different from the reversing currents observed in coastal areas. Open-ocean tidal currents continually change direction. Instead of slack-water periods with no current, such as we find in coastal areas, there are periods in which the current is at a minimum. In the Northern Hemisphere the current usually changes direction in a clockwise sense. Tidal currents in the open ocean are generally weaker, typically about 30 centimeters per second (slightly more than 1 kilometer per hour), than in coastal waters.

A log anchored by an elastic tether near Nantucket Shoals Lightship during a simple semidaily tide would move in a clockwise sense in an elliptical path, returning to its initial point after 12 hours, 25 minutes, as indicated in Fig. 9-17. The absence of wind or nontidal currents is again assumed. If currents or winds are present, the log moves around the elliptical path at the same time that the whole water body is being moved.

In areas with mixed tides, where there is a substantial inequality, the current ellipse is more complicated (Fig. 9-18) because there are

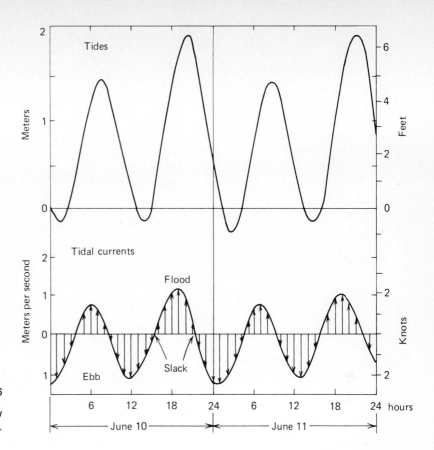

Figure 9-16

Tides and tidal currents at The Narrows, New York Harbor.

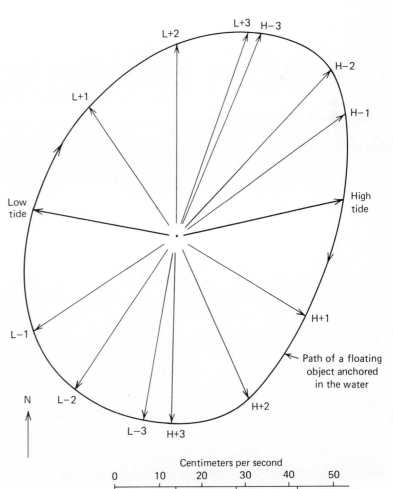

Figure 9-17

Average tidal currents during one tidal period at Nantucket Shoals Lightship off the Massachusetts coast. The arrows indicate the direction and speed (shown by arrow length) of the tidal currents for each hour during a tidal period. The outer ellipse indicates the movements of a log (tied by an elastic line) in the water in the absence of wind or other currents. (After Marmer, 1926.)

Figure 9-18

Average tidal currents during one tidal day at San Francisco Lightship. The radiating arrows indicate the direction and speed (shown by arrow length) of the tidal currents for each hour of a tidal day. The outer ellipse indicates the movements of an object in the water (tied by an elastic line) in the absence of wind or nontidal currents. (After Marmer, 1926.)

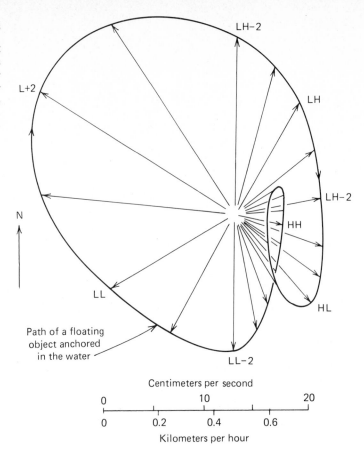

two different ellipses. A log placed in such a current pattern moves through two elliptical paths and returns to its original position after 24 hours, 50 minutes, again assuming no wind or nontidal currents.

TIDAL EFFECTS IN COASTAL AREAS

Tides in the coastal ocean affect estuaries and their tributaries. The tide enters an estuary as a progressive wave and travels upstream. In many estuaries the wave is gradually damped (reduced in height) by friction of the channel and opposing river flow. But in some estuaries the tide wave reaches the end of the estuary and is reflected back. The net result is a standing wave type of tide, where the tide range at the end of an estuary exceeds that close to its entrance. In the Hudson River, for instance (Fig. 9-19), the tidal range is about 1.1 meters at New York Harbor, decreasing to about 0.5 meter at West Point. At Troy, New York (the uppermost extremity of tidal effect), the tidal range is slightly greater than 1.1 meters.

Chesapeake Bay on the mid-Atlantic coast of the United States has relatively simple tidal currents associated with a tide that behaves primarily like a progressive wave. We shall follow the changing tidal currents, beginning with the time when waters are slack at the entrance (Cape Henry), as shown in Fig. 9-20(a). From just inside the entrance, in the southern part of the bay, up to a second slack-water area, about midway up the bay, tidal currents are ebbing. In the northern part of the bay, beyond the second area of slack water, tidal currents are flooding and we can see that areas of slack water separate sections with ebb and flood currents.

Two hours later both areas of slack water have moved into the bay about 120 kilometers, as shown in Fig. 9-20(b). By this time tidal currents are flooding at the bay entrance and the pattern we observed previously has been displaced northward.

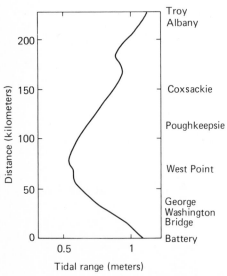

Figure 9-19

Changes in tidal range on the Hudson River.

254

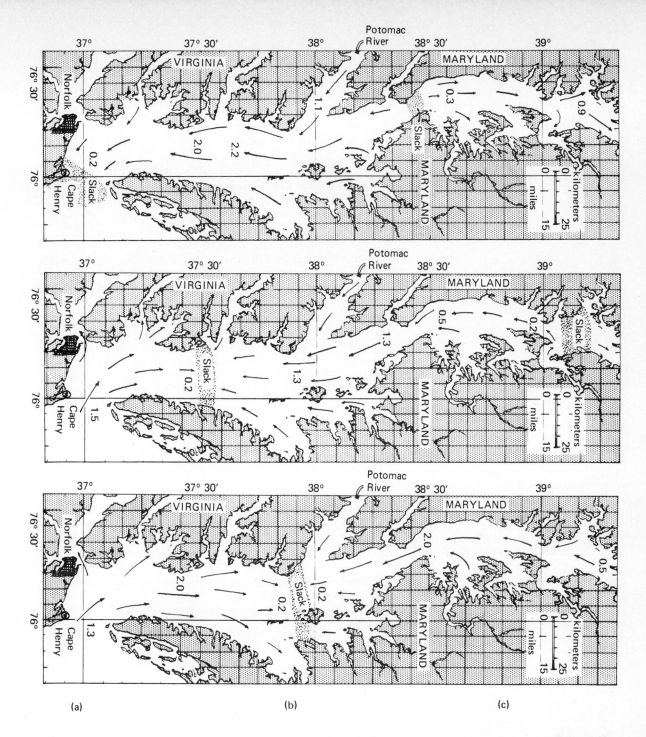

Figure 9-20

Tidal currents in Chesapeake Bay: (a) slack water before flood begins at Cape Henry; (b) 2 hours after flood begins at Cape Henry; and (c) 4 hours after flood begins at Cape Henry. Arrows indicate current direction and the numbers give maximum current velocities in kilometers per hour during spring tides. (After U.S. Naval Oceanographic Office, 1968. Publ. 700, Sec. 1.)

Four hours later the southernmost slack-water area has reached the entrance to the Potomac River and flood-tidal currents at the mouth have diminished somewhat in strength, as shown in Fig. 9-20(c). A final look at Chesapeake Bay 6 hours after slack water before flood shows that the area at the bay mouth is again experiencing slack water, but this time it is slack water before ebb. The arrows in the initial picture are now essentially reversed, flood currents being substituted for ebb currents.

Because of its narrowness, there is little tidal flow across Chesapeake Bay. The reversing tidal currents are therefore relatively uncomplicated, flowing generally along the axis of the bay. In a wider channel, the current pattern would probably be complicated by currents flowing across the channel.

Tidal currents in Long Island Sound are nearly the same over most of the sound at any time. For instance, slack water occurs nearly simultaneously over the entire sound, and flood or ebb currents do also, although current strength varies between areas (Fig. 9-21). The area near the western end of the sound is an exception. An area of slack water separates the small area of opposing tidal currents that have advanced into Long Island Sound from the tidal system in New York Harbor, which connects with the sound through narrow channels.

The situation in which simultaneous slack water is followed by ebb or flood currents throughout the water body is typical of a standing wave type of tide. This type of tide in Long Island Sound results in part from the natural period of the sound, which is apparently about 6 hours.

For our final example, let us consider tidal currents in the North Sea (illustrated in Fig. 9-22), an area whose tidal currents are relatively well known and rather complicated. This basin is wide enough to permit currents across as well as along the basin axis.

The tide enters the North Sea through the Strait of Dover at the south end and from the large opening at the north. In the English Channel–Strait of Dover the tide behaves essentially as a progressive wave, but in the North Sea the tide is an amphidromic system. The area of slack water nearest England corresponds to the center of the amphidromic system. Some of the currents parallel the coastline (for

Figure 9-21

Tidal currents in Long Island Sound after the slack before flood at the Race, the eastern end of the sound. Arrows indicate current direction; the numbers give the current speed in kilometers per hour during spring tides. (After U.S. Coast and Geodetic Survey, 1958. *Tidal Current Charts: Long Island Sound and Block Island Sound,* 4th ed. Serial 574. Washington, D.C.)

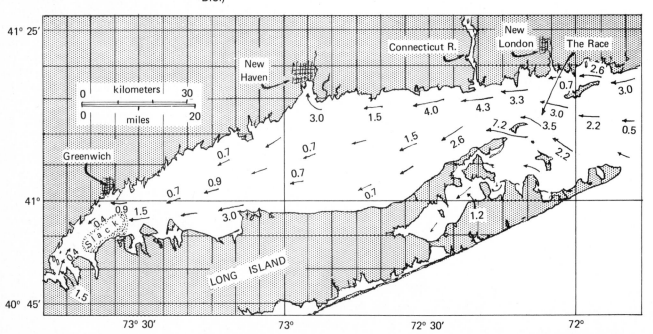

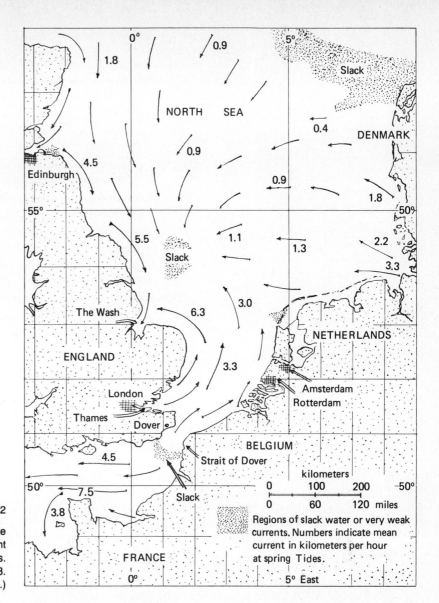

Figure 9-22

Tidal currents in the North Sea. Arrows indicate current direction; numbers give the current speed in kilometers per hour during spring tides. (After U.S. Naval Oceanographic Office, 1968. Publ. 700, Sec. 1.)

example, the coast of Belgium and the Netherlands), as we discussed in the simple reversing currents; other tidal currents are directed primarily in toward the coast (The Wash, on the east coast of England) or directly off the coast (North Germany–Denmark). (Compare this with the simple rotary tide discussed earlier in which there is no slack water, only tidal currents whose strength and direction change constantly.)

Each of the three areas discussed exhibits different aspects of tidal phenomena and illustrates some aspects of the tidal currents associated with them. Collectively they show some of the complexities resulting from interactions between tide-generating forces and the shapes of ocean basins.

Tidal currents are affected by many nonperiodic processes in the coastal ocean. In estuaries large river flow modifies tidal currents, altering their timing so that current prediction is more difficult than predictions of tidal height. Both tides and tidal currents are affected by winds. Times and heights of tides in bays and coastal areas are commonly changed by storms. Thus tide tables often fail to predict the tides as they may be observed under a variety of unusual conditions.

257

1. Diagrammatically show the three types of tidal curves. Label the principal features of each.
2. Define tidal range; daily inequality; spring tide; neap tide.
3. Describe the equilibrium theory of tidal generation. What are the major assumptions? What features of the tide does the theory fail to predict?
4. Explain the dynamical theory of the tide generation. How does it differ from the equilibrium theory?
5. Explain why the tide acts like a forced shallow-water wave in the deep ocean.
6. Show (diagrammatically) how tides and tidal currents are related in the simplest case.
7. Explain the difference between a rotary and a reversing tidal current. Draw a simple diagram illustrating water flow in each.
8. Discuss the differences between progressive wave and standing wave types of tides.

SUMMARY OUTLINE

Tides—periodic rise and fall of the ocean surface
Tidal curves—record of changing sea level
    Tidal day—time between successive transits of the moon over a given point, 24 hours 50 minutes
    Tidal period—time between successive high (or low) tides

Types of tides
    Semidaily tide—two nearly equal high and two nearly equal low tides per tidal day
    Daily tide—one high and one low per tidal day
    Mixed tide—two unequal high tides and two unequal low tides per tidal day
    Tidal range—vertical distance between high and low tide
    Spring tide—greatest tidal range
    Neap tide—least tidal range

Tide-generating forces and the equilibrium tide
    Equilibrium tide would apply to static ocean completely covering a smooth earth; only tide-generating forces are considered
    Unbalanced forces (gravitational attraction of the moon and centrifugal force from earth–moon revolution) acting on the earth's surface cause the tides
    Vertical and horizontal components; horizontal is the stronger
    Equilibrium theory explains daily and fortnightly inequalities
    Variable position of the sun, moon, and Earth causes spring and neap tides

Dynamical theory of the tide
    Includes ocean's response to tide-generating forces
    Tidal constituents may be considered separately, as responsible for partial tides

    Partial tides combined for tidal predictions
    Tides as long-period waves, in shallow water; wave length is one-half Earth's circumference
    Natural period of each basin favors a standing wave; forms amphidrome system affected by Coriolis effect
    Factors controlling tides in nearly enclosed basins:
        Range of tide at ocean entrance
        Natural period of basin
        Size of ocean entrance

Tidal currents—horizontal motions with little vertical component
    Comparable to water movements associated with progressive waves
    Reversing tidal currents near coasts
        Flood current—water moves landward
        Ebb current—water moves seaward
        Slack water—little or no current
    Complex relationship between tides and tidal currents
    Open-ocean rotary currents flow continuously, changing direction
        Particles move in elliptical paths
        Current rose—formed by vectors indicating direction and speed

Tidal effects in coastal areas
    Chesapeake Bay tidal currents behave like a progressive wave
    Long Island Sound has a standing wave type of tide
    North Sea has an amphidromic system set up by tidal currents
    Tidal currents modified by river flow, winds, storms

SELECTED REFERENCES

DARWIN, G. H. 1962. *The Tides and Kindred Phenomena in the Solar System.* W. H. Freeman, San Francisco. 378 pp. Originally published in 1898; classic, exhaustive treatment of the subject; elementary to intermediate.

DEFANT, ALBERT. 1958. *Ebb and Flow.* University of Michigan Press, Ann Arbor, 121 pp. Descriptive and mathematical; elementary to intermediate.

RUSSELL, R. C. H., AND D. H. MACMILLAN. 1954. *Waves and Tides.* Hutchinson, London, 348 pp. Elementary.

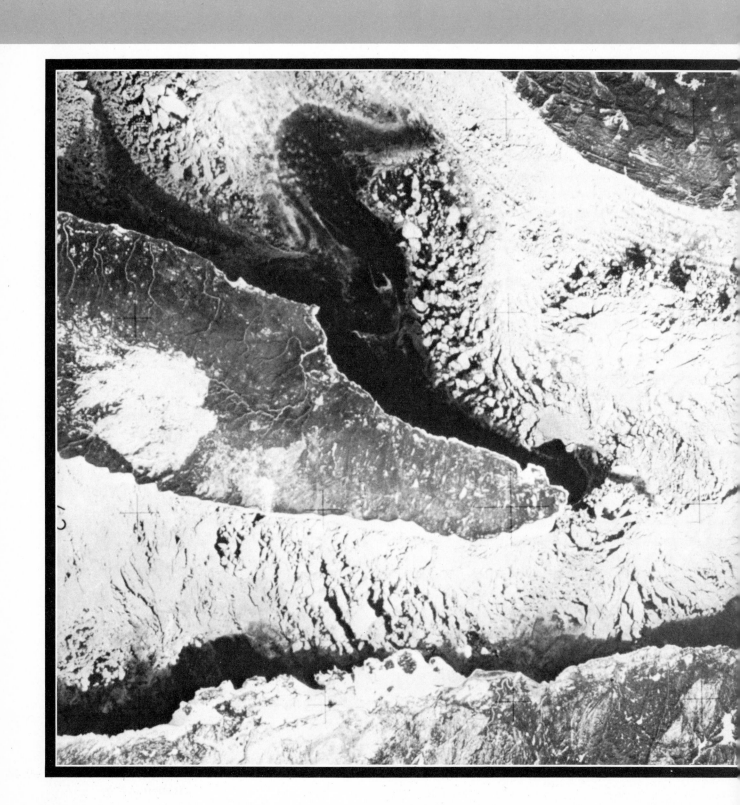

1O ESTUARIES AND THE COASTAL OCEAN

A view of the Gulf of St. Lawrence area of Canada as seen from the Skylab space station in earth orbit. The elongated island mass is the mainland of Quebec. The rounded coastline in the southwest corner of the photograph is Quebec's Gaspe Peninsula. The St. Lawrence River which drains the five North American Great Lakes empties into this body of water. Note the evidence of much ice and snow. (Photograph courtesy of the National Aeronautics and Space Administration.)

In the open ocean, where depths typically range from 4 to 6 kilometers, many processes are bounded only by the air–water interface and the current–deflecting continents. The coastal ocean, however, is only about 70 meters deep on the average. It is usually confined to continental shelf waters, but there are exceptions. Where shelves are narrow on the United States West Coast, for instance, boundary currents can define the limits of the coastal ocean. Its properties thus may extend seaward of the continental shelf break, with a portion of the continental slope serving as a floor.

The shallow bottom and nearby shorelines affect oceanic processes so markedly that coastal ocean areas act as separate oceanographic entities (Table 10-1). Like marginal seas, coastal ocean regions

TABLE 10-1

Environmental Characteristics of Some U.S. Coastal Ocean Regions*

COASTAL REGION	SHORELINE FEATURES	CONTINENTAL SHELF	BOUNDARY CURRENT
Gulf of Maine	Rocky, embayed by drowned river valleys; fjords	Rocky, irregular, 250–400 km wide	Labrador
Middle Atlantic	Smooth coast, embayed by drowned river valleys and lagoons	Smoothly sloping, 90–150 km wide	Gulf Stream
South Atlantic	Smooth, low lying, marshy; many lagoons	Smoothly sloping, 50–100 km wide	Gulf Stream
Caribbean	Irregular; mangrove swamps, coral reefs	East side 5–15 km wide; west side 300 km wide	Gulf Stream
Gulf of Mexico	Smooth, low lying; barrier islands, marshes, lagoons	Smoothly sloping, 100–240 km wide	Seasonally variable
Pacific Southwest	Mountainous, few embayments	Irregular, rugged, average 15 km wide	California
Pacific Northwest	Mountainous, few embayments, some fjords	Irregular (Oregon) to smoothly sloping (Washington), 15–60 km wide	California
Southeast Alaska	Mountainous, many embayments and fjords	Rocky and irregular in places, 50–250 km wide	Alaska
Hawaiian	Mountainous, few rivers, few embayments	None	None

*After U.S. Department of Interior, 1970. National Estuarine Pollution Study.

are partially isolated by shoreline features. In the United States, for instance, Cape Code separates the Gulf of Maine from the mid-Atlantic region and Cape Hatteras separates the mid-Atlantic and South Atlantic regions (Fig. 10-1). Each region (Table 10-2) has distinctive features.

In this chapter we examine the effects of wind and temperature changes, evaporation, and dilution on the coastal ocean.

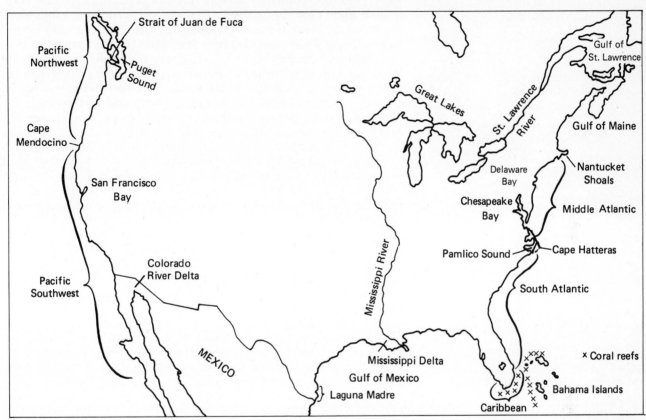

Figure 10-1

Some major North American coastal oceanic regions and major estuarine systems.

TABLE 10-2

Characteristics of U.S. Coastal Ocean Regions*

COASTAL REGION	TEMPERATURE AVERAGE (range) (°C)	SALINITY AVERAGE (range) (‰)	TIDE†	RIVER DISCHARGE	
				WATER (km³/yr)	SEDIMENT (10⁶ tons/yr)
Gulf of Maine	8 (18–0)	30.3 (31–29)	SD	64	?
Middle Atlantic	12 (23–2)	31.5 (31.8–31.3)	SD	95	15
South Atlantic	24 (30–10)	35.5 (36.0–32.5)	SD	137	58
Caribbean	27 (30–22)	35.7 (36.3–34.9)	SD	10	?
Gulf of Mexico	23 (30–12)	32.3 (37.0–30.3)	D	712	360
Pacific Southwest	15 (20–13)	33.6 (33.8–33.3)	M	21	24
Pacific Northwest	10 (13–7)	30.8 (32.5–28.5)	M	113	127
Southeast Alaska	5 (13 to −1)	31.9 (32.1–31.2)	M	?	?
Hawaiian	27 (29–23)	35.0	M	?	?

*Data from U.S. Department of Interior, 1970. National Estuarine Pollution Study.

†SD—semidiurnal tide
 M—mixed tide
 D—diurnal tide

TEMPERATURE AND SALINITY IN COASTAL OCEANS

Large changes in temperature and salinity occur in shallow, partially isolated coastal waters, well beyond anything observed in the open ocean. Such changes are partly due to the proximity of the land. Winds blowing from a continent over the coastal ocean, for instance, affect water temperatures and salinities to a marked degree. On the U.S. Atlantic Coast winds coming from the continent are much hotter than the ocean in summer and much colder in winter, and they influence coastal water temperatures. Dry, cold winds from continents, having lost much of their moisture through precipitation, cause evaporation when they blow across coastal waters. In contrast, winds blowing across the ocean toward a continent, as in the northwestern United States during winter months, have less effect on the coastal ocean.

Salinity extremes occur in coastal waters, usually in isolated embayments with little rainfall or river discharge where mixing with the ocean is severely restricted. In the western Mediterranean and northern Red Sea, for example, evaporation is high and surface-water salinities exceed the average for their latitude by more than 4‰ (see Fig. 6-14). As we know, evaporation (Fig. 6-12) occurs from all ocean surfaces. But along coasts excess freshwater is discharged from continents to be mixed back into the ocean. For this reason, the lowest ocean salinities also occur in coastal waters, near major rivers, such as the Amazon or Congo. Isohalines tend to parallel ocean boundaries, reflecting the fact that the greatest net evaporation occurs near centers of ocean basins.

Highest and lowest ocean surface temperatures occur in coastal waters. Surface-water temperatures exceed 40°C in the Arabian Gulf and in the Red Sea during the summer whereas the surface temperature of the open ocean rarely (if ever) exceeds 30°C. Again, water in such enclosed seas is restricted from mixing laterally with a larger body of water, in contrast to the open ocean, where temperature is distributed by surface currents. Secondly, the limited fetch in small ocean areas limits wave action and thereby inhibits vertical mixing. Furthermore, the intertidal zone heats up markedly and warms the water brought in on flood tides. Water ebbing from a tidal flat on a summer's day is generally much warmer than waters offshore.

Lowest surface-water temperatures are controlled by the point of initial freezing of seawater; generally this is about $-2°C$, depending on salinity (see Chapter 5). In winter, sea ice forms in high-latitude coastal areas, particularly in shallow bays and lagoons where salinities are low due to river discharge and cooling is rapid because of the large surface area and the small volume of water.

Time scales are short in the coastal ocean and warming of surface waters during the day can be detected fairly readily, especially in areas protected from winds and waves. Surface-water temperatures are highest in midafternoon and lowest at dawn, as indicated in Fig. 10-2.

Marked seasonal temperature differences in the coastal ocean occur particularly at midlatitudes, where air temperature changes are greatest. Coastal waters are mixed by storms and cold water can mix all the way to the bottom during winter. Offshore waters, even in winter, never get as cold as continental shelf waters because surface cooling is distributed throughout the water column, to great depths. Frequently a temperature front near the edge of the shelf marks the limit of coastal ocean circulation.

In spring, when surface waters are warmed, the depth of the warmed layer depends on mixing. As the summer progresses, the warmed mixed layer deepens and a pronounced thermocline develops, with marked temperature differences between surface and bottom waters (see Fig. 10-3). During autumn surface layers cool by radiation, evaporation, and mixing with deeper waters, caused pri-

Figure 10-2

Variation in near-surface water temperatures during the course of a day. At midnight (a) the surface zone is nearly isothermal. Cooling continues throughout the night so that by dawn (b) the surface layer is cooler than the waters beneath. After the sun comes up, the surface layer is warmed and by noon (c) the surface layer is distinctly warmer than waters immediately below the surface. Warming continues until late afternoon, when the surface temperature is highest. Later the surface layer begins to cool and is slightly cooler at dusk (d). During the night the surface zone cools until it is again isothermal around midnight. Daily temperature changes are usually confined to the upper 10 meters or less. [Redrawn from E. C. LaFond, 1954. Factors affecting vertical temperature gradients in the upper layers of the sea. *Scientific Monthly* 78(4): 243–253.]

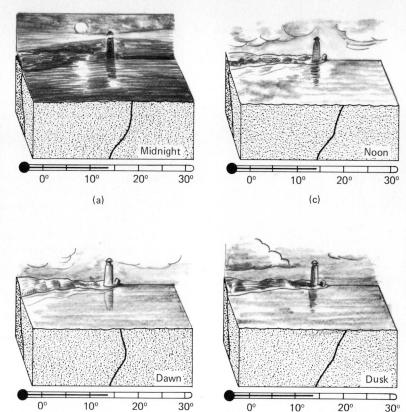

Figure 10-3

Water temperatures at three different depths at an oceanographic research tower located about 1.6 kilometers offshore from Mission Bay, California, in waters about 18 meters deep. [After J. L. Cairns and K. W. Nelson, 1970. A description of the seasonal water. *Journal of Geophysical Research* 75(6): 1127–1131.]

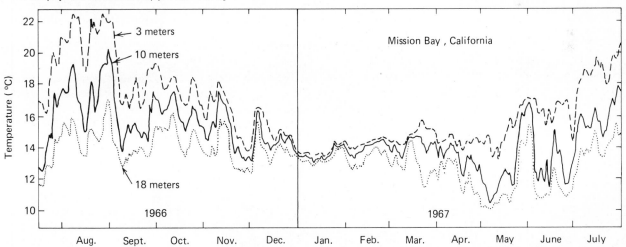

marily by storms. By the time winter begins coastal waters are thoroughly mixed, with almost no temperature or salinity differences between surface and bottom waters (note January conditions in Fig. 10-3).

COASTAL CURRENTS AND UPWELLING   Coastal currents are horizontal water movements that generally parallel the shore. They are primarily geostrophic, driven by winds (especially storms) and river discharge. Such currents may form, disappear, or change flow direction within a matter of hours or days.

Coastal currents are strongest when runoff is large and winds are strong. Along the West Coast of the United States, for example, surface waters are blown shoreward and held there by continuing winds, depressing the pycnocline. A sloping sea surface results, creating geostrophic currents that parallel the coast and are quite variable.

Coastal circulation is bounded seaward by the boundary currents that move water around the major ocean basins. Along United States coasts, nearshore currents often set in the opposite direction to the boundary current offshore. When winds diminish or river discharge is low, coastal currents weaken or disappear. Such conditions are common in late summer when coastal waters tend to form large, sluggish eddies rather than distinct coastal currents. With the onset of winter storms, these eddies disappear, the surface layers are mixed, and coastal currents are reestablished within a day or two.

Water discharged by rivers or from rain or snow eventually moves across coastal currents to mix into the open ocean. Because currents tend to parallel the coast, transfer of freshwater across the system can be slow. In other words, the *residence time* is long. For example, within the mid-Atlantic Bight between Nantucket Shoals off Cape Cod, and Cape Hatteras, North Carolina, the freshwater necessary to account for the lowered salinity is equivalent to about 2½ years of discharge from the region's rivers. In comparison, in the Bay of Fundy it takes about 3 months for freshwater to be replaced, about 3 to 4 months in Delaware Bay.

When winds move surface waters away from the coast, subsurface waters move upward to replace them, as seen in Fig. 10-4. The resulting wind-induced upwelling is common along the western sides of the continents, in the eastern boundary-current areas. Coastal upwelling occurs along the California–Oregon coast, the western coast of South America, the Somali coast of East Africa, southwestern Africa, northwestern Africa (Morocco), and the western coast of India.

Figure 10-4

Schematic representation of subsurface circulation in an area of coastal upwelling. Nearly horizontal lines show equal density surfaces. [After H. U. Sverdrup, 1938. On the process of upwelling. *Journal of Marine Research* 1(2): 155–64; and T. J. Hart and R. J. Currie, 1960. The Benguela Current. *Discovery Reports* 31: 123–298.]

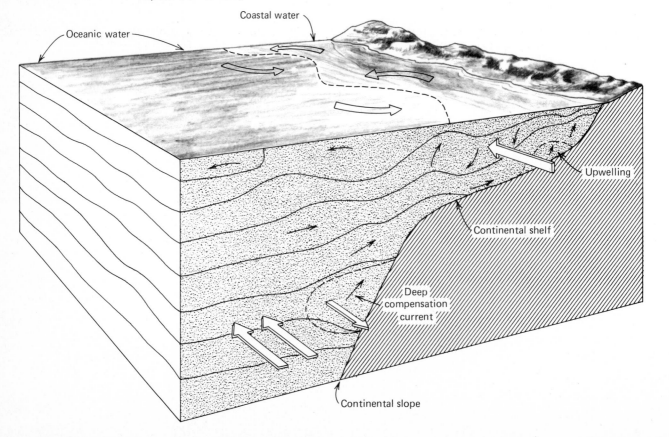

Coastal water

Oceanic water

Upwelling

Continental shelf

Deep compensation current

Continental slope

During upwelling, deeper, much colder waters are brought to the surface, usually from depths of 200 meters or less. They form cold surface or near-surface water masses near the coast and cause significant cooling of adjacent land areas (Fig. 10-5).

Upwelling is a major cause of the high biological productivity of coastal regions. Phosphate and nitrate ions, depleted in sunlit surface waters by phytoplankton growth, move vertically during upwelling at a rate of about 2 meters per day. In doing so, they resupply food for active plant growth, which, in turn, feeds small animals and eventually fishes. Areas of pronounced upwelling have photosynthetic-carbon production rates from two to ten times that of average open ocean waters. In fact, a map of upwelling regions nicely outlines the most biologically productive ocean areas (see Fig. 12-22) and the

Figure 10-5

Upwelled water off the California–Oregon coast shows up as light-colored swirls in this infrared satellite photo taken in September 1974. Puget Sound is at the top and San Francisco Bay is near the bottom. (Photograph courtesy National Oceanic and Atmospheric Administration.)

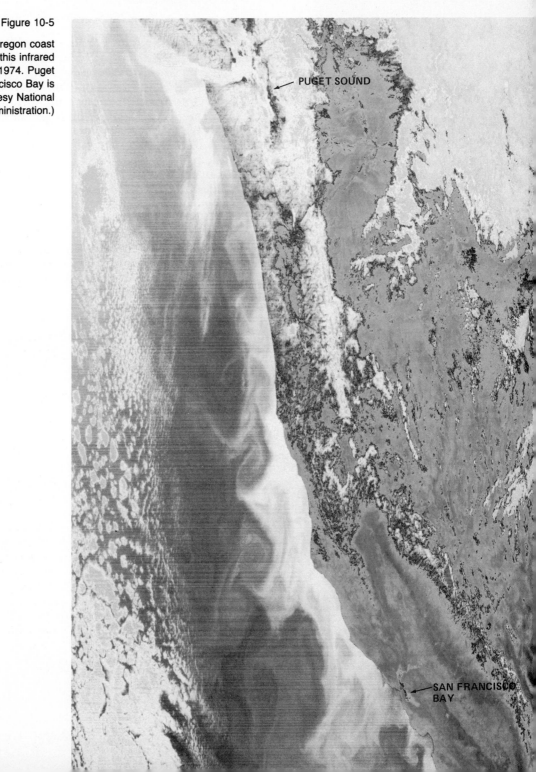

world's richest fishing grounds. More than half the world's fish production comes from upwelling areas.

The cooling effects of upwelling are most noticeable along arid coasts in subtropical areas, such as southern California. There little rainfall or river runoff is present to form a stable, low-salinity surface layer with a pycnocline that inhibits denser waters from coming to the surface. But near small rivers, as along the Oregon coast, upwelling occurs when winds move the low-salinity surface layer seaward and the cold, upwelled waters underneath are exposed. Along the Washington State coast, upward movement of subsurface water also occurs; but an extensive low-salinity surface layer formed by the discharge of many large rivers prevents the deeper waters from reaching the surface. For this reason, upwelling is not readily discernible along the Washington coast.

ORIGIN OF ESTUARIES, FJORDS, AND LAGOONS

Coastal oceans and estuaries formed during the past 20,000 years as sea level rose from the continental shelf break to its present level, following the most recent glacial stage. The sea has been near its present level for only about 3000 years and coastal ocean features are still adjusting as sea level continues to rise at the rate of about 15 centimeters per century (6 inches per century).

We can observe coastal evolution by comparing the recently glaciated and hence relatively young Canadian and New England coasts with regions farther south. On New England and northwest U.S. coasts, glaciers removed the soil cover and carved deep *fjords*. The ancient continental margin was partly destroyed, leaving a narrow, irregular continental shelf. These regions have remained almost unaltered for thousands of years.

In the mid-Atlantic region, beyond the reach of glaciers in the last ice age, there are *coastal plain estuaries* where the drowned mouths of ancient rivers are now below sea level. Southern coastlines tend to have a wide coastal plain from which abundant sediment is discharged to estuaries and the nearshore ocean. Shallow, lagoon-type, bar-built estuaries and deltas and smooth, heavily sedimented coastlines are the rule, both above sea level and on the continental shelf. On the Pacific Coast, where mountain building has been active within recent times, coastal plain estuaries or shallow bays *(tectonic estuaries)* are cut off from the ocean by young mountain ranges. Characteristics of major estuarine types are summarized in Table 10-3.

From the Strait of Juan de Fuca northward, the Pacific coast has been deeply eroded by glaciers. This erosion formed narrow, deep, steepsided fjords and fjordlike estuaries along the Canadian and Alaskan coasts, like those in Fig. 10-6, perhaps the most spectacular of all estuarine systems. Fjords occur in other high-latitude coastal areas, such as Norway and Tierra del Fuego in South America. Glaciers moving down mountain valleys carved gorges hundreds of meters deep and often deposited boulders and clay at the mouths of the valleys. This glacial debris forms an underwater sill at the entrance to many fjords, restricting movement of bottom waters.

Coastal plain estuaries assumed more or less their present size and shape about 3000 years ago, when rapid melting of glaciers ceased and the sea reached its present level. About 30,000 years earlier, these rivers entered the ocean through ancient estuaries that now lie tens of kilometers offshore and 100 or more meters below present sea level. Coastal plain estuaries are typically broader than any part of the river upstream and they gradually deepen seaward except where mud and sand have been deposited. Some, such as the Hudson or Columbia rivers, have associated submarine canyons extending across the former coastal plain, now the submerged continental shelf.

TABLE 10-3

Characteristics of Some North American Estuarine Systems

ESTUARINE SYSTEMS	ESTUARY AREA (square kilometers)	ESTUARY VOLUME (cubic kilometers)	MEAN WATER DEPTH* (meters)	ANNUAL FRESH WATER DISCHARGE (cubic kilometers per year)	LAND AREA DRAINED (thousands of square kilometers)	MAJOR RIVERS
DROWNED RIVER VALLEYS						
Chesapeake Bay System						
(Maryland–Virginia)	11,000	67	6.1	65	110	Susquehanna
Potomac River	1280	7.3	5.7	12	36	Potomac
James River	650	2.3	3.5	10	26	James
Raritan Bay						
(New York–New Jersey)	230	1.1	4.5	1.8	3.4	Raritan
New York Harbor	159	1.2	7.5	19.4	34.7	Hudson
Long Island Sound						
(New York–Connecticut)	3180	62	19.4	21	40.7	Connecticut Housatonic
LAGOONS						
Pamlico–Albemarle Sound						
(North Carolina)	6630	23.9	3.6	7.8	51	Neuse Pamlico
Laguna Madre						
(Texas)	158	1.1	0.9	−0.85‡	nd†	
FJORDLIKE						
Strait of Juan de Fuca						
(Washington–British Columbia)	4370	490	112	nd	nd	
Puget Sound						
(Washington)	2640	185	70	36.5	37.6	Skagit Snohomish
Strait of Georgia						
(British Columbia)	6900	1025	156	145	270	Fraser
TECTONIC						
San Francisco Bay						
(California)	1190	6.2	5	40	161	Sacramento San Joaquin

*Mean depth = volume/area.
†nd = no data.
‡Evaporation exceeds river runoff plus rainfall.

Figure 10-6

The rugged topography of a small inlet on Canon Fjord, Ellesmere Island, in the Canadian Arctic was cut by a glacier, a remanent of which (the large, white mass) can be seen at the upper left. The icebergs in the water come from such glaciers. Note the small floes of sea ice in the foreground. (Photograph courtesy National Film Board of Canada.)

A *lagoon* is a broad, shallow estuarine system. Flow of water between estuary and coastal ocean is restricted by a barrier beach offshore, generally paralleling the shoreline. Lagoons are invariably shallow because barrier beaches form only in waters a few meters deep where wave action suspends sediment and transports it along the coast. Barrier beaches enclosing a lagoon are interrupted by narrow inlets through which tidal currents move water in and out of the lagoon (Fig. 10-7).

On the U.S. Gulf Coast and on the Atlantic Coast south of New England, long stretches of barrier beach separate lagoons from the coastal ocean. Often a single stretch of beach isolates several lagoons and estuaries. All parts of such a system communicate with the coastal ocean through the same set of narrow inlets.

In New England and the Pacific Northwest, where the coastline has been scoured by glaciers, lack of sediment moving along the coast prevents formation of extensive bars or spits. On much of the Pacific coast the continental shelf is too narrow and steep to permit barrier beaches to form.

When all inlets to a lagoon are closed by storms or changes in wave or sediment conditions, the lagoon is transformed into a lake. Freshwater flowing into such a lake from local rivers would raise its level until it overflowed its boundaries if not for the fact that sandy beaches are permeable to water. Seepage through barrier beaches keeps lake levels fairly constant.

Estuaries and lagoons make up 80 to 90% of the Atlantic and Gulf coasts but only 10 to 20% of the Pacific. Mountain building around the Pacific has left little low-lying coastal plain and few rivers have cut through the mountains to reach the sea. Furthermore, the deserts behind mountain ranges in this region yield little water; many areas discharge no water to the ocean. The large tectonic estuarine systems of San Francisco Bay and the Strait of Juan de Fuca (Fig. 10-1) formed when former river valleys sank below sea level because of mountain building.

Figure 10-7

Pamlico Sound, North Carolina, is formed by barrier islands (the light-colored narrow strips) partially isolating drowned river valleys from the adjacent coastal ocean. Note the plumes of turbid water coming out of the inlets and the cloudlike water masses moving along the coast. The puffy white masses at right are clouds. (Photograph courtesy NASA.)

ESTUARINE CIRCULATION

Estuaries are semienclosed arms of the ocean, where freshwater from the land mixes with seawater. Their characteristic circulation patterns are governed by the amount and rate of freshwater entering the estuary, the size and shape of the basin, and the effects of winds and tides. In an ideal estuary, low-salinity water from the river flows in at the head of the bay and spreads out over the seawater beyond because it is less dense (Fig. 10-8). There is a more-or-less horizontal pycnocline zone with a marked density discontinuity separating the two layers.

This oversimplified picture is modified by several factors. First, friction between seaward-moving freshwater and the underlying seawater causes currents, as shown in Fig. 10-9. Water is entrained (dragged up from below) and mixed with the surface layer. This newly mixed water cannot reenter the lower seawater layer because its reduced salinity makes it less dense. Consequently, ocean water flows into the estuary along the bottom to replace that drawn seaward by the surface flow. Each volume of freshwater ultimately mixes with several volumes of seawater and surface salinities increase in a seaward direction. Landward flow along the bottom is much greater in volume than the flow of river water into the estuary. The subsurface saltwater in such an estuary often forms a wedge with its thin end pointed upstream; so in its simplest form this is called a *salt-wedge estuary*.

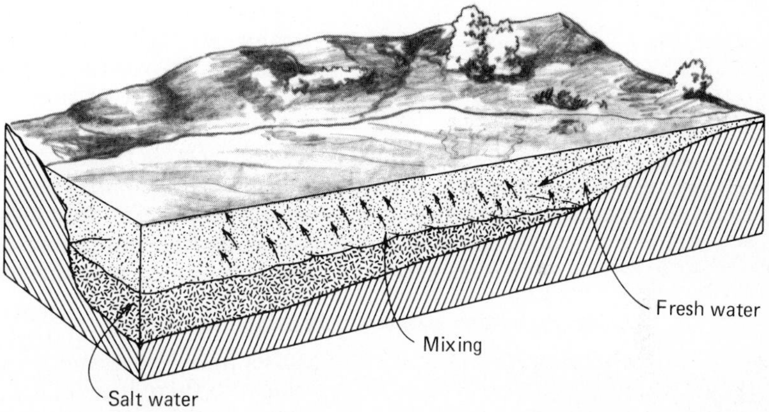

Figure 10-8

Schematic representation of a simple salt-wedge estuary showing the two-layered structure with a landward flow in the salty subsurface layer and a seaward flow in the less saline surface layer. [Redrawn from D. W. Pritchard, 1955. Estuarine circulation patterns. *American Society of Civil Engineers Proceedings* 81(717): 11 pp.]

Fresh water

Mixing

Salt water

Figure 10-9

Variation in salinity and current speed with depth in a moderately stratified estuary.

Moderately stratified estuary

Seaward   Landward

Depth

Salinity
(parts per thousand)

Velocity

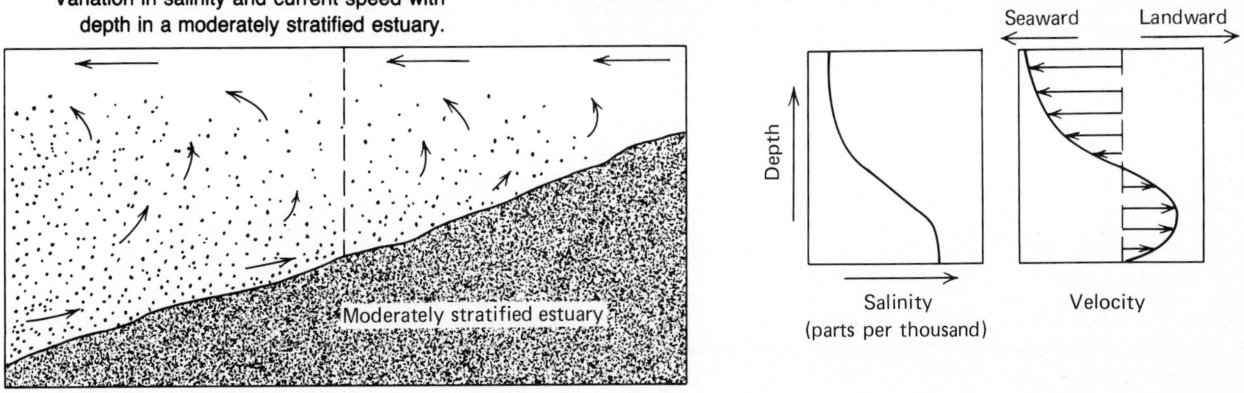

If the frictional drag between the two layers is great enough, internal waves form on the pycnocline. Such waves at the boundary between the two layers can break, like the surf seen at the beach. This process greatly increases the rate of mixing between the layers. The more mixing that occurs in the estuary, the greater is the landward flow in the subsurface layer over the continental shelf (see Fig. 10-10).

271

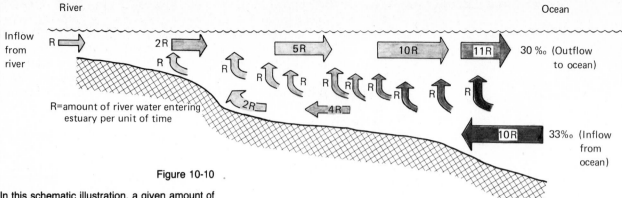

River   Ocean

Inflow from river

R=amount of river water entering estuary per unit of time

30 ‰ (Outflow to ocean)

33‰ (Inflow from ocean)

Figure 10-10

In this schematic illustration, a given amount of river water, designated as R, draws 10 times as much seawater upward from below and carries it downstream, toward the ocean. To replace the outflow, 10 R amount of seawater moves into the estuary. Thus the effect of estuarine circulation is to draw subsurface, high-salinity water into the surface layer. Note that it is essentially an upwelling process.

Simplified salt-wedge stratification, such as described, is rarely observed in estuaries. Certain systems closely approximate it, however, especially where river flow is large and tidal range is low. Salt-wedge estuaries are found where a large river discharges through a relatively narrow channel, such as the lower passes of the Mississippi River or the Columbia River during floods. Forces associated with large river flows may be much stronger than the tides so that they dominate the estuarine circulation. A nearly ideal salt-wedge estuary is formed, one in which relatively little saltwater is mixed into the upper layer. Nearly freshwater is discharged into the coastal ocean, where it mixes with ocean water over the continental shelf. The Amazon is a particularly striking example of such a river.

As river discharge into an estuary decreases, the importance of tidal effects increases. A single estuary may act like a salt-wedge estuary during floods, but during low-flow periods of the year it is usually influenced much more by tides and tidal currents, departing substantially from the pattern of a simple salt-wedge stratification. It then becomes a *moderately stratified estuary*.

Tidal currents cause mixing between layers so that the waters become only moderately stratified, as Fig. 10-11 illustrates. Tidal flow causes *turbulence* (random movements) throughout the water column, which, in turn, increases mixing and so more saltwater is transferred from the subsurface to the surface layer. Some freshwater from the surface also mixes downward. Consequently, salinity decreases in a landward direction in both the surface and subsurface layers. As in the salt-wedge estuary, however, there is a net landward flow in the subsurface layer, replacing saltwater lost from the system, and a net seaward flow in the surface layer, removing both freshwater and saltwater from the estuary, as Fig. 10-12 indicates.

Figure 10-11

Schematic representation of water movements in a moderately stratified estuary. There is a net landward movement in the subsurface layers and a seaward flow in the surface layers, although the difference between the layers is not as pronounced as in the salt-wedge estuary. (After Pritchard, 1955.)

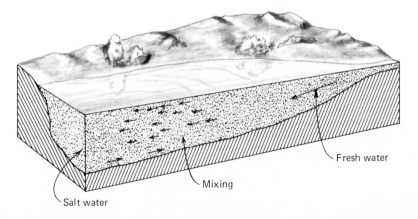

Fresh water

Mixing

Salt water

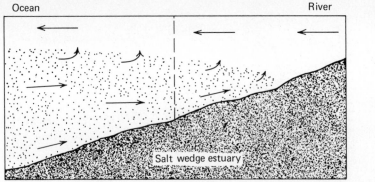

Figure 10-12

Variation in salinity and current speed in a salt-. wedge estuary.

Consider the volume of flow in these two layers. River water has almost no salt as it enters the head of the estuary, whereas seawater moving in along the bottom has a salinity of 35‰. A mixture of equal volumes of fresh and saltwater gives a salinity of 17.5‰; mixing 2 volumes of saltwater with 1 volume of river water results in a salinity of about 23.4‰; as shown by Fig. 10-13. By the time surface-water salinity has reached 34‰, 1 volume of river water has mixed with 39 volumes of seawater. In other words, at this point, landward flow in the subsurface layer is 39 times the volume of river water discharged at the head of the estuary. The volume of surface water moving seaward in the surface layer will be 40 times the original river discharge (see Fig. 10-10).

Because of their greater depth, fjords have more complicated circulation systems. Many fjords have a *sill* or submerged ridge at the entrance deposited by the glacier that cut the fjord. This sill cuts off most of the deeper water from communication with the adjacent ocean, as can be seen in Fig. 10-14. Alberni Inlet on Vancouver Island, British Columbia, for example, has a sill that comes to within 40 meters of the surface. The basin behind the sill is 330 meters deep and so nearly seven-eighths of its volume is partially isolated from communication with the ocean.

Figure 10-13

Relationship between salinity of water masses formed in an estuary and the relative amounts of seawater (salinity 35‰) and freshwater (salinity 0‰) involved in the final mixture.

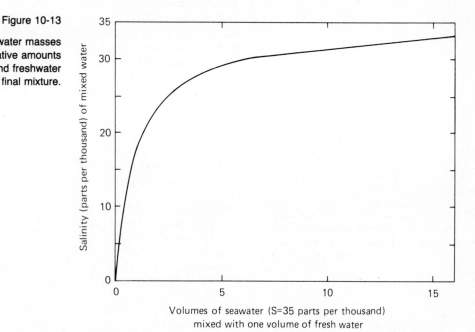

Figure 10-14

Schematic representation of circulation in a simple fjordlike estuary. Note the water mass formed by mixing surface and subsurface waters while flowing over the sill. [After M. Waldichuk, 1957. Physical oceanography of the Strait of Georgia, British Columbia. *Journal of the Fisheries Research Board of Canada* 14(3): 321–386.]

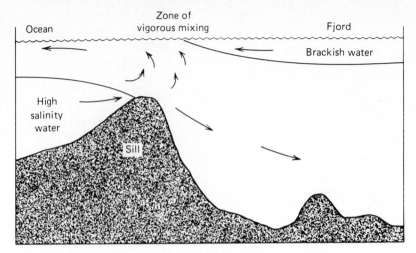

Freshwater flowing into the fjord forms a low-salinity surface layer that moves seaward, in a typical estuarine circulation. This layer, however, often extends no more than a few tens of meters below the surface and usually involves only the waters above the top of the sill.

The deeper waters of a fjord are almost completely isolated from the surface circulation but are strongly affected by conditions at the fjord mouth, outside the estuary. Since river flow, tides, and winds have little effect on this deep water, its circulation is controlled primarily by density differences. If the water at or slightly above the sill outside the estuary becomes more dense, it can flow into the fjord and displace the deeper water there, which then moves out slowly. Strong winds can affect the deep waters by setting up seiches (standing waves), which cause the waters to "slosh" back and forth.

Tides dominate the currents in most estuaries. Any current seen in an estuary is probably of tidal origin. To observe estuarine circulation, it is necessary to make current measurements extending over many tidal cycles. During this period water flows upstream on the flood current a distance nearly equal to its downstream travel on the ebb current. When averaged over many tidal cycles, tidal currents cancel one another, leaving a nontidal estuarine circulation. A current meter placed in the surface layers would indicate a net flow seaward, one in the deeper layers a net flow landward.

Where tidal effects are relatively strong, the estuary tends to be less stratified. In an estuary with small river flow but large tides and tidal currents, the waters may be mixed almost completely from top to bottom.

The rise and fall of tides controls the timing of river water discharge into the ocean. Most freshwater is discharged on the ebb tide and enters the ocean as a series of pulses. River discharge forms cloudlike parcels of low-salinity water, which, when turbid with suspended sediment, are easily recognizable from the air (Fig. 10-15). Each tends to overrun the previous one, forming a complex of irregular fronts (Fig. 10-16). As the cloudlike masses move away from the river mouth, waves and tidal currents continue the mixing over the continental shelf. Boundaries are obscured and eventually a large lens or *plume* of low-salinity water forms. Such a plume can be traced for many tens or hundreds of kilometers off the mouths of major rivers (Fig. 10-17), especially where there are no other large rivers discharging in the area.

Having examined the basic processes acting in estuaries and the coastal ocean, let us consider some examples.

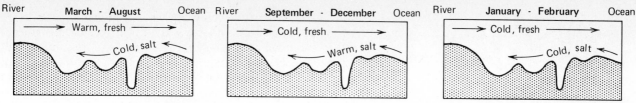

Figure 10-21

Schematic representation of changes in temperature relationships between surface and bottom waters of Chesapeake Bay. (After R. C. Seitz, 1971. *Temperature and Salinity Distribution in Vertical Sections along the Longitudinal Axis and Across the Entrance of the Chesapeake Bay.* Chesapeake Bay Institute, The Johns Hopkins University, Baltimore, Md. p. 25.)

often freeze during January and February, with ice forming first in shallow, protected bays and inlets and then gradually extending toward more open water. In especially cold winters ice covers large areas from shore to shore in the upper estuary.

Strong winds have a pronounced effect on the Bay. In addition to the waves they generate, storm winds can also change sea level locally, sometimes causing flooding of low-lying coastal areas. Winds can totally change circulation in Chesapeake Bay, but currents set up by river discharge and tides are usually dominant influences in the absence of strong winds.

LAGUNA MADRE

One of the few *hypersaline* (above-average salinity) *lagoons* in the United States is Laguna Madre, a series of long, narrow lagoons on the Texas coastal plain between the Rio Grande and Corpus Christi (Fig. 10-22). A sandy barrier island (Padre Island) parallels the coast for approximately 160 kilometers and separates the lagoon from the Gulf of Mexico. Circulation in this lagoon is quite different from the typical estuarine system. At each end of Padre Island narrow inlets connect Laguna Madre with the ocean. Before being protected by jetties, they were frequently closed and reopened by storms and longshore currents.

A hurricane in 1919 formed large shallow areas, which now divide the lagoon into two smaller basins. The deeper basin, Baffin Bay at the northern end, averages about 1.6 meters. Much of Laguna Madre is extremely shallow (average depth 0.9 meter); many areas are covered by only a few centimeters of water at high tide.

Because the inlets at each end are quite narrow, there is relatively little tidal interchange with the open ocean. Also, this section of the Gulf of Mexico has a tidal range of less than 0.5 meter. Consequently, the tidal range in Laguna Madre is only a few centimeters. Wind effects are usually more important than the tide. Wind-caused changes in water level are predicted to exceed 1 meter, based on observed wind velocities.

Winds also control the currents in Laguna Madre. Winds from the north cause currents to set southward, with flow out of the southern inlet. From the south they cause essentially the opposite current direction, as indicated in Fig. 10-22. Note that the shallow areas in the center of Laguna Madre greatly influence currents set up by winds from the south. Back-and-forth seichelike water movements are sometimes set up in the lagoon by winds blowing steadily from a single direction. These winds stir up sediments and so lagoon waters are often turbid.

Laguna Madre is located in a semiarid region where evaporation usually exceeds rainfall. Runoff comes from a few small streams or irrigation ditches. Infrequent but intense local storms may bring more than 10 centimeters of rain in a few hours, providing large amounts of freshwater. Between storms, salinity of waters in the lagoon is usually substantially higher than in the adjacent ocean (Fig. 10-23) and may exceed 80‰. Such erratic freshwater inputs cause extreme salinity variations. The highest salinities recorded are between 110

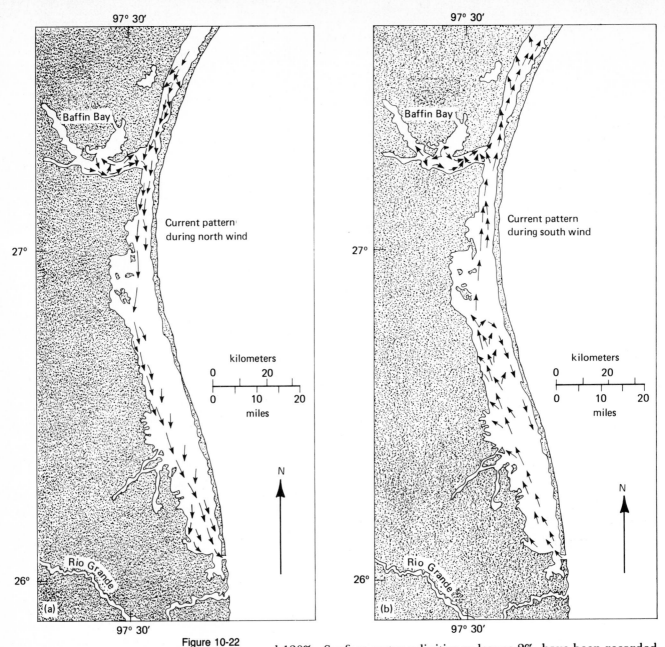

Figure 10-22

Postulated current patterns in Laguna Madre during north (a) and south (b) winds. (After G. A. Rusnak, 1960. Sediments of Laguna Madre, Texas. In F. P. Shepard et al. (Eds.), *Recent Sediments, Northwest Gulf of Mexico,* pp. 82–108. American Association of Petroleum Geologists, Tulsa, Okla.)

and 120‰. Surface-water salinities as low as 2‰ have been recorded immediately following a cloudburst, but within a relatively short time salinity is increased by evaporation and mixing.

Surface-water temperatures are also extremely variable. In winter cold air from the north chills surface waters. Because of the extreme cooling and lack of escape channels for fish in shallow bays, mass mortalities occur among temperature-sensitive fish during unusually cold weather. Substantial fish mortalities *(fish kills)* occur in summer when water temperatures and salinities become intolerable to some species. Dredging of ship channels and inlets has apparently reduced fish mortality by providing escape channels and shelter from the highly variable surface water conditions.

Vigorous wind mixing permits relatively little stratification within this shallow bay. Near inlets, mixing is caused by tidal currents. Excess evaporation sets up a circulation system opposite to the estuarine pattern. The lowest-salinity water comes from the adjacent ocean and thus occurs at the surface. As it flows into the lagoon, ocean water evaporates and sinks with the increased density. Salty, denser subsurface waters flow out into the ocean.

Dredging of channels and construction of tidal inlets, jetties, and other engineering works have modified water circulation and biology in Laguna Madre. The estuary was substantially altered in 1949 when dredging of the Intra-Coastal Canal was completed. This channel increased water exchange between the northern and southern portions of the lagoon and prevented the near drying up of one or the other of these basins, as had happened before dredging. A bridge and solid-fill causeway at the northern end of the lagoon have greatly restricted passage of water from one side to the other.

Laguna Madre is highly productive of fish. In the early 1940s slightly more than half the total fish production for Texas was taken there. In part, it was a result of the deterioration of such nearby areas as Galveston Bay, once an excellent fishing ground. There is also some evidence that the high salinity of the waters causes certain fish to grow larger.

Bay waters receive wastes discharged by urbanized and industrialized areas, as well as runoff from agricultural land, which brings silt, pesticides, and excess fertilizer into the water. Declines in Laguna Madre fish production during the 1960s were ascribed to the cumulative effects of these waste discharges, especially pesticides.

NORTHEAST PACIFIC COASTAL OCEAN

High rainfall and river runoff make the entire North Pacific Ocean an area of net dilution, with a margin of especially low-salinity water bordering its coasts (see Fig. 6-14). These conditions cause a basically estuarine circulation north of 40°N, with net flow of surface water toward the center of the basin.

Flowing nearly due east from Asia, the high-latitude Subarctic Current (Fig. 7-3) brings relatively cool water of 33 to 34 parts per thousand salinity (see Fig. 6-14) toward the North American continent. There it is deflected to form the eastern boundary currents— Alaska Current setting northward and California Current setting southward parallel to the continental margin. The point of divergence shifts seasonally; it is about 50°N in summer and about 43°N in winter.

Because current speeds are low and flows rather diffuse, no sharp boundary separates the coastal current system from the eastern boundary current, in contrast to the situation prevailing in the Gulf Stream system along the Atlantic Coast. Along the Pacific Northwest coast of the United States, the 32.5‰ isohaline is taken to separate the low-salinity coastal waters from the California Current offshore.

During the rainy winters an almost continuous band of low-salinity water parallels the coast from central Oregon to the Strait of Juan de Fuca (see Fig. 10-1). In summer and autumn, however, the Columbia River is the largest source of freshwater along the entire West Coast of the United States. Relatively little mixing occurs in its estuary so that these waters are still rather fresh as they enter the ocean. The effluent forms a distinctive water mass, the Columbia River plume, having salinity less than 32.5‰.

Within 50 kilometers of the river mouth, the plume is mixed by tidal currents and waves; farther offshore, currents and regional winds move it parallel to the coast much as a column of smoke is moved in the air by gentle winds. During the winter it is held near the coast by southwesterly winds (Fig. 10-24), forming a band of north-ward-moving, low-salinity water derived from all the coastal rivers— the *Davidson Current.* This coastal current is recognizable from southern California to the Strait of Juan de Fuca at the U.S.–Canadian border.

From July until October Columbia River discharge forms a low-

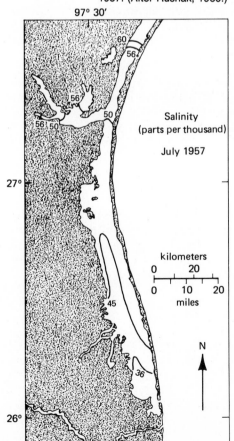

Figure 10-23

Surface-water salinity in Laguna Madre, July 1957. (After Rusnak, 1960.)

Figure 10-24

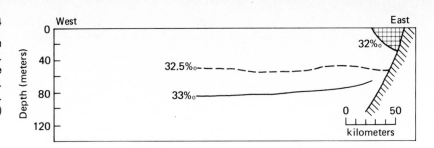

Distribution of salinity with depth near the mouth
of the Columbia River, January–February 1962.
Note that the low-salinity waters (shown by the
cross hatching) are held close to the coast.
Location of the profile is shown in Fig. 10-27.
(After Duxbury et al., 1966.)

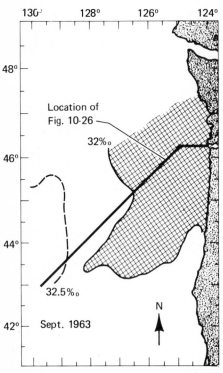

Figure 10-25

Distribution of surface salinity in September
1963. Note the large area covered by surface
water with salinity less than 32‰, derived
primarily from the Columbia River. [After T. F.
Budinger, L. K. Coachman, and A. C. Barnes,
1964. *Columbia River Effluent in the Northeast
Pacific Ocean, 1961, 1962; Selected Aspects
of Physical Oceanography.* University of
Washington, Department of Oceanography
Technical Rept. No. 99 (ref. M63–18), 78 pp.]

salinity plume extending southwestward (Fig. 10-25) as a layer about
30 meters thick (Fig. 10-26). In October 1961, for instance, it covered
an area of 200,000 square kilometers, extending 800 kilometers southward. Following major summer floods, the plume may be even larger.
(See Fig. 6-14, where it appears as a southward dip of the 33 ‰
contour in the East Pacific.) It persists until the first winter storm,
usually in October, after which surface waters are rapidly mixed and
the plume exists only near the river mouth (Fig. 10-27).

Over the continental shelf, the landward flow of subhalocline
waters (part of the estuarinelike circulation) produces near-bottom
currents; these currents are studied using *seabed drifters*—plastic devices (usually brightly colored for ease in spotting) designed to float
near the ocean bottom. Drifters are dropped from ships or low-flying
aircraft in groups held together by a "collar" made of rock salt, which
dissolves after its weight has dragged the group of drifters to the
bottom. Position and time of release of each group are noted, as well
as the serial number of each drifter.

Drifters are recovered by beachcombers or fishermen. For a
small reward, the finder sends the data and location of recovery with
the serial number of the drifter to the laboratory that released it. The
apparent direction and distance traveled by the drifter and its apparent speed are calculated by comparing release and recovery points
and elapsed time.

Recovery of seabed drifters released on the Washington–Oregon
coast continental shelf indicates a net northerly movement of near-
bottom waters at depths exceeding 50 meters. In waters less than 50
meters deep, near-bottom currents move toward the coast. This feature appears to be a result of near-bottom water movements and wind-
induced upwelling in the summer (Fig. 10-28).

Estuarine circulation causes net landward movement of near-
bottom waters toward the Columbia River within about 10 kilometers
of its mouth. A similar situation prevails within 70 kilometers of the
Strait of Juan de Fuca. Less water is discharged from that system, but
the landward subhalocline flow is from 5 to 20 times greater due to
extensive mixing in the large, silled Strait of Juan de Fuca–Strait of
Georgia system. More seawater is carried out in the surface layers for
each volume of freshwater entering the estuary and therefore more
seawater must come in to replace it. Strong currents are set up by the
flow of large amounts of water through the strait and the associated
near-bottom currents affect a large part of the continental shelf.

Figure 10-26

Distribution of salinity with depth near the
Columbia River mouth, September 1963.
Location of profile is shown in Fig. 10-25.
Waters with salinities less than 32‰ form a
relatively thin layer (less than 30 meters thick.)

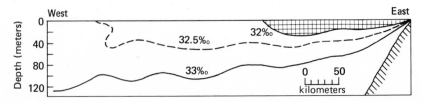

In the Columbia River estuary the relatively large amounts of freshwater discharged through the narrow, deep-river mouth preclude much mixing in the estuary itself. Most of the initial mixing occurs over a relatively large area outside the estuary, setting up an ill-defined estuarine circulation. Near-bottom currents associated with this weak circulation are thus observed within only about 10 kilometers of the river mouth.

Despite the relatively few people living along the Washington–Oregon coast, the harbors and the coastal ocean itself have been subject to contamination. Waste chemicals and paper mill wastes cause locally severe pollution problems. Indeed, marine life in some of the harbors has been nearly destroyed, for short periods, by large accidental releases of wastes.

This particular coastal area is unique because of the large volume of radioactive wastes that it received. From the 1940s to the early 1970s plutonium-producing reactors on the Columbia River about 600 kilometers upstream from the river mouth released large volumes of slightly radioactive wastes to the river. These radionuclides came primarily through corrosion in the pipes used to cool the reactors. The warm and radioactive cooling waters were discharged to the river. In 1957 these waste discharges accounted for more than 95% of all radioactivity released to the environment by the Western world, excluding fallout from nuclear weapons testing in the atmosphere. These radioactive substances were used to trace the movement of the Columbia River plume and the sediment from the river.

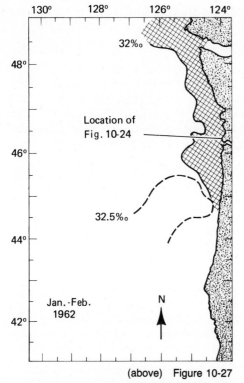

(above)   Figure 10-27

Distribution of surface salinity in the Pacific Northwest, January–February 1962. Note that low-salinity surface waters form a narrow band along the coast. [After A. C. Duxbury, B. A. Morse, and N. McGary, 1966. *The Columbia River Effluent and its Distribution at Sea, 1961–1963.* University of Washington, Department of Oceanography Technical Rept. No. 156 (ref. M66–31), 105 pp.]

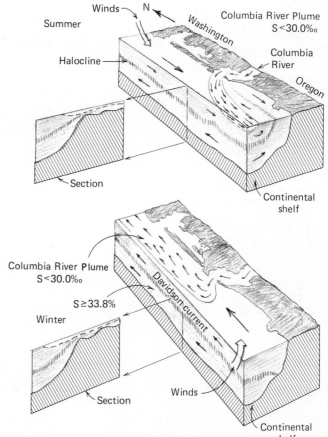

(right)   Figure 10-28

Block diagrams showing generalized surface and subsurface circulation in the Pacific Northwest region, near the mouth of the Columbia River in winter and summer. Compare the surface and subsurface currents for the two seasons. Also note the effects of upwelling along the Oregon coast in summer and the downwelling in winter.

REVIEW QUESTIONS

1. Explain why temperature and salinity variations are greater in coastal ocean areas than in the open ocean.

2. Describe the daily cycle of surface-water temperature in near-shore waters.

3. What are the principal causes of coastal currents?

4. Describe coastal upwelling and the processes that cause it.

5. What are the four types of estuaries? Discuss the formation of each type.

6. Describe the net, nontidal estuarine circulation. What are the principal factors causing the circulation?

7. Describe the annual cycle of temperature and salinity in the Chesapeake Bay.

8. Contrast the circulation in Laguna Madre to the normal estuarine circulation.

9. Discuss the annual changes in coastal currents off the Oregon–Washington coast. What causes the changes in the surface currents and the upwelling?

SUMMARY OUTLINE

Coastal ocean—average 70 meters deep, generally above continental shelf
    Coastal oceanic regions bounded by shoreline features

Temperature and salinity
    Winds from continent evaporate, heat, cool
    Extremes of temperature, salinity
        Highest temperatures, salinities in enclosed basins
        Lowest salinities near rivers
        Rapid cooling, sea ice, in shallow areas
        Large seasonal temperature changes

Coastal currents and upwelling
    Geostrophic currents set up by winds, river discharge
    Boundary currents are outer margin
    Upwelling on western sides of continents, especially caused by winds, currents
    Brings deep water to surface; important for productivity

Origin of four types of estuaries
    Fjords—northern coasts altered by glaciers
    Coastal plain estuaries—drowned river mouths
    Lagoons—shallow; sedimentation forms barrier islands, with inlets
    Tectonic estuaries—mountain-building regions; estuaries separated from continent

Estuarine circulation—freshwater mixes with seawater
    Salt-wedge estuary
        Less saline upper layer from river; denser deep layer from ocean
        Seawater entrained by freshwater flowing out
        Deep, denser water circulates toward surface; a type of upwelling

Moderately stratified estuary—greater volume of seawater circulated
    Tides, tidal currents cause mixing
    Relatively less river flow, less well-defined pycnocline
    Fjords—sill at mouth, marked density stratification
    Tides usually dominate estuarine circulation
    River discharge enters coastal ocean with tidal pulses, forms low-salinity plume

Chesapeake Bay—coastal plain estuary, many tributary rivers
    Volume of estuary discharged annually
    Reversing tidal currents
    Salinity varies with depth
    Temperature distribution controlled by local weather because of shallowness

Laguna Madre—a hypersaline lagoon
    Shallow, small tidal range
    Winds control currents
    Net evaporation except after storms; little density stratification
    Reverse of estuarine circulation—more saline water flows toward ocean

Northeast Pacific coastal ocean—area of net dilution
    Estuarine-type circulation due to heavy river discharge
    Boundary currents—Alaska setting northward; California setting southward
        Point of divergence shifts seasonally
        Boundary with coastal currents diffuse
    Low-salinity Columbia River plume
        Forms northward-setting Davidson current—winter
        Moves southwestward—summer

Generally northward near-bottom currents studied using seabed drifters
Net landward bottom-water flow near coast caused by estuarinelike circulation

Strait of Juan de Fuca—estuarine circulation; extensive mixing in estuary
Columbia River—heavy river flow; ill-defined estuarine circulation outside estuary

SELECTED REFERENCES

CONOMOS, T. J. (Ed.). 1979. *San Francisco Bay: The Urbanized Estuary.* Pacific Division, American Association for the Advancement of Science. San Francisco, CA. 493 pp. Collection of technical papers on San Francisco Bay, how it functions and how it has been altered by human activities.

GROSS, M. G. (Ed.). 1976. *Middle Atlantic Continental Shelf and the New York Bight.* American Society of Limnology and Oceanography, Special Symposium Volume 2. Allen Press, Lawrence, Kansas. 441 pp. Collection of technical papers on the effects of human activities on the coastal ocean off the middle Atlantic portion of the United States.

KETCHUM, B. H. 1972. *The Water's Edge: Critical Problems of the Coastal Zone.* MIT Press, Cambridge, MA. 393 pp. Studies of management of coastal zone resources and man's impact on coastal ocean.

LAUFF, G. H., (Ed.) 1967. *Estuaries.* American Association for the Advancement of Science, Washington, D.C. Publ. 83, 757 pp. Collection of technical papers on many aspects of estuaries and coastal ocean processes, emphasizing biology.

PICKARD, G. L. 1975. *Descriptive Physical Oceanography,* 2nd ed. Pergamon Press, New York, 214 pp. Elementary treatment of general oceanography, including coastal processes.

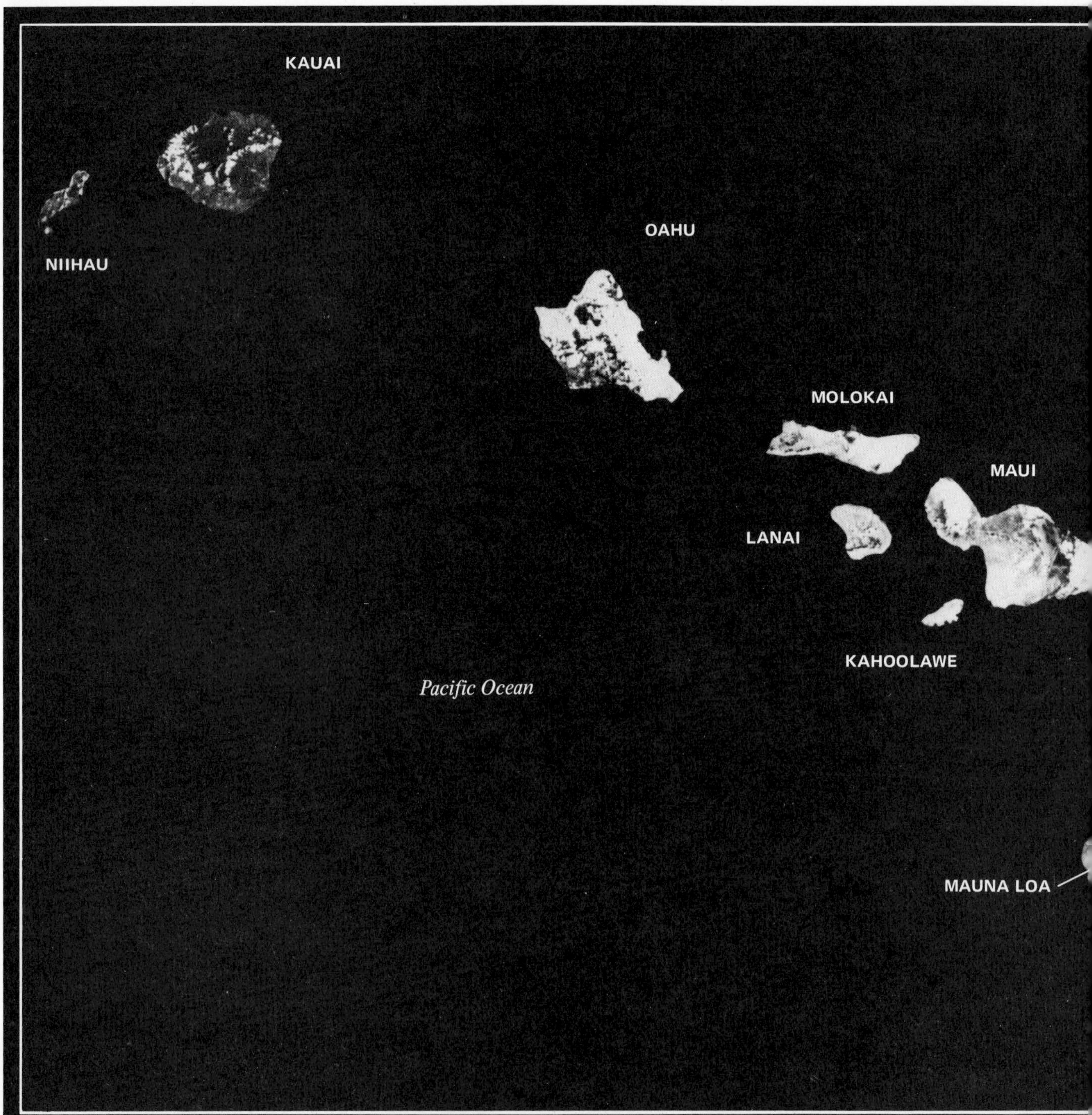

KAUAI

NIIHAU

OAHU

MOLOKAI

MAUI

LANAI

KAHOOLAWE

Pacific Ocean

MAUNA LOA

HALEAKALA

HAWAII

KILAUEA

The Hawaiian Islands were formed by volcanic eruptions. The active volcanoes are at the southeastern end of the chain. Mauna Loa on the island of Hawaii is the world's largest volcano. The islands are progressively older to the northwest. Beyond Niihau only isolated pinnacles or submerged banks remain to mark the locations of extinct volcanoes. The oldest islands, Midway and Kure, have only coral reefs at the surface. White areas are cloud covered. (Photograph courtesy of the National Aeronautics and Space Administration.)

The *shoreline*—where land, air, and sea meet—is the most dynamic part of the coastal zone. Here continents respond to the same forces that affect the coastal ocean: tides, winds, waves, changing sea level, and humans. The shore, as shown in Fig. 11-1, extends from the lowest tide level to the highest point on land reached by wave-transported sand. In this chapter we study the processes that shape the part of the ocean most familiar to us.

FORCES ACTING
ON THE COASTAL ZONE

The broad zone (known as the *coast* or *coastal zone*) directly landward from the shore is also strongly affected by the processes that shape the shore. Usually these forces act most effectively under extreme conditions, such as storms or exceptionally high tides. Violent storms or exceptional tides strike most coasts every 10 to 100 years. Southern New England, for example, was hit by destructive hurricanes in 1635, 1815, 1938, and 1954. The intervals involved are short in the billion-year history of a continent. Storms of such strength have a major effect on the coast, particularly on beaches.

Processes affecting shores act on time scales ranging from a few seconds to tens of thousands of years. Among the more obvious forces are waves, which break, rush up the beach face, and then retreat in a matter of seconds. Along open coasts there are few days without waves to move materials on the beach face, dissipating the energy originally imparted to the sea surface by winds.

The rise and fall of the tides submerges or exposes parts of the beach every 6 or 12 hours. To a great extent, it controls the type and vertical dimensions of the beach, as well as the types of plants and

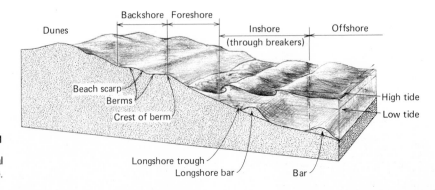

Figure 11-1

Profile of a typical beach and adjacent coastal zone.

288

animals that live there. Tidal currents are important in inlets and coastal embayments but have relatively little influence on open, straight coasts.

Storm surges—relatively sudden large changes in sea level resulting from strong winds blowing across a shallow and partially enclosed body of water—also cause large changes in the shoreline and the coastal zone landward because of the flooding of normally protected areas. Although major storms may come only every few 10 or 100 years, the changes they made often remain visible for decades. For instance, some inlets on the south shore of Long Island broke through barrier islands during hurricanes in the 1930s, changing previously isolated and brackish water ponds and lakes into lagoons and estuaries that are now strongly affected by the tides. Today the inlets are kept open by jetties and continued dredging in order to prevent accumulations of sediment from filling them in and forming solid barrier beach once more.

Other processes controlling the shore act so slowly that their effects are difficult to detect over a human lifetime or even in historical records. Such changes occur over periods ranging from 100 to 10,000 years; even areas with the longest historical records have little reliable information about coasts and their features that predate the classical historians, who wrote about 2000 years ago. To detect and interpret long-term changes, it is necessary to rely on the record preserved, often indistinctly, in elevated beaches, such as those on the West Coast of the United States, an area that is generally rising because of continuing mountain building, or in Scandinavia, where the continent is still rising because of the unloading of the crust caused by melting of the continental ice sheets over the past 20,000 years. In parts of the Mediterranean area some ancient Roman port cities are now submerged as a result of the sinking of the crust, also under the influence of large-scale crustal movements; others have been elevated.

The most profound influence on coasts as we now see them was the repeated advance and retreat of the continental glaciers and resulting sea level changes (see Fig. 11-2). When sea level stood at its lowest, about 20,000 years ago, the shore was near the present continental shelf break. As the glaciers waned and water was released, the shoreline moved back across the continental shelf until it reached its present level about 3000 years ago. If the remaining ice sheets (Antarctica, Greenland) melt, the sea surface may eventually stand as high as 50 meters above its present level. The shoreline we see today will be submerged and a complex of beach ridges and lagoon or estuarine deposits will mark its former location.

The present shore is the result of processes acting on different time scales. The beach itself has the shortest "memory," recording distinctly only what happened when the most recent waves rushed up

Figure 11-2

Changes in sea level caused by growth and melting of continental glaciers during the most recent glacial stage. [After J. R. Curray, 1965. Late Quaternary history, continental shelves of the United States. In H. E. Wright and D. G. Frey (Eds.), *The Quaternary of the United States,* pp. 723–735. Princeton University Press, Princeton, N.J.]

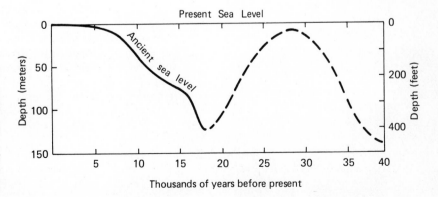

and then retreated or the level of the last high tide; perhaps a recent exceptionally high tide will have left its mark. The beach, as a whole, usually records effects of storms and seasonal changes. In general, it is difficult to detect records of events that occurred more than a few years ago on a beach; even the extensive changes resulting from a storm or storm surge in winter are usually quickly obliterated during the following summer unless humans intrude by dredging inlets or building seawalls, groins, or jetties in order to inhibit erosion and reduce sand movements.

COASTLINES

Most coastlines of the world exhibit signs of the recent rise of sea level, modified by wave action and by deposition of sand and gravel to form beaches or mud to form deltas and marshes (see Fig. 11-3).

One major type of coastline, called a *primary coast,* is formed mainly by terrestrial processes; the ocean has not been at its present level long enough to reshape the land features along its margin. An example of such a coastline is the drowned river valleys of the United States' East Coast, where the former valley of the Delaware River is now Delaware Bay and the former valley of the Susquehanna River and its lower tributaries form Chesapeake Bay. The original shapes of these valleys are virtually intact although largely underwater and somewhat modified by sediment deposition. In glacially carved regions movements of the great ice sheets created deep valleys with U-shaped cross sections. Now filled by seawater, the U-shaped bottoms are concealed, but the valleys cutting back into mountains form typical Norwegian fjords. Fjords also occur in Canada, Alaska, and southern Chile.

Not all processes acting on the coastline have cut away the land surface; other forces were active in leaving behind deposits that now dominate the coastline. Long Island, in New York State, is part of two terminal moraines formed during Pleistocene glaciation. The moraine is a deposit of sand and gravel originally scoured from the continent, carried toward lower latitudes, and left behind at the terminus of the glacier when it melted and retreated. The delta at the mouth of a sediment-laden river is another example of coastline modification caused by deposition of material eroded from the land.

Spectacular but rare coastline features are formed by volcanoes. In the Hawaiian Islands (Fig. 11-4) volcanic cones and lavas come directly down to sea level. In other areas the volcanoes have been worn down to sea level, and in still other areas the sea has broken into the top of the volcano, producing a round bay, usually steep sided and deep, making an excellent harbor. The harbor at Pago Pago (American Samoa) is thought to have such an origin.

A *secondary coast* is shaped by marine processes or by marine organisms. Such coastlines are often cut in rocks or sediments soft enough so that even the limited time of the present sea level stand has been sufficient to cut back the coast. Where bluffs of unconsolidated sand and gravel rise above the shore, they are often cut back by waves to provide a wave-straightened coastline, as illustrated in Fig. 11-5. The materials derived from the erosion of these bluffs are deposited near shore or form small beaches or *baymouth bars* across bays and inlets near the bluffs. Sand moves along the coast by the action of longshore current, forming *barrier islands* and *spits* that separate such inlets and bays from the ocean.

BEACHES

Beaches are the most familiar shore features. Sand beaches, barrier islands, and bays border the Atlantic Coast of the United States from Long Island, New York, to Key West, Florida; in the Gulf Coast

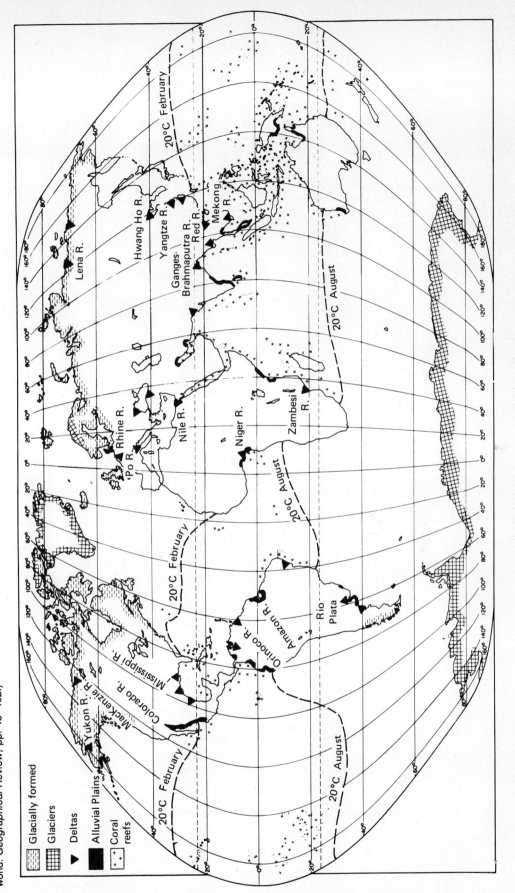

Figure 11-3

Coastlines of the world. (After J. T. McGill, 1958. Map of coastal landforms of the world. *Geographical Review*, pp. 48–402.)

Glacially formed

Glaciers

Deltas

Alluvial Plains

Coral reefs

Figure 11-4

The coastline at Waikiki Beach on the island of Oahu in the Hawaiian Islands. Diamond Head in the background is a remnant of an extinct volcano. The original volcanic features of this island have been altered by erosion and by humans, as demonstrated in the large amount of dredging visible in the yacht basins in the foreground. (Photograph courtesy Hawaii Visitors Bureau.)

region, barrier islands and lagoons are also a characteristic feature. On mountainous coasts, beaches are usually less extensive, being generally restricted to low-lying areas between rocky headlands.

A *beach* is a sediment deposit in motion. At any time the motion of sand grains in the surf may be obvious while the rest of the beach appears quite stable. But a visit to a beach following a major storm will show substantial changes, demonstrating that large segments of the beach have moved. Higher or more protected parts may move only during exceptionally powerful storms; nonetheless, the whole beach does move and has justly been called a "river of sand."

To most of us, the word beach brings to mind a sand beach composed of grains with diameters between 0.062 and 2 millimeters. On midlatitude coasts, beach sands are usually derived from the weathering of silicate rocks. The mineral quartz is a common constituent of such beaches. In fact, some beach sands are so pure that they are mined to obtain quartz sand pure enough for making glass. In tropical and subtropical areas where silicate rocks are rare or absent, beach sands come from broken carbonate shells and skeletons of marine organisms. Such beaches are often white or slightly pink— as in the Bermuda Islands, where the broken shells of a red foraminifer (*Homotrema rubrum*) color the sands. In the Hawaiian Islands the famous black-sand beaches are formed from recently erupted lava (volcanic glass) that make up the islands.

Not all beaches are made of sand. Where wave and current action is especially vigorous, sand may be washed away faster than it is being brought in, leaving behind gravel or even *cobbles*. Still other beaches consist of mixtures of gravel and sand where wave action is not strong enough to remove the sand completely. Sand–gravel beaches occur on the north shore of Long Island, in New England, and on many Pacific Coast shores.

292

Figure 11-5

The coastline on the northern shore of Long Island has been modified by wave erosion and sediment transport. Sand and gravel eroded from bluffs at the points (sharp bends in the coastline) move along the coast and smooth original irregularities. In the left center a small inlet connects a salt marsh (known as Flax Pond) to Long Island Sound. Sand and gravel beaches separate the marsh from the sound. On the right side of the photograph a relatively large, shallow harbor has been created by a long beach that separates submerged former valleys from the sound. (Photograph courtesy National Ocean Survey.)

Beaches are accumulations of locally abundant materials not immediately removed by waves, tidal currents, or winds. In areas like Long Island, beach sands and gravels are derived from erosion of glacial deposits, originally containing unsorted gravels, sands, and clays. Only the gravels and sands remain on the beaches; silt and clay-sized particles are usually washed out of beach areas by even weak waves or tidal currents. Fine-grained sediments tend to accumulate in areas with little wave action or tidal currents, either on the continental shelf at depths below about 30 meters or in lagoons, bays, or tidal marshes.

Most beaches consist of fine sand and tend to slope gently (see Fig. 11-1), with a hard-packed *foreshore* (the zone between high and low water). As the average grain size of the material forming the beach increases, the slope of the beach also increases. Beaches composed of fine sand (⅛ to ¼ millimeter) have average beachface slopes of about 3°. Pebble (4 to 64 millimeter) beaches typically slope about 15° and cobble (64 to 256 millimeter) beaches typically slope about 24°. Even on a single beach, we can see the slope increase as the grain diameter increases. Changes are easily observed as we go from areas with weak currents, where fine sands can accumulate, to areas where stronger currents remove fine sands, leaving gravels and a more steeply sloping

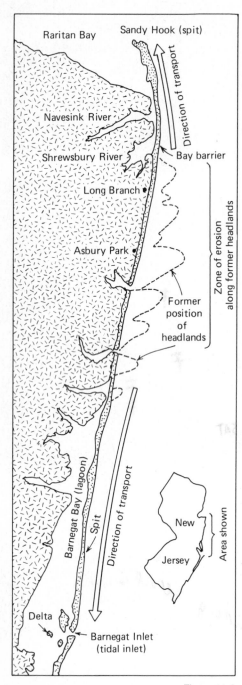

Figure 11-6

Barrier beaches occur along the New Jersey coast. The beaches are formed by sand moving north and south from an area of headlands in the vicinity of Long Branch and Asbury Park. (After D. L. Leet and S. Judson, 1971. *Physical Geology*, 4th ed. Prentice-Hall, Englewood Cliffs, N.J. 704 pp.)

beach. Fine-sand beaches are usually hard packed, permitting easy walking and even driving of cars on them. Coarse-sand or gravel beaches are much looser because the larger grains do not pack as compactly as fine sands.

Beaches typically form near a sediment source—at the base of a cliff or near a river mouth. Sediment is moved onto beaches by waves and currents, replacing materials either moved out into deeper water or transported along the coast. The Columbia River is a conspicuous example; near its mouth are some of the largest beaches and sand dunes on the Washington–Oregon coast. In most other coastal areas, including much of southern California, beaches tend to be small and located near the rivers that supply their sand.

On the North Atlantic Coast of the United States almost no river-borne sediment escapes the many large estuaries to enter the ocean. So, many Atlantic Coast beaches are formed from erosion of nearby cliffs or from sands deposited offshore during times of lower sea level. In many cases, a beach forms where transport of sand along the coast is interrupted by an obstruction, such as a headland downstream.

Not all beaches occur on coastal plains. If the land along a coast is low lying and slopes gently toward the ocean, sand deposits form parallel to the coast and a short distance offshore. If submerged, they are called *longshore bars;* but where large enough, they form *barrier islands,* which typically have a shallow lagoon or bay between the island and the mainland. Such *barrier beaches,* formed by the onshore movements of sands and the longshore movements of currents, are the most common type of beach occurring along the low-lying coastlines of the world (Fig. 11-6). When connected to the mainland, usually at some *headland* (a point of land that juts out from the coast), they are known as *barrier spits.* At small indentations, a barrier of sediment may build completely across the mouth of a bay—then the barrier is known as a *baymouth bar.* A variety of beaches can occur within a single bay.

Two theories have been advanced to explain why barrier beaches form where they do. One is that they develop on a base of ancient, submerged sediment. The other is that sand moving past headlands tends to be deposited rather than being moved farther downcurrent, forming spits that bridge the mouths of bays along the coast, as shown in Fig. 11-7. In any case, barrier bars result from a dynamic balance between longshore currents, wave attack, sediment supply, and bottom topography. They tend to be long and straight, broken at intervals by inlets through which water is borne in and out of the bay or lagoon by tidal currents. The force of these currents keeps the inlets open; thus they act in opposition to the longshore currents, which tend to deposit sediment across the channel.

The barrier beach is normally being washed away and replaced at a fairly constant rate. It is extremely sensitive to changes in the force with which waves break on it or changes in the amount of sand carried toward it by those waves. In a few hours a major storm can move large quantities of sand to form new inlets or close old ones.

Offshore from a beach we usually find one or more submerged low sand ridges on the ocean bottom, called *longshore bars,* parallel to the shore and generally situated in a few meters of water (see Fig. 11-1). At extreme low tides the tops of these bars may be exposed. Separating the longshore bar from the beach proper is the *longshore trough,* which normally remains filled with water at low tide. On coasts with a restricted tidal range several such bars and troughs may develop. Coasts with large tidal ranges generally have only one set of bars, at or near the low tide mark. Bars can usually be identified, even when submerged, by the fact that waves break on them; on many coasts several sets of bars can be spotted from the lines of breakers offshore.

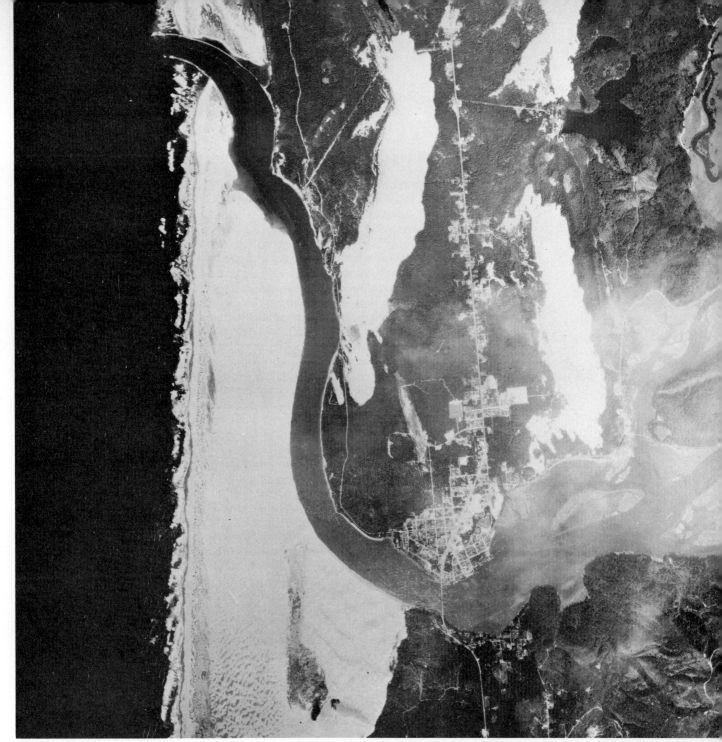

Figure 11-7

A baymouth bar formed by longshore currents moving sediment northward along the Oregon coast at the mouth of the Siuslaw River. The light-colored spit is an area of active dune migration with no vegetation. Darker areas on the spit are depressions between dunes; some have small lakes in them. Dark-colored areas landward of the spit are heavily vegetated and sand there is completely stabilized and not readily moved by wind. (Official U.S. Geological Survey photograph.)

As we leave the offshore portion of the beach and come onto the beach proper, there is a sandy area dipping gently seaward, the *low tide terrace*. This part of the beach is exposed at low tide and submerged at high tide. A small scarp or vertical face often occurs at or near the upper limit of the low tide terrace. It is commonly left by a recent cycle of more intensive wave action that has caused erosion into the beach profile formed during a preceding cycle. The seaward-dipping portion of the beach, collectively known as the *foreshore,* leads up to the *berm crest* or *berm,* the highest part of the beach. Several berms may be present on a beach at any time. Berm crests usually form during storms and represent the effective upper limit of wave action during that storm. As a rule, the highest berm on a beach is formed during winter storms and is referred to as the *winter berm.*

The berm slopes gently downward toward the base of the cliffs or dunes behind the beach. These sands are usually not moved often, as indicated by the large trees that grow there (and often by the accumulations of beer cans and other human artifacts). Where wind action is especially strong, the dry sands are winnowed, removing the lighter particles and leaving behind the heavier mineral grains. On some beaches the heavy minerals remaining give the surface a deep red or purple color, especially noticeable near the base of sand dunes.

Where a beach is not backed by cliffs, dunes are often formed from beach sands blown by the prevailing onshore winds, as illustrated in Fig. 11-8. When not actively gaining or losing sediment, dunes may be colonized by salt-tolerant plants or trees, to become stabilized through time; this process is shown in Fig. 11-9. Dunes protect low-lying lands behind them; in the Netherlands, for example, dunes form a vital part of the defense against flooding from the North Sea. Access to the dunes is often limited or prohibited in beach park areas to preserve the slow-growing plants that protect dunes against active erosion by storm winds (see Fig. 11-9).

On the Atlantic Coast of the United States there are large areas of dunes—for instance, near Cape Canaveral, Florida, and near Provincetown, Massachusetts, on Cape Cod. On the West Coast dunes are associated with the beaches near river mouths (see Fig. 11-7) and also occur on isolated beaches along the rest of the coast. Because they are controlled by the winds, dunes can form obliquely to the coast. Several intersecting sets of dunes may co-occur in the same region.

BEACH PROCESSES

Waves dominate beach processes. Currents and turbulence generated by waves stir up sediment and currents caused by the waves and tides transport sediment parallel to the coast (illustrated in Fig. 11-10). Transport generally takes place between the upper limit of wave ad-

Figure 11-8

Typical sand beach with active dunes, stabilized dunes, and salt marshes. Such a beach-dune complex is found on the south shore of Long Island and at many points along the U.S. Atlantic Coast. (Redrawn from F. D. Larsen, 1969. Eolian sand transport on Plum Island, Massachusetts. *Coastal Environments of Northeastern Massachusetts and New Hampshire.* University of Massachusetts Coastal Research Group, Amherst, MA. pp. 356–367.)

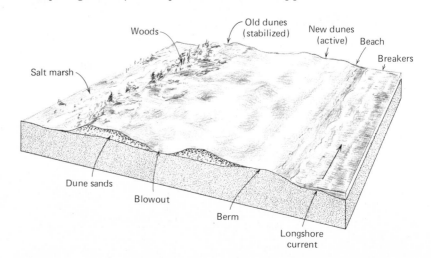

Figure 11-9

Beach grasses stabilize the dunes at Lighthouse Point near Gainesville, Florida. (Photograph courtesy Florida News Bureau.)

Figure 11-10

Incoming waves are refracted and change directions as they enter shallow water. Striking the beach at an angle, they cause a longshore current in the surf zone that moves sand on the beach and along the bottom. Coastal currents are outside the breaker zone.

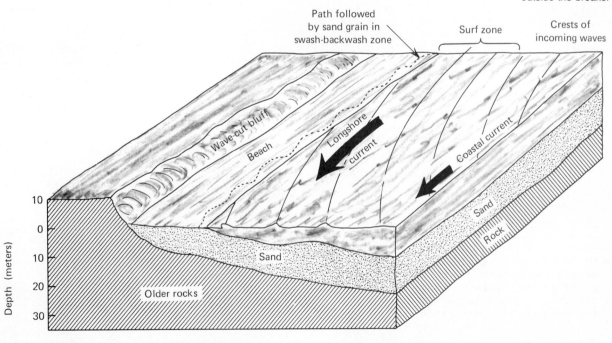

vance on the beach and depths of about 15 meters. Large amounts
of sand are transported in suspension; relatively little is transported
along the bottom.

Beaches change seasonally. During periods of low, long-period
swell, sand is moved back onto the beach, usually building up the
beach in height and width. Longshore bars migrate shoreward, filling
in the troughs, and a new berm forms, usually at a level lower than
the preceding one.

During periods of high, choppy waves (mainly in winter) beaches
are, in general, cut back. The beach foreshore becomes more gently
sloping and a beach scarp forms as erosion proceeds. Strong longshore
currents caused by the waves develop deep channels. Bars develop
because of the offshore movement of sand from areas seaward of the
breakers. Most of the sand removed from the beach is deposited
nearby in the offshore zone to be moved back onto the beach during
the next period of smaller waves.

Waves rarely approach the beach at right angles. Even though
they are refracted on entering shallow water so that crests are more
nearly parallel to the coast, the process is rarely complete and most
waves approach the shore obliquely (see Fig. 11-11). Wave energy
acting parallel to the coast causes longshore currents that move in the
same direction as the waves when they approach the coast. The cur-

Figure 11-11

Waves strike the beach north of Oceanside,
California; their approach at an angle moves
sediment along the beach parallel to the
shoreline and also causes longshore currents.
(Photograph from R. L. Wiegel, 1964.
Oceanographical Engineering. Prentice-Hall,
Englewood Cliffs, N.J. p. 372.)

rent is strongest in the band between the surf zone and the beach. The strongest currents are predicted to occur when the waves approach the shore from a 45° angle. This situation rarely happens; crests of most waves usually deviate less than 20° from being parallel to the beach when they strike the shoreline. Along some trade wind coasts the amount of sand moved by longshore currents exceeds 4 million cubic meters per year. On the south shore of Long Island these currents move about 500,000 tons of sediment westward every year. A comparable quantity of sand moves along New Jersey beaches.

Each wave hitting the beach causes an uprush of a relatively thin sheet of water or *swash* onto the beach face. The water rises until all the energy of the oncoming wave is dissipated or until the water moved by the wave percolates downward into the sand. Any water remaining on the surface runs back down the slope of the beach face.

Because waves rarely strike the beach head on (see Fig. 11-11) but rather at some angle, the swash rises obliquely across the beach face. When the water with its entrained sediment runs back, it goes directly down the slope of the beach. Sand moving along the beach as a result of wave effects is called *littoral drift*. Its direction can change during a single day or over a season; this is a small-scale phenomenon associated with waves hitting the beach. Longshore currents are large-scale phenomena that also result from waves striking the coast.

Wave-induced sediment movements can be surprisingly fast— up to 25 meters per hour and 1 kilometer per day. A more typical rate would be 5 to 10 meters per day on the average. Direction of net movement along the beach is determined by the direction of the strongest and longest acting waves. On most beaches it is the direction from which storm winds come.

Within an individual wave there is substantial movement of water as it breaks. Water at the surface and along the bottom moves toward the beach, carrying with it materials floating or dragged along the bottom, which is why a beach acts as a convergence zone, collecting all sorts of debris as well as sediment. Return flow of the water occurs at mid-depth within the water column, as illustrated in Fig. 11-12. In general, this flow occurs through several meters of water and is not a strong current.

Rip currents are another manifestation of the movement of water toward the beach in the surf and its return flow. After moving toward the beach in the breaking waves, water tends to flow parallel to the beach for short distances until it enters a narrow stream of return flow through the breaker zone—the *rip current*. In rip currents the most rapid flow is relatively narrow and has speeds of up to 1 meter per second until it reaches a distance of perhaps 300 meters from the coast. In this section of the current it often forms a channel deeper than the surrounding ocean and is therefore recognizable from above by the different color of the water and its high sediment content.

Seaward of the breaker zone, the current becomes more diffuse and spreads out, forming a "head" to the current. The water is caught up at this point in the general flow toward the beach. Rip currents and their associated water movements form a cell-like, nearshore circulation system within the breaker zone.

Waves striking certain beaches cause other types of wavelike water motions, called *edge waves*. They, in turn, form *beach cusps*: rather uniformly spaced tapering ridges, with rounded embayments between them, as illustrated in Fig. 11-13. The regular spacing ranges from less than 1 meter to several tens of meters and is related to wave height: higher waves are associated with wider spacing of cusps. These cusps form and reform quickly, especially on fine-sand beaches.

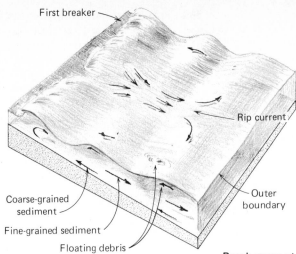

Figure 11-12

Schematic representation of breaking waves and water movement at and below the water surface. Water depth at the outer boundary is approximately half the wavelength for storm waves. [Redrawn from R. L. Miller and J. M. Zeigler, 1964. A study of sediment distribution in the zone of shoaling waves over complicated bottom topography. In R. L. Miller (Ed.), *Papers in Marine Geology,* pp. 133–153. Shepard Commemorative Volume. Macmillan, New York.]

First breaker

Rip current

Outer boundary

Coarse-grained sediment

Fine-grained sediment

Floating debris

Figure 11-13

Beach cusps at El Segundo, California, photographed from the air. (From Wiegel, 1964, p. 34.)

Sediment budgets are useful in determining major sources and losses (sinks) of sand along a stretch of beach. The major sand sources are usually near the beaches, either rivers or erosion of sea cliffs [see Fig. 11-14(a)]. Major losses of sand to the beach occur through longshore transport out of the particular segment, offshore transport (down submarine canyons), and losses due to wind transport to form sand dunes or to deposit in marshes behind the beach.

This approach has been used extensively in southern California where the coast was divided into four cells [shown in Fig. 11-14(b)]. Each cell consists of a river (or rivers) providing sand, littoral drift along the shore, and a submarine canyon that comes in close to the shore and traps the sand flow, carrying it offshore. Each cell is separated from adjoining cells by a stretch of rocky coast devoid of large beaches.

As another example of the processes that control beach development and sediment movement, let us consider the New Jersey coast (see Fig. 11-6). This 200-kilometer stretch of coast consists of barrier islands separated from the mainland by bays, lagoons, and tidal marshes. Along the coast beaches are interrupted by three rocky headlands and ten major inlets leading to lagoons or bays. Tides here are about 1.5 meters; northeasterly storms are common and there are occasional hurricanes.

At the northern end of the coast, near Sandy Hook, the net littoral drift is northerly because nearby Long Island shelters this stretch from waves approaching the coast from the north and northeast. Over a period of about 100 years, sand accumulation on Sandy Hook amounted to about 400,000 cubic meters per year. Near Cape May, at the southern end of this stretch of coast, the net littoral drift is directed southerly and amounts to about 150,000 cubic meters per year. The total amount of sand in transit is substantially greater than the net littoral drift. For example, the total littoral drift at Cape May is estimated to be about 900,000 cubic meters per year, equivalent to about 1 million tons per year.

Between the northern and southern portions of the New Jersey

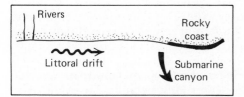

(above)   Figure 11-14(a)

Simple littoral cell consisting of a source of beach sand (rivers), littoral drift that carries sand along the beach, and sink or submarine canyon that transports sand out of the cell. The rocky coast marks the boundary of the cell. [Redrawn after D. L. Inman and J. D. Frautschy, 1966. Littoral processes and the development of the shoreline. *Proc. Coast. Eng. Specialty Conference, ASCE (Santa Barbara, CA.),* pp. 511–536.]

(right)   Figure 11-14(b)

The southern California coast includes four littoral cells. The beach sand moves from the sources (rivers) to the submarine canyons (sinks) where sand is transported offshore into deep water, out of reach of littoral processes. (Redrawn after D. L. Inman and J. D. Frautschy, 1966.)

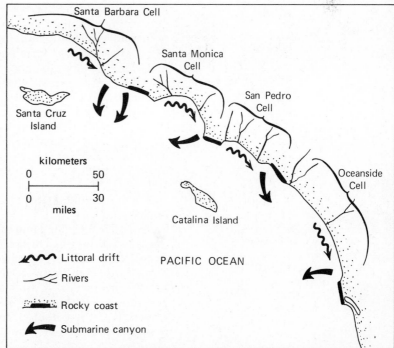

beaches there is a nodal point—located about 60 kilometers south of Sandy Hook—where the net littoral drift is zero. Sand accumulating at Sandy Hook comes from erosion of this section of headland, for there are no major rivers in the region. Between 1838 and 1953 this section of coast was cut back 150 meters at a rate of about 1.5 meters per year. The materials eroded were about two-thirds sand, which was added to local beaches; about one-third was silt and clay-sized material, which moved seaward and was lost from the coastal area. Extensive construction of seawalls and groins in this section reduced the rate of littoral drift by only about 12%.

Inlets and lagoons behind barrier islands act as traps for sediment moving along the coast. Sand is stirred up and put in suspension by waves and then moved into the inlets or lagoons by flood-tide currents. When they slacken or when sediment encounters the dense vegetation of the tidal flats, it settles out. Without resuspension by waves, the ebb tide currents fail to move this sediment back out of the inlet. Thus each of the New Jersey inlets accumulates about 200,000 cubic meters—equivalent to 250,000 tons—per year, filling in navigation channels and necessitating dredging.

DELTAS

Sediments carried downstream by most rivers and streams are deposited either in the estuary or at the river mouth. For many streams, the only sign of their sediment load near the river mouth is a small sandbar, moved by the tides and waves. The bulk of this sediment is eventually resuspended by wave action and moved along the coast by longshore or tidal currents, to be carried onto nearby beaches or deposited on the continental shelf beyond the influence of waves.

Some rivers, however, carry far more sediment than can be dispersed along the adjacent coast. In this case, sediment usually is deposited at the river mouth, forming a *delta.* This is a fairly common occurrence; well-known examples include the Mississippi (as shown in Fig. 11-15), Rio Grande, Fraser, Nile, Tigris–Euphrates, and Rhine deltas. Delta lands are flat, well watered, and often exceedingly fertile because sediment deposition during flood periods provides a continuous supply of rich soil. Many primitive agricultural societies developed and flourished on river deltas, the Egyptian culture on the Nile delta being a familiar example. Lower Mesopotamia, site of the early Sumerian civilization, was an ancient delta and all over the world deltas are valued as farmlands. (Their low elevation above sea level makes them vulnerable to flooding by storm surges.)

The relative capacity of the coastal ocean to transport sediment is determined by the tidal range, the strength of tidal currents, the wave energy dissipated on the coast, and the strength of longshore currents. Strong tidal currents set up by a large tidal range cause scouring of recently deposited sediment. Waves resuspend sediment in the surf zone; they also cause *longshore currents,* which move sediment along the coast in a direction determined by prevailing wave approach.

The Columbia River on the West Coast of the United States, discharging between 10 and 30 million tons of sediment per year into the ocean, lacks a delta due to its relatively large tidal range and strong waves at its mouth. On the other hand, the nearby Fraser River in British Columbia, with approximately the same sediment load, has formed a delta in the protected waters of the Strait of Georgia because strong wave action is inhibited there by the limited fetch for wave generation. The Mississippi River has formed the largest delta in the United States as a result of its large sediment load (about 300 million tons of sediment per year) and the low tidal range in the Gulf of Mexico.

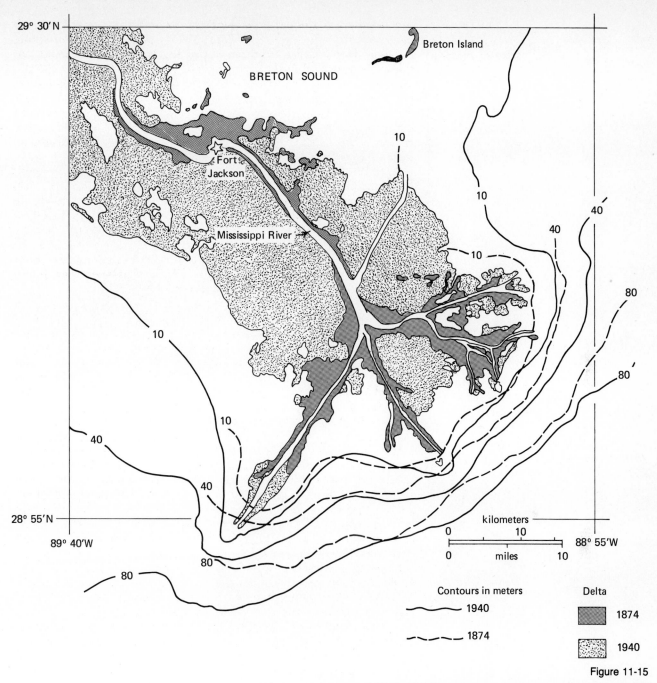

29° 30' N

Breton Island

BRETON SOUND

Fort
Jackson

Mississippi River ➤

kilometers

28° 55'N

89° 40'W

88° 55'W

Contours in meters

——— 1940

– – – 1874

Delta

▓ 1874

░ 1940

Figure 11-15

Mississippi River delta and its growth between 1874 and 1940. Note the
characteristic "bird-foot" shape resulting from each distributary having been built
seaward, with flooding between the distributaries.

Some river systems supply so much sediment that a delta forms
in spite of a large tidal range and extensive wave action. The region
where the Ganges–Brahmaputra rivers in India discharge at the head
of the Bay of Bengal, shown in Fig. 11-16, is an example. These two
rivers, which reach the ocean through the same delta complex, carry
about 700 million tons of sediment each year, forming a large delta.

The first step in delta formation, illustrated in Fig. 11-17, is
filling an estuary with sediment. The Mississippi, for instance, probably filled its estuary soon after sea level reached its present position.
Rivers with large estuaries and relatively moderate sediment loads,

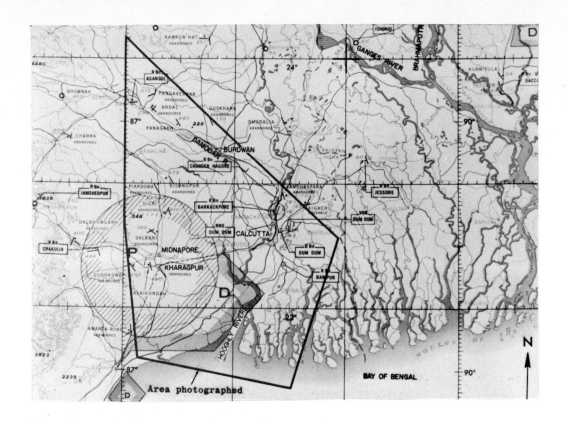

Figure 11-16

A portion of the delta at the mouth of the
Hoogly River as photographed from a satellite.
(Photograph courtesy NASA.)

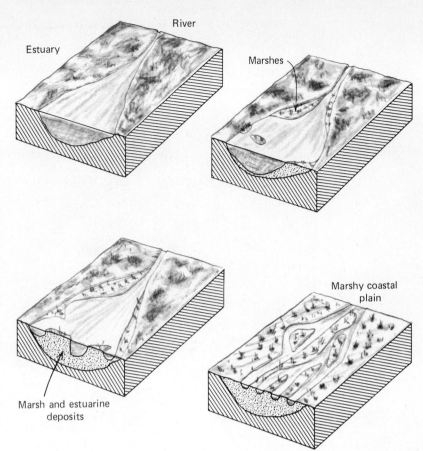

Figure 11-17

Stages in the development and filling of an estuary by sediment deposits, converting a river valley to a salt marsh.

such as the Amazon River, are still filling their estuaries. Until the estuary is completely filled, little or no sediment can be deposited at the river mouth or can form a delta.

As a delta forms, the river builds channels across it called *distributaries,* as shown in Fig. 11-18, through which water flows on its way to the ocean. Often a series of these channels extends across the delta as long, radiating, and often branching fingers. An active distributary continually builds its mouth farther seaward until the distance to the sea is so great that the river flow can no longer maintain that channel. At this time the river shifts course, often during flood, so that the flow cuts a new channel through a different set of distributaries to reach the ocean and the whole process begins again. The Mississippi River Delta has several abandoned distributaries, each with its own subdelta, forming a complex *lobate delta* (see Fig. 11-15). Distributary abandonment is not always sudden but may occur gradually as one channel becomes too shallow to carry a large amount of water and another gradually receives more of the flow so that it becomes enlarged.

Many deltas occur in areas where the land is subsiding. For example, the Mississippi Delta region as a whole is subsiding at a rate of about 1 to 4 centimeters per year. Furthermore, the sediment beneath the delta continually compacts and expels water from the sediment. Consequently, the delta surface is constantly subsiding except in those areas currently receiving sediment.

When a distributary is active, adjacent areas are occupied by ponds or lakes with many marshes, as shown in Fig. 11-19. Areas between distributaries receive less sediment and the deposits there are finer grained and contain bulky plant debris (peat). These deposits

Figure 11-18

Part of the modern delta of the Mississippi River is pictured here, looking south from the "Head of Passes," where one distributary separates into three. The left-hand pass is a natural artery; the other two are kept open for navigation by dredging and by construction of jetties to direct the river flow—thus increasing its velocity so that less sediment is deposited in the channels. (Photograph courtesy U.S. Army Corps of Engineers, New Orleans District.)

compact (and subside) more than the coarse-grained material deposited along the distributary banks. The banks, therefore, subside less than the adjacent marsh and eventually stand higher, forming natural levees on which towns and roads are built.

As the marsh subsides below sea level, it changes into a shallow bay. Seawater in these bays is diluted by groundwater discharged through the delta so that bay waters are usually brackish. These areas, even though their appearance may be unprepossessing to an outsider, are highly productive of marine life. About 90% of commercially important coastal fish and game fish depends on these bays or similar marsh areas during some critical stage of their life cycle.

Deltas are rarely formed by rivers draining areas that were covered by ice during the last glacial period. Lakes, gouged out by ice, are common in these areas and act as traps to prevent the escape of river-borne sediment to the sea. An example is the St. Lawrence River, which drains the Great Lakes. These lakes act as effective sediment traps.

After lakes in the river's drainage basin and the estuary at the river's mouth are filled with sediment, the stream may then be able to transport enough sediment to the ocean to build a delta. The shape of a delta is controlled by the balance between erosion and sediment

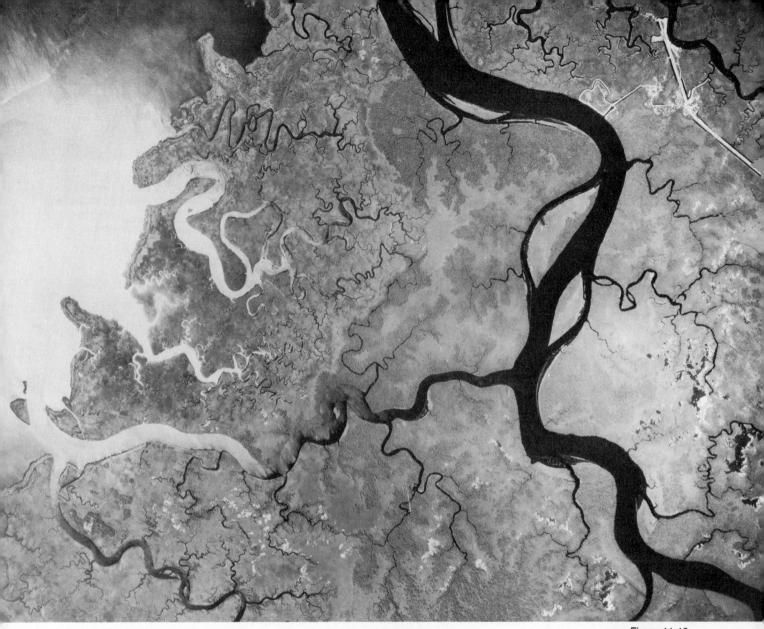

Figure 11-19

Distributaries of the Suwannee River, Florida, showing the tidal marsh and cypress swamps lying between them. Note the meandering tidal creeks in the marsh. A road and dredged channel are conspicuous in the upper-right portion of the photograph. (Official U.S. Geological Survey photograph.)

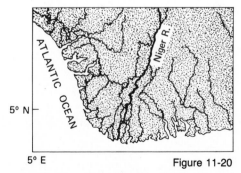

Figure 11-20

Delta of the Niger River on the western coast of Africa.

transport along the coast, on the one hand, and the rate of sediment supply on the other. Where coastal processes predominate, the margin of the delta tends to be rounded, as in the Niger delta, as shown in Fig. 11-20. Where the sediment supply from the river exceeds the capacity of the coastal ocean to transport it, the delta shape is primarily controlled by river processes rather than coastal processes. The Mississippi delta's "birdfoot" shape results from each distributary having built seaward, with flooding between the distributaries, as previously discussed (see Fig. 11-15).

307

SALT MARSHES

Salt marshes are low-lying portions of coastline that are submerged by high tides but protected from direct wave attack. Their surfaces are nearly flat and generally overgrown by salt-tolerant plants, adapted to periodic submergence.

Marshes form where sediment is available from rivers or from resuspension in waves or strong tidal currents. In some areas, sediment depositing in marshes comes from local sources, such as erosion of nearby headlands; this is the source of sediment for many New England marshes (see Fig. 11-21). Size and shape are determined by the general outline of the depression in which the marsh forms; vertical extent is controlled by the tidal range. The upper limit of a marsh is generally controlled by the spring tides in an area, which is the highest level to which ocean water can periodically transport sediment.

Figure 11-21

Small salt marshes in nearly filled bays on the north shore of Long Island (Stony Brook Harbor and West Meadow Creek). Both marsh areas have been modified by dredging. (Photograph courtesy National Ocean Survey.)

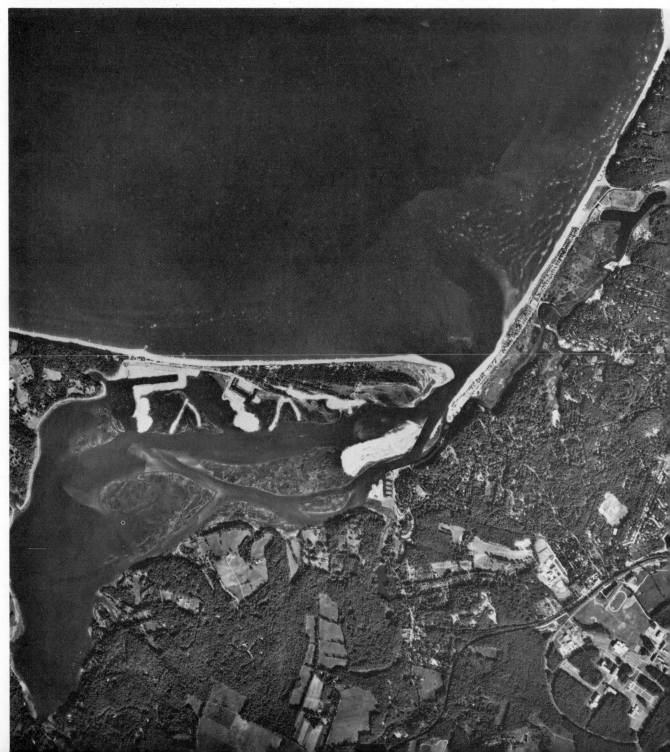

Most of the marsh area consists of nearly flat-topped banks of sand or mixtures of sand and silt. The tops of the banks, known as *tidal flats,* are commonly exposed at low tide and submerged at high tide. If a marsh is well protected, the flats are usually covered by dense growths of marsh grasses. Where the marsh is big enough and open enough for wind waves to be set up across its surface, tidal flats may be completely barren of plants.

Cutting through the tidal flats are many meandering channels, through which seawater enters and drains from the marsh in response to the tidal cycle. The largest and deepest of these channels contain water even at low tide and plants rarely grow in them. The bottom material is generally shifting sand or gravel because strong tidal currents resuspend the finer-grained materials, to be deposited on the flats. These large channels connect with smaller branching and meandering channels that cut back into the marsh and are often exposed at low tide.

Sediment is transported into marshes by strong currents in the tidal channels. As rising water moves out into the smaller channels and then onto the tidal flats, current velocities decrease. At some point the current velocity is too low to keep sediment in suspension so that particles settle out and are deposited. When the tide goes out, current velocity is inadequate to resuspend the sediment grains, which therefore remain where they settled out of the water. Plants on tidal flats also retain sediment and so marshes act as effective traps for fine-grained sediment moving in the coastal ocean. Many marshes mark the locations of former shallow estuaries, now filled by sediment.

Some of the largest and best-studied marshes lie behind the Friesian Islands on the Dutch North Sea coast and the adjacent North German coast along the edge of the Rhine delta. In the Netherlands much of the marshland that formerly bordered that delta has been reclaimed by building dikes to exclude ocean waters and then draining the saltwater to make fields suitable for agriculture.

Salt marshes, or *wetlands,* are important for many marine organisms. In temperate climates salt-tolerant plants and grasses produce an abundance of food for large and small animals. Much plant debris escapes to be deposited with the sediment accumulating on the nearby ocean bottom.

In tropical climates grasses play a less conspicuous role and *mangroves* (see Fig. 11-22) dominate marshes along the estuarine border. These large treelike plants have extensive root systems. The roots form dense thickets at the water level that provide shelter for both marine and land animals—a zone intermediate between land and water. Many forms of life are specially adapted to survive in this environment; the mangrove oyster, for instance, attaches itself to roots and branches that are exposed at low tide, presenting the spectacle of oysters growing on trees.

Mangrove roots trap sediment and organic matter and eventually the swamp is filled in—to be replaced by a low-lying tropical forest. Over a 30- to 40-year period, 1500 acres of new land were created in Biscayne Bay and Florida Bay by colonization of shallow-water areas by mangrove seedlings.

REEFS AND ATOLLS

On tropical and subtropical coasts coral reefs are common where there are no rivers to bring large amounts of sediment to the ocean. Also, the ocean waters must be warmer than 18°C throughout the year. In such areas (Fig. 11-23), carbonate-secreting animals and calcareous algae grow on the shallow bottom, depositing calcareous skeletal material.

Figure 11-22

Mangroves growing out into the water at Ten Thousand Islands, Florida. (Photograph courtesy Florida News Bureau.)

If the platform on which they are growing slowly sinks, the organisms can grow upward and thus keep the upper part of the reef in shallow waters. In this situation, a characteristic sequence of reef forms can be found. First, a *fringing reef* [Fig. 11-24(a)] grows along the shore of an island with gaps at river mouths. As the island sinks, the land area becomes smaller. Rain, wind, and wave action erode it, further reducing the size of the island. But while the coastline retreats, the fringing reef offshore grows upward, becoming a *barrier reef* [Fig. 11-24(b)]. Eventually a lagoon separates the reef from the island. The Great Barrier Reef off Australia is an example.

Finally, the island may totally disappear, leaving only the reef, now called an *atoll*, surrounding the lagoon [Fig. 11-24(c)]. Atolls are commonly irregular in shape, often elliptical in outline (Fig. 11-25).

Lagoons are typically about 40 meters deep and connect to the ocean through passes—interruptions in the reef large enough for ships to navigate. Often small islands of sand or reef debris are on top of the reef or near it within the lagoon. Small *patch reefs*, their tops covered with coral and algae, grow up from the lagoon floor (Fig. 11-26).

A typical reef—for instance, off the West Indian island of Jamaica—can be divided into three zones: back-reef, reef crest, and

310

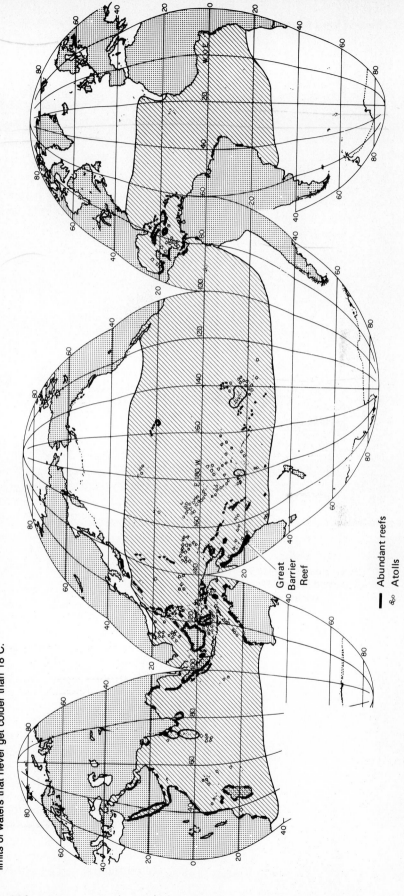

Figure 11-23

Distribution of coral reefs and areas of abundant atolls. The shaded area shows the limits of waters that never get colder than 18°C.

Great
Barrier
Reef

▬ Abundant reefs
&o Atolls

Figure 11-24

Stages in the transition from fringing reef to
barrier reef to atolls.

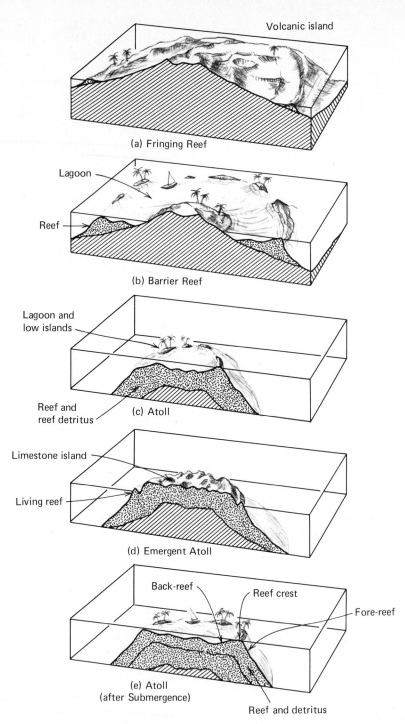

Volcanic island

(a) Fringing Reef

Lagoon

Reef

(b) Barrier Reef

Lagoon and
low islands

Reef and
reef detritus

(c) Atoll

Limestone island

Living reef

(d) Emergent Atoll

Back-reef

Reef crest

Fore-reef

(e) Atoll
(after Submergence)

Reef and detritus

forereef [Fig. 11-24(e)]. The *back-reef* region consists of the lagoon
and the protected inner side of the reef. The region rises to a high
point, the *reef crest,* which is exposed to the air at low tide. It extends
from about sea level to depths of a few meters and includes a buttress
zone that is commonly cut by a series of grooves alternating with spurs
of reef growing seaward. This *spur-and-groove* structure absorbs wave
energy and protects the reef from damage by breakers. Carbonate
sediment and debris, swept by waves from the reef crest, slide down
the channels of the steep reef front to collect in deeper, more protected waters.

Figure 11-25

Six atolls in the Tuamoto Archipelago, part of French Polynesia, in the central South Pacific Ocean (16°S, 145°W). The white bands are reefs surrounding lagoons. Islands occur on the reefs, often at the sharp bends in the atoll, and can be recognized because they appear darker owing to the vegetation on them. [Faint lines transverse to the reef mark the deep-water passes through which ships enter the lagoon.] The white masses are clouds in the upper right. (Photograph courtesy NASA.)

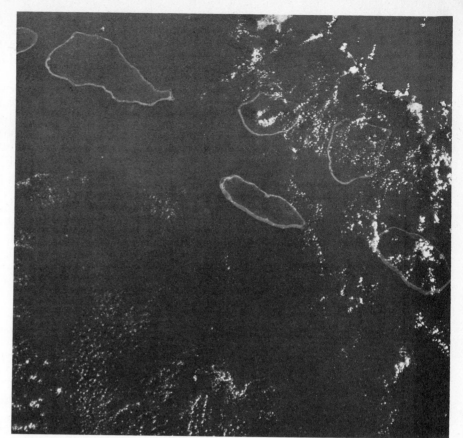

Figure 11-26

Edge of a patch reef in the lagoon at Midway Island, an atoll in the central North Pacific. The irregular surface consists of coral and encrusting calcareous algae. The lagoon floor is about 5 meters deep and is covered with carbonate sand and larger pieces broken off the patch reef. (Photograph courtesy U.S. Geological Survey. J. I. Tracey, Jr., photographer.)

Reef crest and buttress zones are near-surface features, extending to depths of 7 to 10 meters. Below them the seaward slope or *forereef* is typically subdivided into a gently sloping *forereef terrace* (about 7 to 15 meters depth) bordered by a steep, pinnacled *forereef escarpment*, shown in Fig. 11-27. At about 30 meters depth the reef front levels off to a seaward-dipping, sandy *forereef slope* studded with coral pinnacles. Below about 50 to 60 meters, the reef drops off sharply to depths of several hundred meters.

The reef itself is a carbonate framework built of coral skeleton bound together by encrusting calcareous algae. It forms an open, cavernous structure. Loose carbonate sediment collects in the internal spaces and is itself cemented, forming a nearly solid carbonate mass.

Reefs usually grow for millions of years. Vertical accretion of calcium carbonate on the forereef terrace and slope of a typical West Indian reef proceeds at a rate of 0.5 to 1 meter per thousand years and horizontal accretion of material on the deep forereef has averaged about 0.2 meter over the same period.

Figure 11-27

Sketch of forereef, forereef slope, and deep forereef from the perspective of a diver at 50 meters depth. Forereef terrace and escarpment are seen at upper left, merging with forereef slope, and large pinnacles rise above it to become promontories of the deep forereef. Note the diver in the center of the sketch to indicate vertical scale. (Sketch by T. F. Goreau.)

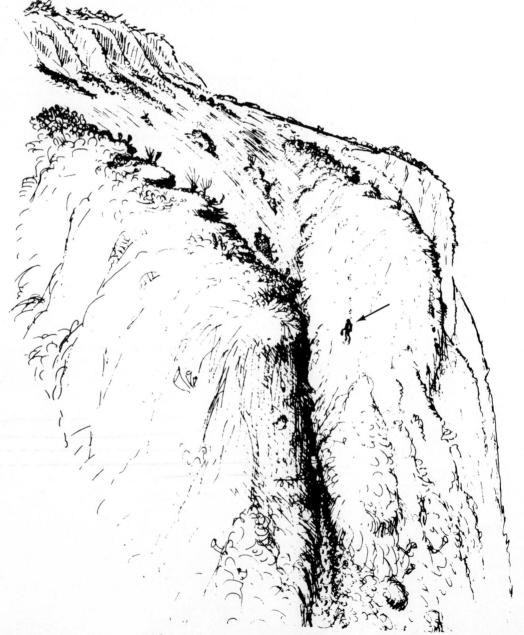

MODIFICATION OF COASTAL REGIONS

Coastal areas have been substantially modified by human activities, especially near urban centers, where population pressures are greatest. Improvement of shipping facilities in estuaries and harbors, filling in of marshes for industries, airports, and housing, prevention of beach erosion, and use of coastal areas for waste disposal account for most coastal modifications in the United States. Because of their importance to shipping, estuaries have perhaps been more extensively modified than any other coastal feature.

Dredging deeper (and usually straighter) channels for safer passage of ships is perhaps the oldest form of coastal alteration commonly undertaken in this country. Most estuaries have bars—shallow sand deposits—at their mouths, whose greatest depth in channels would be around 3 to 5 meters below sea level. These shallow channels are kept open by river flow and tidal currents, but they are far too shallow to accommodate deep-draft vessels; furthermore, they tend to shift location frequently. Dredging them to depths of up to 15 meters has improved navigation in large harbors, but it has also caused other problems to the estuary.

Deeper channels act as sediment traps for sands moving along the coast, as well as for river-borne sediment, so that a continuous program of dredging is required to keep the channels open. Alteration of normal circulation patterns affects salinity distributions; for instance, enlargement of inlets may permit larger amounts of seawater to enter and water may flow through man-made channels toward remote areas that would normally be relatively fresh. Fish and bottom life are affected by such changes.

Beaches have also often been subjected to conflicting uses and have been greatly altered in the process. In many areas, beaches are mined to produce sand for construction and land-fill operations, making them unsuitable for recreational use. In other cases, changes in sand transport has caused erosion of beaches (shown in Fig. 11-28). When a harbor or inlet, for instance, is dredged for improvement of navigation facilities, the resulting deep channels or basins often act as traps for sediment moving down the coast. Consequently, the sand supply to beaches downcurrent is cut off.

Figure 11-28

A scarp cut by waves. The boardwalk has been undermined and snow fences have been placed in an effort to trap wind-blown sand. The dark patches and streaks at lower left are heavy mineral grains left behind by the waves. Some lamination is visible along the scarp. (Photograph courtesy Fire Island National Seashore, National Park Service.)

Efforts to prevent accumulation of sediment in inlets have met with limited success. Seawalls have been built to stem erosion and groins have been built to inhibit sediment movements along beaches. Other efforts might include construction of jetties to keep sand out of the inlets, perhaps combined with pumping of sand past the inlet. Sand may also be dredged from bays and used to replenish beaches. Such approaches are expensive and some are undesirable because of side effects, such as destruction of large marsh areas by dredging.

Once damaged, a beach is not easily restored. Several approaches have been used to protect or rebuild beaches. One of the most direct has been to construct *groins* (low stone walls), such as those in Fig. 11-29, built at regular intervals along a beach to retard sand movement. On the side from which sand is moving, the beach is widened; but the beach downcurrent is usually narrowed unless particular care is taken in constructing and placing the groins. In some instances where groins have been improperly placed, erosion has actually increased. On the New Jersey coast, where groins were used extensively, they have reduced the rate of sand movement by only about 12%.

The other approach to beach stabilization is *replenishment,* a procedure whereby large volumes of sand are brought in and put on the beach. In some instances, the sand is dredged from shallow, protected bays behind the beach. The effects are noticeable but transitory because the sand moves downcurrent as part of the regional movement of sediment along the beach. Often a single storm will remove the newly added sand.

Salt marsh areas were commonly bypassed in the early development of coastal areas in the United States. Unsuited for most agricultural purposes and difficult to build on, they were often used initially for pasture, then as waste-disposal sites, and later for building purposes. About 20% of Manhattan Island in New York City is built on "reclaimed" marsh and shallow harbor areas. Continued population growth has often resulted in accelerated use of salt marsh areas as they come to represent the only available open land. LaGuardia and Kennedy airports in New York City are built on filled marsh and estuarine areas, as are San Francisco's International, Boston's Logan, and Washington, D.C.'s National Airport; Fig. 11-30 illustrates how this was done for Newark International Airport. Once dredged and filled, salt marshes are attractive sites for industrial development because of their proximity to water transport. Housing developments have also taken their toll of salt marsh areas.

In addition, coastal marshes are dredged and filled to form numerous narrow peninsulas and boat channels, offering access to waterways for small boats. In the process, the marshes are altered or destroyed and circulation within any remaining wetlands is greatly modified. Frequently the dredged basins and channels provide poor circulation and the resultant accumulations of wastes, including untreated sewage from housing developments and boats, have caused severe local problems, such as unpleasant odors and fish kills.

Even where marshes are not developed for housing or other purposes, they are often modified by ditching as part of mosquito-control measures. Ditches are dug to improve drainage in the interior of the marsh and thereby eliminate mosquito-breeding areas. Part of the original marsh surface is destoryed and circulation locally altered in the process. Erosion may be accelerated because of improper location of drainage channels.

Through the combination of all these alterations, about 23% of estuarine systems (including salt marshes) in the United States had been severely modified and about 50% moderately altered in 1970. In 1968 the United States government alone spent more than $870 million on coastal-channel and harbor improvement projects. Many

Figure 11-29

Groins have been constructed along the ocean beach (left side of the picture) at Sandy Hook, N.J., in an attempt to halt the flow of sand along the beach.

Figure 11-30

Former marshland on the western side of Newark Bay, N.J., has been developed by dredging channels and filling low areas to build terminals for handling containerized freight (center of photograph) and Newark International Airport (left side). Relatively undisturbed marsh is visible in the background. (Photograph courtesy Port Authority of New York and New Jersey.)

millions more were spent by state and local governments and private concerns.

In the past, planning and engineering of large-scale coastal modification projects depended primarily on experience accumulated over many years of building (and sometimes rebuilding) coastal installations. During the Middle Ages, for instance, Dutch engineers were constructing dikes to reclaim shallow areas of the North Sea.

Coastal engineers use various types of models to predict the effects of new structures on the ocean. Where processes are well understood and computation facilities adequate, it is possible to construct mathematical models—expressions describing in mathematical terms the nature and effects of physical processes. Using computers, the effects of projected structures can then be predicted. Designs can be altered to achieve the desired objectives or to meet other criteria.

Although the use of mathematical models has become increasingly common since the development of computers, engineers have successfully calculated the effects of tides and winds on coastal structures for the past several decades. The great Dutch Afsluitdijk ("enclosure dike"), for instance, has protected the former Zuider Zee estuary since 1932. Extensive studies of tides and current forces of the probable strength and frequency of storms preceded its construc-

317

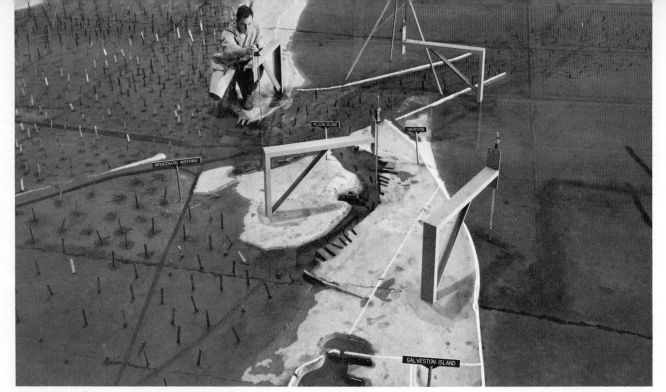

Figure 11-31

Hydraulic model of Galveston Bay showing the inlet in the barrier beach. The model was constructed on a scale of 1:100 vertically and 1:3000 horizontally. Tides and tidal currents are reproduced in the model by a tide generator located in the Gulf of Mexico portion of the model, on the right-hand side. The strips in the model on the bay side are necessary to adjust flow in the model to make it reproduce conditions observed in Galveston Bay. (Photograph courtesy U.S. Army Corps of Engineers, Waterways Experiment Station.)

tion. Much valuable agricultural land has been reclaimed from shallow coastal areas behind the dike by first pumping out the saltwater and then permitting rainfall to flush away salts remaining in the soil (see also Chapter 15).

Yet many problems cannot be handled by mathematical techniques. The processes are either too poorly understood to be described mathematically or too complicated to be solved by existing computers. For these problems, *hydraulic models,* such as the one shown in Fig. 11-31, have been extremely useful. The model simulates an area on a greatly reduced scale so that physical processes can be modeled and studied relatively inexpensively. Furthermore, the model permits the collection, in a few days, of data that would require years to collect in the field.

Hydraulic models are carefully constructed to duplicate the original feature. Then careful observations are made in the field and the model is adjusted to reproduce physical processes accurately. Hydraulic-model studies of estuarine systems have been especially useful in studying such diverse phenomena as the effects of breakwaters on waves in harbors, movement of wastes in estuaries, and sediment movement in harbors.

REVIEW QUESTIONS

1. Draw a profile of a beach and label the major features.
2. Contrast a primary and secondary coastline. Discuss the processes causing each.
3. Discuss the effects of the continued rise in sea level on shoreline features.
4. Describe the filling of an estuary to make wetlands or deltas.
5. What causes sediment movement along beaches?
6. Describe a typical rip current and the processes that cause them.
7. Discuss the role of plants in building and maintaining wetlands.
8. List and briefly discuss the ways in which human activities modify shorelines.

SUMMARY OUTLINE

Shore—extends from low tide to high tide levels

Coast—broad zone extending landward from shore

Processes affecting shores are effective over long as well as relatively short times

Storm surges—relatively sudden sea level changes, usually caused by storm winds

Coastlines

Primary coast—formed by terrestrial forces: drowned river valleys, volcanoes

Secondary coast—shaped primarily by marine processes or organisms; barrier beaches, coral reefs

Deltas—large deposits of river-borne sediment at the river mouth

Modified or prevented by tidal currents or waves

Steps in delta formation: filling of estuary, formation of distributaries, shifting of distributaries

Shape controlled by river processes and coastal processes

Beaches—deposits of loose sedimentary material moved by waves

Usually sand (sometimes gravel) derived from local sources, such as rivers or bluffs; nature of material depends on coastal processes

Barrier beaches—islands or spits built by action of waves and currents

Beach coastlines—generally smooth outlines of preexisting topography

Offshore—sandbar and trough

Onshore—low tide terrace, beach face, berm crest (high point, formed by storms)

Dunes—commonly behind large sand beaches

Beach processes

High, choppy waves erode beaches, usually in winter

Long-period, low swell builds beaches, usually in summer

Waves approaching beach obliquely cause longshore currents and beach drift

Water movement within breaking wave is toward land at surface and along bottom, seaward at mid-depths

Rip currents—return flow of water moved toward beach by breakers

Salt marshes—low-lying area covered by vegetation, usually submerged only at highest tides

Tidal flats—covered by high tide, bare at low tide; may or may not have vegetation cover

Tidal creek channels—flooded at all times, avenues for drainage

Sediment moves into marsh during flood current, not readily eroded by ebb current

Vegetation also traps sediment—for instance, mangrove swamps, marsh grasses

Coral reefs—wave-resistant platforms built in shallow water by carbonate-secreting organisms

Corals, encrusting calcareous algae form framework, sediment is trapped

Grow in shallow, warm, sunlit waters of near-normal salinity

Atolls result from submergence of platform, most common in western Pacific

Back-reef—inner surface, lagoon

Reef crest—exposed at low tide, spur-and-groove structure below

Forereef—coral pinnacles

Modification of coastal regions

Estuarine modification

Includes dredged navigation channels, stabilization of harbor entrances with jetties, waste disposal

Alters estuarine circulation patterns and sediment flow

Salt marsh modification

Used for waste disposal, filled for construction of houses and airports

Navigation-channel dredging and mosquito-control drainage alters circulation

SELECTED REFERENCES

Bascom, Willard. 1980. *Waves and Beaches: The Dynamics of the Ocean Surface*, rev. ed. Doubleday Anchor Books, Garden City, N.Y. 366 pp. Elementary, well written.

Bird, E. C. F. 1969. *Coasts*. MIT Press, Cambridge, MA. 246 pp. Wave, current and wind effects on the structure of coastline; elementary.

Johnson, D. W. 1965. *Shore Processes and Shoreline Development*. Hafner, New York. 584 pp. A classic study, first published in 1919.

Komar, Paul D. 1976. *Beach Processes and Sedimentation*. Prentice-Hall, Englewood Cliffs, N.J. 429 pp.

Steers, J. A. 1969. *Coasts and Beaches*. Oliver & Boyd, Edinburgh. 136 pp. Elementary discussion of beaches and beach processes; examples primarily from Great Britain.

Van Veen, Johan. 1962. *Dredge, Drain, Reclaim: The Art of a Nation*, 5th ed. Martinus Nijhoff, The Hague. 200 pp. Elementary discussion of Dutch reclamation work.

Zenkovich, V. P. 1967. *Processes of Coastal Development*. Interscience, New York. 738 pp. Thorough treatment of shore processes, with emphasis on Russia and recent Russian research.

12 OCEANIC LIFE

Communities of unusually large organisms live near active hydrothermal vents. Brilliantly red-plummed worms living in white, plasticlike tubes are among the most spectacular. They have no mouth or gut, but grow to be nearly 3 meters long and 5 centimeters in diameter. (Photograph by Kathleen Crane, Woods Hole Oceanographic Institution.)

Conditions for life in the ocean are markedly different from those on land. On land we live at the bottom of an "ocean" of air. Some spores and seeds, as well as bacteria and viruses, indeed remain for relatively long times in the atmosphere and some spiders can float attached to a fine thread. Birds and flying insects often use updrafts to support themselves for awhile, but even they alight from time to time. Yet the inhabited portion of the atmosphere is relatively thin. The tallest trees extend only about 70 meters into the air and few animals penetrate into the soil more than a meter. Most terrestrial organisms live in a zone about 20 meters thick. Many marine organisms spend at least part of their lives attached to a solid surface.

The open ocean, on the other hand, is three dimensional. Many organisms float or swim in a seemingly boundless ocean. Life exists at all depths—from sunlit surface waters to the dark ocean bottom. The ocean averages about 4 kilometers deep and covers about 70% of the earth's surface, a living space about three hundred times larger than that inhabited on land.

The density of seawater, like that of living protoplasm, is around 1.025 grams per cubic centimeter whereas the density of air is much less—around 0.0012. This means that the effective weight of animals and plants in the sea is only a small fraction of what it would be in the air. Support requirements for organs and appendages of the body are quite different in the two environments. The heavy internal skeletons of land animals, developed to hold us upright against the pull of gravity and to facilitate motion and leverage, are unnecessary in the ocean, as are the rigid cellulose structures characteristic of large land plants.

One problem unique to life in the open ocean is sinking, which has no counterpart on land. Most marine organisms are adapted for survival in a particular layer of water that satisfies their specific light and temperature requirements. An organism less dense than seawater tends to float toward the surface. But most marine plants and animals are slightly denser than seawater and tend to sink unless they can swim or are equipped with some means of maintaining buoyancy.

Oceanic environments are far less variable than on land. Seawater temperatures change slowly because of seawater's large heat capacity. Although surface-water temperatures vary widely with latitude (Fig. 12-1), daily and seasonal fluctuations are much smaller than in land areas at the same latitudes (see Chapter 6). Thus marine organisms are rarely subjected to large or rapid temperature changes, as many land forms are.

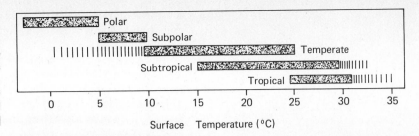

Figure 12-1

Plants and animals live at all temperatures in the world ocean. Note that temperature variations are greatest in mid-latitude waters and least at high latitudes. [After J. W. Hedgpeth (Ed.), 1957. *Treatise on Marine Ecology and Paleoecology.* The Geological Society of America, New York. p. 364.]

Most marine animals are cold blooded, meaning that their internal temperatures are essentially the same as the surrounding water. Within the temperature tolerance range for a species, such processes as growth and development, decomposition, photosynthesis, oxygen consumption, and metabolic rate accelerate with increased water temperature. As a rule, cold-blooded organisms can more easily tolerate temperature changes toward low extremes than toward the high end of their tolerance range. Pronounced cooling is likely to bring about a period of quiescence in which organisms require little or no food and less oxygen than normal. Extreme heating causes death.

Most marine organisms in the open ocean have body fluids whose salinity is close to the seawater in which they live. Thus they need not develop mechanisms to protect themselves from losing water, as land forms do, nor are they in danger of taking in too much or too little salt. But salinity is often variable in intertidal, estuarine, and coastal environments. Some fishes and invertebrates have ways of excreting water, others of excreting salt, and yet others of protecting their internal salinity with outer shells or scales. In estuaries, where there is a marked salinity change going from freshwater to seawater, the different kinds of plants and animals are distributed according to their tolerance for a fresh, brackish, or high-salinity environment (Fig. 12-2). Note that the fewest species live in brackish water, where salinity is most variable.

Another important aspect of the ocean as an environment for life is that it is relatively homogeneous. Seawater carries dissolved nutrient salts, oxygen, and carbon dioxide throughout the ocean basins, although they are not uniformly dispersed. Most marine plants need no roots to absorb nutrients and water from soils, nor leaves to exchange gases with the atmosphere and to catch the sun's light. In

Figure 12-2

Distribution of plants and animals according to salinity tolerance in the Tees River estuary, on the northeast coast of England. (After W. A. Alexander et al., 1935. *Survey of the River Tees.* Water Pollution Research Technical Paper No. 5. Her Majesty's Stationery Office, London. p. 66.)

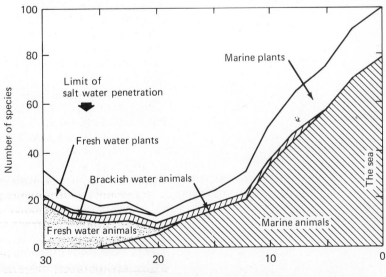

a typical marine plant these functions are carried out within a single cell that absorbs nutrients, light, and water from the surrounding waters.

There are three *principal marine life styles: drifting, swimming,* and *attached.* The drifters *(plankton)* are organisms that have limited ability to swim. Included here are *bacteria, phytoplankton* (one-celled plants), and *zooplankton* (animal plankton), most of which are microscopic. The strong swimmers, or *nekton,* include most adult fishes, as well as squid and whales. *Benthos,* or bottom-dwelling organisms, include large plants that occur only in shallow waters and bottom-dwelling animals at all depths.

Marine communities are classified according to habitat as *intertidal, neritic* (coastal ocean), *pelagic* (free swimming), *benthic* (bottom dwelling), and *abyssal* (deep ocean) (Fig. 12-3). Perhaps the most important distinction in the living environment is between the *photic,* or sunlight surface zone, and the *aphotic,* essentially dark waters below. In the photic zone phytoplankton produce the carbohydrates and proteins that support almost all marine life at all depths in the ocean.

The depth of the photic zone depends on water clarity (see Chapter 6) as well as the intensity of sunlight. In the clearest open ocean about 50% of light penetrating the surface remains at 18 meters depth, with about 1% extending to 100 meters. But nearshore—for example, off Cape Cod or in the English Channel—phytoplankton grow so abundantly in the nutrient-rich waters that they block the sunlight. As a result, only about 50% of the light penetrates to 3.5 meters and 10% to 8 meters. At 17 meters depth only 1% of insolation remains. The 1% level in insolation is usually considered the bottom of the photic zone.

Figure 12-3

Classification of marine environments.

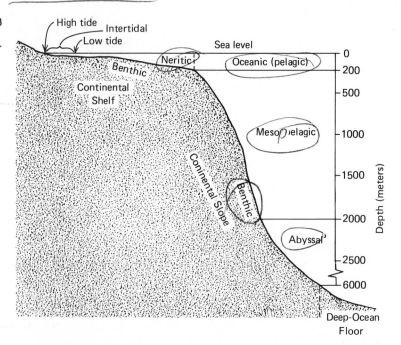

LIMITING FACTORS IN MARINE ECOSYSTEMS

Plants and animals in the ocean, as on land, depend on one another to provide the conditions that support life. Members of a community exchange matter and energy with other organisms and with the nonliving environment. In a self-sustaining community, or *ecosystem, autotrophic* organisms (plants) produce food from inorganic compounds, using energy from the sun. *Heterotrophic* organisms (animals) consume plant food and *decomposers* (bacteria) break down organic matter into

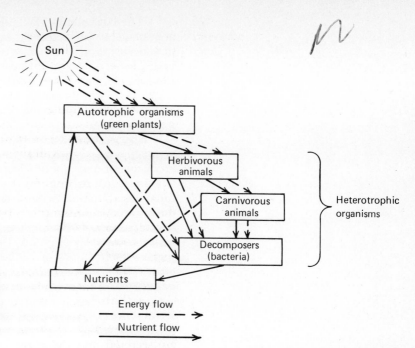

Figure 12-4

Flow of matter and energy through the marine food web. Nutrient salts are returned to solution in seawater to be recycled. Energy, originally derived from the sun and stored by plants in the form of carbohydrates, is dissipated as work and heat.

its inorganic components (Fig. 12-4). Dynamic patterns of growth and interaction may cause the amount of a particular element to vary from time to time, but the basic components remain in balance so that the ecosystem as a whole maintains a steady state.

Population size is controlled by *limiting factors*. In other words, population growth is limited by the availability of some factor, such as food, living space, or another resource. Factors that limit phytoplankton production control marine ecosystems, for most life in the ocean depends on food production by these microscopic floating plants. Phytoplankton growth is limited by three factors: availability of *light*, *nutrients*, and *trace organic materials* in the water.

During winter lack of sunlight limits plant growth in mid- and high-latitude waters. When sunlight becomes stronger in early spring, the phytoplankton *bloom* locally—that is, reproduce rapidly. Under favorable conditions they may become so abundant that plants growing near the surface cut off light from those below. In this case, light again becomes a limiting factor, for growth of plants is then limited to the topmost surface waters. In clear water photosynthesis takes place to depths of 100 meters or more.

Scarcity of nutrients, nitrates, or phosphates in near-surface waters limits plant growth over much of the ocean. It is particularly true in the centers of ocean basins for two reasons. First, in very deep waters dying plants and animals sink below the photic zone before decomposition releases all the nutrients into the water. The rate of nutrient return from deep to surface waters in these areas is slow.

Secondly, in the tropics, where light levels are high throughout the year and there are no seasons, phytoplankton can use available nutrients throughout the year and so they never build up in the water. Outside the tropics, in contrast, nutrient levels rise during the winter when photosynthesis is light limited.

Besides phosphorous and nitrogen, many other elements are essential to marine plants; some are typically present in quantities that limit the growth of a particular species of organism. Lack of silicon for example, can limit growth of *diatoms*, a common high-latitude phytoplankter, or of *silicoflagellates* (these organisms are described in Chapter 13).

The last column, *(S/P)* in Table 12-1, is a measure of the relative abundance of each element in a kilogram of seawater(s) compared to its concentration in 100 grams of dry plankton tissue *(P)*, a figure that varies widely for the different elements. *S/P* is small for nitrogen, phosphorus, and silicon, normally available in seawater as NO_3^- (nitrate), PO_4^{3-} (phosphate), and $Si(OH)_4^{4-}$ (silica). Distribution of these nutrients is affected by uptake and release by organisms; thus they behave as nonconservative elements in seawater.

Plankton exhibit remarkably constant atomic ratios of carbon, nitrogen, and phosphorus (see Table 12-2). The latter two elements are incorporated in animal protein and skeletal materials and a small amount is lost through undecomposed shells and skeletons incorporated in sediments. Most nutrients, however, are returned to seawater during decomposition of organic matter by bacteria. Much of the decomposition occurs in near-surface waters and the nutrients released are quickly recycled. The remainder of the decomposition occurs in deep waters or sediments below the pycnocline. These nutrients move with the subsurface currents. They are returned to the photic zone by upwelling or other vertical transport processes.

A third factor limiting phytoplankton growth is the presence (or absence) of *organic materials in trace amounts*. Such substances as vitamins, antibiotics, toxins (poisons), hormones, and trace nutrients are secreted into the water by bacteria, plants, or animals. One or more of these substances may be necessary in some cases (in other cases, detrimental) to the success of another plant or animal species. Recently upwelled waters are often relatively unproductive because organic compounds, such as vitamins, are lacking.

TABLE 12-1

Abundance of Biologically Important Elements in Plankton and Seawater*

ELEMENT	PLANKTON† *(P)* (grams per 100 grams of dry weight)	SEAWATER *(S)* (grams per cubic meter, 35‰)	RATIO OF ABUNDANCE IN SEAWATER RELATIVE TO PLANKTON *(S/P)*
O	44.0	8.57×10^5	19,600
C	22.5	28	1.25
Si	20	3	0.15
K	5‡	380	72
H	4.6	1.08×10^5	23,500
N	3.8	0.5	0.13
Ca	0.8	400	500
Na	0.6	1.05×10^4	17,500
S	0.6	900	1500
Cl	0.5‡	1.9×10^4	38.000
P	0.4	0.7	0.17
Mg	0.32	1.3×10^3	4100
Fe	0.035	10^{-2}	0.29
Zn	0.026	10^{-2}	0.39
B	0.01‡	5	500
I	0.003‡	6×10^{-2}	20
Cu	0.002	3×10^{-3}	1.5
Mn	0.00075	2×10^{-3}	2.7
Co	0.00005	3×10^{-4}	6
V	0.00005	2×10^{-3}	40

*After H. J. M. Bowen, 1966. *Trace Elements in Biochemistry*. Academic Press, London.

†Mainly diatoms.

‡Concentrations poorly known.

TABLE 12-2

Atomic Ratios of Carbon, Nitrogen, and Phosphorus in the Elementary Composition of Plankton*

TYPE	CARBON	NITROGEN	PHOSPHORUS
Zooplankton	103	16.5	1
Phytoplankton	108	15.5	1
Average	106	16	1

*After A. C. Redfield, B. H. Ketchum, and F. A. Richards, 1963. The influence of organisms on the composition of seawater. In M. N. Hill (Ed.), *The Sea*. Interscience, New York, Vol. II, pp. 26–77.

PHOSPHORUS AND NITROGEN CYCLES

Nitrate and phosphate concentrations vary widely because of currents and the processes that take up nutrients as plants grow and then release them after the organisms die (see Fig. 12-5). Most nutrients are released soon after an organism's death and are quickly recycled in the near-surface waters of the photic zone. Other nutrients incorporated in organic matter sink into deep water. Because decomposition also occurs below the photic zone, organic nitrates, phosphates, and silicates are released there but not taken up by photosynthesis. In tropical and subtropical surface layers, as well as in temperate environments following a phytoplankton bloom, inorganic nitrate and phosphate concentrations may be undetectable chemically.

In shallow coastal waters nutrients recycle fairly quickly. Winter storms mix the entire water column; so when light increases in early spring, a bloom takes place. Seasonal variations in phosphorus content of subpolar coastal waters are shown in Fig. 12-6. During the winter, when phytoplankton populations are small, dissolved inorganic phosphorus concentrations are as high as 0.002 milligram of atomic phosphorus per liter or more. The first spring bloom, in about February or March, causes total soluble phosphorus levels to drop sharply while the particulate phosphorus (in plants and later in animals) rises correspondingly. When nutrient levels drop off, the organisms die and release nutrients or, in some cases, a zooplankton bloom results from the abundance of plant food and the animals release nutrients during metabolism. In any case, there is often a second bloom in April or early May, again causing a drop in soluble phosphorus and a rise in particulate organic matter.

Zooplankton are an important source of dissolved phosphorus

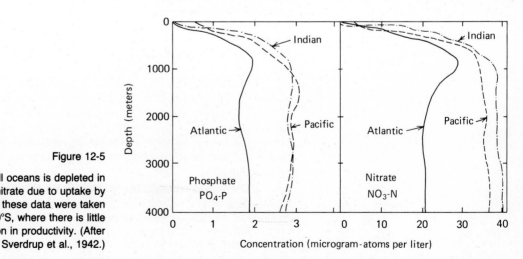

Figure 12-5

The photic zone in all oceans is depleted in phosphate and nitrate due to uptake by phytoplankton. Most of these data were taken between 30°N and 30°S, where there is little seasonal variation in productivity. (After Sverdrup et al., 1942.)

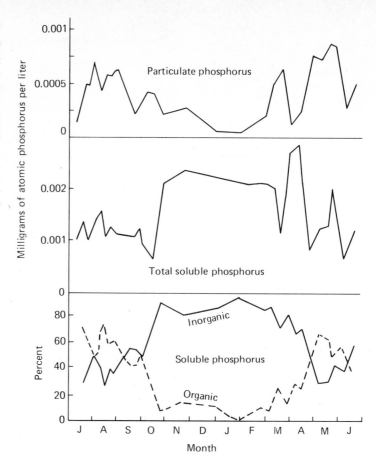

Figure 12-6

Seasonal variations in three forms of phosphorous in a coastal embayment (Departure Bay, British Columbia). Particles containing phosphorous—for instance, plants and animals—are abundant in surface waters during spring and fall blooms. Inorganic soluble phosphorus content of surface waters is highest between blooms—that is, during the winter months. The curve for organic soluble phosphorus shows that it is present when plants and animals are in the water; thus it is close to the curve for particulate phosphorus. (After Parsons and Takahashi, 1973. *Biological Oceanographic Processes.* Pergamon Press, Oxford. p. 37.)

in many environments. For instance, the shrimplike copepod, *Calanus,* in a North Atlantic study, retained 17% of its dietary phosphorus for growth, eliminated 23% in solid wastes, and excreted 60% as soluble phosphorus. Another study showed that during 10 weeks of a phytoplankton–zooplankton bloom in the North Sea, inorganic phosphate remained at or above 0.0006 milligram of atomic phosphorus per liter. But in the absence of zooplankton, phosphate levels dropped to 0.0001 milligram of atomic phosphorus per liter after 2 weeks.

Dissolved and particulate organic phosphorus released by phytoplankton and animals is utilized by bacteria and certain heterotrophic phytoplankton. Because many bacteria are consumed by one-celled animals, the organic phosphorus→bacteria→microzooplankton chain (Fig. 12-7) provides a pathway by which organic phosphorus reenters food webs directly without being changed to inorganic phosphate.

Phosphates dissolve readily from organic compounds, aided by bacterial and phytoplankton enzymes that are abundant in seawater. But nitrogen cycles more slowly because energy is required to "fix" it in organic form, reduce it, and finally oxidize it back to nitrogen. Zooplankton facilitate this process by releasing soluble nitrogen compounds, such as urea and ammonia, which are readily taken up by phytoplankton.

Under conditions of nitrate or phosphate depletion, marine plants can survive with greatly lowered internal concentrations of these nutrients for short periods. Phytoplankton can extract phosphate from seawater rapidly and "hoard" much greater amounts than needed for growth. For this reason, they can continue to grow for several generations after the supply in the water has been depleted.

328

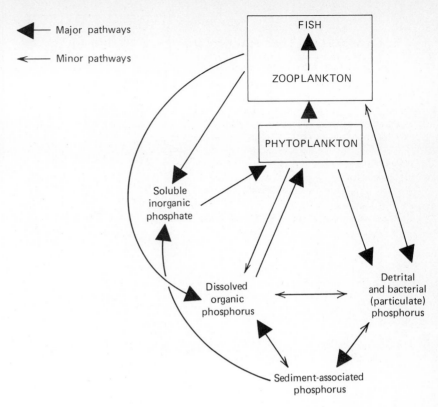

Figure 12-7

Major pathways in the phosphorus cycle of the sea. Fish and zooplankton release soluble organic and inorganic phosphates and phytoplankton release dissolved and particulate organic phosphorus. Plants take up soluble inorganic phosphates and some are able to utilize dissolved organic phosphorus. Some organic phosphorus enters sediment and is eventually released, after decomposition, as inorganic phosphate. (Modified from Parsons and Takahashi, 1973.)

Nitrate uptake and release in temperate environments generally follow a pattern similar to that for phosphorus except that nitrates nearly disappear from surface waters during summer. But in tropical and subtropical oceans the scarcity of nitrates probably limits productivity at all times.

Bacteria play an important part in the nitrogen cycle (Fig. 12-8) because they oxidize ammonia to nitrites and subsequently to nitrates. Reduction of nitrates and nitrites also occurs, as Fig. 12-8 shows, because these reactions are reversible. Some nitrate is lost by formation of nitrogen gas through denitrification, but certain algae and bacteria also "fix" dissolved nitrogen—that is, assimilate it directly from seawater, especially in tropical regions.

Figure 12-9 summarizes seasonal variations in concentrations of limiting nutrients in the English Channel. In spring and late summer intense diatom blooms deplete the nutrients, in the water; the curve for silicon shows this situation especially well. During winter, when lack of sunlight and mixing of surface waters below the photic zone inhibit plant growth, nutrient levels are at a maximum.

ORGANIC GROWTH FACTORS

Scarcity of dissolved organic compounds, such as *vitamins*, can also limit plant growth. Organisms that secrete these substances precondition waters for other organisms that require them; thus the production of such "micronutrients" often determines succession of species.

One or more vitamins—specifically B_{12}, thiamine, and biotin—are required by many phytoplankton species. Bacteria are the main producers of B_{12}, but other bacteria require it and some phytoplanktors secrete it. Complex relationships arise between producers and consumers of vitamins. The diatom *Nitzschia punctada*, for instance, requires B_{12} but synthesizes biotin and thiamine. A related species, *N. closterium*, requires both B_{12} and thiamine; *N. putrida* does not

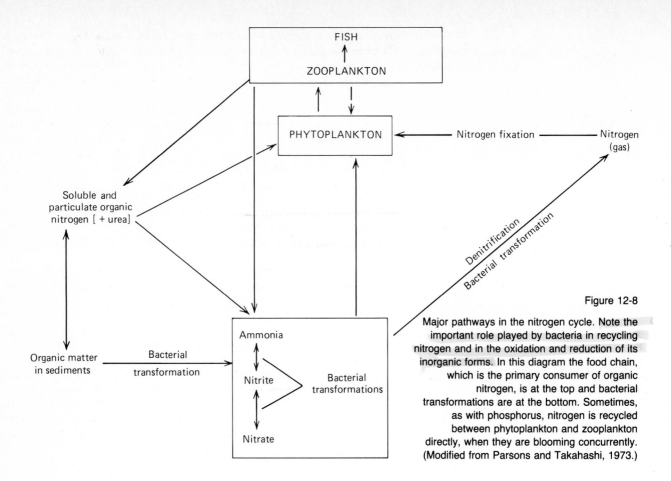

FISH

ZOOPLANKTON

PHYTOPLANKTON ← Nitrogen fixation ———— Nitrogen (gas)

Soluble and particulate organic nitrogen [+ urea]

Organic matter in sediments — Bacterial transformation →

Ammonia

Nitrite

Nitrate

Bacterial transformations

Denitrification
Bacterial transformation

Figure 12-8

Major pathways in the nitrogen cycle. Note the important role played by bacteria in recycling nitrogen and in the oxidation and reduction of its inorganic forms. In this diagram the food chain, which is the primary consumer of organic nitrogen, is at the top and bacterial transformations are at the bottom. Sometimes, as with phosphorus, nitrogen is recycled between phytoplankton and zooplankton directly, when they are blooming concurrently. (Modified from Parsons and Takahashi, 1973.)

Figure 12-9

Seasonal variation in nitrate, silicate, and phosphate in the English Channel. The curve for silicate reflects the fact that, in the channel, diatoms (yellow green algae with siliceous shells) bloom profusely in spring and again in late summer. Their shells are voided by animals rather than being incorporated into animal tissue. Channel waters are so shallow and well mixed that there is little difference in nutrient concentration with depth in the water column at any time. (After J. E. G. Raymont, 1963. *Plankton and Productivity in the Ocean.* Macmillan, New York. 625 pp.)

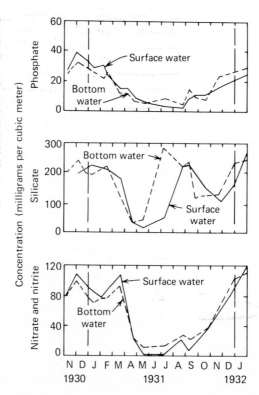

require any. A dinoflagellate plankter, *Gymnodinium brevis*, requires all three. In general, about half or less of the diatom, green algae, and blue green algae species are known to have vitamin requirements whereas among other groups of phytoplankton the vitamin-requirers dominate.

In the open ocean vitamin B_{12} concentrations are commonly higher below 50 meters than in surface waters. In some areas lack of it may limit plant growth. Coastal waters normally contain an adequate supply, although it may be temporarily depleted by a heavy diatom bloom. B_{12} is absorbed in large quantities by detritus and is thus ingested by filter-feeding animals.

Soluble organic *chelating agents* (e.g., amino acids, nucleotides, and hydroxy acids) can be growth factors. They form compounds with certain metals that otherwise cannot be utilized by phytoplankton. At the normal pH of seawater, in the absence of chelating agents, iron, manganese, and some other trace elements are insoluble, and hence unavailable to plants. Different organisms require different concentrations of trace metals and chelating agents and, in general, coastal waters are less likely to be deficient in these substances. Iron, for example, may be growth limiting where organic chelators are lacking in tropical and subtropical oceanic waters. Chelating agents can also reduce the toxicity of metals, such as copper, by making the metal unavailable to organisms.

These growth factors can have still other effects. The characteristic form of certain marine plants depends, for instance, on bacterial byproducts. The sea lettuce *Ulva* loses its normal broad-leafed shape and grows into long, tubular filaments in the absence of certain bacteria. The microorganisms, in turn, seem to be stimulated by the seaweed.

FOOD WEBS, TROPHIC LEVELS, AND BIOMASS

Phytoplankton, the main *primary producers* of the ocean, are eaten by the smallest marine animals, the herbivorous zooplankton. These *primary consumers* are, in turn, eaten by *secondary consumers,* either carnivorous zooplankton or fish. The position of an individual or group of individuals in a *food web*—who eats it and whom it eats—(Fig. 12-10) defines its *trophic level.* Plants are the first trophic level, herbivores the second, primary carnivores the third, secondary carnivores the fourth, and so on. At each trophic level, dissolved and suspended organic matter (*organic detritus*) is released from living organisms into the water. (This process should not be confused with the release of inorganic nutrient salts during animal metabolism.)

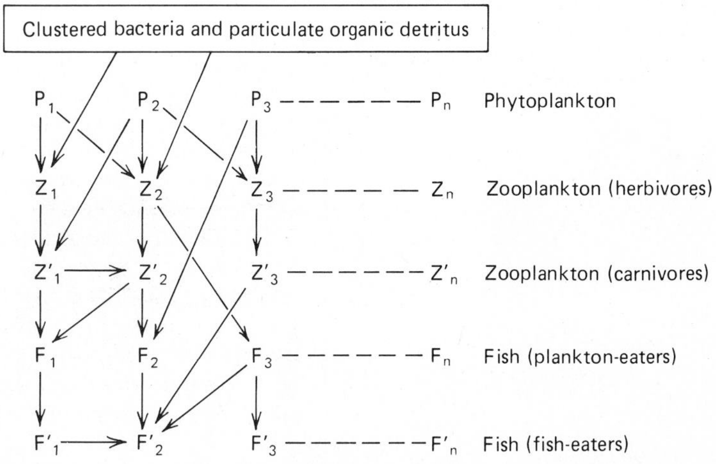

Figure 12-10

A theoretical pelagic food web. Many species (P_1, P_2, P_3, etc.) of organisms at each of five trophic levels are shown. A nearly infinite number of food relationships is possible, although at any given time and place abundance of one organism may dominate a particular trophic level. Note that an organism (e.g., F_2) may feed on at least three trophic levels. (Adapted from Parsons and Takahashi, 1973.)

Clustered bacteria and particulate organic detritus

P_1 P_2 P_3 — — — — — P_n Phytoplankton

Z_1 Z_2 Z_3 — — — — — Z_n Zooplankton (herbivores)

Z'_1 → Z'_2 Z'_3 — — — — — Z'_n Zooplankton (carnivores)

F_1 F_2 F_3 — — — — — F_n Fish (plankton-eaters)

F'_1 → F'_2 F'_3 — — — — — F'_n Fish (fish-eaters)

In most ocean areas relationships among primary producers and consumers are complex, as shown schematically in Fig. 12-10. For discussion purposes, it is often useful to consider simpler relationships in the form of a *food chain* in which each level is eaten only by the next higher trophic level. One of the most conspicuous is the *grazing food chain:*

Phytoplankton→Zooplankton→Fishes→Birds, Humans

Primary producer→Herbivore→Carnivores

Detritus is produced at all levels of food webs; it sinks below the surface, where it supports life at all depths so that energy produced at the surface is transported through food webs to all depths in the ocean. When they die, organisms are eaten by scavengers or decomposed by bacteria, which are themselves a food source for some marine animals.

The detritus food chain is important in the flow of energy from primary producers to fishes and similar organisms or to the sediment deposits. The detritus food chain is much less conspicuous than the grazing food chain and generally involves microscopic organisms living in the sediments. A simple detritus food chain is

Phytoplankton→Detritus→Bacteria, fungi→

→Ciliates, Invertebrates→Fishes
(worms)

Sediments (burial)

Ultimately nearly all the chemical energy stored during photosynthesis is used up in respiration. Recall that energy, unlike matter, is a one-way flow, as illustrated in Fig. 12-4.

Most of the food energy available to a plant or animal cell is used by the cell itself in growth, reproduction, or other functions. Animals use more energy than plants because they actively seek and consume prey. To calculate how much energy is available to the trophic level, *ecological efficiency (E)* is defined as follows:

$$E = \frac{\text{amount of energy extracted from a trophic level}}{\text{amount of energy supplied to a trophic level}}$$

As a general rule, we assume E to be 10 to 20% in marine ecosystems. A 10% ecological efficiency means that it takes 500 grams of phytoplankton to support a population of herbivorous zooplankton that weighs 50 grams. This zooplankton population provides a fish with enough food to increase its body weight by 5 grams.

The value of E is determined by a variety of factors, such as the amount of detritus recycled in a given food chain and the amount of work an animal must do to locate and capture its prey. In general, energy transfers in open-ocean food webs are less efficient than coastal-ocean food webs.

To quantify the amount of organic material in an ecosystem, we use the concept of *biomass*. Biomass is the weight of plants or animals, expressed as weight of organic carbon, per volume of seawater. We might find, for example, that the biomass of primary producers at a location is 880 milligrams of carbon per cubic meter of seawater (mg C/m³). A cubic meter of water taken from the same location might contain 90 milligrams of carbon per cubic meter of herbivorous zooplankton, measured by filtering the water through a fine sieve and picking out zooplankton known to consume only plant food. The same water might also contain 10 milligrams of carbon per cubic meter of carnivorous zooplankton.

Biomass can also be expressed as total weight of organisms in the water under a specified area (e.g., a square meter) of seawater, or grams per square meter. Bottom-dwelling populations are usually measured in terms of weight per square meter of the bottom. Sometimes plant biomass is given as plant pigment (e.g., chlorophyll) units per volume of water.

PHOTOSYNTHESIS

Marine plants, like most of their terrestrial counterparts, are *autotrophs;* they contain *chlorophyll* or other light-absorbing pigment that enables them to make their own food—typically *carbohydrates,* complex organic substances in whose bonds chemical energy is stored. In this process, known as *photosynthesis,* carbon dioxide combines with water and 120 kilocalories of radiant energy are absorbed per mole of carbohydrate formed. The end product is a sugar containing chemical energy with oxygen being released as a byproduct. Thus

Photosynthesis

(chlorophyll)

$$6 \ CO_2 \ + 6 \ H_2O \ + 120 \ kcal \rightarrow \ C_6H_{12}O_6 \ + \ O_2$$

carbon water energy ← carbohydrate oxygen
dioxide

Respiration

This simplified statement shows that during *photosynthesis* oxygen is released and energy is stored, but it does not describe the sequence of chemical steps involved.

The process is reversed during plant or animal *respiration* when oxygen is taken up and chemical bonds are broken. Carbohydrate is *oxidized* and energy released. All *heterotrophic organisms* (animal and most bacteria) obtain energy from breaking down and rearranging chemical-energy-containing materials originally synthesized from inorganic compounds by autotrophic organisms.

Plants also utilize part of the energy they produce. During respiration plants and animals take up oxygen and expend energy on growth, reproduction, and work. The net production of food by a green plant depends on its photosynthetic processes exceeding the amount of energy used in respiration.

The first stages in photosynthesis involve absorption of solar energy by chlorophyll. These stages only take place in sunlight and are called the *light reaction.* Subsequent parts of the synthetic process are carried on irrespective of light conditions. Respiration takes place in light and dark.

There are two basic steps in the light reaction. The first involves synthesis of a substance that will later accept oxygen (defined as a *reduced* substance). Hydrogen is added to a complex organic molecule, *nicotinamide adenine dinucleotide phosphate* (NADP), yielding NADPH:

$$NADP + H \rightarrow NADPH$$

The added hydrogen is now available to create other reduced substances, meaning that NADPH has *reducing potential.*

In the second step, energy-containing molecules are created. Energy from the sun is conserved in a chemical form by means of a cyclical electron-transport chain whereby ATP (adenosine triphosphate) is formed from ADP (adenosine diphosphate) and phosphate (PO_4^{3-}).

$$ADP \quad + \quad P_i \quad \rightarrow \quad ATP$$

adenosine   inorganic   adenosine
diphosphate   phosphate   triphosphate

This process is called *phosphorylation.*

The chemical energy contained in ATP and the reducing power of NADPH now remove oxygen from CO_2, forming carbohydrates, as well as synthesizing proteins and fats. Collectively, this part of the process is referred to as the *dark reaction.*

With cells, plant and animal alike, energy is transferred as shown in Fig. 12-11. A certain amount of energy is conserved in the internal recycling of ATP and ADP, but some is permanently lost, as shown by the two black arrows that represent energy lost from the system.

Figure 12-11

The cyclic transfer of energy within single cells. Conservation of energy is achieved by the ADP–ATP recycling process. In plants the organic "food fuel" is derived from compounds formed by photosynthesis within the cell itself. In animals food must come from consumption of a plant or another animal. (Adapted from W. D. Russell–Hunter, 1970. *Aquatic Productivity.* Macmillan, New York. p. 15.)

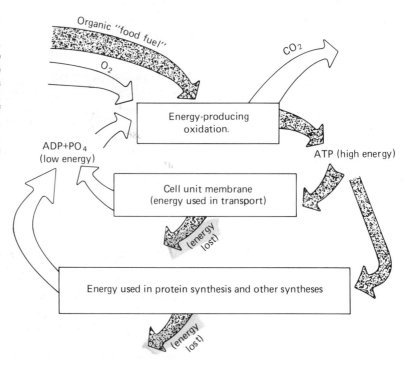

MEASURING PRIMARY PRODUCTIVITY

The net amount of food produced in an area during a given time is the *productivity* of the area's plant population. Productivity is thus a rate phenomenon whereas *standing crop,* or biomass, is a measure of the amount, by weight, of plant (or animal) material in a given area at any time. Productivity of a population; for example, may be high even though its biomass is quite small. Such is the case when grazing by zooplankton sharply depletes a rapidly reproducing, highly productive phytoplankton crop, as shown in Fig. 12-12.

Productivity can be estimated by approximating the average standing crop of a population and multiplying that figure by its rate of generation (doubling time) over short intervals. If the average standing crop of zooplankton is 500 milligrams of carbon per square meter and the population doubles once a month, then its productivity is 6000 milligrams or 6 grams of carbon per square meter per year (assuming constant production throughout the year).

Productivity is hard to measure accurately in the ocean. One way

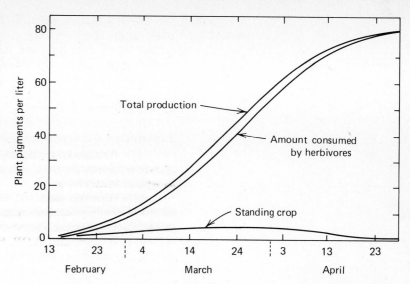

Figure 12-12

Standing crop represents only a small part of total plant production. Data based on a study in the English Channel showing that of 85,000 plant pigment units per cubic meter produced (according to predictions from PO_4^{3-} uptake), only 2500 units per cubic meter, or about 3%, were measured as a standing crop. Difference was assumed to be due to grazing. (After Raymont, 1963.)

to quantify *primary production* (plant productivity) is to introduce radioactive $^{14}CO_2$ into a sample of seawater and monitor its rate of uptake by phytoplankton. This measures *gross productivity*, which is defined as the mass of carbon fixed per unit area per unit time, usually expressed in grams or kilograms of carbon per square meter per year. Net productivity is calculated from gross productivity by correcting for losses due to respiration, which are usually 10 to 50%.

Another method for measuring productivity is to determine the amount of oxygen produced by the plants in seawater because oxygen is given off by plants in direct proportion to the amount of organic carbon synthesized. In this "oxygen-bottle" technique, sealed bottles of seawater containing phytoplankton are lowered to selected depths below the surface. After a known period, the oxygen content of the water in the bottles is measured and compared with the amount present when the bottles were sealed.

In the photic zone, production of organic matter decreases with depth until a point of no net increase is reached. It is known as the *compensation depth,* a point where the organic matter (and oxygen) produced by photosynthesis exactly equals the amount utilized in respiration. At greater depths, respiration exceeds photosynthesis and no plant production occurs. Finally, no organic matter is produced at all and plants eventually die. The concept of compensation depth, then, is used in reference to metabolism of an individual plant sample at a specific depth, which is determined by the extent of light penetration.

Marine and land plants contribute about equal amounts annually to global food production. The productivity of phytoplankton is several orders of magnitude greater than that of land plants, but the biomass of plankton in the world ocean appears to be several orders of magnitude less than that of terrestrial producers, such as trees, grass, and agricultural crops.

FACTORS IN PLANT PRODUCTIVITY

The most important factor in the productivity of an ocean area is the rate of nutrient supply to the photic zone. For this reason, most continental shelves are highly productive; they are shallow and well mixed and the estuarine circulation tends to retain nutrients in the coastal ocean. These factors and the close coupling between the water column and nutrient regeneration in the sediments lead to high rates of nutrient resupply to the photic zone.

335

Photosynthesis at mid- and high latitudes is inhibited during local winter, however, due to lack of sunlight. Plants can grow in near-surface water, but they are frequently carried below the sunlit zone because of mixing by storms in the absence of a well-stabilized surface zone. Thus the plants cannot produce effectively because they use as much (or more) energy through respiration than they produce through photosynthesis.

In spring increased sunlight permits phytoplankton to grow and reproduce rapidly in areas where the waters are well supplied with nutrients. But there can still be no productivity increase as long as plants are carried below the photic zone, causing net respiration to exceed production.

The concept of *critical depth* expresses the relationship between the depth of the surface layer in which photosynthesis equals the respiration and the depth of the surface mixed layer. In spring when a thermocline is formed by warming at the surface plants are no longer mixed below the critical depth. Production equals or exceeds respiration and phytoplankton can reproduce. Very rapid reproduction causes a bloom.

As light increases in spring (as early as January in Long Island Sound), phytoplankton begin to bloom. At this time the compensation depth is very near the surface because sunlight is still relatively weak. The critical depth (always below the compensation depth because it represents a lower limit for the growing population as a whole) may occur at some point within the mixed surface layer. In this case, plant growth is retarded because part of the population is continually being mixed below the critical depth. As sunlight becomes stronger, the critical depth may extend to the bottom in shallow areas. In deeper waters a thermocline may form at or above the critical depth. In either case, the plant population expands rapidly as soon as the critical depth exceeds the depth of the mixed surface layer, as found in the North Sea (Fig. 12-13).

Differences between the biological properties of coastal and offshore waters are illustrated by the sequence of blooms off the New England coast. In shallow, nearshore waters the spring diatom increase begins sometime between late January and early March. Farther offshore, on Georges Bank, diatoms do not bloom until late March or April, and this process occurs up to a month later in continental slope waters and the open Gulf of Maine. Blooms near the coast are smaller than offshore; and there may be additional blooms in late summer or fall, which does not usually happen in deeper water. Another difference is that the number of individuals in nearshore areas never falls as low as in continental slope and Gulf of Maine waters. In both areas, heavy grazing by zooplankton terminates the blooms within a few weeks.

The different oceanographic conditions in these waters explain the differences in productivity. Near the coast, river discharge lowers surface salinities and forms a stable surface layer during much of the year. Phytoplankton production begins to exceed consumption soon after there is an increase in sunlight (Fig. 12-14). Since some photosynthesis goes on all winter, nutrients never accumulate to the extent that they do offshore and the spring flowering is less dramatic. There is some nutrient enrichment of near-bottom waters during the summer as animals feed, release organic detritus into the water, and die. Detritus and decomposition products sink to the bottom but are recycled from time to time through the shallow waters, bringing nutrients to the surface and causing late summer blooms.

Farther out on the continental shelf, in deeper waters, the spring bloom does not occur until a thermocline develops. Then as organic matter decomposes, it sinks far below the photic zone. Nutrients re-

Figure 12-13

Observations at a weather station (66°N, 2°E) in the North Sea show that plankton do not appear in large numbers until the critical depth exceeds depth of the mixed layer. (After R. W. Wimpenny, 1966. *The Plankton of the Sea.* American Elsevier, New York.)

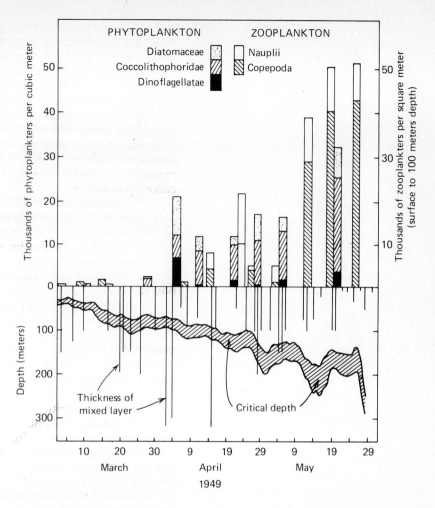

Figure 12-14

The seasonal cycle of phytoplankton off Woods Hole, Mass. [Redrawn from G. A. Riely, 1969. Seasonal fluctuations of the phytoplankton. In D. J. Reish (Ed.), *Biology of the Oceans,* p. 72. Dickenson Publishing Co., Belmont, CA.]

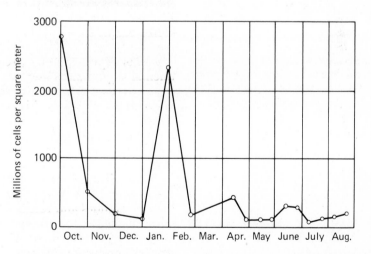

leased are not recycled to the surface until winter storms break down the thermocline and a new annual cycle begins. Thus one bloom is the rule in the open ocean rather than the two that occur closer to shore.

Red tides are particularly striking examples of plankton blooms. They occur when dinoflagellates become so abundant that they discolor the water, often giving it a "tomato soup" color extending for hundreds of yards or even many kilometers. In some instances, red tides are accompanied by cases of paralytic shellfish poisoning in which neurotoxins produced by certain dinoflagellates are concentrated by shellfish. Often there is no discernible effect on the organisms themselves, but when eaten by humans, the toxins cause symptoms ranging from tingling in the hands and feet to paralysis and even death. Shellfish beds are usually closed to harvesting during red tides.

Red tides result from a redistribution of organisms due to several coincident circumstances. First is stratification of the water column, caused by heavy rains or river discharge. Then in the presence of strong sunlight dinoflagellates reproduce rapidly and swim toward the light. They are concentrated from a large volume of water into the surface layer.

In addition to paralytic shellfish poisoning, decomposition of red tide organisms at high temperatures uses up the dissolved oxygen in bottom waters, killing bottom-dwelling fishes and other organisms.

Red tide outbreaks have been noted in New England coastal waters in May and June and again sometimes in late August and early September when nutrient levels are high in a stable surface layer. In southern California red tides are restricted by the longshore current to a narrow band of waters along the coast or to local embayments.

Frequently there is a direct relationship between increased insolation and increased productivity. But not all plants respond to light level in the same way. Figure 12-15 shows how three types of marine phytoplankton respond to increased light intensities. In the study from which these data were taken, light saturation occurred between 500 and 750 foot-candles for green algae, between 2500 and 3000 foot-candles for dinoflagellates, and at intermediate light levels for diatoms. Above these intensities, photosynthesis decreased. "Sun-type" species that flower in the summer, or at shallow-water levels, commonly photosynthesize most efficiently at high light intensities. During the winter and in deep waters, however, "shade-type" communities flourish because they photosynthesize best at lower light intensities. They can evidently utilize weak sunlight more efficiently than the "sun-type" species can.

Figure 12-15

Relative photosynthesis—light curves in some marine phytoplankton. (After Parsons, Takahashi, and Hargrave, 1977.)

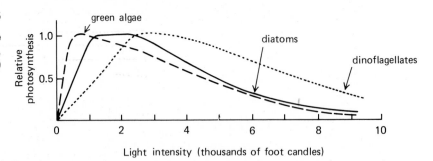

NONLIVING ORGANIC MATTER

Most of the ocean's organic content is not living but is dissolved or occurs as very small, nonliving particles (Table 12-3). Similar data appear graphically in Fig. 12-16, although here no distinction is made between living and nonliving matter. Note the tiny fraction of particulate matter (greater than 10^{-3} millimeter), the larger proportion of *colloidal*-sized particles (10^{-3} to 10^{-6} millimeter) and the very large percentage of dissolved organic matter (particle size less than 10^{-6} millimeter). Open ocean waters contain about 0.4 to about 2 milligram of carbon per liter (mg C/l) of dissolved material, and particulate organic matter ranges from about 0.01 milligram of carbon per liter in deep water to about 0.1 to 0.5 milligram carbon per liter near the surface. This enormous reservoir of nonliving organic material reenters food webs largely through bacterial activity.

TABLE 12-3

*Relative Abundance of Various Forms of Organic Matter in Seawater**

FORM OF ORGANIC MATTER	RELATIVE ABUNDANCE (percent)
Dissolved organic matter ($< 10^{-6}$ mm)	80
Colloidal organic matter (10^{-6}–10^{-3} mm)	18
Nonliving particulate organic matter ($> 10^{-3}$ mm)	1.8
Phytoplankton	0.2
Zooplankton	0.02
Fishes	0.0002

*Data from J. H. Sharp 1973(b). Size classes of organic carbon in seawater. *Limnology and Oceanography* **18**: 441–447; and Raymont, 1963.

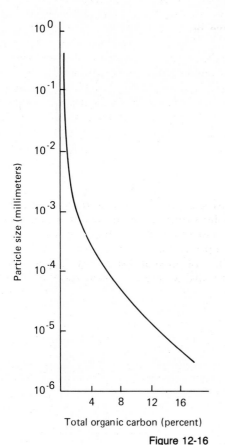

Figure 12-16

Generalized size distribution of organic carbon in seawater. Only about 2% is in particles larger than 1 micron (10 millimeters). Only about 18 to 20% is in particles larger than 10 millimeter. [After Sharp, 1973(b).]

Sources of dissolved organic compounds are as follows.

1. Decomposition of dead plants and animals. Up to 50% of an organism's total mass may dissolve in seawater as cell membranes are decomposed by bacterial enzymes. Subsequent bacterial action releases more amino acids, peptides (a type of protein derivative), carbohydrates, and fatty acids into the water.

2. Secretion by plants. Sugars, amino acids, fats, organic phosphates, vitamins, enzymes, toxins, and a variety of other organic solutes are released by living phytoplankton. Less than 10% of the carbon assimilated during photosynthesis is usually released into the water as dissolved organic compounds during active plant growth and reproduction. But under conditions of stress, such as in very high light intensities, 50% or more of the photoassimilated carbon may be released. This situation occurs near the end of a bloom or when a population is, for some reason, inhibited from reproducing. Amounts released by healthy populations generally vary inversely with the availability of nutrients. A greater proportion of organic matter (12 to 27%) is secreted on a daily basis in nutrient-poor waters of the open ocean than in more fertile coastal waters (6 to 12%).

3. Excretion and exudation by animals. Animals excrete some dissolved organic waste products during normal metabolic activity. Certain animals also release dissolved materials while moulting, or sloughing off one shell to grow another.

Principal sources of particulate organic detritus are

1. Dead organisms and parts of them.
2. Solid wastes excreted by animals.
3. Dissolved and colloidal organic matter that adheres to solid particles or surfaces in seawater. This topic is discussed in more detail later.

Organic carbon concentrations in the North Atlantic are highest near the surface because there is more life there than in deeper waters (Fig. 12-17). For the same reason, shallow and nearshore waters contain proportionately more particulate organic carbon than the open ocean (Table 12-4). In open-ocean surface waters, 10 to 50% of the organic matter may be alive—hence particulate. But in the deep waters below the photic zone, more than 90% of organic matter is nonliving. So the particulate fraction in open ocean water from surface to depth is only about 1% of total organic carbon and usually less than that in deep waters.

TABLE 12-4

*Particulate Organic Carbon as a Percent of Total Organic Carbon for Various Marine Environments**

AREA	PERCENT
North central Pacific Ocean (subtropical)	0.7
Central western North Atlantic Ocean (includes Gulf Stream, Caribbean)	1.5
Strait of Georgia, North Pacific (nearshore)	13
Chukchi Sea, Arctic Ocean (nearshore, shallow)	24

*Data from Sharp 1973(a).

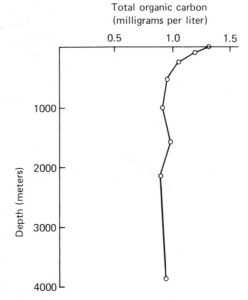

Figure 12-17

Depth curves of total organic carbon from two studies in the western North Atlantic Ocean. [Data from J. H. Sharp, 1973(a). Total organic carbon in seawater. *Marine Chemistry* 1: 211–229.]

BACTERIA IN FOOD WEBS

Bacteria are less than 2 microns in diameter (< 0.002 millimeter) and have negligible biomass. But their role in marine food webs cannot be evaluated on the basis of weight, as it is for plants and animals, because decomposition rather than energy transfer is their most important function.

A total of 2.5 to 3.0 × 10^{10} metric tons of CO_2, for instance, is "fixed" annually by photosynthesis. An equivalent amount is therefore released through respiration during the same period, with decomposition of organic compounds releasing inorganic nutrients. Bacterial activity accounts for roughly 90% of this turnover. Populations of bacteria turn over quickly so that at any one time the bacterial biomass may represent only about one-tenth of annual production. Populations can double in a few hours to a few days, but predation usually keeps their number constant.

In general, there are thousands of living bacteria per milliliter (cubic centimeter) of coastal water, although as few as ten and as many as several million have been counted in different water masses. In the open ocean we normally find a few hundred per milliliter, but below 100 meters depth there may only be a few per liter. Some species can live free in seawater; others must be attached to a surface in order to grow and reproduce.

At surface temperatures and pressures, bacteria have a high

340

oxygen consumption and metabolic rate, but at the low temperatures and high pressures of the deep ocean they seem to be less active. This fact was inadvertently demonstrated in 1968 when the Woods Hole Oceanographic Institution's submersible research vessel, *Alvin,* sank with the hatch open in 1540 meters of water and remained on the bottom for 10 months. When recovered, the crew's lunch of apples, bologna sandwiches, and bouillon in thermos bottles was found in a plastic box, wrapped but soaked with seawater. The food smelled and tasted fresh when it was first opened, but after 4 weeks of storage at 3°C in sterile seawater it became putrid. Additional experiments showed that organic material, wrapped to prevent water or air from circulating past it, decomposes 10 to 100 times more slowly at deep-ocean pressures than at the surface, even at very low temperatures. Studies of digestive tract fauna in deep-ocean animals, however, indicate that bacteria play the same role in decomposing undigested material as they do at sea level.

Bacteria are important sources of concentrated protein for some filter-feeding animals and for organisms that scrape accumulated detritus from underwater surfaces. Individual microbes usually cannot be captured directly from the water because of their small size, but colonies of bacteria become locally concentrated on suspended particles and other submerged surfaces in the ocean [Fig. 12-18(a)]. Bacteria collect on the surface-active or oily organic residues that adhere to sediment particles, piers, and ships' hulls, forming a greasy film. Dissolved and colloidal organic compounds also clump together in seawater, aggregating into amorphous masses called "marine snow." Such organic masses attract bacteria. Any solid object in the water soon becomes coated with a film of organic matter containing bacterial microcolonies and their metabolic products. Many benthic organisms colonize surfaces where bacterial slime has accumulated, a process known as *fouling* [Fig. 12-18(b)]. Some organisms may simply be seeking a firm surface on which to attach themselves; others live by scraping off and consuming the film with its aggregated organic components [Fig. 12-18(c)].

Figure 12-19 summarizes major pathways by which materials are recycled in marine ecosystems. Organic matter derived from byproducts of plant and animal metabolism is shown cycling to the right of the living components, products of decomposition to the left. Note that bacteria decompose organic material and release carbon dioxide and inorganic nutrients; these reenter the water to support photosynthesis. Some inorganic nutrients are also returned directly to the environment by animal excretions. The term *heterotrophic uptake* refers to direct assimilation of dissolved substances from seawater by bacteria and certain kinds of heterotrophic phytoplankton. For example, some bacteria quickly take up and assimilate dissolved organic matter that has been exuded by photosynthesizing plants, such as amino acids and glucose.

DISSOLVED OXYGEN IN THE OCEAN

Bacteria utilize oxygen to decompose organic matter and thereby release nutrients. For this reason, curves showing distribution of oxygen in the ocean (Fig. 12-20) are usually a mirror image of nutrient distributions, at least in the top 1000 meters (see Fig. 12-5). Note in Fig. 12-20 that in the North Atlantic, where newly formed deep waters are moving toward the bottom, oxygen levels are high throughout the water column. But in most of the ocean, although surface layers are saturated with oxygen, its concentration diminishes with depth for a few to several hundred meters just below the pycnocline due to respiration and decomposition of animals and bacteria.

Oxygen demand at great depths is limited. Low temperatures

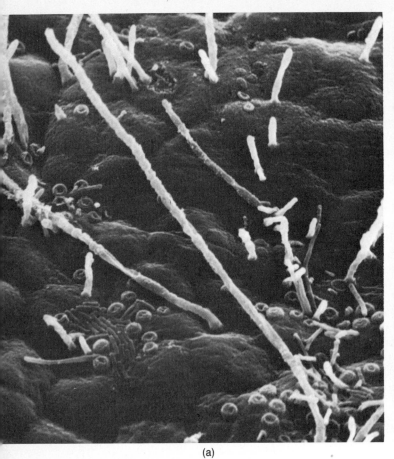

(a)

(b)

Figure 12-18

Bacteria colonize submerged surfaces, creating a slimy matrix that supports many benthic organisms. (a) Rod-shaped, ring-shaped, and filamentous bacteria attached to the red seaweed *Rhodymenia palmata*. The longest filament here is about 0.05 millimeter (50 microns) in length. (b) A "garden" of diatoms and bacteria grows on fiberglass rods under water. An area about 0.37 millimeter square is shown in this photograph. (Photograph courtesy Dr. J. M. Sieburth and University Park Press, Baltimore, Md.) (c) The tunicate *Molgula*, a common fouling organism in Chesapeake Bay, feeds on organic detritus from the side of an aquarium. (Photograph courtesy Michael J. Reber.)

(c)

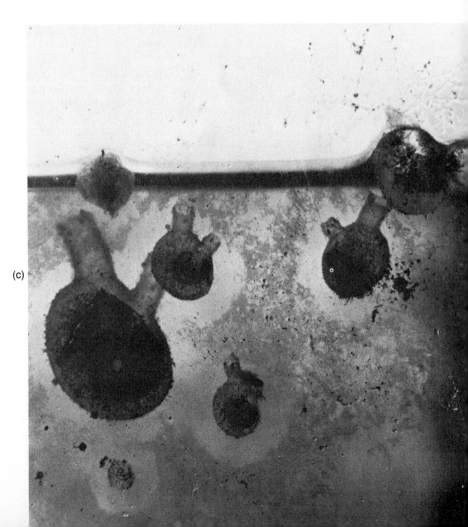

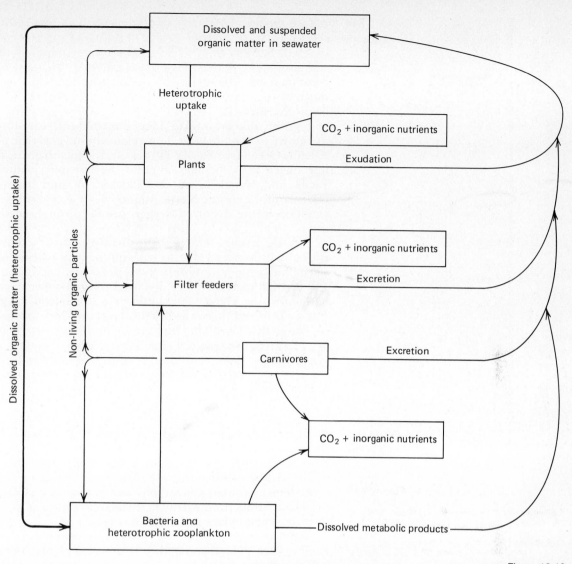

Figure 12-19

Pathways of transfer and nutrient remineralization from dissolved and suspended organic matter in the marine environment. Compare with Fig. 12-7. Note that both plants and animals contribute dissolved and suspended organic matter to the reserves in seawater; some of it reenters the food web via filter-feeding animals and heterotrophic plants and the rest is the decomposed bacteria. (Modified from Parsons, Takahashi, and Hargrave, 1977.)

Figure 12-20

Characteristics oxygen profiles of major ocean basins. Positions: 12°N, 137°W; 0°N, 80°E, 9°S, 5.5°W; 55°N, 45°W. (After Dietrich, 1963.)

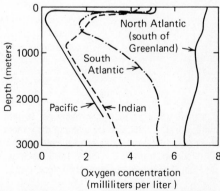

reduce the metabolic rates of organisms; high pressures also seem to reduce bacterial activity so that the rate of oxygen consumption is lowered. Scarcity of food keeps deep-water populations sparse. Decomposition of detrital material is minimal, for most of it has been consumed near the surface. Consequently, deep-ocean waters contain much of the oxygen with which they were saturated when they sank below the surface.

After sinking, it takes many hundreds of years for deep waters to return to the surface. The relatively oxygen-rich profile of the Atlantic Ocean (see Fig. 12-20) reflects the fact that surface waters are introduced at both poles whereas the northern boundaries of the Pacific and Indian oceans are blocked by land barriers. South of Greenland, where the North Atlantic deep water forms, there is little variation in the dissolved-oxygen profile throughout the water column.

As deep water returns to the surface, its dissolved oxygen content is gradually reduced due to consumption by marine organisms. In warm and temperate waters there is a zone, varying in depth from 150 to 1000 meters, where dissolved oxygen concentrations are at a minimum (see Fig. 12-20). Little or no dissolved oxygen from the surface is mixed downward to this layer and little oxygen remains in the water that has risen from great depths.

In areas of restricted circulation, such as some fjords and deep-ocean basins, deep waters may be devoid of dissolved oxygen; in summer many relatively shallow estuaries also have no oxygen in near-bottom waters.

FACTORS CONTROLLING PRODUCTIVITY

What controls productivity in the ocean? Ocean waters contain vast amounts of nutrients and ample sunlight for plant growth occurs over most of the earth. Furthermore it has been shown that productivity of the richest ocean areas (shallow-water estuarine and coral reef ecosystems) compares favorably with intensively cultivated agricultural land; gross productivity in these areas can be as high as 20 grams of carbon per square meter per day (Fig. 12-21).

But there are important differences between production of organic matter on land and in the ocean. Land plants can grow for years before soils are depleted of nutrients whereas phytoplankton can only

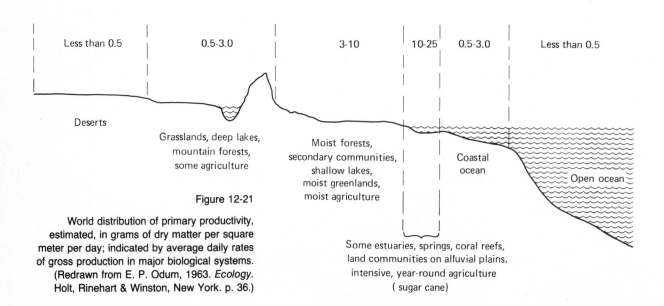

Figure 12-21

World distribution of primary productivity, estimated, in grams of dry matter per square meter per day; indicated by average daily rates of gross production in major biological systems. (Redrawn from E. P. Odum, 1963. *Ecology.* Holt, Rinehart & Winston, New York. p. 36.)

345
distribution of productivity
in the ocean

sustain maximal production for a few days before nutrients in the waters must be replenished. Highly productive ocean water contains only about five parts usable nitrogen per ten million parts of seawater, enough to produce perhaps 5 grams of dry organic matter. Over most of the ocean, nitrogen concentrations average only about one-fourth as much. But in fertile soil about 30,000 times as much nitrogen is available for plant growth, on the order of 0.5%.

One key to productivity of ocean waters is the rate at which nutrients can be renewed. Another is the quantity of light absorbed by plants. Primary productivity in polar regions is light limited throughout the year because the growing season is short and a heavy ice cover absorbs much of the available insolation. Lack of stability in the surface zones, where plankton are carried below the critical depth, limits production in temperate latitudes. But, in general, there is ample light in surface waters and productivity is controlled by the rate at which nutrients are recycled to the photic zone.

DISTRIBUTION OF PRODUCTIVITY IN THE OCEAN

Figure 12-22 shows relative productivity of surface waters in the ocean. Note that the northernmost Arctic appears to be an area of low productivity due to its limited insolation. Note also that the centers of ocean basins have low productivity because nutrients sink out of the photic zone and are not rapidly renewed. It has been estimated that subsurface waters move upward through the open ocean thermocline (at about 1000 meters depth) at rates between 0.5 and 1.6 centimeters per day. Nitrogen concentrations at this depth are usually about 20 milligrams atomic nitrogen per liter so that 3 to 9 grams of carbon per square meter per year might be a normal productivity if not for the fact that zooplankton excrete nutrients in the surface layer. In fact, primary productivity in central ocean basins averages around 50 grams of carbon per square meters per year (Table 12-5).

Figure 12-22

Distribution of primary productivity in surface waters of the world ocean. (Redrawn from O. J. Koblentz–Mishke, V. V. Volkovinsky, and J. G. Kabanova, 1970. Plankton primary production of the world ocean. *Symposium on Scientific Exploration of the South Pacific.* National Academy of Sciences, Washington, D.C. pp. 183–193.)

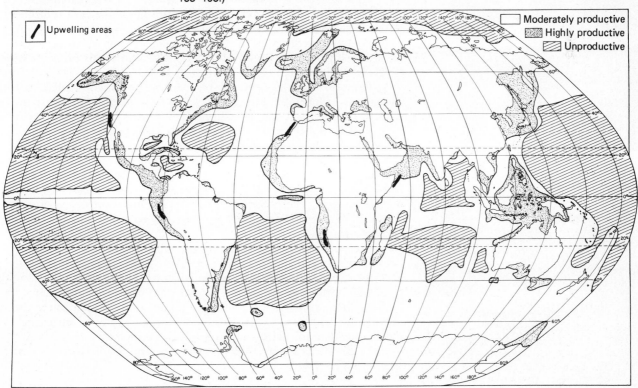

Upwelling areas

Moderately productive
Highly productive
Unproductive

TABLE 12-5

*Primary Productivity of Ocean Areas**

PROVINCE	PERCENTAGE OF OCEAN	AREA (millions of square kilometers)	MEAN PRODUCTIVITY (grams of dry carbon/m²/yr)	TOTAL PRODUCTIVITY (10⁹ tons carbon per year)
Open ocean	90	326	50	16.3
Coastal zone†	9.9	36	100	3.6
Upwelling areas	0.1	3.6	300	0.1
Total	100.0		–	20.0

*After John H. Ryther, 1969. Photosynthesis and fish production in the sea. *Science* 166:72–76.

†Includes offshore areas of high productivity.

Nutrient concentrations are much higher on continental shelves and slopes and upwelling regions because vigorous vertical mixing returns nutrients to the photic zone from deeper waters. In the Pacific Ocean, upwelling (Fig. 12-22) occurs along much of the eastern margin, especially the Peruvian–Chilean coast and California, as well as off Japan, Kamchatka, India, and western Canada. Similar processes occur in the Atlantic off West Africa, northeast Brazil, and the southeast coast of South America. Regions of divergence and upwelling in the Indian Ocean are caused by monsoon winds; in the open ocean upwelling is associated with divergences along the equator (see Fig. 12-22) and at the border between the Antarctic and midoceanic current gyres. Strong vertical water movements also occur in parts of marginal and inland seas, such as the Bering Sea and Arabian Sea.

Relatively unproductive waters are most widespread in the Pacific Ocean, in the Arctic, and at subtropical latitudes. The Indian Ocean has the largest relative proportion of productive areas, particularly in temperate and equatorial latitude belts.

A total of 25 to 30 billion tons of inorganic carbon is apparently fixed by marine plants every year. Including the carbon utilized by autotrophic organisms in respiration, net primary productivity of organic carbon is around 15 to 20 billion tons per year. It would appear to be the limit of primary productivity unless nutrients can somehow be artificially recycled on an enormous scale.

How do these figures compare to the production of food for human consumption? The answer depends on distribution of productivity throughout the world ocean. About 90% of the ocean surface, or 326×10^6 square kilometers, is open ocean. If annual productivity here averages about 50 grams of carbon per square meter per year, this is a total of 16.3×10^9 (16.3 billion) tons. Coastal waters, defined as those less than 180 meters deep (but excluding areas of active upwelling), account for another 9.9% of the ocean surface, or 36×10^6 square kilometers. These areas are about twice as productive as the open ocean and contribute about 3.6×10^9 tons of organic carbon annually.

Coastal upwelling regions are the most productive of all. Where prevailing winds drive surface waters away from coasts, causing active upwelling, nutrient-rich waters can support productivity of up to 10 grams of carbon per square meter per day. However, only about 1% of the world ocean (3.6×10^5 square kilometers) experiences these conditions. So even though productivity may average 300 grams of organic carbon annually, these areas only contribute about 0.1×10^9 tons to the annual total of 20×10^9 tons for the hydrosphere as a whole.

347
distribution of productivity
in the ocean

Table 12-5 shows that most of the ocean's organic carbon production takes place in open ocean waters. But most major fisheries (Chapter 15) are located in continental shelf waters, particularly in areas of upwelling. In terms of food production, the map of productivity distribution (Fig. 12-22) indicates the richest ocean areas. The explanation lies in the relative efficiencies of oceanic and neritic food chains.

Open-ocean food webs are typically long and complex. Recall that only the smallest phytoplankters (less than 0.005 millimeter) grow in the nutrient-poor waters of open oceans, and they can only be consumed by very small herbivores, on the order of 0.1 millimeter. These tiny animals are preyed on by primary carnivores in the 1 millimeter size range and they, in turn, by secondary carnivores that are about ten times as large, or about a centimeter long. (Examples of these organisms are described in Chapter 13.) In many cases, one or two larger invertebrate animals or fishes form additional links in an open-ocean food chain before it reaches a top carnivore, such as a mackerel or tuna.

Energy is lost at each trophic level in the chain, the more so because food is so scarce in open oceans. Predatory animals grow more slowly and expend more energy hunting for food than do coastal predators. It is estimated that ecological efficiencies do not average more than 10% at any level of an open-ocean food chain.

In contrast, areas of upwelling are characterized by a large proportion of phytoplankton (diatoms) that aggregate into clumps or long filaments. They can either be eaten directly by schooling fishes or there can be a small herbivorous consumer in the food web. In any case, highly productive ocean areas commonly support short food chains of rapidly growing organisms whose ecological efficiencies may be as high as 20% on the average. Little energy is wasted in hunting, and the yield of food to humans is much higher for the same amount of primary production. About half the world's fish catch comes from upwelling areas (Fig. 12-22, Table 12-6).

Coastal areas where upwelling is not common have intermediate productivity. Large phytoplankton are often present, especially during the short intense blooms of temperate zones. Density of even very small phytoplankton, however, is higher in coastal waters than in the open ocean and long food chains undoubtedly exist along with very short ones, depending on local conditions. Ecological efficiency in coastal waters is assumed to average around 15%.

Results of these calculations are summarized in Table 12-6. They indicate that the food potential of an ocean area is typically proportional to the rate at which nutrients are recycled to surface waters.

TABLE 12-6

*Estimated Fish Production of Ocean Areas**

PROVINCE	PRIMARY PRODUCTIVITY (10^9 tons of organic carbon)	TROPHIC LEVELS	ECOLOGICAL EFFICIENCY (percent)	FISH PRODUCTION (millions of tons, fresh weight)
Oceanic	16.3	5	10	1.6
Coastal	3.6	3	15	120
Upwelling	0.1	1½	20	120
Total	20			240

*After Ryther, 1969.

OPEN-OCEAN BIOLOGICAL PROVINCES

Distributions of open ocean organisms are controlled by surface currents and the abundance of light and nutrients. The major surface currents outline the principal biological provinces in the open ocean. Within each of these provinces, temperature, light, and nutrient concentrations—the chief factors controlling distribution of marine organisms—are relatively constant. Each gyre is partially self-contained and supports a relatively distinct assemblage of organisms.

The major open ocean provinces include the subpolar and subtropical gyres and the equatorial zone (shown in Fig. 12-23). Coastal ocean regions, boundary currents, and the very high latitudes are excluded from this discussion. The subpolar gyres are highly variable in light intensity, rich in nutrients, have relatively few species, but contain high biomasses. The surface zone is relatively thin because of divergence within the subpolar gyres bringing nutrients into the photic zone, a form of upwelling.

Subtropical gyres are relatively constant in temperature and light intensity, low in nutrients, and have large numbers of species but a low biomass. The surface zone is relatively thick because of convergence within this gyre. Nutrients are only slowly replenished by diffusion from below and by inward transport of nutrient-rich waters on the margins of the gyre.

The equatorial zone is characterized by east–west currents causing high variability in species abundance, large numbers of species, and intermediate-to-high biomass, partly because of upwelling along the equator.

Distributions of biomass are mostly controlled by water movements and the availability of light and nutrients rather than the swimming efforts of the plant or animal plankton. Areas high in nutrients generally maintain high biomasses whenever light is sufficient. Areas low in nutrients neither produce nor support high biomasses regard-

Figure 12-23

Schematic representation of the major open-ocean surface currents in a Northern Hemisphere ocean basin. Physical processes in each gyre have an effect on the biological characteristics of the corresponding biological province. [Adapted from J. L. Reid and others, 1978. Ocean circulation and marine life. In H. Charnock and G. Deacon (Eds.), *Advances in Oceanography*, pp. 65–130. Plenum Press, New York.]

Physical processes

Surface Current Patterns

Characteristics
of Biological
Provinces

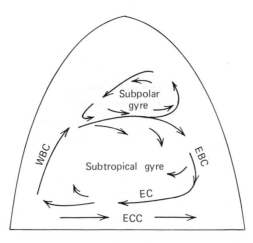

Subpolar grye
  Divergent gyre
    Large Seasonal changes, especially
      in light intensity
    High nutrients, supplied by upward-
      moving subsurface waters

Subtropical gyre
  Convergent gyre
    High salinity (Evaporation)
  Low nutrients, resupplied by
    vertical diffusion

Equatorial zone
  Divergent area
    High rainfall
    High nutrients, supplied by
      upward-moving subsurface
      waters, upwelling on equator

WBC Western Boundary Current
EBC Eastern Boundary Current
EC   Equatorial Current
ECC Equatorial Counter Current

Subpolar
  Relatively few species
  Large biomass

Subtropical
  Relatively large number of
    species
  Small Biomass

Equatorial
  Large number of species
  Large biomass
  Substantial east-west
    variability

less of light intensity. Phytoplankton tend to be relatively widespread and few species are restricted to only one gyre. Apparently phytoplankton tend to be spread by currents from one gyre to the next. A single "seed cell" is sufficient to establish a new phytoplankton population. Also, phytoplankton tend to be resistant to unfavorable conditions, thus facilitating their survival for many months. In fact, the decreasing light intensity with depth in the open ocean is more of a barrier to phytoplankton growth than lateral differences. In other words, phytoplankton cells can more readily survive moving hundreds of kilometers horizontally than sinking a few hundred meters below the photic zone.

Zooplankton species, on the other hand, exhibit much more restricted distributions. Many open-ocean zooplankton species occur in a single gyre. And some species have vertical migrations in and out of the photic zone to maintain themselves in a gyre and to avoid predation.

The subtropical gyres are the most extensive of the major oceanic habitats. They share a large fraction of their zooplankton forms. A few species are restricted to particular current systems, such as the West Wind Drift or the equatorial zone. Mixing of zooplankton from different systems occurs in the western boundary currents, which have the largest number of species.

REVIEW QUESTIONS

1. Contrast conditions for life in the ocean with those on land.

2. Which factors limit production of organic matter in the ocean? Where is each factor most important?

3. Describe the processes that cause the observed distribution of phosphate and nitrate in deep-ocean waters.

4. Illustrate the different trophic levels in a simple food chain.

5. Describe the basic process involved in photosynthesis. Contrast photosynthesis and respiration.

6. Discuss the seasonal cycle in phytoplankton abundance in mid-latitude surface waters. Indicate the factors controlling each of the features in the distribution.

7. Describe a red tide. Discuss probable causes.

8. On an outline map of the world, show the most productive areas. Indicate the location of major upwelling zones.

9. Describe (using diagrams) the characteristic dissolved profiles in the deep Pacific, South Atlantic, and North Atlantic (south of Greenland). Explain the observed differences.

10. Discuss the factors limiting the amount of fishes now caught in the ocean. Explain how the amount caught could be increased.

SUMMARY OUTLINE

Marine environment

Relatively homogeneous—contains gases, nutrients, water

Small temperature fluctuations—organisms mainly cold blooded

Plants typically one celled

Major classifications of organisms: bacteria, phytoplankton, zooplankton, nekton

Major environments: intertidal, neritic, pelagic, benthic, abyssal

Photic zone is region of food production

Limiting factors in marine ecosystems
    Ecosystems self-sstaining: include autotrophs, heterotrophs, decomposers
    Light can limit plant production
    Availability of nitrates, phosphates, silicates limits productivity—at low latitudes in open ocean, centers of ocean basins, during heavy blooms in temperate climates
    S/P ratio is small for elements that limit plant growth
    Organic trace materials—for instance, vitamins, toxins—can promote or inhibit growth

Phosphorus and nitrogen cycles
    Nutrients in organic matter sink below photic zone, quickly recycled in shallow water
    Inorganic soluble phosphorus high when particulate phosphorus is low, quickly recycled
    Zooplankton release nutrients during metabolism
    Nitrogen cycles more slowly than phosphorus, requires energy to "fix"
        Bacteria oxidize and reduce nitrogen compounds

Organic growth factors
    Vitamins synthesized by some bacteria, phytoplankton
    Chelating agents—make metals (e.g., iron) available to plants

Trophic level and biomass
    Primary producers, primary and secondary consumers form food web
    Position in food web defined as trophic level—"who eats whom"
        Energy in food webs used up, not recycled
    Ecological efficiency a measure of energy transfer between trophic levels
    Biomass—weight of organisms in population

Photosynthesis—solar energy stored as chemical energy in carbohydrates
    Bonds broken, energy released in respiration
    Net production is energy remaining after respiration, cellular metabolic processes
    Light and dark reactions in photosynthesis; light reactions absorb insolation
        NADPH created for reducing power
        ATP created from ADP and phosphate; cyclical energy-transport chain

Measuring primary productivity—a rate function
    Productivity equals average standing crop times rate of generation
    Measured by takeup of $^{14}CO_2$, or by oxygen produced
    At compensation depth, production equals respiration

Factors in plant productivity—recycling of nutrients to surface; stable surface layer
    Coastal zone: a productive area

Critical depth: when mixed layer is no deeper than photic zone, blooms can occur
Red tides caused by water stratification, rapid multiplication at surface
Some species photosynthesize best at lower light intensities.

Nonliving organic matter
    Most organic matter nonliving, very small particles
    Sources of dissolved organic matter
        Decomposition of dead organisms
        Secretion by plants
        Excretion and exudation by animals
    Sources of particulate organic matter
        Dead organisms
        Solid wastes excreted
        Adherence of dissolved material to surfaces

Bacteria in food webs
    Have low biomass, but important as decomposers
    Decomposition less effective in very deep water
    A concentrated source of protein for filter-feeders, scavengers
    Bacterial slime accumulates, followed by fouling, on surface-active materials

Dissolved oxygen in the ocean
    Abundant at surface, exchanged with atmosphere
    Least abundant where respiration, decomposition is high
    Abundant in deep, open ocean because of sinking polar waters, abyssal circulation
    Deficient in bottom waters of basins where surface production is high, resupply of bottom water limited

Productivity of the world ocean
    Ocean surface waters can quickly be depleted of nutrients
    Rate of nutrient renewal, abundance of light, stability of surface zone are primary factors
    Nutrients renewed quickly where deep waters cycle to surface: shallow regions, divergences, upwelling areas
    Food production in the ocean
        Most carbon produced in open ocean areas
        Oceanic food chains long, coastal food chains short
    Ecological efficiency greatest in coastal, upwelling areas, short food chains

Open-ocean biological provinces
    Distributions of open ocean organisms controlled by surface currents and abundance of light and nutrients
    Major provinces include: subpolar, subtropical, and equatorial
    Phytoplankton widespread laterally, primarily restricted to surface zone
    Zooplankton more restricted to single gyres but more capable of vertical movements

SELECTED REFERENCES

BATES, MARSTON. 1960. *The Forest and the Sea.* Random House, New York. 277 pp. Compares life in the ocean and in a tropical rainforest.

CUSHING, D. H., AND J. J. WALSH. 1976. *The Ecology of the Seas.* Saunders, Philadelphia. 467 pp.

PARSONS, T. R. M., M. TAKAHASHI, AND B. HARGRAVE, 1977. *Biological Oceanographic Processes*, 2nd ed. Pergamon Press, Oxford. 332 pp. Review of recent research in biological oceanography.

RAYMONT, J. E. G. 1963. *Plankton and Productivity in the Oceans.* Pergamon Press, New York. 625 pp. Comprehensive survey of plankton research literature, including aspects of physical oceanography as they relate to marine biology.

RUSSELL–HUNTER, W. D. 1970. *Aquatic Productivity: An Introduction to Some Basic Aspects of Biological Oceanography and Limnology.* Macmillan, London. 306 pp. Modern treatment of plankton and productivity in lakes and in the ocean, emphasizing North Atlantic forms. Good sections on plankton and world food production.

SIEBURTH, J. M. 1979. *Sea Microbes.* Oxford University Press, New York. 491 pp. Intermediate-to-advanced treatment of marine microorganisms; assumes biological background.

STEELE, J. H. (Ed.). 1970. *Marine Food Chains.* University of California Press, Berkeley and Los Angeles. 552 pp. Collection of papers dealing with relationships between plankton, productivity, and energy transfers in the ocean and estuaries. Intermediate to advanced in difficulty.

WIMPENNY, R. S. 1966. *The Plankton of the Sea.* American Elsevier, New York. 426 pp.

13 PLANKTON AND FISH: Marine Food Webs

Towed rapidly behind a moving ship, the Isaacs-Kidd mid-water trawl collects marine creatures living in waters a mile or more deep. The trawl is named after its inventors, John D. Isaacs, professor of oceanography at Scripps Institution of Oceanography, University of California, San Diego, and Lewis W. Kidd. Use of the trawl has revealed many new species of fishes in the mid-depths and has caught large quantities of other fishes once thought very rare. (Photograph courtesy Scripps Institution of Oceanography, University of California, San Diego.)

The open ocean provides a living environment for organisms that swim or float that is unlike anything known to us on land. The boundaries for most organisms are often subtle changes in water temperature and salinity or the presence or absence of a necessary food supply or even changes in light intensity. The average organism living in this unbounded environment is small, about the size of a mosquito. The small size, the lack of boundaries, and the fact that even familiar processes act in somewhat unfamiliar ways at the small sizes afford distinct advantages and limitations to open ocean organisms. In this chapter we explore the open ocean waters and the organisms that inhabit them. And we investigate the relationships among plant productivity, grazing by herbivores, and predation by fishes, leading eventually to organisms eaten by humans.

CHARACTERISTICS OF PLANKTON

Food in the open ocean is produced by minute floating plants—most of them single celled—living in the sunlit surface waters. Small size is an advantage to a minute floating plant, which depends on diffusion in the water to supply nutrients and remove wastes. This situation is in contrast to that on land, or in very shallow water, where air and/or water move past the rooted plant, bringing gases and nutrients and carrying away wastes. Each water parcel, with whatever is suspended in it, moves as a unit. Where nutrient concentrations are low, as in open ocean waters, the availability of nutrient ions within water parcels limits growth. So small size is an advantage because a large ratio of surface area to body mass provides a relatively larger area across which dissolved substances can be exchanged with the water.

Some phytoplankton can move a little relative to the water around them. Many have tiny, whiplike flagellae with which they propel themselves by lashing motions while others sink slowly through the water and depend on turbulence and currents to keep them in the photic zone.

An object's rate of sinking in the ocean is determined by two factors: its density relative to the water around it and the drag or resistance offered by the medium. Phytoplankton protoplasm is generally a little heavier than seawater, having a density range of 1.02 to 1.06 grams per cubit centimeter, but still dense enough to sink. The second sinking rate determinant, resistance of the medium, is greater for small objects than large ones.

An object with a large surface-to-volume ratio sinks more slowly than one of the same density but with a smaller ratio of surface area to volume. The smaller the object, the larger the ratio.

$$\text{Ratio of surface to volume} = \frac{\text{area}}{\text{volume}} = \frac{4\pi r^2}{4\pi r^3/3} = \frac{3}{r}$$

Thus an object having a radius of 0.1 millimeter has a surface/volume of 30 whereas an object of the same shape and density with a radius of 0.01 millimeter has a surface/volume ratio of 300 and sinks more slowly. A very slow rate of sinking is an advantage because phytoplankton must remain in the photic zone to grow and reproduce and yet motion through the water increases nutrient supply. Mixing in the surface zone returns part of a sinking population to the surface. In addition, many flagellated forms respond positively to light and swim upward.

PHYTOPLANKTON

Major phytoplankton forms are described in Table 13-1. The smallest *ultraplankton* (including bacteria) are less than 0.005 millimeter in diameter. Next largest are the *nannoplankton,* ranging from 0.005 to 0.07 millimeter, followed by *microphytoplankton* (0.07 to 1.0 millimeter), which are in the same size range as many zooplankton.

TABLE 13-1

Dominant Forms of Marine Phytoplankton

TYPE AND CHARACTERISTICS	LOCATION	COLOR AND APPEARANCE	METHOD OF REPRODUCTION
Diatoms: silica and pectin "pillbox" cell wall, sculptured designs; of major importance for coastal ocean productivity; has floating and attached forms	Everywhere in surface ocean, especially in colder waters, upwelling areas, even in polar ice; some heterotrophic below photic zone; some form "resting spore" under adverse conditions	Size: 0.01–0.2 mm Yellow green or brownish; single cells or chains of cells; radial or bilateral symmetry; many have spines or other flotation devices	Division, splitting of nuclear material; average reduction of one cell wall thickness at each division (Fig. 13-1); when limiting size is reached, cell contents escape, form new cell
Dinoflagellates: next to diatoms in productivity; many heterotrophic, inject particulate food; some have cellulose "armor"; very small open ocean species are naked	In all seas, and below photic zone; some parasitic; warm-water species very diverse; some have resting stage for protection; sometimes abundant in coastal areas as red tide (see Chapter 12)	Size: 0.005–0.1 mm Usually brownish, one celled; have two whiplike flagellae for locomotion; many are luminescent	Simple, longitudinal, or oblique divisions; daughter cells achieve size of parent before dividing
Coccolithophores: covered with calcareous plates, embedded in gelatinous sheath; important source of food for filter-feeding animals	Mainly in open seas, tropical and semitropical; sometimes proliferates near coasts; some heterotrophic forms at depths to 3000 meters	Size: 0.005–0.05 mm Many flagellated; often round or oval single cells; when present in great numbers, they give the water a milky appearance	Some individuals form cysts from which spores arise to develop into new individuals
Silicoflagellates: very small, have silica skeleton; some heterotrophic forms	Widespread in colder seas worldwide, especially in upwelling areas	Size: about 0.05 mm Single celled, one or two flagellae; starlike or meshlike skeleton	Simple cell division
Blue Green Algae: small, relatively simple cell structure; cell wall of chitin	Mainly inshore, warmer surface waters, tropics	Size: filaments to 0.1 mm or more Blue green or red rafts of mottled filaments; can cause a colored "bloom" in water	Simple division of each cell into two

Nannoplankton and the very small flagellated ultraplankton are most important in coastal and open-ocean equatorial waters, where they provide 50 to 80% of total standing crop and CO_2 uptake—perhaps because very small organisms absorb nutrients more efficiently than larger ones in low-nutrient waters. In coastal waters proportionately more of the larger microphytoplankton are present than in open seas, but nannoplankton still contribute most of the primary production on an annual basis. Where the currents are strong or upwelling replenishes nutrients, the larger forms are apparently favored. In the open ocean there may be less turbulence to bring phytoplankton back to the surface layers. Here the swimming abilities of many smaller forms may provide an important advantage over the larger forms, which slowly sink out of the photic zone.

Microphytoplankton (or *netplankton* because they are caught by nets) have large standing crops at higher latitudes. They proliferate during spring blooms when more than 99% of plants in an area may belong to a single species. Diatoms (Fig. 13-1) are usually the most important. Rate of division may exceed once a day and population typically increases 500 to 2000 times the winter "seed crop." Microzooplankton, locally present as consumers of the year-round nannoplankton population, cannot ingest the larger diatoms. But when larger zooplankters grow and reproduce in response to the abundance of food, their grazing quickly reduces phytoplankton biomass (see Fig. 12-12).

Diatoms (Table 13-1) have protruding, gelatinous threads and form long chains, especially in nutrient-rich coastal and upwelled waters. They grow by division (Fig. 13-1), some species forming *auxospores* seasonally, others every 2 or 3 years. A particular population may be recognizable over a long time period by its gradually diminishing shell size; some North Sea water masses can be traced by their characteristic diatom populations.

Often a single species dominates an area for 2 or 3 weeks, or a season, and is then succeeded by another. Diatoms seem most buoyant when they are young and growing rapidly so that, as one population ages, it sinks to lower levels and another species that had previously been represented by a small number of individuals begins to bloom in near-surface waters. This process may be repeated several times during spring and summer.

Diatoms form resting spores when environmental conditions become less favorable, remaining in this condition for long periods if necessary. After death, their glasslike shells sink to the ocean floor.

Dinoflagellates are second to diatoms in abundance (Fig. 13-2). They are sometimes considered one-celled animals, for many are heterotrophic. Some live on dissolved or particulate organic matter absorbed or ingested from seawater. Many require less light than diatoms and can tolerate lower nutrient concentrations. Dinoflagellate blooms commonly exceed diatom production in some areas partly because a scarcity of silicon can limit diatom growth but does not affect dinoflagellates. Changing light intensity also affects the succession of species in a particular area. In the mid-Atlantic coastal ocean, for example, nannoplankton dominates during November and December, but diatoms are the important producers during the January-to-March blooms.

The so-called armored dinoflagellates are covered with small plates of cellulose that form the organism's cell wall. Usually two flagellae are set in grooves at right angles in the cell wall for propulsion. This adaptation, together with a sensitivity to light, permits the organism to remain at its preferred light level by swimming upward. Several species are bioluminescent; they emit chemically produced light, particularly when agitated. In tropical waters breaking waves are often lit by the phosphorescence of these tiny organisms.

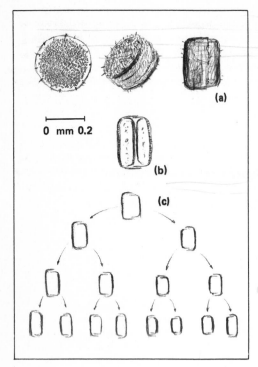

Figure 13-1

Reproduction in diatoms. There is a "pillbox" shell (a), perforated with pores for exchange of metabolic products. When a diatom grows large enough to divide, the "lid" and "box" separate (b), and each gets one-half the cell contents. A new "box" is then secreted over the exposed protoplasm so that the original shell half becomes the new "lid." The daughter cells become progressively smaller in this type of division. When a certain size limit is reached, both old shells are discarded and the resulting bare auxospore doubles or triples in size before forming a new set of shells.

0 mm 0.2

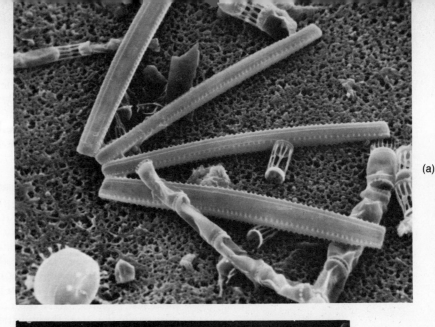

(a)

Figure 13-2

Variously shaped diatoms (a), collected on a plankton net, include rod-shaped, jointed, spool-shaped and "pillbox"-type species. The longest measures about 0.08 millimeter (80 microns). (b) The dinoflagellate *Peridinium* shown here is about 0.4 millimeter (400 microns) long. Its locomotive flagellum (not visible in this photograph) is attached at the groove near the upper left. (Photographs courtesy Dr. J. M. Sieburth and University Park Press, Baltimore, Md.) (c) This living coccolithophore, *Cyclococcolithina leptopora,* was photographed through an electron microscope. It was collected in the North Pacific Ocean and measures a little over 0.01 millimeter (10 microns) in diameter. (Photograph courtesy Dr. Susumu Honjo.) (d) The silicoflagellate *Distephanus,* whose skeleton is shown containing organic debris, is only about 0.03 millimeter (30 microns) in diameter, including the spines. (Photograph courtesy Dr. J. M. Sieburth and University Park Press, Baltimore, Md.)

(b)

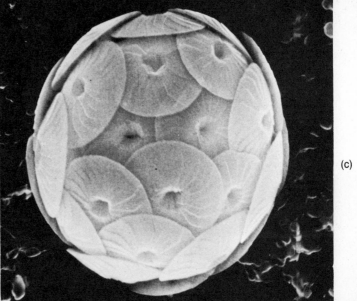

(c)

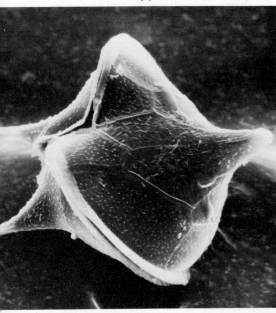

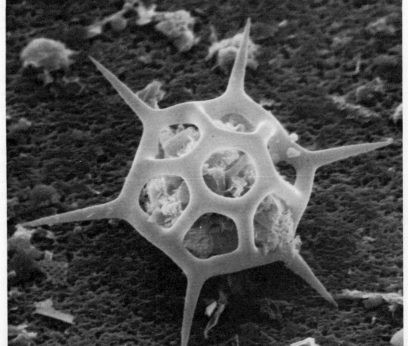

(d)

Coccolithophores [Fig. 13-2(c)], another important group of flagellates, are covered with tiny, calcareous plates that are important contributors to marine sediments in certain parts of the deep ocean. Less common are *silicoflagellates* [Fig. 13-2(d)] and numerous other types of flagellated nannoplankton. Near coasts, filamentous *blue green algae* may be locally abundant and sometimes green algae, although they grow in shallow water.

ZOOPLANKTON

Most lower-trophic-level consumers in marine food webs are zooplankters. As herbivores and primary carnivores, they transfer matter and energy from phytoplankton to higher-level predators consumed by humans. And as conspicuous members of pelagic ecosystems, their life cycles and behavior have been extensively studied.

Zooplankton gather food and reproduce in a variety of ways. Some swim well enough to pursue prey. But they are mostly filter feeders, bearing tiny hairs or mucous surfaces to capture floating food particles. Because these animals are usually limited to food particles of a particular size, their distribution depends largely on the availability of organisms that they are adapted to catch.

Another important factor in zooplankton distribution is the narrow temperature range (generally only a few degrees) in which they can reproduce. Adult populations have greater temperature tolerance and may be borne far out of their breeding range by currents, making them available to predators in a wider area. The greatest number of plankton species breed in tropical waters, with a steady decrease toward higher latitudes. But the number of individuals in an area is normally a function of its productivity; there may be 500 or 1000 times as many animals per square meter of near-surface water in the North Atlantic coastal ocean, for example, than in the tropical Atlantic open ocean. Again, currents can carry nutrient-rich waters through a relatively sterile area so that abundant zooplankton are sometimes found in a region of normally low productivity.

Not all zooplankton remain free floating throughout life. The ones that do, known as *holoplankton*, are generally more important in marine food webs than the *meroplankton*, which live attached to the bottom as adults.

HOLOPLANKTON

Many small consumers of nannoplankton are single-celled *protozoans*, of which *Foraminifera* [Fig. 13-3(a)], *Radiolaria* [Fig. 13-3(b)], and *Tintinnidae* (Fig. 13-4 and frontispiece) are important examples. Foraminiferans live nearly everywhere in the ocean. Many have delicate, porous shells or "tests," which are conspicuous constituents in calcareous sediments (Chapter 4). Thin extrusions of protoplasm called *pseudopodia* (false feet) extend through holes in the shells to capture food particles by surrounding them. Digestive juices are secreted onto the food to dissolve it so that it can pass directly into the cell. Wastes are excreted from body surfaces. Protoplasmic strands connect outside the animals' body to form a branching network around and through the tests and the animal grows. Additional calcium carbonate is laid down around newly extruded pseudopodia so that successive interconnected chambers are added to the shell.

Radiolarian protozoa have an internal capsule of organic matter and a siliceous or strontium sulfate skeleton. The skeleton is arranged as a sphere or as concentric spheres, with radiating pieces. Their protoplasmic strands do not form a net but project in all directions as long, sticky filaments. These trap tiny particles, which are then borne by a protoplasm toward the center of the body to be digested. Individuals range from 0.1 to more than 10 millimeters in diameter. Reproduction is by division into many small flagellated cells.

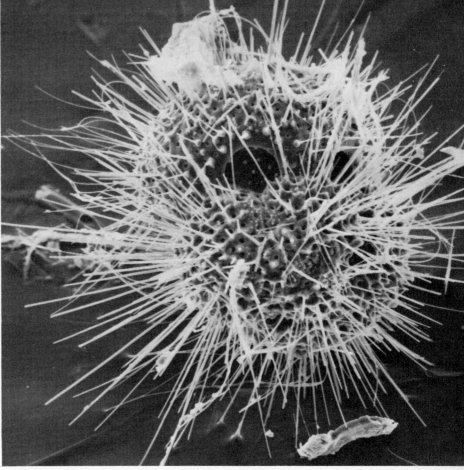

(a)

Figure 13-3

(a) A foraminiferan, *Globigerinoides ruber,* from Bermuda waters. Two apertures, through which food is ingested, are visible just above center. Excluding the spines, this organism is about 0.3 millimeter (300 microns) in diameter. (Photograph courtesy Dr. J. M. Sieburth and University Park Press, Baltimore, Md.) (b) This delicate skeleton of an acantharian—a close relative of the radiolaria—is made of crystalline strontium sulfate. The skeleton dissolves in seawater within a few hours after the animal dies. Thus it is not found in sediments. (Drawn by Ernst Haeckel, Report of scientific results, *H. M. S. Challenger. Zoology* 18: 1803, 1887.)

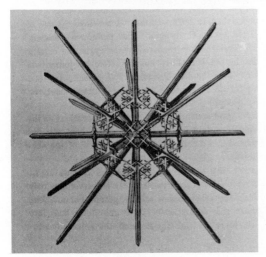

(b)

Tintinnids ("bell animals") and other tiny, ciliated protozoans are common throughout open oceans, where plants are small and populations sparse. Tintinnids are enclosed in a goblet-shaped or tubular hard shell of protein a few tenths of a millimeter long. The mouth is surrounded by a circle of hairlike cilia (Fig. 13-4) whose waving motions propel the organisms and entrap food particles; some forms also have tentacles around the mouth. Tintinnid, radiolarian, and foraminiferan shells or capsules are sometimes colonized by phytoplankton, with which metabolic products are evidently shared—another adaptation for increasing the food supply in a nutrient-poor environment.

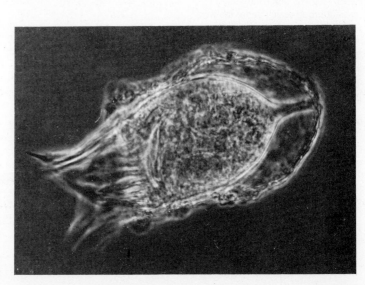

Figure 13-4

This cultivated specimen of *Tintinnoposis lohmani* (on the order of 0.1 millimeter in length) has clearly visible cilia and lorica (outer sheath) surrounding the goblet-shaped body. (Photograph courtesy Kenneth Gold.)

Rotifera are tiny roundworms that are seasonally common in coastal and open oceans. A double wheel of beating cilia on the head serves to propel the animal and to sweep microorganisms into its mouth.

In contrast to the small, sparsely distributed animals of open oceans, the characteristic zooplankters of productive waters are relatively large and complex. *Crustacea* (Fig. 13-5) are the most numerous; they constitute 70% or more of the zooplankton, both in bulk and numbers. *Copepoda* (Fig. 13-6) and *Euphausiidae* (Fig. 13-7) are most important in marine food webs.

Crustaceans have been called the "insects of the sea." In fact, both insects and crustaceans are arthropods, meaning "jointed-feet," characterized by segmented bodies and appendages. Each pair is usually specialized for a particular function, such as feeding, movement, sensation, or reproduction. There is a stiff, chitinous outer shell. The characteristic, free-swimming *nauplius* larva commonly molts many times (Fig. 13-7) before assuming the adult shape.

Copepods are present throughout the ocean and may be the most numerous animals in the world; certainly they are the most numerous marine herbivores. In the northwestern Pacific they average 15,000 individuals per cubic meter in surface waters. In Arctic waters there may be nearly twice as many of a single species in a cubic meter of seawater; even at 500 meters depth well over a thousand individuals may occur in a cubic meter. Depending on temperature and availability of food (usually diatoms), large copepods can double

Figure 13-5

Plankton crustacea. Clockwise, beginning at the bottom:

Amphipods hatch in brood-pouches, emerging as tiny adults. Some groups are locally very abundant in the plankton. The *Hyperiid* type shown here has characteristically huge compound eyes. The species illustrated is *Themisto,* an important food of the North Sea herring.

Cumaceans are more common in shallow coastal waters than in the deep ocean. They often remain on the ocean floor during the day, rising to surface layers at night to feed.

Mysids are a shrimplike form. Many species live near continental margins, but a few are found at depths of 4000 meters or more. Metamorphosis, from egg through nauplius stages to adult form, takes place in a brood-pouch under the front legs of the female.

Copepods (meaning "oar-footed") occur at all depths, in all parts of the ocean. They have been termed the "insects of the sea" because of this abundance and variety. The eggs hatch into free-swimming nauplii, which often moult many times before attaining adult form.

Euphausiids are relatives of the edible shrimp, which is a bottom dweller. All euphausiids are planktonic and are found everywhere in the ocean. Many are filter feeders, especially in high-latitude waters; others have grinding jaws to accommodate larger food particles.

Many *isopod* species are land-dwellers. The species *Gnathia,* of which the larva (left) and adult male are shown here, lives as a fish parasite during the larval stages but is free swimming in adult life.

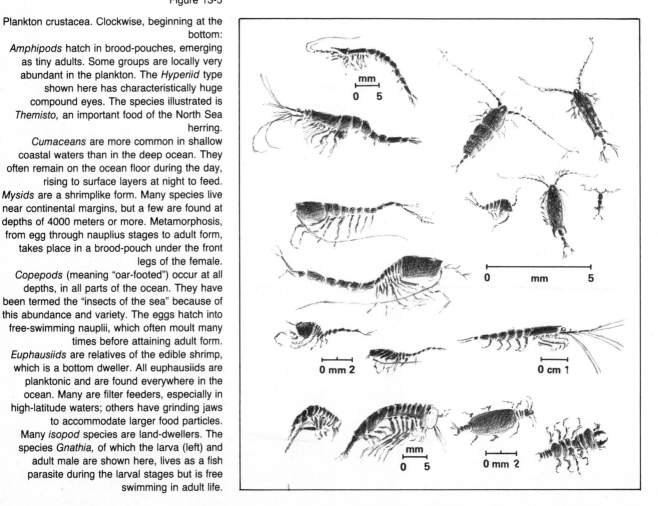

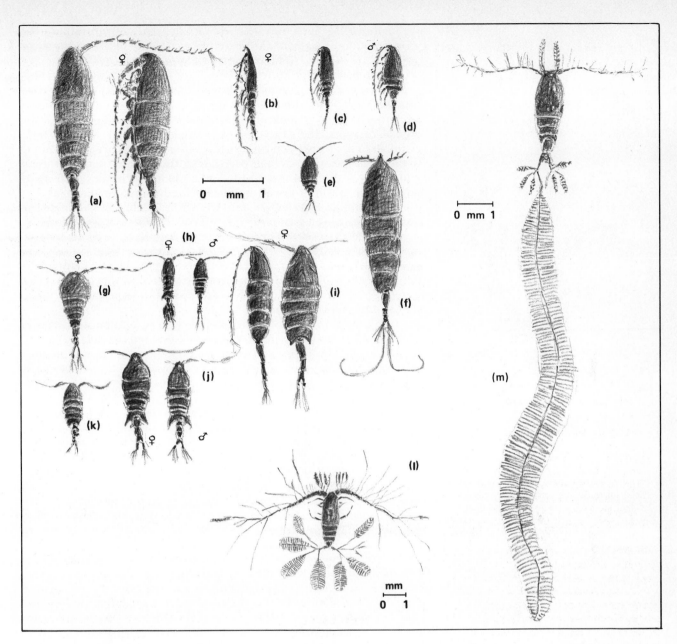

Figure 13-6

Calanoid copepods: (a) *Calanus finmarchicus,*
(b) *Rhincalanus,* (c) *Pseudocalanus,* (d)
Paracalanus, (e) *Microcalanus,* (f) *Euchaeia,* (g)
Temora, (h) *Eurytemora,* (i) *Metridia,* (j)
Centropages, (k) *Isias*—all found in the vicinity
of Great Britain. (Redrawn from Newell, 1967.)
(l) and (m) are tropical varieties of *Calocalanus.*
(Redrawn from A. Hardy 1967. *Great Waters.*
Harper & Row, New York. 541 pp.)

Figure 13-7

Life cycles of *Euphausia superba:* Egg (a), early
and late nauplius stages—(b), (c), (d)—the
intermediate *mysis* stage, and nature adult are
shown.

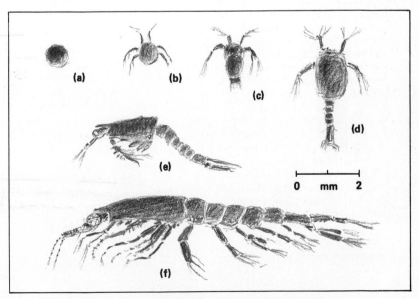

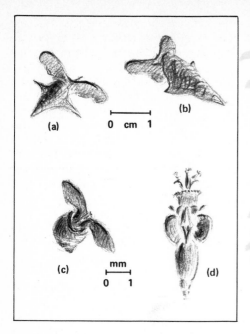

Figure 13-8

Marine pteropods: *Clio*—(a) and (b)—found throughout the world ocean, *Spirialis* (c), and *Dexiobranchaea* (d). (Redrawn in part from Hardy, 1965.)

Figure 13-9

These chaetognath worms include several species of *Sagitta* common to North Atlantic waters. They are sometimes used as tracers for water masses, for different species are found in waters having different nutrient compositions. (Redrawn from G. E. Newell and R. C. Newell, 1967. *Marine Plankton: A Practical Guide.* Hutchinson, London. 221 pp.)

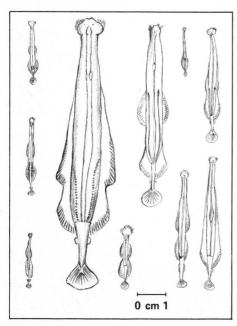

their numbers a few times in a year; smaller species reproduce even more frequently. Daily food consumption for older larvae ranges from 50% of body weight to much more when food is plentiful.

Copepods range in length from about 0.3 millimeter to about 8 millimeters and have feathery, curved bristles that form a filter chamber behind the mouth. By constant movement of appendages near the head, two opposing currents are set up. One moves the animal forward; the other forces a stream of water into the filter chamber. Tiny plants and fine particles thus trapped are passed along to the mouth. Copepods in northern seas, their bodies rich in proteins and fats, are eaten in great numbers by herring, whales, and many other organisms that are an important food source for humans.

Shrimplike *euphausiids,* also known as "krill," are another important crustacean group. Dense swarms of these animals feed on diatoms and themselves constitute the chief food of large, filter-feeding whales in Antarctic waters, as well as of many fishes on the high seas. Euphausiids are larger than copepods, being up to 5 centimeters long, and they are found on or near the bottom as well as in surface waters. The group includes herbivorous and, especially in warm waters, carnivorous species.

In Antarctic and colder temperate waters, euphausiids mature slowly and live up to 2 years. It has been estimated that a large baleen whale (see Fig. 13-24) eats an average of 850 liters of euphausiids per day. *Euphausia superba* may produce 10^{11} kilograms annually, nearly twice the world's commercial fishery production for 1978 (0.6×10^{11} kilograms).

Another important type of crustacean, especially in nearshore waters, is *Cladocera* ("water fleas"), such as *Podon* and *Evadne* (see Fig. 13-29). They sometimes occur in tremendous concentrations in coastal ocean (e.g., the North Sea) and in such estuaries as Chesapeake Bay. They are small—around 1 millimeter long—and possess a single large eye formed by fusion of the two larval eyes.

Besides the crustaceans, certain other types of zooplankton are locally conspicuous. *Pteropoda* ("wing footed") are small pelagic snails that occur in dense swarms in all seas. Carbonate-shelled species are common in tropical oceans, where their remains form "pteropod oozes." In all members of this group the characteristic snail foot is modified into fins (Fig. 13-8). Rows of beating cilia set up currents that bring food particles against a mucous secretion and then bear the food toward the mouth. In surface waters pteropods eat phytoplankton; in deeper currents they are carnivorus. Northern species ordinarily do not have shells; some resemble a little slug with wings, 2 centimeters or less in length.

Some small marine carnivores are strong swimmers, although usually classed as plankton. Among the most important are the *arrowworms* or *Chaetognatha* (Fig. 13-9), whose name means "bristle jawed." Adults, 2 to 8 centimeters in length, are active and abundant from the surface to great depths in all seas. They are transparent and possess chitinous spikes embedded in muscle around the mouth. These spikes can open out like a fan and then turn inward as seizing jaws to consume their prey. By fixing their long bands of longitudinal muscles, arrowworms dart rapidly through the water in pursuit of small zooplankton. There is no larval stage; small worms hatch directly from fertilized eggs.

Different North Atlantic water masses can be distinguished by determining the species of arrowworm that lives in them. *Sagitta setosa* is found only in North Sea water masses and *Sagitta elegans* in more fertile oceanic waters. The distribution of these water masses changes from year to year, affecting pelagic species as well as the benthic animals that may develop more successfully in one or the other type

Figure 13-10

The adult jellyfish *Chrysaora* measures about 10 centimeters across the bell and is found worldwide. (Photograph courtesy Michael J. Reber.)

of water. They can be distinguished by chemical analysis, but the quickest and easiest method of determining whether a water sample is *setosa* or *elegans* water is to capture and identify the arrowworms that live in it.

The animals discussed so far are all important members of food chains that lead to fish and from there to human predation. But other zooplankters seem to be dead ends in food webs exploited by man— for example, the well-known jellyfish. They have a continuous, two-layered body wall surrounding a digestive cavity that has only one opening, around which are tentacles bearing stinging cells. The tentacles capture swimming or drifting particles and move them into the mouth, through which wastes are also eliminated. The animal moves by rhythmic pulsations of its bell (Fig. 13-10).

This group includes the spectacular colonial *siphonophores* of the open sea like the Portuguese man-of-war pictured in Fig. 13-11. Both jellyfish and siphonophores paralyze their prey with stinging cells that consist of a barb attached to a sack of poison. Dangling tentacles entangle the food and sweep it toward the mouth, inside the bell. Some forms swim upward and then sink with bell and tentacles extended, trapping prey beneath. A siphonophore is not a single animal, but a colony of individuals that live together and function as one. Some species (e.g., *Physalia*) have a gas-filled float, but most have one or more swimming bells. Prey-catching, reproductive, and digestive polyps trail below like a fisherman's drift net.

Physalia are common along Mediterranean shores and in nearly all tropical waters, where their stinging cells are feared by swimmers. The poison works instantly and can be fatal to a child. But *Physalia* has enemies of its own. A small, blue-shelled snail, *Janthina*, about 2 to 3 centimeters long, floats at the water surface by means of gas-filled bladders; two of these snails can devour a 10-centimeter long *Physalia* colony in 24 hours.

Figure 13-11

The Portuguese man-o'-war *(Physalia)* has a purple, air-filled bladder that floats at the surface. This animal is colonial in the sense that beneath the float hang clusters of polyp individuals, some of which entrap, paralyze, and engulf their victims, digesting them and absorbing their juices on the spot. Other polyps serve a reproductive function, producing sexual medusae. The man-o'-war fish, immune to *Physalia's* poison, feeds on discarded scraps and acts as a lure to attract victims. (Photograph courtesy Wometco Miami Seaquarium.)

Figure 13-12

Comb jellies (ctenophores) include various forms, three of which—(a) *Bolinopsis,* (b) *Meroe* and (c) *Pleurobrachia*—are shown here. They swim by motions of the fused cilia, of which typically there are eight rows on the body. Medusae, on the other hand, swim by pulsations of the bell.

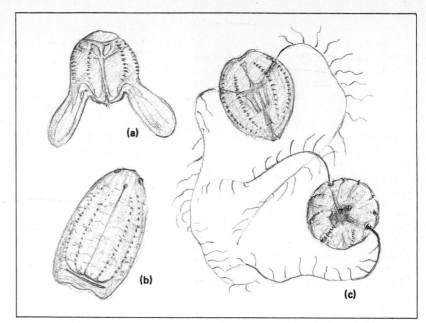

The delicate *Ctenophora* (Fig. 13-12) look something like jellyfish. Small and jellylike, they are sometimes known as sea walnuts or comb jellies. Some have trailing tentacles for capturing prey. Voraciously carnivorous, ctenophores often occur in great numbers and may substantially reduce populations of crustaceans and young fishes.

Ctenophores are bilaterally symmetrical, rather than radially symmetrical like the coelenterates, and they lack the stinging cells characteristic of the former group. They swim by moving the bands of fused cilia that forms their "combs." All are luminescent and can sometimes be seen at night in surface waters when disturbed.

Tunicates are primitive relatives of the *Vertebrata* (animals having backbones). In the plankton they are represented by *sea squirts,* the transparent, barrel-shaped *salps* and *doliolids* illustrated in Fig. 13-13. Significant numbers occur in tropical waters, where they feed on tiny plankton by pumping water through their bodies. Sometimes these animals form long chains that lie in masses at the surface of quiet waters. Large tunicates do not seem to be widely eaten by fishes and like siphonophores and ctenophores may represent a dead end in marine food webs harvested by humans.

MEROPLANKTON   Planktonic larval forms of benthic animals are locally important in coastal waters, where they are an important food for deep-water fishes, such as cod and flounders. There are more than 125,000 species of benthic animals, most of which have a free-swimming larval stage that lasts a few weeks. Eggs and sperm of benthic animals are discharged in great clouds to fertilize in the water (Fig. 13-14). These number in the tens of millions per individual per year, but mortality due to predation and other hazards is high. The number of eggs produced just about balances their loss during development, and the number of benthic adults remains roughly the same over a period of years unless the environment changes.

The ability of maturing larvae to find suitable bottom material on which to settle is an important factor. Currents, for example, may carry worm larvae into rocky-bottomed areas where there is no sediment in which to burrow and so the larvae die. A barnacle, on the other hand, needs a solid surface to grow on so that landing on soft, claylike sediment would prevent larvae from completing their meta-

364

Figure 13-13

Tunicates (sea squirts): these animals pump water through their bodies by contraction of the muscular bands encircling them. *Doliolum* (a) has a life cycle of alternative sexual and asexual generations—the asexual is illustrated here. Reproduction is by budding—little groups of cells are formed on the underside of the mature individual (upper left), whence they migrate along the body of the "parent." They then attach themselves in rows to an appendage at the front end, the *cadophore,* shown enlarged at lower left. As the cadophore grows longer, the buds grow to maturity. Some remain attached as a colony; others break off to live as individuals. (Redrawn from Hardy, 1965.)

Salpa (b) also exhibits alternation of generations—the asexual stage is shown here, in solitary and aggregate forms (the individual at lower right is an aggregating type). At the top of the group is a "nurse" salp with a cluster of buds on its underside that will break off to form sexually reproducing individuals. (Redrawn from Hardy, 1965.) (c) At top—solitary (right) and aggregate forms of *Salpa democratica;* below— solitary (bottom) and aggregate forms of *Salpa zonaria.* (Redrawn from Newell, 1967.)

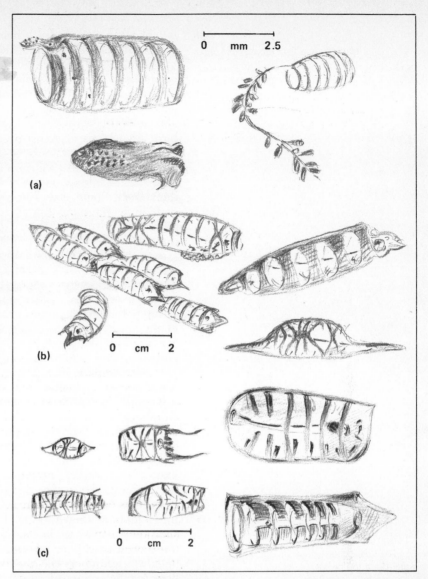

Figure 13-14

Oysters (male on left, female on right) expelling sperm and eggs. (Photograph courtesy Michael J. Reber.)

morphosis into the adult form. Some larvae can resume a swimming or floating existence if the first attempt at finding a suitable bottom material fails. After awhile they seek the bottom again and some can crawl about in search of an appropriate place to attach themselves.

Predation causes larval mortality of 90% or more. Jellyfish, comb jellies, larval fish, arrowworms and many other carnivorous plankton and nekton consume large numbers of benthic larvae, few of which have any defenses against predators.

Among benthic *invertebrates* (animals without backbones), worms are some of the most widely distributed. *Polychaeta* (bristle worms) are segmented animals named for the rows of chitinous bristles along their sides. Some, like *Tomopteris* (see Fig. 13-29), are permanently planktonic, but many have free-swimming larvae (Fig. 13-15). At first the young worm is toplike or rounded, as in Fig. 13-15(a). The mouth is at the "equator," below a ciliated girdle, with the anus at the bottom. Waving cilia send the animal spinning through the water and currents set up by the spinning motion move food particles toward the mouth. This type of larva is called a *trochophore*. As development continues, other ciliated rings and segments appear below the mouth, as shown in Fig. 13-15(b). In some forms, the trochophore becomes the head of the adult worm; in others, the girdle and other larval structures are suddenly sloughed off when the animal settles on the ocean floor and completes metamorphosis, as shown in Fig. 13-15(c).

The *Mollusca* include snails and slugs (*gastropods*), clams and oysters (*bivalves*), and squid and octopus (*cephalopods*). Most of this group are benthic as adults, with planktonic larvae. The first stage is a trochophore, which usually remains in the original egg capsule, shown in Figs. 13-16(a) and (b). In the later, free-swimming stage, the ciliated girdle is extended at the sides to form large lobes—shown in Fig. 13-16(c)—which not only support the increased weight of the larva (now known as a *veliger*) but also prevent the typical spinning motion of the trochophore so that the animal swims straight ahead. The lobes extend out beyond the body of the larva and the internal organs twist into a loop, as shown in Fig. 13-16(g). Mouth and anus use the same shell aperture; when the head and foot are withdrawn for safety inside the shell, a hard plate (the *operculum*) attached to the foot covers that opening tightly. As the shell develops, the snail sinks to the ocean floor to take up benthic life. The development of the clam (illustrated in Fig. 13-17), which does not need to twist its body to fit into a single shell, is a variant on the same theme.

Figure 13-15

Larval stages of generalized polychaete worm. The trochophore stage (a) spins through the water like a top by waving its tuft of cilia; the gut can be seen within. In (b) and (c) the larva metamorphoses to adult form. (Redrawn from Hardy, 1965.) (d) Photograph of an adult benthic bristleworm. (Photograph courtesy Wometco Miami Seaquarium.)

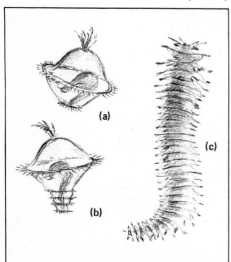

(d)

Figure 13-16

Stages in the metamorphosis of a marine snail from a trochophore (a) and (b) through a free-swimming veliger (c) to the adult form (d).

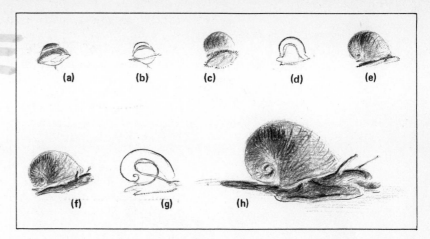

Figure 13-17

Metamorphosis of a generalized bivalve, showing the trochophorelike larva (a) and (b) with a shell forming at the sides. It has been drawn as if transparent in order to show formation of the gut. Nearing the adult state (c), the foot is visible where the shell is partially open.

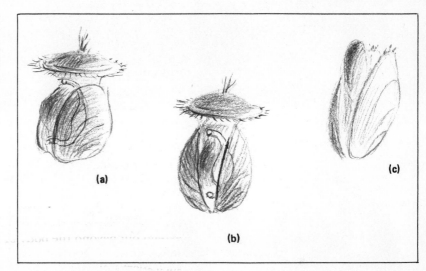

Benthic crustaceans, such as barnacles, develop in much the same way as their pelagic relatives, the copepods and euphausiids (see Fig. 13-7). Nauplii of the common barnacle *Balanus balanus* (Fig. 13-18), about 0.5 millimeter in length, may number 13,000 larvae at a time, of which perhaps 30 survive to attachment. Very few live 2 years to reach sexual maturity.

Some fishes, such as herring and sand eel, attach their eggs to rocks or vegetation, some lay them in "nests" near the shore, and some lay them in gelatinous masses. But most fish eggs are released and fertilized in waters near the continental margins. These eggs drift with the plankton; when sufficiently developed, they hatch and begin to feed. Depending on temperature and species, it may be as soon as 1 or 2 days after release. They then remain in the planktonic stage for weeks. Success of the young fish in a particular *year class* (i.e., those spawned during a single season) is affected by several environmental factors. One of the most important is that the young fish must find enough food to survive before the yolk sac, shown in Fig. 13-19(a), is completely absorbed. This factor can depend, for example, on currents carrying the fish into a region where detritus particles or recently hatched invertebrate larvae are the right size for the developing fish to eat. If the larval stages are transported offshore where food is scarce, the entire year class may die.

Figure 13-18

Developmental stages of the acorn barnacle *Balanus*. The larval "nauplius" (upper left) swims freely, metamorphosing into the secondary "cypris" stage (upper right and center, showing inner and outer views). Having found an appropriate location for its stationary adult life, the animal forms a hard shell that can be opened for feeding or closed for protection. Cutaway views (bottom) show the animal inside its shell, closed and open.

(left & right, bottom)   Figure 13-19

(a) Development of the Atlantic fish *Sardina pilchardus;* the tiny fish is first seen developing inside the egg. (b) Larvae of (from the top) mackerel, anchovy, rockfish, croaker, halibut.

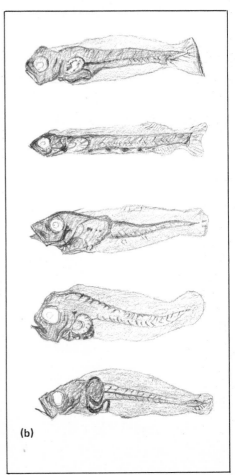

(a)

0   mm   5

(b)

A varied and distinctive fauna inhabits the mesopelagic zone, the depth zone from 200 to 1000 meters. Only about 1% of light penetrates to the top of this zone even in very clear waters. It is close to the lower limit for photosynthesis; but as far down as 1000 meters, some animals respond to diurnally changing light levels by migrating vertically every 24 hours. Swarms of planktonic euphausiids and copepods feed on small particles from above and are, in turn, preyed on by squid and fishes. Food requirements are less in this environment because animal metabolism is slower in colder waters.

This dimly lit zone has relatively more bioluminescent animals than any other part of the ocean. *Stomiatoids* are an abundant group whose form and habits illustrate some typical adaptations for life in dark, deep waters (Fig. 13-20). *Lantern fish* and *hatchetfish* (Fig. 13-21),

Figure 13-20

Stomiatoid fishes are slim, generally dark or silvery color, and range from a few centimeters to a few tens of centimeters long. Probably the most numerous fishes in the ocean, they are especially common at mid-depths of 500 to 2000 meters. Populations may not be dense in any one place, in the manner of coastal herring of anchovies, but they are common through a vast area of the ocean. *Melanostomiatoids* (a) characteristically bear chin barbels, often associated with a luminous lure. These fleshy organs are sensitive to water movements, which cause the fish to snap in the direction of the disturbance. Stomiatoids have big mouths and sharp, strong teeth and many can swallow victims larger than themselves. Luminous dots along the body are common. The midwater *Cyclothone* (bristlemouths) are believed to be the largest single genus (group of species) among marine vertebrates. They are small (2 to 8 centimeters) and rapacious, with mouths that open nearly 180° to capture prey. Illustrated (b) *C. microdon* and (c) *C. Pallida. Idiacanthus niger* (d), about 20 centimeters long, is shown after having swallowed a fish longer than its own gut, the prey being doubled up to fit inside the predator. *Idiacanthus fasciola* (e) is commonly found at depths of 900 to 1800 meters, in low-latitude waters. Females are black, reaching a length of 30 centimeters; males are pale and seldom exceed 5 centimeters in length, with lights on their cheeks, but lacking the characteristic barbel of the female. The larvae, whose eyes are on the ends of long stalks, drift at the surface while feeding on plankton. When they reach a length of about 5 centimeters, they sink to 900 meters or lower, and the eyestalks resorb to a normal position.

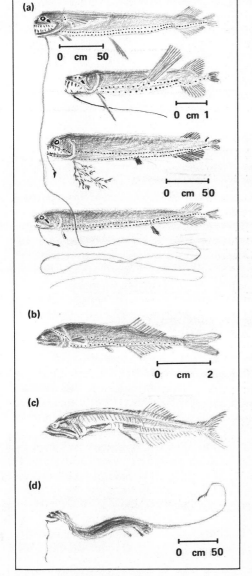

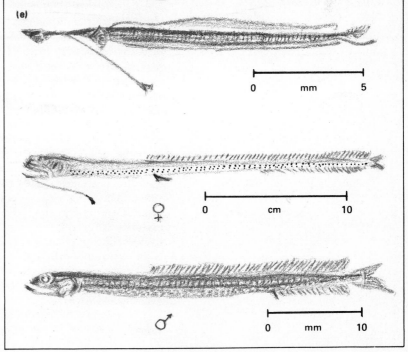

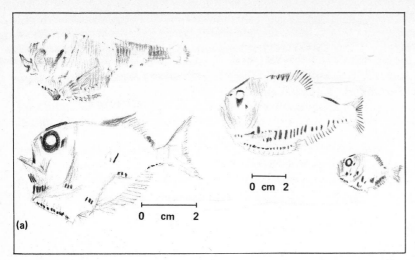

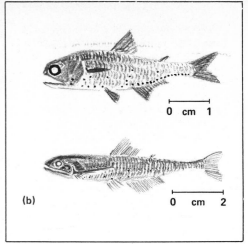

Figure 13-21

Bioluminescent deep-ocean fishes.
(a) Four species of hatchetfish, all small—the largest species is *Argyropelacus gigas*,
at top. These fish are common at depths of 100 to 500 meters, although some have
been taken at depths to 2000 meters. They are characterized by luminous organs and
by their distinctive "hatchet" shape. (b) Second in numbers only to the stomiatoids
among deep-water fishes, there are at least 170 species of lantern fish, in all but the
very coldest seas. Great masses often migrate to the surface at night to feed. Light
organs dotted along the sides and a bright light near the tail are characteristic.
Although sunlight and even moonlight repel them, lantern fish often experience a
reversal of the normal reaction in the presence of a very bright light so that they swim
toward a searchlight on the water, like moths.

both common from 100 to 500 meters depth, have patterns of light
organs that identify particular species. This adaptation permits the
fish to recognize potential breeding mates.

Below 1000 meters there is no sunlight at all and waters are
uniformly cold (Chapter 6). Animals at these depths are widely distributed; the same kinds may be found in almost every part of the
ocean. A few filter-feeding crustaceans are supported by the detrital
rain from above, but predation is the rule. Many species are adapted
to eat anything that will go into their mouths, including animals twice
their own size, as shown in Figs. 13-20 and 13-22. Meals are apparently
rare; a fish at these depths may eat only every few months.

Animals living in absolutely dark waters are usually black or
reddish, either color making them invisible. Lighted lures are common
(Fig. 13-22), for many marine animals are attracted by them.

Squid (Fig. 13-23) are predators in the deep ocean as well as in
surface waters. These swimming mollusks have no external shell but
are stiffened internally by a bladelike chitinous "pen." Encircling their
mouths are long, prehensile, tentacle-bearing suckers. All marine
mollusks have a *mantle*, or tough protective membrane through which
water circulates, but in squid this mantle is highly developed into a
muscular, torpedo-shaped sheath. By powerfully contracting the mantle, they force water out of a siphon, propelling themselves rapidly
backward. Some small species can leap out of the water and glide. It
is an extremely efficient system of propulsion, making squid one of
the fastest animals in the ocean and one of the hardest for humans
to catch. They also have very good eyes, an asset in hunting and
evading predators.

It is difficult to estimate the abundance of these elusive and wide-
ranging animals, but one indication is their high rate of consumption

Figure 13-22

Deep-sea angler fish. The many species of these fishes are most abundant at depths of 1500 to 2000 meters, where the ocean is dark and food is extremely scarce. In life cycle and in body structure, some angler fish illustrate two adaptations for survival at great depths: a movable, lighted fishing lure that dangles in front of the huge mouth and the male is parasitically attached to the female to guarantee mating.

In the species *Ceratius holboelli* (a), the male swims freely until adolescence, at which time he is only a few millimeters long. Locating a female, he attaches to her belly skin with his teeth and grows to a length of several centimeters. Their fertilized eggs float to the surface, where the larvae hatch and feed on copepods until adolescence, when they return to deep water. Sexual maturity is not attained until after the fish have joined together permanently. Adult females are about 1 meter long. Not all angler fish have the "attached male" adaptation, but all females possess some sort of attractive light organ to lure prey. The female ceratioid angler *Gigantactis* (b), living at depths to 5000 meters, carries a maneuverable "fishing rod" many times the length of her body. Other species, such as *Linophryne* (c), have luminous "barbels" on their chins, or a light organ in their mouths, just behind the teeth. A common adaptation at these depths is the ability to extend jaws and belly so they prey much larger than the predator can be swallowed. *Malanocetus johnsoni* (d) is illustrated before and after having eaten a fish twice its length.

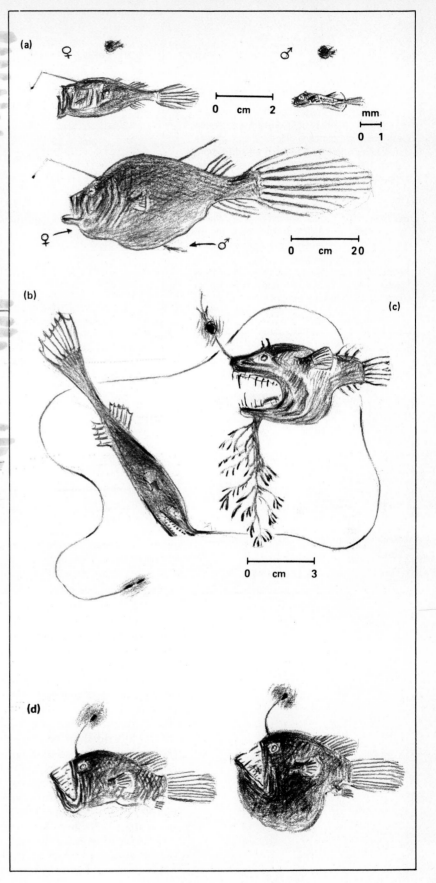

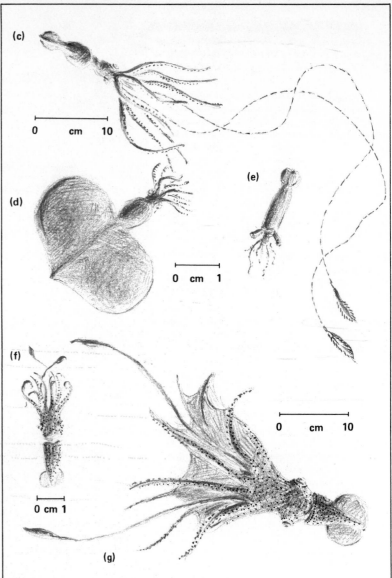

Figure 13-23

Oceanic squids. There are many species of these active, predatory cephalopods in the open ocean, ranging from luminous deep-sea species a few centimeters long to the giant squid more than 16 meters long. They live in all oceans, sometimes to depths of 3500 meters. Illustrated are: (a) and (c) two species of the deep-sea octopod *Cirrotheuthis;* (b) the common nearshore squid *Loligo;* two deep-water Atlantic squids—(d)*Octopodotheuthis,* and (e) *Taonidium;* and two midwater species—(f)*Caliteuthis reversa,* and (g) *Histioteuthis bonnelliana. Histioteuthis* has been captured at about 1000 meters depth, and *Calliteuthis* as deep as 1500 meters, but both are sometimes seen at the surface.

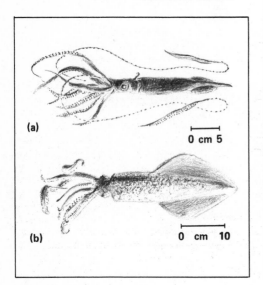

by *sperm whales* (Fig. 13-24), whose principal food they form. The annual take of squid by sperm whales may equal the world commerical fish production.

Many zooplankters, as well as squid and fishes, migrate vertically on a 24-hour cycle, responding to changing light levels. Echo sounders have detected a movable "deep-scattering layer" or dense aggregation of small animals that rise to feed at about 200 meters or less during the night and then return to depths of 500 to 1000 meters during the day. The precise makeup of a deep-scattering layer is difficult to determine. It is known, however, that some copepods, for example, can swim upward at speeds of 17 to 30 meters per hour and the larger euphausiids three or four times as fast. Fishes having gas-filled swim bladders that resonate when sound waves pass through them also make up a sizeable fraction of this migratory population.

Besides making the rich variety of foods near the bottom of the photic zone available to deep-water organisms, vertical migration helps zooplankton to maintain their position. Surface currents may move them away from their preferred location while they are feeding

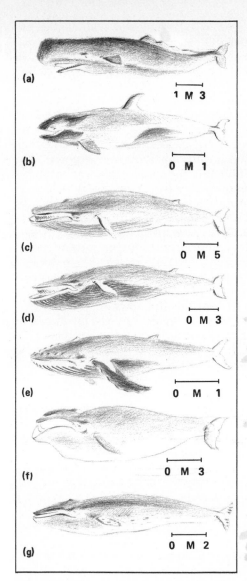

Figure 13-24

Sperm whales (a) are the largest of the toothed whales. Males average 15 meters; females are smaller. Their principal food is squid, which they hunt from the tropics to subarctic latitudes in all seas, diving to depths of 1000 meters or more and remaining below for as long as 1 hour. Giant squids up to 10 meters in length have been found intact in the stomachs of sperm whales. There is a reservoir of *spermaceti oil* in the animal's large, square head. Its function is unknown, although naturalists and scientists have speculated that it might be involved in the exchange of gases during long dives or perhaps in sound production.

Before a dive, a whale fills its lungs with air; while it is underwater, most of the fresh gases are exchanges for carbon dioxide or other products of its metabolism. Large amounts of oxygen are stored in the myoglobin of its muscle tissue, as well as in blood. The heart beats very slowly while a whale (or other diving animal) is submerged, which helps to reduce oxygen demand. Small lungs and a flexible thoracic cavity permit compression without damage as the volume of air decreases with depth.

Sperm whales are sometimes attacked by packs of *killer whales* (b). These are not really whales at all but members of the dolphin and porpoise family, most common in cold water.

Baleen whales (whalebone whales) are the largest animals that have ever lived. They are mammals and thus breathe air and bear live young. A horny sieve that hangs in plates from the palate is the "whalebone"; it is used to strain copepods and krill from seawater. They have no teeth and plankton or small fishes are their only food.

Blue whales (c) are the largest of all, reaching lengths of up to 31 meters. A 27-meter blue has been found to weigh 120 tons. Baleen species also include: (d) the *finback,* which attains a length of about 20 meters; (e) the *humpback,* about 15 meters at maturity; (f) *right whales,* 17 meters or more; and (g) the smaller *California gray whale,* about 12 meters. Baleen whales live in all seas, but particularly in the Arctic and Antarctic, where during the short southern summer (with a peak in February) krill have been reported so thick that a large ship was "slowed to half speed by them." The greatest numbers of baleen whales have been observed in the Antarctic when plankton are dense. During the rest of the year they range widely throughout the oceans.

at night. By encountering countercurrents that flow below the surface layer, they are carried in the opposite direction to remain in nearly the same location.

Vertical migration seems to be an important mechanism for transporting food to the deep ocean. Deep-water animals consume food near the surface and then carry down food that might otherwise have remained and been recycled in surface waters.

FOOD FISHES Fishes are active predators. The typical streamlined shape of most pelagic species testifies to the value of speed for capturing prey and avoiding enemies. Most can swim rapidly for short periods, about 10 times their body length in a second. Over extended periods, as during migration, a medium-sized fish, such as salmon, herring, or cod, can travel hundreds of kilometers in a few days. Large oceanic fishes that feed at high trophic levels, such as the *shipjack tuna* (Fig. 13-25), can swim 100 or more kilometers per day for weeks at a time while hunting schools of smaller fish. Estuarine and bottom-dwelling species, how-

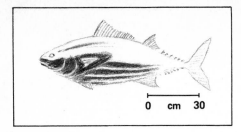

Figure 13-25

Skipjack tuna, an inhabitant of tropical and temperate waters, is related to the great bluefin sport fish (which may be up to 5 meters in length) and also to the albacore tuna and the yellowfin. Tunas maintain a body temperature several degrees higher than the surrounding ocean, perhaps due to an unusually high metabolic rate related to their adaptation for high-speed, long-distance swimming.

ever, tend to hide near rocks or in soft sand and their bodies are adapted for concealment rather than speed.

Of the *demersal fishes*, those that live on or near the ocean floor, the most important to man are the *cod* and its relatives (Fig. 13-26). They are mostly first-level carnivores, feeding on invertebrates, small fishes, and larvae. Another group, the *flatfishes* (Fig. 13-27), are modified to lie on the ocean floor. They are elliptically shaped, although their larvae are bilaterally symmetrical. As flatfishes mature, the eyes migrate to one or the other side of the head so that as adults they can conceal themselves on the bottom with only their eyes exposed. The upper side, left or right, depending on the species, tends to be darker than the lower. This is a common protective adaptation in pelagic as well as demersal fishes; organisms viewing a fish from below will be looking toward the light and so a light-bellied fish is less visible. But looking into deeper water, a dark-backed fish is more difficult to see.

Herring (Fig. 13-28) are commercially the most important pelagic fishes (see Table 15-1). Collectively known as *clupeoids*, where diatoms are abundant. Many clupeoid fishes have specially modified gill rakers for straining phytoplankton out of the water; they also feed on copepods and other zooplankters. Clupeoids and other low-trophic-level feeders have high oil contents. They are made into oils, fertilizers, or livestock feed, besides being consumed directly as human food.

All these relatively small fishes tend to *school*—that is, to aggregate in such a way that they are uniformly spaced and oriented in a common direction. This behavior is typical of fishes that swim unprotected in the surface layer of open seas and are constantly subject to predation. In general, all members of a school are of one species and more or less uniform size.

Schooling fish orient themselves so as to keep their own mirror image constantly in view. When thus positioned, they seem to respire more slowly, eat more, learn better, and exhibit less nervous behavior. If a school is startled, its members give off an "alarm substance" into the water and then pack more closely together to confront the source of the disturbance. The presence of many individuals within a relatively small area decreases the chances of any one individual being attacked.

Smaller predators also school to hunt; they surround the prey school so that its members become confused and less able to escape.

Figure 13-26

Some commercially important demersal fishes. Cod (a) live in cold and temperate seas, chiefly in the Northern Hemisphere, close to rocky or sandy bottoms. A schooling carnivore, its eggs are planktonic and in millions to ensure survival.

Pollack (b), haddock (c), and silver hake (whiting) (d) are found from Newfoundland to the Caribbean, to 450 meters depth. Related species live in deep waters over continental shelves in both hemispheres and over a wide temperature range. The young are adapted to eat planktonic crustaceans while adults prey on small fishes. Like the rest of the cod family and other less oily fishes, they can be effectively preserved by drying. The squirrel hake is shown in (e).

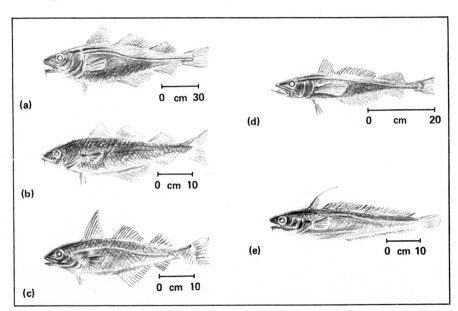

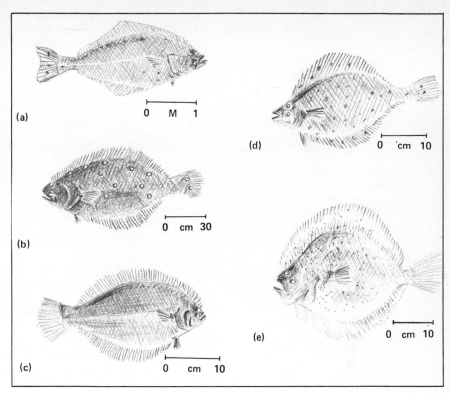

Figure 13-27

Somer important flatfishes. Atlantic halibut (a) are found from the Arctic Ocean to New Jersey; Pacific species live from the Bering Sea to the latitude of California. At lower latitudes, these species seek deeper waters. Halibut sometimes grow to 3 meters in length and are active predators. flatfishes live on large invertebrates and small fishes. Northern fluke (summer flounder) (b) range from Cape Cod to Cape Hatteras and reflected species inhabit all but the coldest seas. This flounder is really a member of the halibut family. Winter flounder (Atlantic sole) (c) are true flounders, as are European plaice (d) and turbot (e).

But some of the very large predators do not school, nor do those fishes that live below the photic zone. Schools may partially disperse toward nightfall, coincident with a period of intense feeding. Some—for instance the Atlantic herring—descend toward the floor of the continental shelf during midday, where they are eaten by demersal fishes, such as cod.

North Sea herring have been extensively studied because of their importance as a fishery. Their life cycle and feeding habits are typical of coastal species (Fig. 13-29). Herring deposit their eggs in thick blankets that may cover 100 square meters or more of the continental shelf. Bottom-living fishes, especially haddock, consume great num-

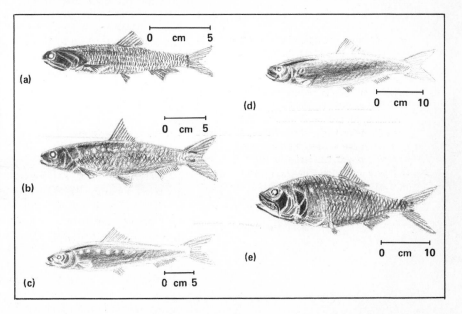

Figure 13-28

The herring family. (a) Anchovies and related species are abundant in warm seas, nearshore and in the open ocean. These valuable fish are widely used for food and as fresh and frozen bait. They are tolerant of a wide salinity range and are a basic food for many larger fishes. (b) Pilchard are found in coastal waters throughout the world, between isotherms 12 and 20°C.
They are of major commercial value off Australia and South Africa. (c) Sardines, (d) herring, and (e) menhaden are distributed from Nova Scotia to Brazil and are a staple in the diet of many seabirds and fishes.

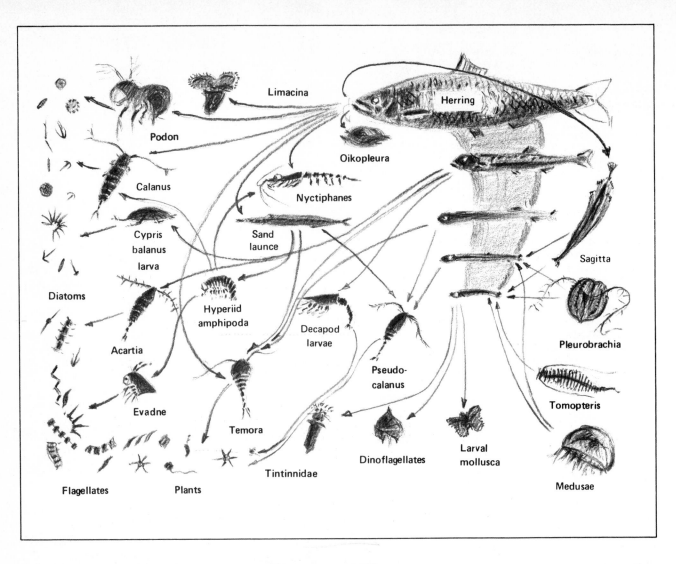

Limacina
Podon
Calanus
Cypris balanus larva
Diatoms
Acartia
Evadne
Hyperiid amphipoda
Temora
Flagellates
Plants
Tintinnidae
Sand launce
Nyctiphanes
Oikopleura
Herring
Decapod larvae
Pseudo-calanus
Dinoflagellates
Larval mollusca
Sagitta
Pleurobrachia
Tomopteris
Medusae

Figure 13-29

Life of the Atlantic herring. (Redrawn from A. Hardy, 1965. *The Open Sea. Part II: Fish and Fisheries.* Houghton Mifflin Co., Boston. p. 62.)

bers of these eggs. Herring develop as tiny, wormlike creatures, curled around the yolk of their eggs. After a few weeks, at about 5 millimeters length, the transparent larvae break free of the egg membrane and rise toward the water surface. Their first food is phytoplankton, later nauplii, very small crustaceans, and young stages of copepods.

Others starve when they do not find suitable food. At a length of about 30 millimeters, they begin to eat some of the smaller adult copepods; and at about 40 millimeters they grow scales and begin to look like young fish instead of tiny eels. They form enormous school—some as much as 5 kilometers long—and travel toward shallow water, often passing into estuaries, now eating mainly estuarine crustaceans. After 6 months or so, the young herring disperse throughout the North Sea. When sexually mature, after a period of 3 to 5 years, they join schools of spawning fish and the cycle is repeated. The same species inhabits the Atlantic Coast of Canada and the United States.

The *mackerel* and related species (Fig. 13-30) are well adapted for strong, continuous swimming in open waters, both near the surface and at depth. Open ocean species feed at high trophic levels and are voracious carnivores.

Mackerel provide an interesting example of migratory patterns in fish. They leave the surface waters about October and aggregate in small areas near the ocean floor. During this period they eat crustaceans and small fishes. In January the mackerel move to the surface

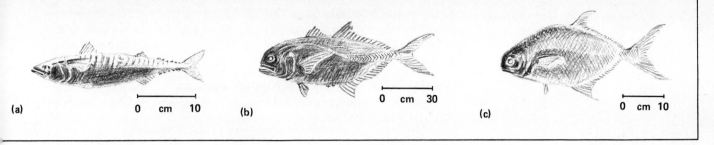

(a)   0 cm 10   (b)   0 cm 30   (c)   0 cm 10

Figure 13-30

Atlantic mackerel (a) are distributed from Labrador to Cape Hatteras and from Norway to Spain. In the western Atlantic they migrate northward in schools along the coast in summer, southward offshore in winter. Preyed on by birds, squids, and other fishes, they themselves eat anything they can swallow. (b) The common jack is related to a number of warm-water species off American shores; jacks range from Cape Cod to Brazil and southward of the Gulf of California. (c) Pompano. Related species include bonitas, tunas, and swordfishes.

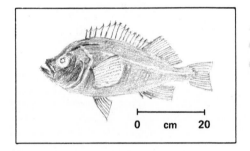

0   cm   20

Figure 13-31

Ocean perch (rosefish, redfish). These live in the North Atlantic with related species worldwide. Some of the many species in this family are referred to collectively as rockfish because they prefer near-bottom waters. The flesh tends to be reddish and many groups have long spines, often poisonous in tropical members of the family. North Atlantic redfish and eastern Pacific rockfish are examples of major fisheries opened after World War II to exploit what used to be considered "trash" (noncommercial) fishes in meeting world protein needs.

in schools and migrate to their April spawning grounds south of Ireland. They spawn near the edge of the continental shelf, gradually moving closer to the land, and during this time they feed on plankton, especially copepods. From June to July they form smaller schools and move close to the shore, changing their diet from plankton to the small fishes that swarm in inshore bays. In the fall they again seek deeper waters.

Ocean perch and related fishes (Fig. 13-31) are taken in large quantities throughout the world for food (see Table 15-1). *Basses, croakers, snappers, kingfish, walleye pike, porgies,* and *groupers* are all members of this varied *percoid* class, especially common in warm nearshore waters. Many sport fishes and also many freshwater species are included among the percoids.

Another group of valuable food fishes are *salmon* and *trout*. Both are active predators, generally confined to northern, fairly cold water, and they typically spend part or all of their lives in freshwater. Salmon spawn in rivers but attain most of their growth in ocean waters (Fig. 13-32).

Figure 13-32   (above)

Chinook salmon (king salmon, spring salmon). Ranging from southern California to northwest Alaska, chinook are especially prized by sport fishermen. They are common along the British Columbia coast. In spring and early summer mature adults leave coastal waters to "run" up large rivers, where they spawn in freshwater and then die. The young may go to sea in the first year or remain for a year or more in the streams. Maximum length is about 1.5 meters. (Photograph courtesy Washington State Department of Fisheries.)

FISH AND COASTAL OCEAN PRODUCTIVITY   Most important sport and food fishes live in coastal waters. More than half are estuarine dependent during at least part of their lives. Estuaries are nursery areas for many species, providing protected environments as well as abundant food at critical life stages for larvae and juveniles. These areas are highly productive, deriving nutrients from three sources; subsurface coastal ocean waters carried in by the estuarine circulation; dissolved minerals leached from soils in the river's drainage basin; and nutrient-rich by-products of agriculture, industry, and human wastes. Furthermore, nutrients tend to be retained within estuaries because phosphates and nitrates released by decomposition in lower layers are returned to the surface by entrainment of subsurface waters in the estuarine circulation (Chapter 10).

Being shallow as well as rich in nutrients, estuarine tidal flats, marshes, and shallow inlets can support dense stands of benthic plants. Included are flowering *seagrasses* that grow in mud and sand, benthic algae, and *epiphytes* (plants, often unicellular, such as diatoms, that grow attached to larger plants). Primary productivity of seagrass communities may exceed 600 grams of carbon per square meter annually. In Atlantic estuaries, for example, the marsh grass *Spartina* and associated algae produce up to 250 tons of plant material per square kilometer (10 tons per acre) in a year, several times the productivity of an average wheat crop. Shore-based vegetation—for instance, in salt marshes—and phytoplankton further increase estuarine productivity.

Fishes, jellyfish, and benthic animals throughout the estuary and coastal ocean are supported by estuarine productivity. Migratory and developing fish populations are major consumers. Food webs in shallow estuaries are based largely on microscopic benthic and epiphytic algae and decaying vegetation. Thus the primary plant food is consumed directly by herbivorous fishes, such as mullet, and by benthic fauna rather than by zooplankton, as is typical in open ocean waters.

Patterns of spawning and development vary widely among estuarine-dependent saltwater fishes and shellfish. Off the Atlantic Coast and the Gulf of Mexico, many coastal fishes and crustaceans spawn in the ocean, but their young soon move into estuaries. Larvae often travel many tens of kilometers, moved passively by strong coastal currents, and enter small inlets where currents are quite strong. Larvae that swim at the surface (such as mullet) may come in on the tides. Bottom species, such as shrimp, whose life cycle is illustrated in Fig. 13-33, are aided by estuarine circulation systems. Once inside, the larvae are distributed within bays by currents; possibly some are sensitive to salinity gradients and swim toward less saline water. In general, low salinities are preferred by fish larvae where waters are warmer. Tolerance for low salinities declines toward higher latitudes.

In spring and summer in the Northern Hemisphere, fish migrations are typically directed inshore and northward; in late summer and fall the pattern reverses. The bottom-dwelling fluke (see Fig. 13-27), for example, winter offshore along the middle-Atlantic coast, sometimes as far as 150 kilometers from land. In spring they move toward the coast to feed, spending late spring and summer in nearshore waters. In early fall fluke start back toward their wintering grounds. They spawn during October and November, 15 to 100 kilometers offshore on the continental shelf, and surface currents carry the buoyant eggs (later the larvae) southward until early spring. They swim toward the coast and spend the summer in shallow estuarine waters; in autumn they reenter the ocean. This type of breeding cycle, with variations, is common among Atlantic and Gulf Coast species.

Some migratory fishes enter Atlantic estuaries to spawn. Striped bass, for instance, lay their eggs in the low-salinity waters of Chesapeake Bay and Albemarle Sound tributaries, and the young drift downstream into the estuary. At maturity (2 or more years), some enter coastal waters and migrate northward as far as New England during the summer. Winters are spent in deep holes or channels and in spring they again swim upstream to spawn.

Productivity in the North Sea (see Figs. 3-20 and 12-23) has been studied in order to determine whether fish yields in the area might be increased. But judging from known levels of primary production,

Figure 13-33

Life history of the Gulf Coast shrimp: (a) shrimp eggs; (b) nauplius larva; (c) late nauplius; (d) mysis; (e) postmysis; (f) juvenile; (g) adolescent; (h) adult.

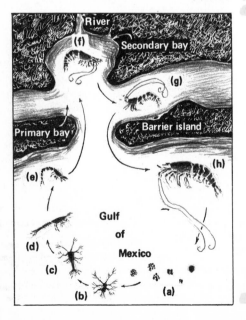

it seems unlikely that present yields can be greatly increased. For example, it has been calculated that net primary productivity in the North Sea averages about 90 grams of carbon per square meter per year. Production of pelagic fish has ranged as high as 2×10^6 tons during peak fishing years in the 1960s and demersal species were caught at a rate of 0.9×10^6 tons annually. Assuming that "natural mortality," or death from causes unrelated to humans, claimed at least half of the pelagic fishes and more than half the demersal production, it has been calculated that 4.0×10^6 tons of pelagic and 1.3×10^6 tons of demersal fishes represent a maximum potential production for the North Sea as a whole. This averages out to 8 grams of carbon per square meter (wet weight) of pelagic fishes and 1.6 grams of carbon per square meter of demersal fishes for each 90 grams of carbon per square meter per year of net phytoplankton produced in the region.

Figure 13-34 is a representation of the North Sea as an ecosystem; it illustrates the food web through which energy passes from phytoplankton to humans. Each gram of organic carbon is considered equivalent to 10 kilocalories (kcal) of potential energy. Each organism in the food web has been assigned a value in kilocalories. calculated either from its biomass or on the basis of its trophic level and presumed ecological efficiency. Copepods, for instance, are assumed to be about 20% efficient in transforming ingested phytoplankton into animal biomass (170/900 = 19%). But only half the herbivores are eaten by pelagic fishes in the direct food chain to humans and pelagic fishes

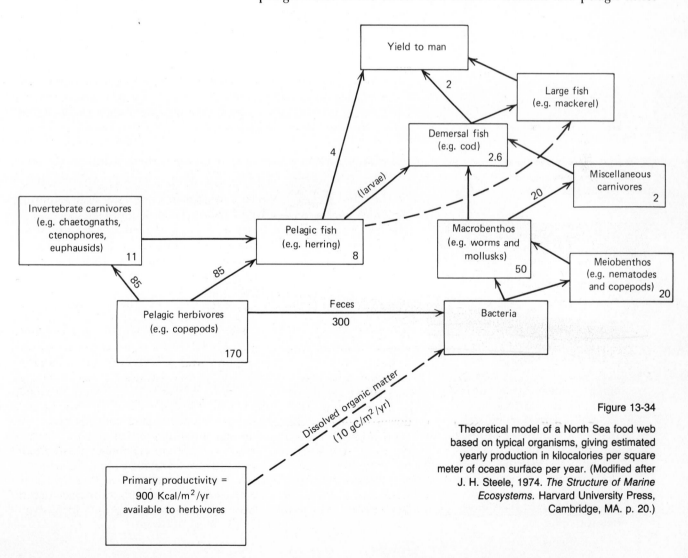

Figure 13-34

Theoretical model of a North Sea food web based on typical organisms, giving estimated yearly production in kilocalories per square meter of ocean surface per year. (Modified after J. H. Steele, 1974. *The Structure of Marine Ecosystems.* Harvard University Press, Cambridge, MA. p. 20.)

are only considered about 10% efficient at building fish tissue from the copepods they eat. Furthermore, even if all chaetognaths (arrowworms) are ultimately eaten by herring, the additional trophic level would use almost all the energy potentially transferable from arrowworm to fish.

On the decomposer–demersal fish side of the model, it is postulated that benthic forms are supported by detrital material from above, which may go at least in part through bacterial transformation before being consumed by bottom dwellers. Bacteria apparently have high ecological efficiency and a very low ratio of biomass to productivity because they live only a short time but reproduce rapidly. They evidently provide substantial food for both *macrobenthic epifauna* (large animals living on the bottom) and *meiobenthos* (minute animals that live in bottom sediments). These organisms probably also have a low ratio of biomass to productivity, especially the very small species. But if efficiencies of around 10% are typical of fish and benthic animals, the model indicates that present populations utilize nearly all primary production in the North Sea. No large increase in fish yields is likely under these circumstances without drastically disturbing energy flows in the food webs.

REVIEW QUESTIONS

1. Explain the advantages and disadvantages of small size to a phytoplankton cell in the ocean.
2. Diagramatically show how cell division by a diatom influences cell size.
3. What is the difference between meroplankton and holoplankton? Give examples of each.
4. Explain why krill are so abundant near Antarctica and why they are important as food for whales.
5. Describe some typical features of animals living in the mesopelagic zone and indicate the utility of these features.
6. Discuss schooling of fishes and indicate its utility to the fishes.
7. List and describe the factors that make coastal and estuarine areas highly productive of fishes and invertebrates.

SUMMARY OUTLINE

Characteristics of Plankton
Small size results in large ratio of surface to volume
Presents relatively larger surface for nutrient, gas diffusion
Retards sinking

Phytoplankton
Major classification—ultraplankton, nannoplankton, microplankton
Small forms dominate in open ocean
Diatoms—dominant producers in coastal, mid- and high latitudes
Characterized by heavy spring blooms, clumping in high-nutrient areas, siliceous shells
Dinoflagellates—next most productive; some heterotrophic

Coccolithophores, silicoflagellates also important; blue green algae in warm coastal areas

Zooplankton—transfer matter and energy from primary producers to predators
Variety of feeding adaptations for trapping small particles
Currents carry them away from narrow temperature ranges of breeding grounds

Holoplankton—permanent members of plankton
Protozoa—consume nannoplankton
Foraminifera—calcium carbonate shell; *Radiolaria*—siliceous; *Tintinnidae;* and *Rotifera* important in open oceans
Crustaceans—abundant in productive ocean areas
Copepoda—most numerous

Euphausiidae—"krill," abundant in Antarctic, eat diatoms, eaten by whales

Chaetognaths—active, predatory worms, abundant

Pteropods—"snail with wings"; herbivorous or carnivorous

"Dead ends" in food chains—Coelenterates (jellyfish); Ctenophores (combjellies); Tunicates (salps, doliolids)

Meroplankton—larval stages only in plankton

Eggs, larvae planktonic, metamorphose to adult form at attachment to bottom

Worms, mollusks—have trochophore larvae

Crustaceans—similar to planktonic forms

Fish larvae—success depends on availability of food, other environmental factors

Life in the deep ocean—mesopelagic zone—200 to 1000 meters

Less than 1% of light penetrates

Many organisms bear patterned lights, lighted lures

Squid—powerful swimmers, high-level predators

Vertical migration by many fishes, invertebrates

Deep-scattering layer: 200 meters deep at night; 500–1000 meters by day

Deep-water organisms get food from surface zone

Food fishes

High-trophic-level feeders—for instance, tuna—swim powerfully

Demersal fishes (cods, flatfishes)—coastal ocean, deep-water, low trophic-level feeders

Herring family—clupeoids; herbivorous, tend to school for protection

Life cycle typical of coastal fishes

Mackerel—migrate throughout wide coastal ocean area

Fish and coastal ocean productivity

Most food and sport fishes depend on coastal, estuarine productivity

Estuaries—high productivity, nursery area; benthic vegetation productive

Different species have various migratory patterns; enter nearshore and estuarine areas at different times of year, different stages of life cycle

North Sea: cannot yield much more than at present

Primary productivity approximately 90 grams of carbon per square meter per year; fish yield approximately 3×10^6 tons

Present population evidently utilizes all energy from primary productivity

SELECTED REFERENCES

BARDACH, J. E., J. H. RYTHER, AND W. O. McLARNEY. 1972. *Aquaculture: The Farming and Husbandry of Freshwater and Marine Organisms.* Wiley-Interscience, New York. 868 pp. Thorough discussion of all aspects of aquaculture.

FRASER, JAMES. 1962. *Nature Adrift: The Story of Marine Plankton.* G. T. Foulis, London. 178 pp. Concise, readable survey of marine plankton forms, their behavior and ecology.

GULLAND, J. A. (Ed.) 1970. *The Fish Resources of the Ocean.* Fishing News (Books) Ltd., West Byfleet, Surrey, England. 255 pp. Survey of world fisheries.

HARDY, ALISTER. 1965. *The Open Sea: Its Natural History.* One-volume edition. Houghton Mifflin, Boston. 355 pp. A classic in the field by a well-known British naturalist; nicely illustrated.

HOLT, S. J. 1969. The food resources of the ocean. *Scientific American* 221(3) (September 1969): 178–194. Survey of contemporary technology and outlook for the future of marine food resources.

IDYLL, C. P. 1964. *Abyss: The Deep Sea and the Creatures that Live in It.* Thomas Y. Crowell, New York. 396 pp. Elementary discussion of marine life with particular emphasis on deep-ocean organisms; also discusses various curiosities, such as sea monsters.

MARSHALL, N. B. 1958. *Aspects of Deep-Sea Biology.* Hutchinson, London. 380 pp. Authoritative treatment of the deep-ocean environment and its ecology.

MARSHALL, N. B. 1971. *Exploration in the Life of Fishes.* Harvard University Press, Cambridge, MA. 204 pp.

RAY, CARLETON, AND ELGIN CIAMPI. 1956. *The Underwater Guide to Marine Life.* A. S. Barnes, New York. 337 pp. Classifies and briefly describes thousands of marine organisms, with line drawings.

WIMPENNY, R. S. 1966. *The Plankton of the Sea.* American Elsevier, New York. 426 pp. Emphasis on the taxonomy of animal plankton; well illustrated.

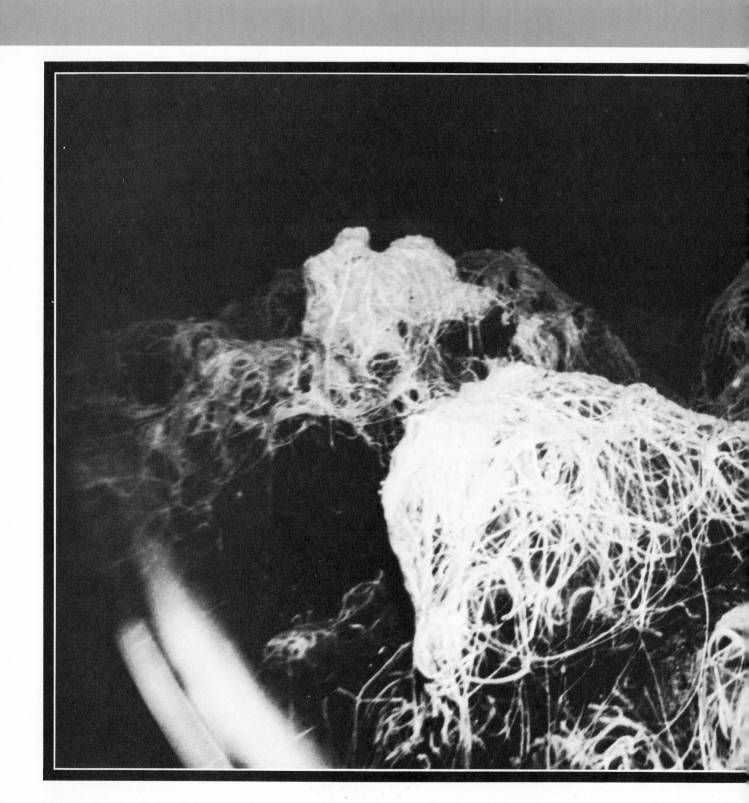

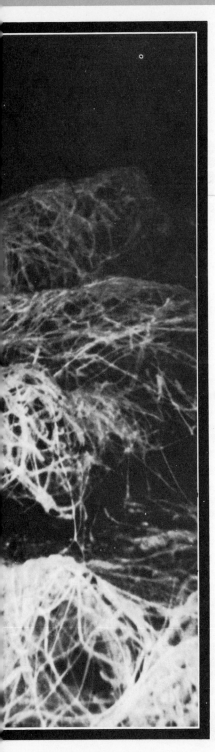

Dubbed "spaghetti" by scientists because they were draped like threads over rocks, these animals were found in the hot springs area in the Galapagos Rift. The animal was identified as *enteropneust,* a type of worm. (Photograph courtesy James Childress, UC-SB.)

Benthic organisms live in a two-dimensional world in contrast to the three-dimensional, nearly boundless existence of plankton and nekton. There are three *benthic life strategies: attachment to a firm surface, movement on the bottom,* or *burrowing in sediment.* Generally they correspond to three *ways of obtaining food: filtering from seawater, predation,* or *swallowing sediment* and digesting the organic matter.

The three lifestyles are combined in many ways. Crabs or worms, for instance, may live in a sand or rock burrow but emerge to hunt for prey or scavenge for detritus. Slow-moving animals with heavy shells, such as snails or sea urchins, feed on attached organisms or detrital particles. But some attached animals, such as sea anemones, are predators; they capture organisms that swim or float past them. In this chapter, we consider life on the bottom of the ocean.

**DISTRIBUTION
AND DIVERSITY
OF BENTHIC LIFE**

Attached plants require sunlight and so can only grow on about 2% of the ocean floor, shallower than 30 meters depth, and many kinds require a firm reef or rock substrate (Fig. 14-1). Fixed seaweeds, although highly productive in shallow seas, thus contribute little to total productivity in the open ocean. A notable exception is *sargassum,* a leafy brown *alga* (seaweed) that forms a unique floating "benthic" ecosystem in the Sargasso Sea (Fig. 14-2).

The makeup of a benthic community is controlled mainly by light, temperature, salinity, and nature of bottom and by whether these features are constant or variable. Environmental stability seems to favor evolution of highly diverse communities in which many kinds of plants and animals coexist. In surf zones and nearshore areas, for example, where waves keep bottom materials in constant motion, benthic organisms are scarce. And in parts of shallow polar oceans the short growing season for plants, together with marked salinity changes due to melting and freezing of sea ice, creates a rigorous climate in which relatively few species can thrive. Quiet waters in nearshore tropical areas, on the other hand, normally have highly diverse bottom communities. Thousands of species can make up a single ecosystem, such as a coral reef community.

Shallow-water ecosystems in temperate climates tend to have relatively few different species but many individuals of each (Fig. 14-3). This lack of diversity may be partly due to the stress of pronounced temperature changes that occurred at midlatitudes during more glacial times. Tropical and deep-ocean environments, on the other hand, have evidently experienced stable temperatures for millions of years.

Figure 14-1

On this rocky Maine coast, brown algae grow thickly along a steep rock face and then cease abruptly except near the bottom, where growth extends farther and farther into the water.

Figure 14-2

Sargassum (sometimes called "gulfweed") is the only large nonattached seaweed. It is found in the south–central North Atlantic, where surface currents converge, an area known as the Sargasso Sea. There it puts out shoots that break off as new plants. Patches of this algae form "floating islands"; they harbor communities of smaller plants and animals that can live nowhere else in the world. Included are many kinds of specialized crabs, snails, copepods, worms, fish, sea horses, and small octopus, as well as the *Sargassum fish* itself. Patches of sargassum collect in windrows along lines of convergence formed by wind-driven Langmuir cells and other circulation systems. Surface films. plankton, and particles of detritus caught there become available to decomposers and consumers in the sargassum community. Large numbers of eggs and larvae of open ocean fishes become trapped in the dense growth, which offers protection from predators as well as a rich supply of food particles. Some of these eggs and young fishes, an energy source that originated outside the sargassum community, are consumed within the community, thus augmenting the local nutrient supply. Seven million tons of "gulfweed" is estimated to float in the approximately 5000 square kilometers of the Sargasso Sea.

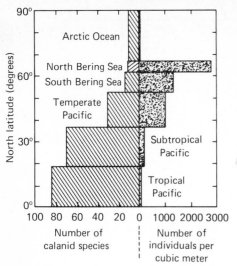

Figure 14-3

The number of species of calanoid copepods (planktonic crustaceans) from the upper 50 meters is greatest in tropical ocean areas. But numbers of individuals are greatest at subpolar latitudes. This relationship also true for benthic organisms. (After A. G. Fischer, 1960. "Latitudinal Variations in Organic Diversity," *Evolution*, 14, 84-81.)

Figure 14-4

Moving south from polar to subtropical waters, the percentage of snail species having pelagic (free swimming) larvae increases with increasing water temperatures. (After Raymont, 1963.)

They are relatively free from seasonal climate changes. Such environments include a greater variety of organisms than do communities where temperature, salinity (see Fig. 12-2), or other conditions are subject to change.

Development of juvenile stages also varies with environment. Benthic organisms commonly have pelagic larvae in tropical and subtropical zones, but in more rigorous polar and subpolar climates nonpelagic larvae predominate (Fig. 14-4).

Abundance of benthic organisms is controlled by productivity of surface waters and by water depth. It is greatest on continental shelves and below areas of upwelling or divergence. Numbers of organisms decrease with increasing water depth and distance from land.

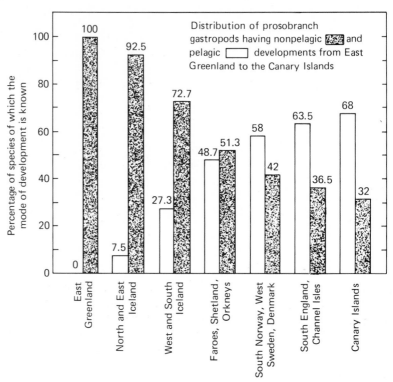

Distribution of prosobranch gastropods having nonpelagic and pelagic developments from East Greenland to the Canary Islands

Benthic populations vary according to fluctuations in year-class size. Variations in temperature, food supply, or number of predators determine the number of larvae that can grow to maturity. This factor, in turn, affects population size, particularly for organisms whose life cycle is not much longer than their generation period. The situation is illustrated by the change in biomass of five species of clams over a 4-year period, as shown in Fig. 14-5. Among relatively long-lived species, where many generations coexist, populations are less variable.

ROCKY SHORE COMMUNITIES

Intertidal areas provide a vertical sequence of living conditions for organisms. Each *zone* or horizontal band is occupied by a particular assemblage of plants and animals, as can easily be seen on a rocky cliff. Organisms are attached at the surface, although many animals hide in moist, protected crevices when the tide is out. Zonation also occurs on sandy and muddy shores, but here many zones occur within the sediment and are more difficult to see.

Conditions in intertidal areas range from nearly always dry (highest high tide level) to nearly always submerged (lowest low tide

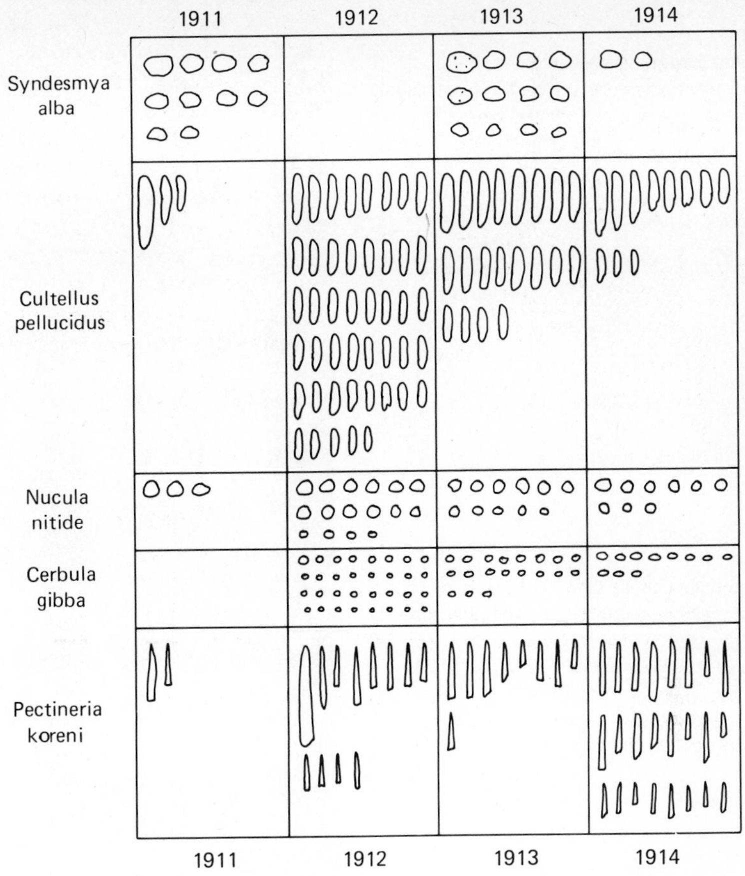

	1911	1912	1913	1914

Figure 14-5

Annual variation in relative abundance of five invertebrate species studied over a 4-year period. (After J. E. G. Raymon, 1963. *Plankton and Productivity in the Ocean,* Macmillan, Pergamon Press, New York 625 pp.)

level). Population zones are often sharply divided instead of being graded into one another. This feature reflects the intense competition for living space. At the line where two populations meet, neither enjoys a marked advantage. Above and below the line, assemblages are determined by differences in the organism's ability to endure air exposure and often by the survival of their predators and diseases under the same conditions. Zonation among barnacles on the Maine coast can be seen in Fig. 14-6.

Where winds, sun, or waves create an unfavorable environment, or where the rock face is steep, attached seaweeds may be sparse or absent. In more protected areas a variety of plants are attached to bare rocks along the seashore. Above the high tide mark, seawater only reaches the rocks at higher spring tides and during storms, but salt spray usually wets the area. Resistance to drying is a prime requirement for plants and animals in this region. Several varieties of blue green algae, lichens (algae growing symbiotically among fungus filaments), and certain small snails are common above the high tide mark. Filamentous green algae known as sea hair sometimes grow in moist, protected spots.

Between the high tide and low tide marks, especially where heavy surf is a factor or where current scouring is intense, firm attachment is a necessity. Barnacles (Fig. 14-6) often dominate the upper, more exposed regions, where the rock is covered by water less than half of each tidal day. In shady, protected areas the barnacle zone extends farther up the rock face than it does on dry, sunny surfaces. These animals feed while covered by water at high tide, filtering small particles from the water (Fig. 13-18). The barnacle shell has a four-part lid that shuts tightly, protecting the animal from exposure.

387

Figure 14-6

Zonation among barnacles, Maine coast. A small species dominates the upper layer, separated by a sharply defined boundary from the larger species below.

Limpets, cone-shaped snails, also live in rocky intertidal areas. They feed by scraping off algae and accumulated detritus. During low tide or when subjected to wave action, some limpets return each to its own depression on the rock face, which closely fits the rim of the animal's shell. The seal is so good that it is nearly impossible to dislodge a limpet by hand. Some species favor attachment on top of another limpet. Lower in the intertidal zone, tube worms and sea anemones compete with the barnacles; closer to the low tide mark, dense mussel beds are common, plus various kinds of attached algae, as seen along the Maine coast in Fig. 14-7.

Preference for a specific light intensity governs distribution of plants. Green algae may be found in areas ranging from somewhat above the low tide mark to a depth of perhaps 10 meters in temperate and tropical zones. Varieties range from lush-looking bright-green sea lettuce, up to 1 meter in length, to delicate, mossy types only a few centimeters long.

Red algae grow worldwide but most abundantly in temperate and tropical seas. They prefer the dim light of very deep water or well-shaded pools. Brown algae flourish in colder water, although some kinds are found on rocky coasts throughout the world. An abundant form in North America is brown rockweed, with its cluster of berrylike floats. Marine algae lack true roots but absorb nutrients from seawater directly. Some benthic algae attach to the bottom by rootlike structures called *holdfasts.*

Figure 14-7

A rocky intertidal area in Maine shows zonation of attached animals and plants: barnacles above, brown algae below, and at the bottom, underwater even at low tide, green and red algae grow in a shady pool. Snails, worms, sea stars, and tiny crustaceans live among the plants.

Figure 14-8

Many varieties of red and green algae shade one another in this tide pool on the Maine coast, providing shelter for small shellfish and worms. The area photographed is about 1 meter across.

Tide pools contain specialized plants and animals. The protected environment afforded by the rocks often permits more delicate organisms to live in these pools and a large variety of plants and animals may live in a small area (Fig. 14-8). Tiny, shrimplike crustaceans (amphipods), swimming worms, and many kinds of snails are common in tide pools, especially where deep crevices or beds of seaweed retain water during low tide. Evaporation, overheating, and oxygen depletion occur on warm, sunny days whereas heavy rains can markedly

Figure 14-9

Warm-water sea anemones with sea stars at the New York Aquarium. The animals at left center and far right are open; the mouth opening is at the center of the "flower." When startled, they close instantly, giving the appearance of a wrinkled stump (right center).

Figure 14-10

The coelenterate *Chrysaora quinquecirrha*, a native of Chesapeake Bay: (a) hydroid polyps, growing on an oyster shell; (b) detail of a typical feeding polyp, about 1 millimeter tall—the mouth, surrounded by 16 tentacles, is at the top of the polyp; (c) reproductive polyp shows budding of ephyrae and may produce a new stack of full-grown, free-swimming *medusa* in 2 to 4 weeks; polyp may give rise to 2 to 12 ephyrae and may produce a new stack up to four times in one summer; (d) full-grown medusa of *Chrysaora,* about 10 centimeters across the bell; their unpleasant sting sometimes closes the bay to swimmers. (Photograph courtesy Michael J. Reber.)

lower salinity within a few minutes. Survival for a tide pool organism requires a tolerance for sudden changes in temperature, salinity, and dissolved-oxygen content of the water.

The most populous zone on a rocky beach is generally located around and below low tide level. Starfish and crabs are common, usually hidden in rocky crevices. Small scavenging snails inhabit protected niches containing stagnant water and decaying debris. Sea anemones (Fig. 14-9), sea urchins, and sea cucumbers may be locally abundant below the low-water mark. Hydroid colonies (Fig. 14-10) grow in quiet but not stagnant waters, as do the nudibranchs (a shell-less snail; see Fig. 14-11) that feed on them.

Lobsters (Fig. 14-12) and crayfish (spiny lobsters, Fig. 14-13) scavenge on subtidal hard bottoms, both nearshore and far out on the

(a)

(b)

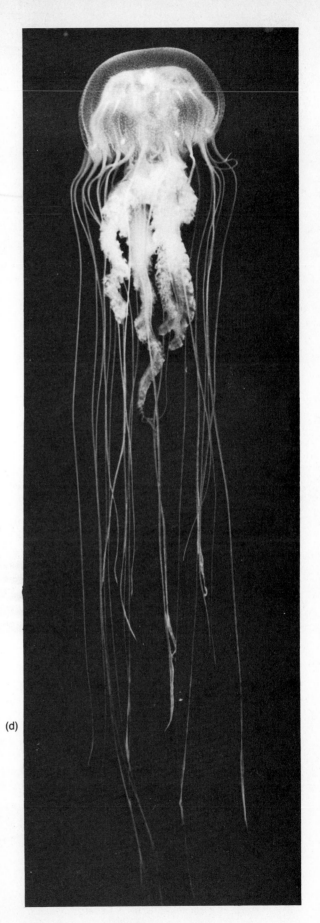

(c)

(d)

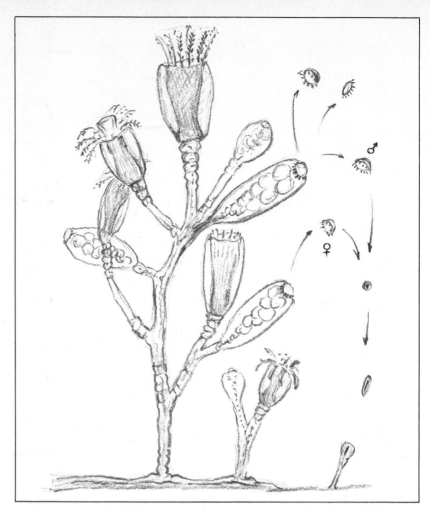

Figure 14-10e

Medusa reproduce sexually to form a new, sessile animal. The hydroid *Obelia,* showing feeding polyps and reproduction cycle, is illustrated in (e).

continental shelf. They walk about at night, searching for worms, mollusks, and organic debris and usually seek shelter during daylight under rocks or among seaweed. In autumn Maine lobsters move offshore to breed, probably seeking warmer deeper waters and thereby avoiding the cold shelf waters in winter.

The octopus (Fig. 14-14) is another common resident of hard-bottom and continental shelves. It prefers crevasses to hide in and usually emerges only at night to feed.

SOFT-BOTTOM COASTAL OCEAN COMMUNITIES

Tidal marshes, beaches, and estuarine shorelines offer a variety of habitats for benthic plants and animals. Abundant rooted plants in marshes contribute plant debris to the fine-grained sediments and protect the marsh from erosion. On open coastlines, where wave action or longshore currents inhibit rooted plant growth and also carry away silt and detritus, coarse, sandy bottoms with low organic content are the rule. In beach and marsh environments, plankton, rooted plants, benthic invertebrates, fishes, birds, and mammals form complex ecosystems (Fig. 14-15).

Figure 14-11

The nudibranch *Craten pilata,* length about 20 millimeters, which feeds actively on hydroid polyps in Chesapeake Bay. This shell-less snail breathes through the gill-like projections along the sides. Because of their noxious taste, they have a few natural enemies. (Photograph courtesy Michael J. Reber.)

Figure 14-12

American lobster, which lives off the East Coast of the United States. (Photograph of a model in the U.S. National Museum, Washington, D.C.)

Figure 14-13

Spiny lobster (crayfish), similar to a true lobster but lacking the large claws, is found in the warm waters of the U.S. Pacific and Gulf coasts. (Photograph courtesy Marineland of Florida.)

Figure 14-14

Octopus. The eyes are on top of the head; below is a siphon through which water is ejected after passing over the gills. There are eight tapered tentacles or arms, all equipped with double rows of suckers. Maximum size is around 50 kilograms with a 9-meter armspread, but many common species are much smaller. (Photograph courtesy Marineland of Florida.)

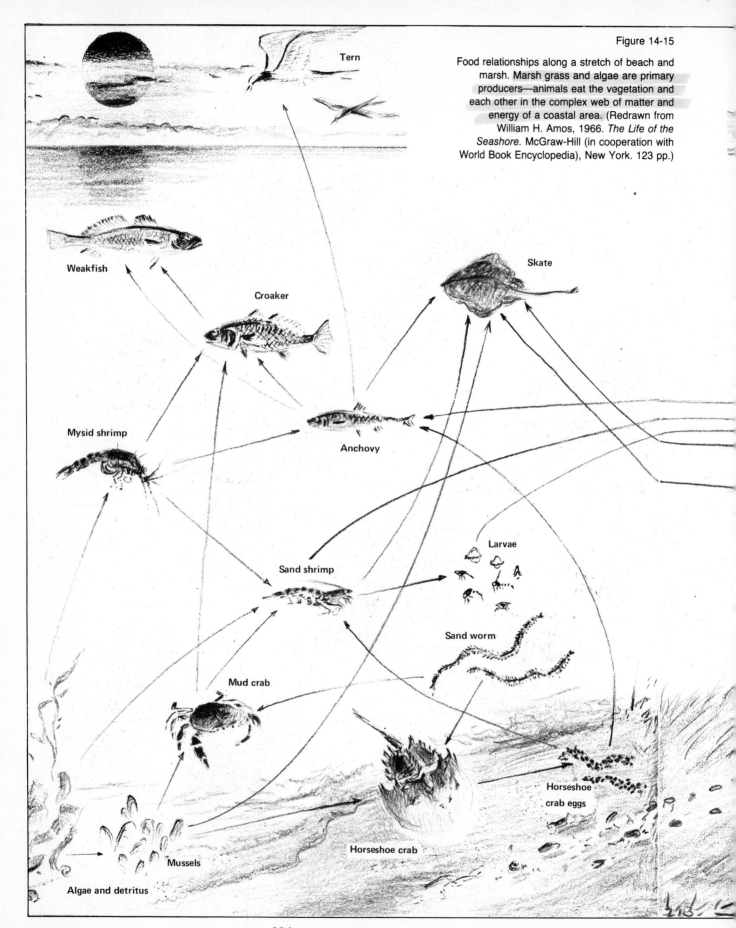

Tern

Weakfish

Croaker

Skate

Mysid shrimp

Anchovy

Larvae

Sand shrimp

Sand worm

Mud crab

Horseshoe crab eggs

Mussels

Horseshoe crab

Algae and detritus

Figure 14-15

Food relationships along a stretch of beach and marsh. Marsh grass and algae are primary producers—animals eat the vegetation and each other in the complex web of matter and energy of a coastal area. (Redrawn from William H. Amos, 1966. *The Life of the Seashore.* McGraw-Hill (in cooperation with World Book Encyclopedia), New York. 123 pp.)

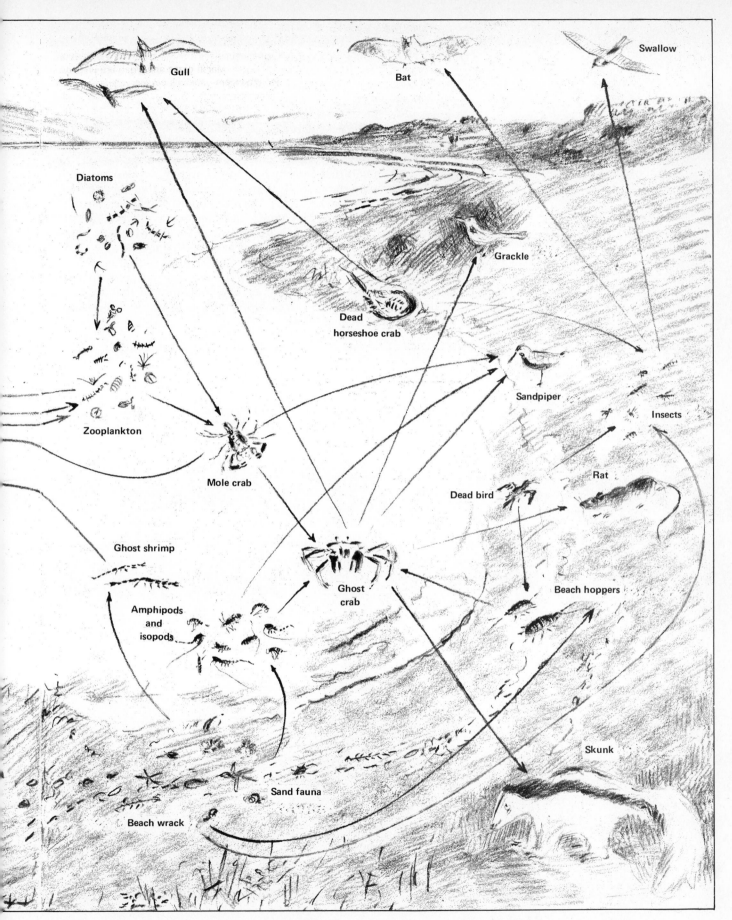

Infauna—animals that live buried in sediment—are much less conspicuous than surface-dwelling animals, or *epifauna*. Many of the former are *selective deposit feeders* [Fig. 14-16(a)], lifting and sucking food particles out of the mud, whereas others pump large quantities of water through filtering devices to remove edible materials. Still others feed *unselectively* on sediment deposits, eating their way through the substrate and digesting their food in the sediment; these include some nudibranchs, sea cucumbers, and many worms, such as the one in Fig. 14-16(b).

Distribution of infauna is governed partly by grain size of the sediments. Animals that feed by filtering plankton and detritus from seawater dominate stable sandy deposits below low tide. They require relatively clear, nonturbid water, which will not clog delicate, mucous-coated filtration devices. Fine muds, easily suspended by bottom currents, are generally not a satisfactory substrate for filter feeders. Muds and clays, however, are well suited to organisms that feed unselectively by ingesting sediments because the smaller particles, with their higher ratio of surface to volume, normally contain more organic matter. Such detritus supports bacteria and very small animals that are food for deposit feeders. These microscopic members of the infauna are discussed in the next section.

Figure 14-16

(a) Benthic bivalves: the cockle is a filter feeder; the clam *Tellina* sucks edible particles from the sediment surface. (After Hedgpeth, 1957.) (b) Sipunculid worms can retract the tentacle-bearing head into the body. Many live in soft ocean sediments and feed by swallowing great quantities of mud and sand, from which the organic component is extracted in the gut. (Photograph of a model in the U.S. National Museum, Washington, D.C.)

(a)

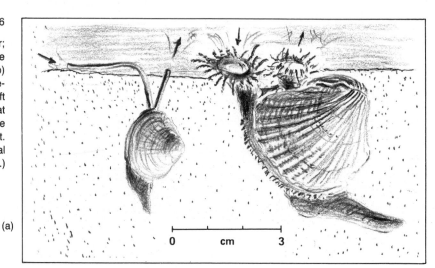

(b)

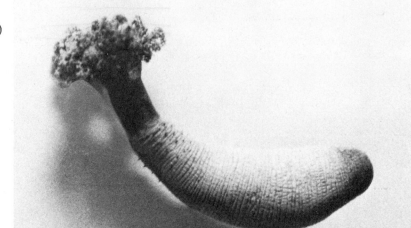

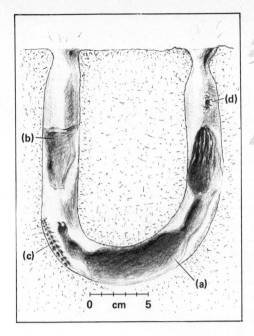

Figure 14-17

The pink echiuroid worm *Urechis caupo* (a) is a common inhabitant of Pacific mud flats at or below the low tide mark. This so-called innkeeper worm constructs a deep, U-shaped burrow with narrow openings by scraping away mud with bristles at each end of its body. Inside it secretes a funnel-shaped, fine-meshed mucous net that fits over its neck like a collar (b). As *Urechis* pumps water through the burrow, food particles are trapped in the net. When the funnel becomes clogged, the worm eats it and constructs a fresh one. Wastes and debris are ejected by a blast of water pumped from the gut, but enough remains in the burrow to support other animals. A worm *Hesperone* (c) lives behind the "host," along with tiny pea crabs *(Scleroplax),* and sometimes small fish, called gobies, shelter in the burrows.

Even for deposit feeders, the most desirable sediments are not necessarily entirely silt- or clay-sized particles, especially where waters contain large amounts of dissolved or suspended organic matter. Excess organic material in the sediment may cause oxygen depletion in the near-bottom waters that is intolerable to most benthic animals. About 3% of organic matter in the sediment appears to be optimal for deposit-feeding bivalves.

Mud-burrowing animals have specialized breathing structures that are not clogged by fine silt. Breathing through the skin—as sea stars do, for example—is a poor adaptation for a muddy environment. One widely used mechanism involves flowing water through the burrow (Fig. 14-17) or in and out of the shell of a bivalve. In this way, oxygen is supplied for respiration, fine particles are swept off the gills, and a food supply is maintained.

For filter feeders, fine sands are more suitable than coarse. Particle size does not affect the animal directly, but coarse-grained deposits are usually associated with strong bottom currents. Fine sands occur where currents are weaker. In intertidal areas deposit feeders generally dominate whatever the substrate because unstable sediment surfaces prevent accumulation of phytoplankton or other particulate food utilized by filter feeders.

Many deposit feeders make fecal pellets that are quite durable (Fig. 14-18). This factor tends to reduce turbidity at the sediment–water interface and so is beneficial to certain benthic organisms. In deep, quiet water the bottom may be nearly covered by fecal pellets. The fecal pellets tend to stabilize the bottom, thereby making it more resistant to erosion.

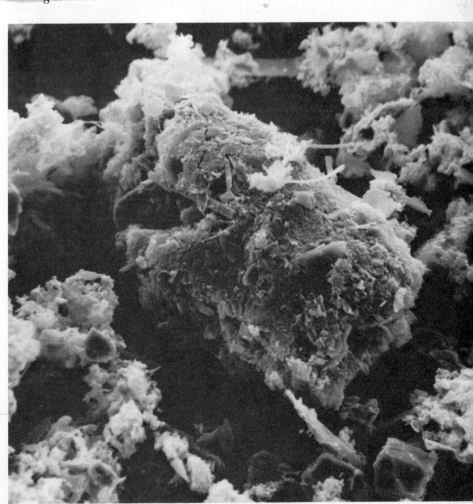

Figure 14-18

Fluffy light-colored particles or organic debris surround this fecal pellet. Darker mineral grains appear at the bottom of the photograph, which shows an area about 0.14 millimeter square. (Photograph courtesy Dr. J. M. Sieburth and University Park Press, Baltimore, Md.)

Figure 14-19

A Long Island, New York, tidal marsh at high tide (above) and at low tide (below), when the surface is set mud rather than water.

In the absence of strong waves or currents, intertidal sediment deposits may be stabilized by marsh grasses, forming *salt marshes* (Fig. 14-19). Such sediments are rich in detritus. Walking through the area at low tide, one sees little evidence of life, but the ground is full of animal burrows. Mussels live buried in the wet mud and so do burrowing worms.

Around the high tide level, fiddler crabs (Fig. 14-20) dig their burrows. These burrows may be up to 1 meter in length, but they are dug obliquely to be shallow so that the crab can remain dry. While

(a)

(b)

(c)

Figure 14-20

Fiddler crab and his burrow: (a) fiddler posed on the sand; (b) about to dart into his burrow, fighting claw held protectively over his head; (c) a freshly made burrow surrounded by lumps of compacted sediment excavated by the crab and deposited outside.

the tide is low, a crab picks through the mud, selecting bits of plant or animal detritus. He enters his burrow when flooding is imminent, sealing the doorway against the rising tide, and remains inside until the water subsides. Fiddler crabs require only enough contact with the sea to keep their gill chambers moist, although they can survive for weeks immersed in seawater.

The burrowing land crab *Cardisoma* and the beach scavenger *Ocypode* (ghost crab) maintain even less contact with seawater than does the fiddler. These animals have enlarged gill chamber linings that function as lungs. Temperate-to-tropical environments seem to favor this adaptation to a semiterrestrial way of life; relatively higher humidities at these latitudes keep the animal's gill membrane moist while it is away from the water.

Dense stands of *eelgrass*, a seed-bearing marine plant, grow in shallow-water sediments, as shown in Fig. 14-21(a). Eelgrass has an underground stem and ribbonlike leaves as much as 1 meter long. Specialized communities of plants and animals find food and shelter in eelgrass beds; its leaves are coated with epiphytic diatoms, blue green algae, protozoans, organic detritus, and small hydroids [Fig. 14-21(b)]. Tube-building crustaceans and worms attach to the leaves and small grazing shrimps and snails scrape off accumulated detritus. This regular cleaning of the eelgrass by animals is essential to the continued growth of the epiphytes, especially the diatoms, for if they become coated with detritus, they die for lack of sunlight. Other common residents of eelgrass communities include tunicates and scallops.

The composition of an eelgrass community depends partly on temperature, depth, and current speed. Where currents are strong, for instance, small crustaceans living in tubes (which they secrete and cement to the blades of eelgrass) are likely to survive while heavier, crawling species are swept away. Infauna grow profusely in the sediments because the grass has a stabilizing effect and turbidity is reduced. Eelgrass and epiphytes are food for many animals in the community and detritus for bottom dwellers is abundant. In addition, sediments are usually firm enough to permit construction of burrows; holes dug in coarse sands tend to collapse, although some worms and other animals are able to build tubes that prevent such accidents.

Below low tide level, in muddy sediments on the continental shelf, there are usually large numbers of deposit-feeding bivalves (such as the clam in Fig. 14-22). Where water is too deep for algae and eelgrass, sediments are often soft and easily eroded, due in part to constant reworking by burrowing polychaetes and mollusks. A few tens of clams per square meter can rework all newly deposited sediment once or twice each year, to a depth of 2 centimeters. Attached animals are unable to colonize the semiliquid substrate and filter feeders are choked by the high concentration of suspended sediment.

The deposit-feeding clam *Yoldia* is equipped to live in this habitat. Its low-density body is often swept up by strong bottom currents and suspended along with silt from the soft bottom. Heavier species, such as a filter-feeding bivalve that tries to settle there, tend to be buried and suffocated by these turbid flows. If *Yoldia* is buried by sediment, it tunnels upward toward the new mud–water interface where oxygen is available. If exposed at the surface by wave action, it just as quickly digs in again.

Where deposit feeders are abundant, filter feeders are scarce. This situation may occur when the activities of the former make the area less easily habitable by the latter. In this case, one organism forces another out of the area—not by competing for food or even for living

(a)

Figure 14-21

(a) Model of an eelgrass community, Woods Hole, Mass. Scallops, shrimps, and crabs are shown in the eelgrass; several kinds of worms live in the soft bottom. The burrow at left contains the polychaete *Chaetopterus,* which maintains a current through the tunnel by flapping leaflike *parapods* ("side-feet"), seen along the middle of its body. The burrow lining is a tough, parchmentlike material, shown in the W-shaped burrow at center. At right the tube-dwelling polychaete *Diopatra,* with gills emerging from the mouth of the tube. (Photograph courtesy American Museum of Natural History.) (b) Bacteria and several species of diatoms (magnified about 200 times) grow as a crust on eelgrass. (Photograph courtesy Dr. J. M. Sieburth and University Park Press, Baltimore, Md.)

(b)

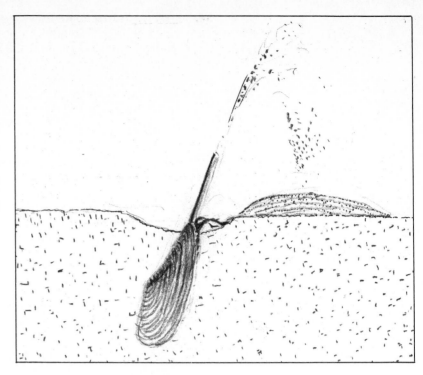

Figure 14-22

Feeding orientation of *Yoldia limatula*. Half buried in mud, the clam uses its feeding palps to collect sediment and bring it into the mantle cavity. The large inedible fraction is forcibly ejected as a cloud of loose sediment, creating mounts of laminated, reworked sediment as the clam feeds. When resting, *Yoldia* commonly burrows a few centimeters below the surface. [After Donald C. Rhoads, 1963. Rates of sediment reworking by *Yoldia limatula* in Buzzards Bay, Massachusetts, and Long Island Sound. *Journal of Sedimentary Petrology* 33(3):723–727.]

space but by creating an undesirable habitat. On the other hand, oxygen introduced by biological reworking of sediment *(bioturbation)* permits bacteria, protozoans, and other small benthic animals to live deeper below the sediment–water interface than is possible in oxygen-deficient muds.

MICROORGANISMS IN MARIN SEDIMENTS

Figure 14-23

Batillipes mirus, one of the 340 known species of marine tardigrades. It is found in shallow coastal waters. (After S. Hinton.)

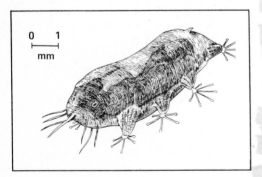

Marine sediments contain numerous one-celled and small multicellular animals called *meiobenthos* (from the Greek *meion*, meaning smaller), about 1 millimeter or less in length. Some are larvae of larger animals. Conspicuous groups of permanent meiobenthos include *harpaticoids* (very small copepods), *turbellarians* (flatworms), *ostracods* (small bivalved crustaceans), and several kinds of roundworms, especially *nematodes*. Some common types do not fit into classifications previously described—for example, the *tardigrade* or "water bear" illustrated in Fig. 14-23. Benthic foraminiferans and ciliate protozoans are especially abundant in sediments all over the world.

Meiobenthic organisms crawl, climb, and swim between sand grains, feeding on bacteria, protozoans, diatoms, and other epiphytes or on the organic matter coating particles. Average biomass of meiobenthic organisms in coastal marine sediments ranges from 0.2 to 2 grams wet weight of silty sand. This is only 3% of total benthic faunal biomass in such regions, although numbers of individuals may range from 50,000 to more than a million per square meter. Relative biomass is often greater in brackish water regions and on beaches. Shallow-water and intertidal muds usually support denser populations than do pure sand deposits or the deeper, finer sediments offshore.

On the continental shelf, 4000 to 32 million meiobenthic individuals live beneath each square meter of ocean bottom. In coastal waters—for example, in the Mediterranean Sea and nearshore waters off South India—meiobenthic biomass may be 10 to 30% or more of the total fauna, depending in part on whether larger (e.g., harpaticoids) or smaller (e.g., nematodes) organisms predominate. In the Norwegian Sea at 1500 to 1800 meters depth, where there are large populations of nematodes, biomasses of 2 grams of carbon (wet weight) per square meter have been recorded. In continental shelf and slope waters, deposit-feeding macrobenthos preying on the microfauna contribute to the diet of demersal fishes. In this way, meiobenthos enter pelagic food webs that include large predators, such as fishes and humans.

Deep-sea sediments often contain as much meiofaunal as macrofaunal biomass, although generally there are fewer individual organisms than in coastal waters. For example, at 3000 to 5000 meters depth, Indian Ocean sediments contain from 20,000 to 110,000 individuals per square meter. This figure may only represent a few tenths of a gram, or less, per square meter, but macrofauna have extremely low density in the deep ocean and average only 0.002 to 0.2 gram per square meter so that meiobenthos here are proportionately quite important.

Meiobenthic activity governs the availability of dissolved oxygen in the uppermost 1 or 2 centimeters of sediments. Where water and oxygen are available due to stirring by the meiobenthos, oxygen-consuming animals occur at depths of a few tens of centimeters or more. Anaerobic organisms (those that do not require dissolved oxygen) inhabit oxygen-deficient sediments. In intertidal environments many organisms move up and down continuously with the tide so as to remain near the air–water interface in flooded sand.

Minute animals have higher metabolic rates and consume relatively more food and oxygen per unit of body weight than larger ones. In general, they also have shorter life cycles and, consequently, populations turn over more rapidly. Oxygen consumption in meiofauna ranges from 200 to 2000 cubic millimeters of oxygen per hour per gram wet weight. This is 20 times higher than the figure for large macrobenthos, with medium-sized animals utilizing intermediate amounts. On the average, the meiofaunal metabolic rate is five times that of larger animals.

Generation rates are quite variable. Some groups reproduce every few days, others as infrequently as once a year, and the majority a few times a year. Standing crop represents perhaps a third of gross production, due mainly to regular predation and death from other causes. Using three generations per year as an average, meiofaunal production is estimated to be about five times the rate for macrobenthos. In subtidal silty sand, populations representing only a few percent of the total standing crop may provide some 15% of the annual contribution to food webs.

Microbenthos (including bacteria and fungi, ciliate and flagellate protozoans, amoebae, benthic diatoms, and filamentous algae) are smaller than 100 microns and range in weight from 10^{-4} to 10^{-9} milligram. Autotrophs live at the surface and heterotrophs mainly within sediments. In nearshore deposits, the microbenthos constitute an even smaller mass than do the meiobenthos, but at the same time they constitute a hundredfold larger number of individuals. Most of their food is used in respiration and reproduction; relatively little contributes to increased biomass.

When large quantities of organic matter are present, a bacterial film often forms at the sediment surface. Biomass in the topmost layer of silt may reach several tens of grams per square meter and numbers of individuals may reach 10,000 to 100 million per gram of mud.

Bacteria contain abundant proteins, fats, and oils. Although supplying little energy as carbohydrates, they are an important food for deposit-feeding macrofauna and meiofauna.

CHEMOSYNTHESIS Some bacteria can utilize such compounds as ammonia (NH_4^+), methane (CH_4), and nitrate (NO_3^-) or such elements as hydrogen, iron, and sulfur as primary energy sources. By a process known as _chemosynthesis,_ they form new organic matter in the absence of sunlight and even in oxygen-deficient waters. Chemosynthesis has lately been recognized as an important source of food for filter-feeding organisms that live near active hydrothermal vents on midocean ridges. Such bacteria are also important food sources in anaerobic sediments and in the oxygen-deficient waters of the Black Sea and in some fjords where the bottom waters are partially isolated from the rest of the ocean circulation.

Anaerobic bacteria obtain oxygen for chemosynthesis and respiration from carbon dioxide, nitrite, nitrate, or sulfate (SO_4^{2-}). In doing so, they reduce these compounds, forming the nitrites, ammonia, methane, and sulfur involved in chemosynthesis.

Reduced sulfur combines with hydrogen, forming hydrogen sulfide gas (H_2S), or with iron to form the iron sulfide that gives anoxic subsurface muds their characteristic black color and rotten egg smell. Breakdown of organic matter releases phosphates (PO_4^{3-}), resulting in precipitation of inorganic calcium and iron phosphates in sediments.

Bacteria also play a role in the distribution of elements not directly involved in plant or animal metabolism—for example, manganese. Manganese carried to the ocean by rivers is not concentrated in seawater or in marine organisms to any great extent. Instead it precipitates as iron–manganese nodules (see Chapter 4) on the deep-ocean floor. Heterotrophic bacteria attach themselves to these nodules and the enzymes they produce seem to accelerate manganese oxidation and precipitation. Energy derived from this process apparently enables the bacteria to form ATP and assimilate organic matter more efficiently so that manganese serves as an energy source for these oxidizing organisms.

REEF COMMUNITIES

A reef develops when benthic organisms build a rigid, wave-resistant structure on the bottom. The reef provides a shallow-water environment, favorable to many organisms, where nutrients are readily recycled. Barnacles, algae, mussels, and other sessile organisms, for instance, colonize a reef, as they do any firm surface in shallow water.

The process by which organisms invade a previously uninhabited area is known as *ecological succession*, or "fouling." It begins with an accumulation of bacterial slime (see Chapter 12). Benthic diatoms and protozoans appear next. They multiply rapidly, utilizing absorbed organic compounds and products of bacterial decomposition. Hydroids and multicellular algae follow and then come the planktonic larvae of barnacles (Fig. 13-18), mussels, and snails. Eventually the ecosystem reaches a balanced state or *climax community* in which no further colonization occurs and ecological succession ceases unless a disturbance of the system causes the process to start afresh.

Oysters are sessile, bivalved mollusks that are abundant in coastal oceans and estuarines worldwide and an important source of food in many regions (see Chapter 15). Each mature female produces millions of eggs yearly (see Fig. 13-14). Most are fertilized but only a small fraction of the larvae (Fig. 13-14) survives the planktonic stage (which lasts a few weeks, more or less, depending on latitude) and settles to the bottom to develop as adults. Larvae select a suitable spot, preferring oyster shells to any other substrate and apparently favoring live oysters over dead shells. An oyster needs 1 to 5 years' growth to mature, during which time most of those that settled together as larvae have been killed by crowding, competition for food, or predation.

Oyster reefs contain enormous numbers of individuals whose shells are cemented to rocks and to one another. Tidal flats of partially enclosed bays and river mouths provide advantageous conditions,

Figure 14-24

Sabellarian polychaete worms (a) secrete a tube of mucus that hardens into a parchmentlike substance. Feathery gills, often brightly colored, are the only part of the worm that emerges from the end of the tube. The gills, which absorb oxygen and also trap plankton and detritus in a mucous coating, have eyespots that are sensitive to changes in light intensity; the passing of a shadow causes them to retract into the tube. Sabellarian worms can build reefs of their own. Colonial species live in surf zones, where the shoreline is rocky enough to support their tubes. Currents supply filterable food particles and sand grains, which they bond with mucous secretion to make wave-resistant structures. As a sabellarian colony grows, new worms settle around and over old ones and a reef is formed that may extend for several meters along a beach or breakwater. This photograph was made at the Wometco Miami Seaquarium. The calcareous tube worm *Hychoides dianthus* (b) is commonly found encrusting oysters or other mollusks.

especially where moving water brings fresh supplies of plankton and oxygen and little silt and few predators exist. (An exception is the Virginia oyster *Crassostria virginica,* which is well adapted to life in silty waters.)

An oyster pumps many gallons of water each hour. Plankton or other food particles in the water are caught on a mucous net that is moved steadily toward the mouth by ciliary action. Some food concentrated in this way is selectively swept out of the shell, before the oyster can consume it, and other filter-feeding animals take advantage of the food supply.

Attached organisms, such as barnacles, mussels, and tube worms (Fig. 14-24), add to the bonding that holds oysters together in clusters, thus increasing the stability of the structure. Crevices between shells shelter small filter feeders. Microorganisms, flatworms, and tiny crabs feed parasitically on the attached animals while certain fishes (gobies and killifish, for example) are adapted for life in the shallow tidal waters over the beds. The latter prey on small crustaceans and soft-bodied animals among the rocks and crevices. Some animals, like the pea crab, live in the shells of live oysters.

Sea stars, snails, crabs, and various fishes prey on oysters. Oyster-drill snails chemically dissolve and physically bore holes through the shell to eat the oyster. Sea stars exert tremendous suction with their tube feet to pull open the valves and then consume the meat by everting their stomachs to surround the oyster.

(a)

(b)

Oysters thrive in low-salinity waters. In Chesapeake Bay, for example, there are large natural beds in intermediate-salinity areas (7 to 18‰) where oysters can survive but their principal enemies and diseases cannot. In the absence of predation, oysters grow faster and are more prolific in higher-salinity waters near the mouth of the estuary than upstream. In the high-salinity water of the seaside along the oceanshore, they survive only in intertidal zones.

Coral reefs occur mainly in tropical and subtropical oceans warmer than 18°C, where carbonates precipitate readily. Reef-building coral is a coelenterate that builds a cup-shaped calcareous skeleton, shown in Fig. 14-25. New individuals bud from the side of the parent. As animals die, subsequent generations build over and around the dead skeletons.

Unicellular algae, called *zooxanthellae*, live in coral tissues. Theirs is a symbiotic relationship in which the algae make photosynthetic products directly available to the host coral and, in return, receive nutrients and CO_2. Zooxanthella photosynthesis apparently alters CO_2 concentrations in coral tissue, greatly increasing the animal's ability to extract $CaCO_3$ from seawater to produce the carbonate skeleton. Calcification rates of reef-building corals are strongly influenced by the amount of insolation received by the algae and are highest in sunlit near-surface waters.

But corals are only the most conspicuous contributors to the framework of a reef. Encrusting algae known as *crustose red algae* are important cement depositors at shallow depths. In addition, red and green *calcareous algae* produce much of the loose carbonate sediment incorporated in crevices in the reef mass.

Coral reefs create a complex shallow-water benthic environment. The ecosystems that flourish on and around reefs are some of the most productive in the ocean, supporting extensive benthic and pelagic communities. They are also highly diversified in the kinds of living space available and in the kinds of organisms that inhabit them. More than 3000 animals species can coexist on a large reef tract. Competition for firm surface space, rather than lack of food, probably limits the population of filter-feeding benthos, at least above 200 meters depth.

On the reef crest, tides and waves periodically expose parts of the reef. There hardy, soft-bodied algae, barnacles, and coralline algae

Figure 14-25

Encrusting coral growing under experimental conditions. (Photograph courtesy Michael J. Reber.)

form rimmed pools that project above sea level 10 centimeters or more. Water splashed into these pools drains or dries out slowly, keeping plants and animals in the pool moist. Coralline algae may also form a purplish red ridge, the *algal* or *Lithothamnion ridge,* at the waterline. Crabs, small fishes, worms, and algae live in the shallow pools.

In the photic zone reef-building corals from the dominant cover, but attached plants greatly exceed the mass of animal material. Besides the coralline algae that add calcareous material to the reef framework, filamentous green algae embedded all over its surface manufacture food during the day, using nutrients released by animals and bacteria. At night coral polyps extend their filter-feeding apparatus (Fig. 14-26) to capture plankton and detritus from the water.

Many kinds of invertebrates and fishes (Fig. 14-27) live in coral reefs, although there are relatively few individuals of any one kind. Some, like the sea star in Fig. 14-28, consume coral polyps and algae; others feed on detritus. Still others—for example, the moray eel and sea anemone—are predators. Parrot fish and other browsers actually graze on the reef structure to eat the algae and animals living in it. They excrete clouds of calcareous silt; this material, along with mollusk shells and calcareous debris from benthic foraminifera, corals, and algae, accumulates in interstices in the forereef structure.

Zonation of plants and animals on the forereef is largely controlled by light and temperature. In shaded environments, such as caves and tunnels and the undersides of ledges, *sclerosponges* deposit a skeleton of slender silica needles and calcium carbonate, the only living organisms known to deposit both minerals.

Evidently they contribute substantially to some reefs, especially below 70 meters depth, where reef-building corals are rare. Sediments trapped by sponges and corals undergo rapid cementation, contributing significantly to the solidification of the reef framework.

Figure 14-26

Branch whip coral, *Leptogorgia:* (a) branches of coral with a red beard sponge in the background; (b) closeup of a feeding polyp showing tentacles extended. (Photograph courtesy Michael J. Reber.)

(a)

(b)

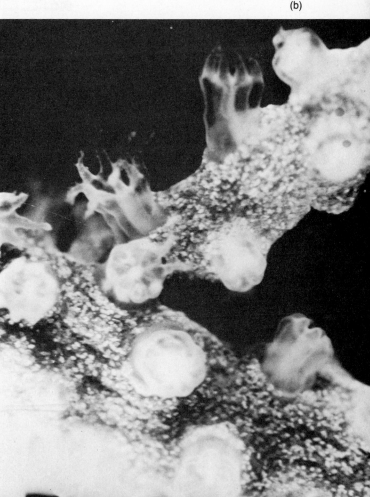

Figure 14-27

Fishes swimming over a coral reef, Midway Island. (Photograph courtesy U.S. Geological Survey.)

Many attached organisms of the deep forereef are restricted to protected areas where they are not covered by falling sediment. Non-reef-building corals, soft sponges, hydroids, crustose red and filamentous green algae, ascidians (benthic tunicates), *crinoids* (see Fig. 14-31), bivalve mollusks, tube-dwelling worms, and sea anemones are among the many kinds of attached organisms below 70 meters depth. Some, such as tube worms and ascidians, contribute skeletal materials to the reef. Small lobsters, crabs, sea stars, and sea urchins move over the surface, grazing on attached organisms. Boring and cavity-dwelling sponges and many kinds of small and large worms live within the reef itself, and fishes prey on the benthic fauna. Thick sediment deposits on flat surfaces contain animal burrows and trails.

Below about 200 meters, the sediment layer is thicker. Shrimps, sea stars, small fishes, worms, sea anemones, and sea cucumbers inhabit holes and mounds in the bottom. Stalked crinoids are also abundant at these depths. Scavengers and grazers include hermit crabs, brittle stars (see Fig. 14-32), and sea urchins. Corals grow where sediment layers are thin, as do unstalked crinoids and encrusting sponges. Attached animals support mobile organisms; for example, brittle stars are often found clinging to corals. Among the pelagic fauna found at these depths are jellyfish, chaetognaths, mysids (see Fig. 13-5), and many kinds of crustacean nauplii.

Artificial reefs can support ecosystems, just as naturally occurring reefs can. Most of the U.S. Atlantic and Gulf Coast continental shelf is flat and sandy, with little rock or natural reef on which communities can build or pelagic populations find shelter. But migratory bluefish, tuna, mackerel, flounder, pollock, and jacks gather around wrecked ships because of the protection they offer, particularly for juveniles. In addition, attached benthic organisms provide food.

408

Figure 14-28

Crown-of-thorns sea star *(Acanthaster planci)* feeds on coral reef at Guam. In the lower photograph, the same area of coral is shown as it appeared a few months later. The coral is dead and its skeleton has been overgrown by algae. (Photographs courtesy Westinghouse.)

To construct artificial reefs, heaps of rock, construction debris, discarded automobile tires, and specially built concrete blocks with holes have been deposited in coastal areas for the benefit of sport fishermen. Junk car bodies have also been tried, but they tend to break up after a few years and the metal thus released may locally be a nuisance—catching fishing gear, for example (Fig. 14-29). In the Gulf of Mexico offshore drilling platforms act as artificial reefs. The ideal artificial reef would be high enough to cause upwelling of nutrient-rich bottom waters and it would have a variety of surfaces and openings to provide diverse habitats for benthic organisms and juvenile fishes.

(a)

(b)

Figure 14-29

These photographs of artificial fishing reefs were taken at Artificial Shoals in Hawaii: (a) coral growth on a reef of junk car bodies; (b) a school of *Chromis verator* in a car body; (c) a reef made of damaged concrete pipe. (Photographs courtesy State of Hawaii, Department of Fish and Game.)

(c)

DEEP-OCEAN BENTHOS

The deep-ocean benthos is highly diverse, partly because of the stability of the environment. Below about 2000 meters depth (see Figs. 6-18 and 6-19), temperature and salinity vary little over most of the ocean. Deep waters were probably 3° to 8°C warmer several million years ago, before the ice ages began, but since then bottom conditions have been nearly constant. Immigration and evolution under these colder conditions have resulted in diversity comparable to that of tropical soft-bottomed communities.

Soft sediments and deposit feeders (Fig. 14-30) dominate the deep-ocean bottom, especially in central ocean basins far from shore. In areas of red-clay deposits, where sediment accumulation is slow and deposits contain less than 0.25% of organic carbon, filter feeders are conspicuous. The crinoid in Fig. 14-31, for example, stands on a long stalk above easily eroded muds. Predatory forms, such as brittle

Figure 14-30

Deep-ocean sea cucumbers: (a, right) *Scotoplanes;* (b, below) *Psychropotes* (drawn from the underside, as if crawling on a sheet of glass). These cylindrical, muscular echinoderms creep along the bottom by contractions of the body. At the anterior end, the mouth sweeps up sediment, detritus, and small animals. (From H. Theel, 1882 and 1886. Report on the holothurioidea of the *Challenger* expedition. *Report of the Scientific Results H.M.S. Challenger, Zoology,* vol. 14, pt. 1, 176 pp., 1882; pt. 2, 290 pp., 1886.)

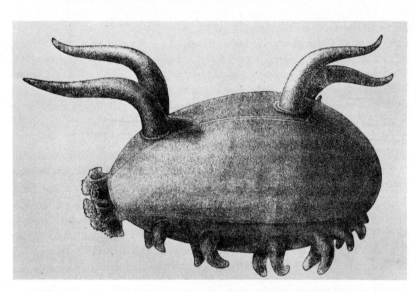

(a)

(b)

Figure 14-31

Sea lilies have cup-shaped bodies and long arms. The skeleton is made of calcareous plates, absent in the vicinity of the mouth. (Photograph of a model in the U.S. National Museum, Washington, D.C.)

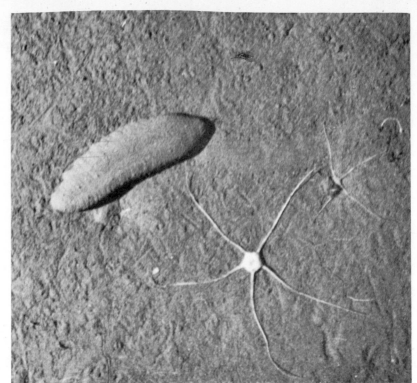

Figure 14-32

Brittle star and sea cucumber, photographed at a depth of about 1500 meters in Hydrographer's Canyon, western North Atlantic. (Photograph courtesy Woods Hole Oceanographic Institution, David Owen, photographer.)

stars (Fig. 14-32), are more abundant where organic carbon content is higher, moving about on long legs that support their bodies above the sediment surface.

Filter-feeding sponges (as in Fig. 14-33), coelenterates (Fig. 14-34), segmented and unsegmented worms (Fig. 14-35), bivalved mollusks and crustaceans—in fact, all the major groups in the shallow-water benthos—occur in deep-water habitats. Uniform coloration (gray or black among the fishes, often reddish in crustaceans) and delicacy of structure are typical in these quiet, dark waters. Smaller animals in deep-ocean sediments include polychaete worms, isopods and amphipods (see Fig. 13-5), bivalves, and *tanaids,* deep-water crustaceans similar in appearance to isopods.

Figure 14-33

Deep-ocean sponges: (a, left) *Toegeria;* (b, right) *Lefroyella.* (From John Murray and A. F. Renard, 1891. Report on deep-sea deposits based on the specimens collected during the voyage of *H.M.S. Challenger* in the years 1872 to 1876. *Report of the Scientific Results H.M.S. Challenger,* vol. 5, 525 pp.)

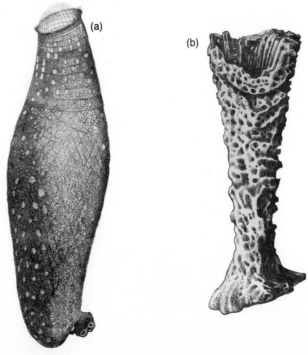

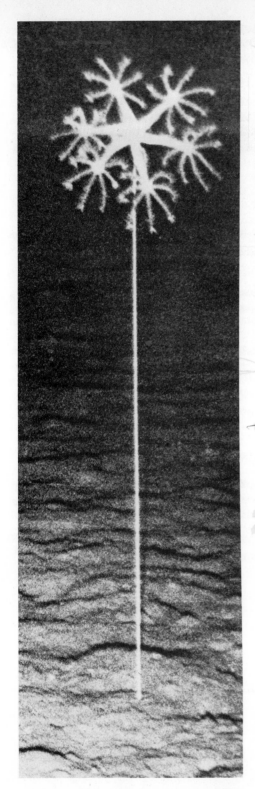

Figure 14-34

The stalked coelenterate *Umbellula,* whose pencil-thick stem appears to be about 1 meter long, was photographed at a depth of more than 5000 meters off the Atlantic coast of Africa by a Naval Oceanographic Office crew aboard the research vessel *Kane.* (Photograph courtesy U.S. Navy.)

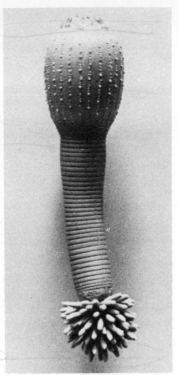

Figure 14-35

The unsegmented worm *Priapus* swallows its prey whole. It lives in bottom sediments of polar regions. (Photograph of a model at the U.S. National Museum, Washington, D.C.)

Deep-sea organisms are smaller than their shallow-water counterparts. Ten times as many individuals are caught if a screen of 0.06-millimeter mesh is used to filter the sediment instead of one that catches only organisms longer than 0.3 millimeter. Meiofauna taken in this way include benthic foraminiferans, nematodes, harpaticoid copepods, and ostracods (Fig. 14-36).

Most of the detritus that falls from surface or mild-depth zones is consumed or decomposed before reaching the deep zone so that little biodegradable material reaches the deep-ocean floor. Inedible remains of large animals, such as sharks' teeth, squid beaks, and whales' earbones, are common, but they are not a source of food for deep-ocean organisms. Eggs of some benthic forms float to the surface zone where food is more plentiful for the larval forms and hatch there before descending to the bottom.

Biomass is low at abyssal depths on the ocean bottom. On outer continental shelves and upper slopes, 5000 to 22,000 individuals may live in each square meter of sediment. The lower slope and upper rise below 500 meters depth average around 1000 individuals per square meter, but only around 30 animals inhabit the same sized area on lower continental rises and abyssal plains. Below the central ocean

Figure 14-36

A generalized ostracod, about 1.5 millimeters in length. (Adapted from M. E. Johnson and H. J. Snook, 1967. *Seashore Animals of the Pacific Coast.* Dover, New York.)

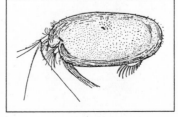

gyres, biomass of the 30 to 200 organisms in each square meter is only a few hundredths of a gram, wet weight, or a few percent of the infaunal biomass in nearshore sediments.

Large fishes, sharks, or whales fall to the ocean floor after death. The carcasses attract large numbers of swimming scavengers, but such events are rare and so the ability to sense food from a distance is an important adaptation in a dark, relatively barren environment.

Near continents, where turbidity currents flowing through abyssal channels bring plant material to the deep ocean, a specialized population exists. One bivalve, for example, bores holes in the coconut husks, bamboo stems, and parts of trees that are carried to the deep ocean. Abandoned holes are then occupied by small mussels, worms, or amphipod crustaceans. Plant debris is eventually decomposed by bacteria, which then make an important contribution to the diet of deposit feeders. But at abyssal depths, as we know, bacterial activity proceeds slowly and may not provide much food for infauna.

BENTHOS OF HYDROTHERMAL VENTS

Hard rock surfaces near active hydrothermal vents on the East Pacific Rise at depths around 3000 meters near the Galapagos Islands support dense populations of benthic animals, primarily filter feeders and associated predators (Fig. 14-37). Unlike benthic communities previously discussed, these organisms depend for their food on bacterial chemosynthesis rather than photosynthesis at the ocean surface. Bacteria in the vents and surrounding areas use hydrogen sulfide and methane as the basis of their food production. Animals lives around the vents and feed on the chemosynthetic bacteria, a food-rich oasis on the deep-ocean bottom. Once a vent system becomes inactive, the chemosynthetic bacteria can no longer grow and the filter-feeding organisms starve. Inactive vents are usually surrounded by the shells of dead animals.

The vents are widely separated and short lived; so the organisms that inhabit them must have effective dispersal mechanisms for their juvenile forms in order to take advantage of new vents as they form, often hundreds or thousands of kilometers away. Furthermore, the hydrogen sulfide that supports the chemosynthetic bacteria is a poison to oxygen-using animals. Thus animals living near the vents must be

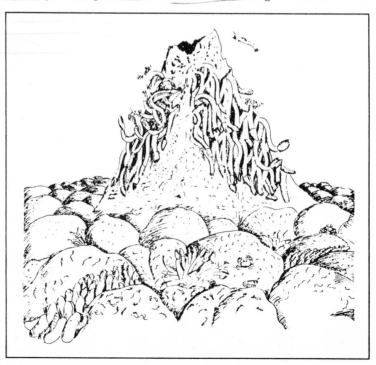

Figure 14-37

Dense growths of red-bodied giant worms grow on an inactive sulfide chimney with clams and crabs shown in the foreground.

able to withstand exposure to hydrogen sulfide by storing oxygen in their blood. The most conspicuous animals are blood red—like liver—because of the hemoglobin in their blood.

Most spectacular are the giant, gutless worms, a type of pogonophoran, which grow to be 3 meters long and 2 to 3 centimeters in diameter. These worms have red gill-like structures that can be withdrawn into the white plasticlike tube when threatened. The worms have no digestive tract and apparently absorb their food directly from the water or possibly from bacteria associated with the worm.

Other pink, polychaetelike worms inhabit the outer surfaces of the cooler vents, the so-called white smokers. They give the chimneys a porous, friable structure. These animals can leave and reenter their tubes. They do not encounter the very hot waters coming from the vents because of the rapid mixing with the cold near-bottom waters.

In addition, clams the size of dinner plates and large mussels grow around the vents. The clams reach their giant size in less than 2 years. Crabs and an eellike fish have been seen feeding on these benthic organisms (see Fig. 14-38).

Figure 14-38

Photograph of giant gutless worms (upper right) and giant clams (lower center) living near an active vent on the East Pacific Rise. The photograph was taken from *Alvin,* the U.S. deep-diving research submarine. (Photograph courtesy Woods Hole Oceanographic Institution).

1. Contrast conditions for life in the benthic and pelagic environments.
2. Describe the three major lifestyles for benthic organisms. Give an example of an animal employing each lifestyle.
3. List the factors that control the makeup of a benthic community. Describe the importance of each.
4. Contrast epifauna and infauna. Give an example of each type of organism.
5. Describe, in general, the role of marsh grasses in a marsh and for the coastal ocean.
6. Describe the role of bacteria in controlling the chemical composition of sediment pore waters.
7. Describe the structure of a coral reef and indicate the roles played by different groups of organisms.
8. Discuss how an artificial reef attracts fishes. How do artificial reefs resemble natural reefs? How do they differ?
9. Discuss how the deep-ocean benthic environment differs from the inshore continental shelf environment.
10. Describe some adaptations of benthic organisms to life on the deep-ocean bottom.
11. Discuss the unusual features of the benthic organisms living around active hydrothermal vents on the deep-ocean floor.

SUMMARY OUTLINE

The benthos: two-dimensional world
    Life strategies for benthic animals and feeding behaviors:
        Attachment to a firm surface—mainly filter feeders
        Free movement on the bottom—mainly active predators
        Burrowing in sediment—mainly deposit feeders

Distribution and diversity of benthic life
    Attached plants rare except in shallow coastal waters
    Benthic community structure controlled by light, temperature, salinity, type of bottom, environmental stability
    Diversity—governed by stability, absence of rigorus conditions
        Greatest in deep-ocean and tropical nearshore waters
        Less in temperate climate communities
    Benthic biomass governed by local productivity

Rocky beaches
    Zonation based on duration of submergence
    Attached organisms dominate in intertidal zone
    Tide pools—specialized environment
    Zone below low tide level most densely populated
    Mobile crustaceans common on hard-bottomed continental shelves

Soft-bottomed coastal communities
    Infauna—filter feeders, selective and unselective deposit feeders, in sediment
    Epifauna—surface dwellers
    Nature of deposits governs community composition
        Filter feeders—coarser deposits, less turbid waters
        Deposit feeders—fine-grained deposits, rich in organic matter
    Fecal pellets can stabilize deposits
    Salt marshes—highly productive; support diverse communities
    Eelgrass community—specialized population; grass roots stabilize bottom
    Infauna of soft, continental shelf deposits tolerate turbidity and avoid burial by sediments

Microorganisms in marine sediments
    Meiobenthos—sediment-dwelling animals, smaller than 1 millimeter; numerous; high energy flow; short lives
        Relative biomass of meiofauna increases with distance from continental shelf
        Organisms stir, oxygenate uppermost sediment layers
    Microbenthos (less than 100 microns)—bacteria, fungi, protozoans
        Bacteria synthesize protein

Chemosynthesis—bacteria reduce inorganic compounds to fix CO_2 as an energy source; can happen in anaerobic environments, sediments, and around hydrothermal vents

> Bacterial activity promotes formation of manganese nodules
> Organic materials decompose slowly in anoxic sediments

Reef communities—build a rigid, wave-resistant structure; nutrients recycled

> Ecological succession (fouling) gives rise to climax community
> Oyster reefs—oysters pump water containing particles for filter feeders
> Coral reefs—subtropical
> Coelenterate polyps and zooxanthella algae form symbiotic relationship
> Crustose algae contribute to reef structure

> Attached plants highly productive
> Sclerosponges contribute siliceous material
> Carbonate sediment incorporated in reef

Artificial reefs—on U.S. mid-Atlantic and Gulf coasts; attract sport fishes

Deep-ocean benthos

> Great diversity due to stable environment
> Soft-sediment bottom—deposit feeders and stationary organisms
> Delicate structures, small size, low biomass characteristic
> Scavengers sense presence of food from great distances

Hydrothermal vents

> Hard rock environment
> Primarily filter-feeding organisms
> Supported by chemosynthetic bacteria
> Large organisms, rapid growth

SELECTED REFERENCES

HEDGPETH, JOEL W. (Ed.). 1957. *Treatise on Marine Ecology and Paleoecology, Vol. I: Ecology.* Geological Society of America, New York. 1296 pp. Detailed, scholarly monographs on habitats and biological conditions in the ocean, with emphasis on biological communities; extensive bibliography.

JOHNSON, M. E., AND H. J. SNOOK. 1967. *Seashore Animals of the Pacific Coast.* Dover, New York. 659 pp. Nontechnical account of marine biology, stressing Pacific Coast organisms.

NICOL, J. A. COLIN. 1960. *The Biology of Marine Animals.* Interscience, New York. 707 pp. A standard reference text, with emphasis on physiology.

RICKETTS, EDWARD F., AND JACK CALVIN. 1962. *Between Pacific Tides,* 3rd ed. Stanford University Press, Stanford, CA. 516 pp. Excellent field guide and reference book, well illustrated and including detailed bibliography.

STODDART, D. R., AND M. YONGE. 1978. *The Northern Great Barrier Reef.* Royal Society, London. 366 pp.

THORSON, GUNNAR. 1971. *Life in the Sea.* World University Library, McGraw-Hill, New York. 256 pp. Elementary biological oceanography, emphasizing benthic environments.

15 OCEAN RESOURCES

Offshore drilling rig exploring for petroleum and natural gas in the United Kingdom sector of the North Sea. (Photograph courtesy CONOCO, Inc.)

$\mathbf{T}$ransportation, defense, and fishing are ancient uses of the ocean. Today the ocean is more and more being exploited for needed materials. Rising demands for fuels, metals, and construction materials have outstripped the resources available on land. This trend is particularly obvious in the case of petroleum and natural gas, for which the continental shelf has become a prime source.

The conference on the Law of the Sea has resolved some questions about who owns the resources of the ocean and thus removed some stumbling blocks for their development. Although many legal problems remain, the pace of ocean exploitation will probably expand greatly. The new legal regime also affects management and protection of such traditional marine resources as fisheries. Under the ancient system of free entry, where anyone with the money to buy a boat could fish as much as desired, fish stocks have been systematically overfished and, in many instances, essentially destroyed as a resource. The new boundaries, 200 nautical miles offshore, permit better control and perhaps better management of fishing activities.

Despite long use of the ocean, it is still a frontier area for resource development. We are even now developing techniques that permit the mining of salt and sulfur from the seabed or drilling to recover gas and oil from greater depths of water. As yet there is little capability to work for long periods at the bottom of the ocean except with expensive, specialized devices. And as in any frontier area, much uncertainty and inflated expectations exist. Conversely, resources that have not been identified are likely to become important within a few decades.

In this chapter we explore some traditional uses of the ocean and consider possibilities for some less developed techniques. But first we must consider the nature of resources based on the rate of their formation and extraction. We distinguish between *renewable* resources, which are replenished through growth or other processes at rates that equal or exceed our rate of consumption, and *nonrenewable* resources. Freshwater and forests are examples of renewable resources; petroleum and metal ores are nonrenewable.

FISHERIES Fisheries (see Fig. 15-1) are a good example of an exploited renewable marine resource involving several groups of fishes [Table 15-1(a)], most of them coastal ocean species. Fishes provide about 3% of human protein consumption worldwide and about 10% of all animal protein consumed. Between 1950 and the late 1970s, worldwide marine fish

(a)

Figure 15-1

Today's commercial fishing methods are not very different from those used for centuries. Bottom-dwelling fishes, such as halibut, are scooped up in a trawl net (a) that is dragged over the sea floor. The photograph shows a trawl catch before it is dumped on the deck of a vessel of the National Marine Fisheries Service in the North Atlantic. Fishes that feed in schools near the surface are often trapped or entangled in a vertically suspended net or in a purse-seine as shown in (b) fishing for menhaden at

(b)

Sealevel, North Carolina. Purse-seines are useful for catching densely schooling fishes in protected waters, but they are also used on the high seas in some parts of the world to catch tunas. The third common fishing method uses hook and line (c); here fishermen land a 30-kilogram yellowfin tuna, using a three-pole, single-line rig in the Gulf of Guinea off Africa. (Photographs courtesy National Marine Fisheries Service.)

(c)

TABLE 15-1(a)

World Commercial Catch of Marine Fish, Crustaceans, and Mollusks by Species Groups (excluding whales and seals), 1978*

SPECIES GROUP	THOUSANDS OF METRIC TONS (live weight)	PERCENT OF TOTAL
Herring, sardines, anchovies, et al.	13,917	23.0
Cods, hakes, haddocks, et al.	10,450	17.3
Miscellaneous marine and migratory (fresh–salt) fishes	8175	13.5
Jacks, mullets, sauries, et al.	6100	10.1
Redfish, basses, perches, et al.	5216	8.6
Mollusks	4384	7.2
Mackerals, snoeks, cutlass fishes, et al.	4057	6.7
Tunas, bonitos, billfishes, et al.	2522	4.2
Crustaceans	2387	3.9
Flounders, halibuts, soles, et al.	1212	2.0
Shads, milkfishes, et al.	838	1.4
Salmon, trouts, smelts, et al.	621	1.0
Sharks, rays, chimaeras, et al.	606	1.0
Total	60,485	100.0

*Data from Food and Agricultural Organization of the United Nations. *Yearbook of Fishery Statistics,* 1978, vol. 46.

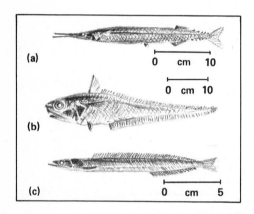

Figure 15-2

Some lesser-known fishes of potential commercial importance. (a) Sauries (garfish) are distributed from Alaska to lower California and south of Japan to the Asian mainland. There are coastal and open ocean species. These schooling fishes feed near the surface at low trophic levels. Related species, such as the halfbeak, are some of the bullet-shaped "flying" fishes reported skittering along the surface in open waters. Pacific sauries have been fished commercially offshore but to a limited extent. (b) Grenadiers, often called "rattails," are found worldwide along lower continental slopes at 200 to 800 meters depth. They are perhaps the most numerous of all fishes on the edges of the deep sea. They feed on benthic invertebrates, also on fishes and plankton at mid-depths. Like many deep-water organisms, they are fragile and easily damaged in fishing. The tapering "rat-tailed" body is characteristic of many deep-water species. (c) Sand lances swim in large schools near northern shores. Their habit of diving into sand and sometimes remaining there after the tide is out gives them their name.

production increased steadily from about 20 million to about 65 million tons. It was hoped that with proper management the oceans would ultimately provide much of the protein needed to feed rapidly increasing populations. But the collapse of California's sardine fishery in the early 1950s and the Peruvian anchovy fishery in the early 1970s has shown that unlimited fish production from the ocean is not possible.

In 1978 the total world production of marine fishes and shellfish was 60 million metric tons (live weight). Another 6 million metric tons of freshwater fishes were harvested. About 60% of the catch was used directly for food while most of the remainder was made into either fish oil or meal for livestock feed. Major fish-producing countries were (in order of tonnage): Japan, the USSR, China, United States, Peru, and Norway. Until the collapse of the anchovetta fishery in 1972, Peru was the world's top fish producer.

In 1979 the 2842 metric tons of fishes landed in the United States were valued at $2.2 billion. Menhaden, a fish used for industrial purposes, accounted for 42% of the volume but only about 5% of the value. Luxury seafoods, including shrimp, salmon, tuna, crabs, and lobsters, were the most valuable products, accounting for 60% of the total value of the U.S. catch [Table 15-1(b)]. Recreational fisheries in the United States take a large and steadily increasing amount of marine fishes each year.

The maximum yield of fishes that can be harvested from the ocean is estimated at about 200 million tons a year, about four times present catches. Achieving this level of production requires harvesting many fishes not now taken (Fig. 15-2). It might also lead to serious reduction—perhaps extinction—of certain species. Heavy fishing has severely depleted several fish stocks; combined with natural climatic fluctuations, such pressure can deplete a stock to the point where the fishery never recovers. An overview of fishery dynamics will illustrate some of the factors that control fish production in the ocean and demonstrate the effects of overfishing.

A fish stock is largest in a virgin, unfished state. Growth of young fishes is balanced by death and predation and there is a relatively large proportion of mature and older individuals. As a fishery is

TABLE 15-1(b)

*U.S. Recreational and Commercial Fisheries Catch, 1979**

TYPE OF FISHERY	THOUSANDS OF METRIC TONS	MILLIONS OF DOLLARS
Recreational fisheries	1500 (est.)	?
Commercial fisheries		
Finfish		
Salmon	243.1	412.8
Tuna	165.3	158.4
Menhaden	1181.2	109.4
Flounders	94.9	82.7
Other	643.1	350.7
Subtotal	2327.6	1114.0
Shellfish		
Shrimp	152.4	471.6
Crabs	221.9	284.2
Lobsters	19.7	85.1
Clams	41.7	79.2
Oysters	21.8	65.6
Other	57.1	134.0
Subtotal	514.6	1119.7
Total	2842.2	2233.7

*Data from National Marine Fisheries Service, 1980. *Fisheries of the United States, 1979.* Current Fishery Statistics No. 8000, National Oceanic and Atmospheric Administration, Washington, D.C.

developed, the number of older individuals is reduced first. Thus the heaviest burden on a fully exploited fishery falls on young adults that have recently experienced their peak growth efficiency (Fig. 15-3). A sardine, for example, grows 120 millimeters in its first year but only 30 millimeters in the third year of life.

When a fish population has grown as large as possible, it is in a steady state and does not increase or decrease. But if the population is reduced—for instance, by heavy fishing—the number of young fishes reaching maturity increases due to the reduction of competition for food and/or living space. Rate of population increase is highest at some intermediate number of individuals, but it approaches zero if the population is reduced to a minimum. In a young fishery, therefore, the number of young fishes may increase rapidly for a few years so that the population as a whole grows larger.

Figure 15-4 shows the relationship of yield to population size. When the harvest reaches about 70% of the maximum potential, rate of increased yield per unit of fishing effort begins to drop rapidly. *Maximum sustainable yield* for the fishery is attained by slightly more than doubling the effort required to reach the 70% mark. Additional effort may increase yields for awhile, but eventually the population shrinks and its rate of natural increase is reduced to zero. Thus *overfishing*, or fishing past the maximum sustainable yield, causes future yields to decline and damages the stock's capacity to recover, even if fishing is later restricted. The growth rate of fishes is also a factor. A shorter time is required to correct overfishing in the case of fast-growing fishes like the anchovy than for a fishery in which individuals mature slowly, such as halibut or whales.

A mature fishery producing the maximum sustainable yield is usually about half the size of the original population. But even if this take is not exceeded, other changes can threaten a fishery. For example, recruitment from a new year class (the individuals hatched in a single spawning season) varies with changes in food supply, predation, and oceanic conditions, such as temperature and current pat-

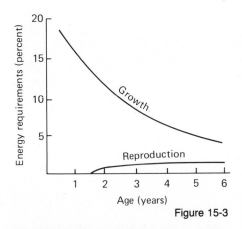

Figure 15-3

The percentage of assimilated energy used by the Pacific sardine for growth and reproduction. Ecological efficiency—that is, ratio of growth to food intake—may achieve a high of 20% or more during its early life. (After J. H. Steele, 1974. *The Structure of Marine Ecosystems,* Harvard University Press, Cambridge, MA.)

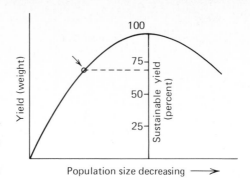

Figure 15-4

Yield curve for a fishery. [After R. Edwards and R. Hennemuth, 1975. Maximum yield: assessment and attainment. *Oceanus* 18(2).]

terns. If fishermen take the same number of fishes year after year or increase their efforts in a poor year to compensate for smaller stocks, overfishing can result. A "bad year," combined with heavy fishing, may reduce a stock to such an extent that a different species takes over the living space and food supply, replacing the commercial species. If there is no market for the successor species, the fishery collapses.

The Pacific sardine fishery off central California illustrates this sequence of events. Once the largest in the Western Hemisphere, it has produced almost nothing since the mid-1960s. Beginning in 1915, this fishery expanded rapidly. Reaching its peak in 1936, it maintained an average production exceeding 500,000 tons until 1944 and then fell off sharply. But even after the near disappearance of sardines, there is no evidence of a drastic reduction in fish biomass of California waters. Instead it seems that the sardine has been replaced by other fishes, especially anchovy. The anchovy standing crop has increased at least twentyfold while the sardine dropped to about 5% of its former biomass. But restrictions on commercial fishing off California made it impossible to develop a market for fish meal and oil from the anchovy, which has not been commercially exploited.

EL NIÑO AND THE PERUVIAN ANCHOVETTA FISHERY

Overfishing combined with natural events—the appearance of an El Niño—to destroy the Peruvian anchovetta fishery in the early 1970s. The results were felt worldwide.

During spring in the Southern Hemisphere off the coast of Peru, subsurface nutrient-rich water upwells, bringing nutrients to support the growth of phytoplankton. Silvery, finger-length anchovetta graze these plants and grow rapidly to maturity. During the 1960s this upwelling area supported the world's largest fishery. In the early 1970s one out of every 5 tons of fish taken from the ocean came from the west coast of South America.

El Niños and overfishing changed this rich but delicately balanced situation within a matter of a few years and nearly destroyed the fishery. El Niño—Spanish for "The Child" because it arrives about Christmas time—is a marked change in the coastal circulation that occurs in some years and not in others. Warm water flows south along the coast and prevents cold upwelled waters from reaching the surface. Phytoplankton growth drops markedly so that the anchovetta either die from lack of food or move elsewhere. Guano birds (cormorants, pelicans, and gannets) that feed on the anchovetta disappear and die (see Fig. 15-5). Heavy rains on shore can cause flooding in the coastal desert areas, resulting in economic hardship among the fishermen and severe disruption of national economies.

Effects of the 1972 El Niño were felt worldwide. In the 1960s and early 1970s the Peruvian anchovetta fishery was a major source

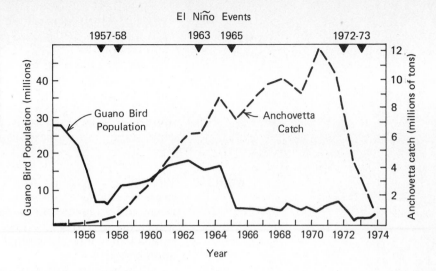

Figure 15-5

Growth and decline of the Peruvian anchovetta fishery and the fluctuations in the guano bird (cormoronts, pelicans, gannets) population in relation to El Niño events. (After Haydee Santander, 1980. The Peru Current System: biological aspects. In *Proceedings of the Workshop on the Phenomenon known as "El Niño,"* pp. 217–227. UNESCO, Paris.)

of protein for animal feed. In 1972 the onset of El Niño caused fish meal production to drop to less than half the 1970 level. The world's protein supply fell between 6 and 9% at the same time that a global grain shortage occurred because of poor crops caused by bad weather. Food prices went up in the United States; there were international incidents because of a U.S. embargo on soybean exports to Japan; and starvation occurred in South America as farmers shifted from their traditional bean crops to grow soybeans for export.

El Niños are caused by changes in the prevailing winds far from the coast of South America. By carefully monitoring the trade winds and understanding their effects on the circulation of the equatorial Pacific Ocean, it is now possible to predict El Niños from 6 months to a year before the event.

During the year preceding an El Niño stronger than average trade winds, blowing from the southeast, increase the amount of water carried westward by the Pacific equatorial currents. These trade winds cause the sea surface to slope upward slightly from South America to the western Pacific. Tide gages show changes in water level around the islands between Samoa and New Zealand after the wind picks up. Six months before an El Niño, tide gages on Christmas Island in the mid-Pacific show a rise in sea level. When the trade winds relax, this water "surges" back toward South America as a wavelike feature. Warm water flows eastward along the equator and then south along the coast of South America. Thus El Niño is the result of a large wavelike feature that involves most of the equatorial Pacific. Using this hypothesis and measuring the strength of the trade winds at selected island stations and changes in sea level, oceanographers were able to predict an El Niño in 1975 and to study its effects along the Peru–Chile coast.

The importance of predicting the occurrence of El Niños can be seen by examining more closely the results of El Niños in the early 1970s, especially the 1972 El Niño that nearly destroyed the Peruvian fishery in conjunction with the continued overfishing. At its peak, the Peruvian anchovetta fishery produced about 12 million metric tons per year. After the El Niño of the 1970s, production dropped to between 1 and 2 million tons per year. Furthermore, other fish species began to replace the anchovetta.

Strong El Niños kill enormous numbers of marine organisms. When these dead organisms decompose, the dissolved oxygen is used up in near-bottom waters, anaerobic bacteria take over, and hydrogen sulfide forms. In the El Niño of 1925—the most severe in the past

400 years—hydrogen sulfide formed and was released to the air in such quantities that it blackened the lead paint on ship's hulls and shoreside houses.

Heavy rainfall in the normally arid coastal areas and resulting floods caused by El Niños have been catastrophic in South America. Archaeologists studying ancient irrigation canals have found evidence that devastating floods around A.D. 1100 destroyed irrigation systems and led eventually to the overthrow of the local empire. A severe El Niño is the most probable cause of such severe flooding.

WHALING

Whaling is an industry that has required international regulation to preserve stocks because whaling grounds extend through international and national waters. Antarctic whales, for example, breed in winter off the African, Australian, and South American coasts. In summer they migrate toward the Antarctic continent to feed on the abundant krill.

The whaling industry dates from before the twelfth century. At first whales were harpooned from small boats launched from shore; later they were hunted far out at sea by ships carrying small boats, a few of which would be launched at once to attack a whale from all sides. In the eighteenth and nineteenth centuries, when whalebone for dresses and oil for lamps were in demand, the seas near Greenland supported a great whale fishery. Sperm whales were hunted by New England whalers for spermaceti, used for cosmetics and candles. This fishery was extended to baleen (whalebone) whales and become worldwide, especially in tropical waters and in the Southern Hemisphere (see Table 15-2). As each new hunting ground was located, it was exploited to near depletion of the whale stock and many stocks became dangerously low. By the mid 1970s right whales had been reduced to about a total of 5000 individuals. There were about 7000 humpbacks, 11,000 grays, and 12,000 blue whales. Fin whales numbered about 100,000 and sperm whales over 600,000. Whales mature slowly. Some live to be 80 years old; so when a stock is depleted, it takes many decades to recover (see Fig. 15-6).

TABLE 15-2

Whales in the Southern Hemisphere*

SPECIES	INDIVIDUAL WEIGHT (tons)	INITIAL POPULATION	POPULATION 1970s
Blue	90	150,000	7500
Fin	49	375,000	75,000
Sei	14	150,000	75,000
Humpback	34	95,000	2200
Minke	6.4	250,000	250,000
Sperm	33	590,000	400,000

*Data from Kenneth R. Hinga, 1979. The food requirements of whales in the Southern Hemisphere. *Deep Sea Research* 26A:569–577.

MARICULTURE

Mariculture, or *aquaculture,* the marine equivalent of agriculture, is a means of extracting more food from the ocean while avoiding some of the uncertainties inherent in harvesting wild stocks. The simplest form of aquaculture is to take young animals from their native spawning grounds and relocate them in another area where they can fatten for the market. This process occurs, for example, with oysters, which

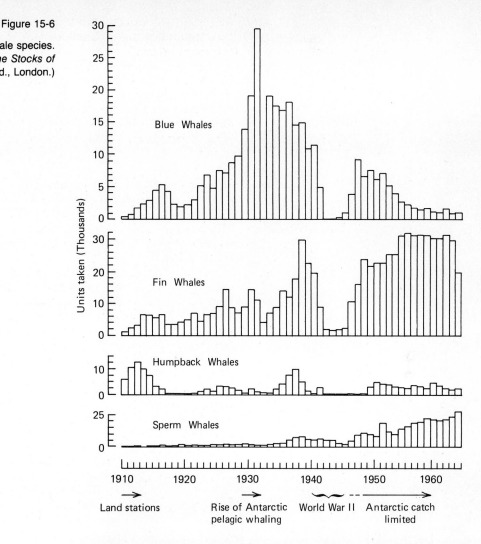

Figure 15-6

World catch of the principal whale species. (After N. A. Mackintosh, 1965. *The Stocks of Whales*. Fisheries News (Books) Ltd., London.)

are taken from spawning areas in Connecticut to fatten in bays on Long Island before harvesting. More complicated forms of aquaculture involve enclosures (fish pens) where feeding is controlled.

The lack of domesticated marine species and ignorance of fish diseases have limited the growth of aquaculture in the United States. Also, the legal status of many marine areas is unclear so that it is often impossible to control access in order to prevent loss or damage to a crop. Despite these problems, aquaculture is likely to become an increasingly important source of some foods, such as shrimp or oysters, as it already is in Asia (Fig. 15-7).

RESOURCES FROM SEAWATER

Water, salt, bromine, and magnesium are extracted from seawater. The ocean is an attractive source of these materials for several reasons. First, seawater is abundant and extraction of metals and salt from seawater is not likely to deplete any significant fraction of the ocean. Furthermore, the annual supply of several substances equals or exceeds present levels of production from the ocean. In the late 1960s 29% of the world's salt came from the ocean, as well as 70% of its bromine, 61% of its magnesium metal, 6% of its magnesium compounds, and 59% of its "manufactured" water. These materials were produced in about 60 countries and were worth more than $400 million.

427

Figure 15-7

In Japan a bamboo raft is used to suspend strings of oysters off the bottom. This
method offers greater protection from predatory starfish and snails. (Photograph
courtesy Consulate General of Japan, N.Y.)

Evaporating basins, located in many relatively dry coastal regions, use solar energy to evaporate seawater. Evaporation of brine is carefully controlled so that only sodium chloride or another desirable component of sea salt is removed at a single stage. Salts recovered from seawater must otherwise be treated to remove magnesium sulfate (Epsom salt) and calcium carbonate. Evaporating ponds for extraction of sea salts operate around southern San Francisco Bay (Fig. 15-8). Sea salt is used by the chemical industry, as is magnesium, a lightweight metal. Bromine extracted from seawater is used as a component of antiknock compounds in gasoline.

Despite the variety and large amount of valuable materials dissolved in ocean water, the extremely dilute nature of seawater makes it prohibitively expensive to produce most of them. Gold is an example; its concentration in seawater is extremely low, about 4 grams per million tons (10^{12} grams) of seawater. This amounts to about 5 million tons of gold in the ocean, but the cost of energy for pumping seawater, added to the cost of chemical treatment, far exceeds the value of the gold recovered. High energy costs further discourage recovery of metals dissolved in seawater. Another problem in many areas is the difficulty of obtaining adequate supplies of undiluted, uncontaminated seawater.

Growth of cities in arid coastal regions makes water from the ocean an increasingly important resource. In obtaining it, we effectively duplicate the natural cycling of water from the ocean through

Figure 15-8

About 5% of the U.S. production of salt comes from evaporating seawater in southern San Francisco Bay. Brine from large evaporating ponds is transferred to "pickle ponds" and then to a crystallization pond, where the salt forms. The photograph shows a harvesting operation after the last of the liquid, known as "bittern," has been removed and the layer of crystallized salt has reached a depth of about 10 centimeters. (Photograph courtesy Leslie Salt Company.)

evaporation (or freezing), using fossil or nuclear fuels instead of solar energy, in order to have water when and where it is needed.

The simplest way to obtain water from seawater is to build a greenhouse. Evaporation from pans of seawater occurs during the day when the sun is shining and condensation takes place at night on the cool surfaces of the shelter. This creates a humid, nearly tropical environment where certain plants can be grown at the same time for local marketing. Such man-made oases are useful for the small production of high-value products like fresh vegetables or fruit but seem to have little potential for large-scale agricultural development.

Recovery of freshwater from the ocean is most easily done by using a simple still (Fig. 15-9). Seawater is fed into a closed container and heated. The water vapor, free of salt, is condensed. The salt remaining behind and most of the original seawater (typically 90%) are discharged as a hot brine, the disposal of which constitutes a potential pollution problem for desalination projects.

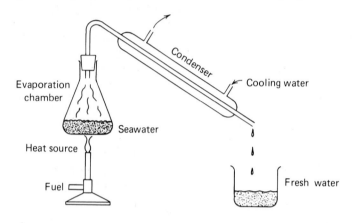

Figure 15-9

Simple single-stage evaporation–condensation apparatus.

Although easily visualized, recovery of freshwater from a single-stage evaporation process like that in Fig. 15-9 is impracticable. Too much heat is lost in the cooling water and high energy costs restrict the process to small-scale or emergency uses.

One way to increase the efficiency is to use the seawater for cooling before it is fed into the boiler (see Fig. 15-10). In this way, energy consumption is reduced. It is also possible to reduce the boiling point by carrying out the process under a slight vacuum, a technique known as *flash evaporation*. A large multistage flash distillation system was built for Elat, Israel. Smaller units are used on ships to make drinking water.

Another way of desalinating seawater is to duplicate the natural freezing cycle in which water and seasalts are separated, as discussed

Figure 15-10

Multistage flash distillation used at Elat, Israel, where water temperatures are relatively high. (After B. J. Skinner and K. K. Turekian, 1973. *Man and the Ocean*. Prentice-Hall, Englewood Cliffs, N.J.)

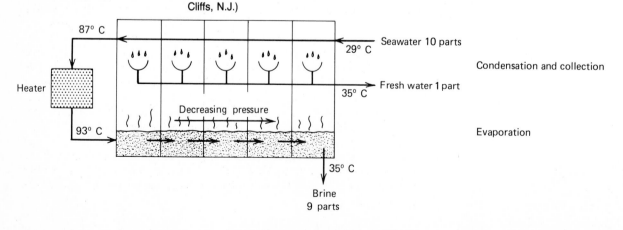

in Chapter 5. In this process, a coolant is added to seawater, which evaporates and removes heat from the liquid surface, causing it to freeze. If the ice crystals are removed and washed free of brine and salt, they can then be melted to obtain freshwater. The coolant is recycled and brines are discharged with approximately twice the salinity of the original seawater. A plant using these principles was designed for Ipswich, England, to produce water at a cost of approximately 50 cents per 100 gallons. Such a process is particularly suited to application in northern latitudes where ocean waters are initially cold, thereby reducing costs.

PETROLEUM AND NATURAL GAS

Petroleum and natural gas are the most valuable resources taken from the ocean. Petroleum comes from altered organic matter, usually phytoplankton, buried in sediments. In ocean areas where productivity is exceptionally high and bottom-water circulation sluggish, dissolved oxygen is used up due to decomposition of uneaten phytoplankton. Growth of consumer organisms is thereby inhibited and organic matter accumulates in the sediment deposits. Eventually bacteria break down some of this material. Higher than normal heat flow from tectonic processes completes the transformation of organic matter to petroleum or natural gas. But if the heating continues too long, all the organic matter may be transformed to natural gas or even totally destroyed.

Over time sediments compact under the weight of overlying deposits. Water and oil are expelled and move from the source sediments into nearby porous rock, where they may accumulate. Ancient reefs and sandstones are especially favorable sites for accumulation of oil when the migration is stopped by cappings of impervious sediments through which the oil and water cannot readily move. Thus the three factors necessary for the formation of oil and gas deposits are

1. Accumulations of sediments rich in organic matter.
2. Heating of the source beds to form petroleum or natural gas from the sedimentary organic matter.
3. Permeable porous rocks to hold the petroleum in an extractable form after its movement has been stopped by an impermeable layer.

Petroleum and natural gas have been produced off California, in the Gulf of Mexico, and in the Arabian Gulf from extensions of well-known oil fields on land (Fig. 15-11). Elsewhere, as in the North Sea between Great Britain and Holland and in Bass Strait between Australia and Tasmania, there were no corresponding land deposits. Other promising areas include the continental shelves north of Alaska, around Sumatra and Borneo in Indonesia, and on the Atlantic and Pacific margins of North America.

Salt domes are common on continental shelves and many form traps for petroleum and natural gas in the uplifted and fractured rocks around the domes (see Fig. 15-16). Such salt domes are important areas for oil and gas production in the Gulf Coast. Similar deposits may also occur in salt domes in deeper waters of the Gulf of Mexico.

It is difficult to predict the petroleum possibilities of the continental slope and rise. Thick sediment deposits and relatively high concentrations of organic matter suggest that petroleum occurs there, too. But difficult engineering and technical problems must be solved before these deep-water deposits can be successfully explored and exploited.

Figure 15-11

Production facilities in the Fateh field, Arabian Gulf. Equipment on the platforms is used to control the production of the wells, to pump the crude oil, and to flare excess natural gas. The tanker in the background stores the oil until it is transhipped for transportation to market. (Photograph courtesy Continental Oil Company.)

Estimates of undiscovered petroleum and natural gas resources are shown in Table 15-3. About 70% of the undiscovered resources is expected to be found on continental shelves and in shallow marginal ocean basins, much of which can be exploited by using available techniques. About 23% is expected to come from the continental slope, which cannot be exploited with present production techniques. Relatively little is expected to come from the continental rise and deep-ocean floor.

432

TABLE 15-3

Estimated Undiscovered Total Offshore Petroleum Resources (oil plus equivalent amount of natural gas)**

OCEAN AREA	RESOURCES	
	(billions of metric tons)	(billions of barrels)
Undiscovered reserves		
Continental shelves, shallow seas	183	1370
Continental slopes	61	460
Continental rises	12	90
Deep-sea trenches and ridges	3.5	26
Total	260	1950
Proved reserves	19	143

*From National Academy of Sciences, 1975. *Mineral Resources and the Environment.* Washington, D.C. 348 pp.

OTHER POTENTIAL ENERGY SOURCES IN THE SEA

The ocean itself, primarily the tide, is a source of renewable energy (Table 15-4). Tides are used for power in France and Russia, and experiments have been made to generate power from waves and from the temperature differences between warm surface waters and the much colder deep-ocean waters.

In areas of recent volcanic activity, water trapped underground in porous or fractured rocks is heated through contact with the heated rocks. Wells drilled into such reservoirs tap either hot water or superheated steam. Both are used as the energy source for steam electric power plants in Tuscany (Italy), New Zealand, northern California, Mexico, Japan, and the USSR. Steam from a large geothermal field is used for heating greenhouses in Iceland and development is underway to harness the power for generating electricity. Similar plants are possible on volcanic islands, such as Hawaii, and exposed portions of midoceanic ridges—Iceland, for example.

For centuries tides have powered mills in coastal regions; for instance, in 1650 Boston had a tidally powered mill to grind corn. Tidal power is now used to generate electricity. Large tidal power plants are operating on the Rance estuary in Brittany (France) and on a bay in the USSR near Murmansk in the White Sea. Some estuaries in England, China, and Korea are possibilities, as is the Bay of Fundy in Canada.

Three factors limit use of tidal power: the need for large tidal ranges, suitable topography, and the timing of power generation. The largest tidal ranges in the ocean are around 15 meters—for example, in the Bay of Fundy (Chapter 9). Even with special turbines designed to work on the ebb and flood tides, the tidal range must exceed 5 meters to be useful. Such ranges are rare (Fig. 15-12), for the tidal range for most coasts is only about 2 meters (see Chapter 9). Fur-

TABLE 15-4

Energy from Fossil Fuels and Other Earth Sources*

SOURCE	ENERGY	
	($\times$ 10^{20} cal/yr)	($\times$ 10^{12} watts)
Combustion by man	0.6	8.0
Dissipated by tides	0.2	2.7
Geothermal losses by earth	3.2	42
Solar energy at earth's surface	6500	86,000

*From National Academy of Sciences, 1975. *Mineral Resources and the Environment.* Washington, D.C. 348 pp.

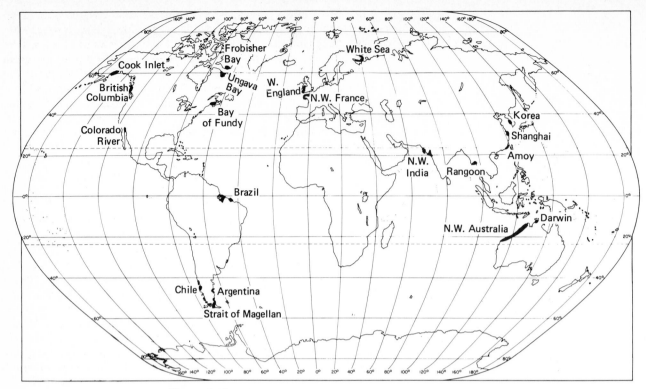

Figure 15-12

Areas whose tidal range exceeds 5 meters.

thermore, many potential sites are in remote areas where power transmission to urban and industrial centers would be expensive.

A second limiting factor for a tidal-power generating station is topography. Most tidal power schemes involve one or more dams. In a typical system, the gates through the dam are opened when the tide is high and then closed, keeping the water behind the dam at the level of the high tide. When the tide has dropped sufficiently, the water is allowed to flow out through turbines, thereby generating power. The larger and wider the opening into the bay or estuary, the more expensive the dam. Many of the most attractive sites are at high latitudes where glaciers have cut deep, narrow embayments and scoured the landscape down to bedrock, providing good foundations for dams.

Finally, there is a timing problem. Tidal power generation is tied to the tidal cycle, which does not usually coincide with peak power demands. The tidal power plant on the Rance estuary, for instance, produces about four times as much power during spring tides as during neap tides. Several ideas have been advanced to solve this problem. One is to use a vast network of power lines so that the electricity can be used somewhere regardless of when it is generated. Another uses several dams to store water at high levels so that one basin serves as reservoir and another as collector. Finally, there is the option of generating electricity and storing it in some way for later use. A fuel like hydrogen could be made and stored in the same way.

Waves are another potential energy source. One wave 1.8 meters high in water 9 meters deep releases about 10 kilowatts of power in each meter of wavefront (Chapter 8). The power dissipated on a beach during a single storm is enormous. Small amounts of power generated by the waves have been used to power whistles or gongs on navigational buoys for many years. The problem is to develop schemes for large-scale extraction of this energy at reasonable costs.

Temperature differences between water masses are yet another potential energy source. Surface waters are heated by the sun to about 25°C over much of the ocean whereas waters a few hundred meters below are 5°C or less. Power can be generated by transfer of heat between deep and surface layers. This step has been done on an experimental basis near Hawaii. Warm surface waters are used to evaporate specially chosen liquids, much as an oil-fired boiler is used to evaporate water in a steam electrical power plant, and waters from below the thermocline can be used to cool the system. Schemes have been proposed to generate power on tropical islands while using the cold, nutrient-rich deep waters to support aquaculture. Still other schemes envision floating power plants in subtropical waters, "grazing the thermocline" to produce power. A small experimental plant operating off the Hawaiian Islands in the early 1980s successfully produced electricity.

MINERALS FROM THE CONTINENTAL MARGINS

Sand, gravel, and shell account for the bulk of the material taken from the ocean and for much of the value of ocean resources produced in the past. Such items are increasingly in demand in urban areas for paving roads, constructing buildings, and filling low-lying areas. As urbanized regions expand, they use up local sand and gravel quarries and often build over potential supplies. So construction firms must go long distances in order to acquire building materials. For coastal cities, nearby bays and harbors have been dredged to provide sand and gravel, especially for landfill operations where composition and grain size requirements are less stringent than in making concrete.

Over the next 10 to 20 years, tens of millions of cubic meters of sand and gravel will be produced from U.S. offshore waters. Such deposits have long been exploited around the North Sea. Carbonate sands have been dredged in island areas where limestone for making cement is scarce on land, as in the Hawaiian Islands and Iceland.

Sand is also used in large quantities to repair and restore beaches damaged by currents or storms. One resort community on the U.S. Atlantic Coast annually uses more than 1 million cubic meters of sand for such purposes.

Material for construction purposes or landfill is commonly obtained by dredging. In open ocean waters dredging is carried out by seagoing ships that drag a special pipe behind the vessel and pump large volumes of water and sand off the bottom (see Fig. 15-13). The water is discharged, but the sand settles out in the large tanks on the vessel, called a hopper dredge. The hoppers are dumped or pumped out at the designated site. Another common type is the hydraulic dredge, which is basically a set of large pumps mounted on a barge. The sand is pumped and then discharged through large pipes to the shore site (or other location) being filled or to sites where the material can be stockpiled (see Fig. 15-14) for later use.

Figure 15-13

Schematic representation of simple hydraulic and hopper dredge operations. In the former, the cutter head is moved across the bottom, stirring and suspending the sediment. This slurry is then pumped into the vessel and out through a discharge pipe to a disposal or storage site where the sand and gravel settle out. The water and fine sediment particles run off. In the hopper dredge, suspended sediment and water are pumped into tanks aboard ship, where the sand and gravel settle out and the water is discharged overboard.

(a)

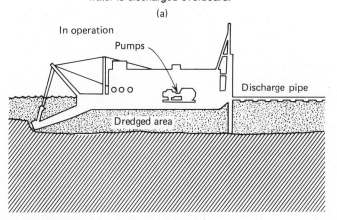

(b)

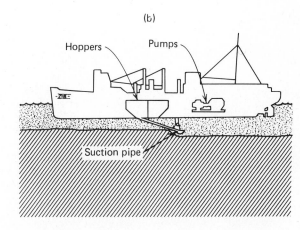

Figure 15-14

A simple hydraulic dredge, with the cutter head shown out of the water, at left. The discharge pipe can be seen at right. (Photograph courtesy Ellicott Machine Corporation.)

The rivers that cut their way across continental shelves during periods of lower sea level, the movements of the shoreline across the shelf as sea level rose, and the constant scouring due to storms, together with tidal current action, have left deposits of valuable minerals on many continental shelves (Table 15-5). These so-called heavy minerals, or detrital deposits, are much denser than normal sands and gravels and they are often concentrated and left behind when other sediments are moved out. Channels formed by rivers that cut across the continental shelf during periods of glaciation, when sea level was lower, are often sites of heavy mineral concentrations. The rivers that flowed through them eroded their rock walls and carved deep valleys. Light and soluble materials were washed away, but heavier grains were left in the river valley. Some detrital deposits formed near beaches, where waves and longshore currents separated particles containing titanium and rare earths from the lighter sand grains, leaving concentrations of the heavy minerals as beach deposits. Where such

TABLE 15-5

Materials Commonly Recovered from Surficial Continental Shelf Sediment Deposits*

	MINERAL		
MATERIAL	NAME	CHEMICAL COMPOSITION	DENSITY (g/cm³)
Gold	Gold	Au	19.3
Tin	Cassiterite	SnO_2	7.0
Chromium	Chromite	$FeCr_2O_4$	4.5
Titanium	Rutile	TiO_2	4.3
Titanium	Ilmenite	$FeTiO_3$	4.7
Rare earths	Monazite	Rare-earth phosphates	5.0
Diamond	Diamond	C	3.5
Sand and gravel	–	–	2.5

*After B. J. Skinner and K. K. Turekian, 1973. *Man and the Ocean*. Prentice-Hall, Englewood Cliffs, N.J. 149 pp.

minerals are sufficiently valuable, they are mined. Gold, tin, chromium, and titanium have been produced from continental shelves (Fig. 15-15).

Among nonmetal ocean-bottom resources (excluding fuels and building materials), sulfur and phosphorite are particularly important. Salt domes are rich sources of sulfur in addition to petroleum, as noted (Fig. 15-16). Sulfur accumulates around a salt plug because an insoluble sulfur-containing compound, anhydrite ($CaSO_4$), is generally present in the original salt deposit. As the plug is forced upward due to pressure from surrounding and overlying denser rock layers, it often encounters groundwater that dissolves some of the salt. Anhydrite, commonly altered to gypsum, and insoluble impurities, such as clay, form a "cap rock" mantle over the salt core. If anaerobic bacteria are also present, in association with petroleum deposits, they act to reduce anhydrite as follows:

$$CaSO_4 \quad + \quad CH_4 \quad \rightarrow \quad CaCO_3 \quad + \quad H_2S \quad + H_2O$$

| anhydrite or gypsum | hydrocarbon (petroleum) | calcium carbonate (limestone) | hydrogen sulfide | water |

Hydrogen sulfide gas is released and later combined with oxygen dissolved in groundwater to yield elemental sulfur:

$$2\,H_2S + \quad O_2 \quad \rightarrow 2\,H_2O + \quad S$$

| hydrogen sulfide | oxygen | water | sulfur |

Sulfur is recovered by forcing hot air and water into a salt dome in order to melt the sulfur and make it rise through pipes to the surface.

Figure 15-15

Areas of present petroleum, natural gas, and mineral extraction from the ocean bottom. Areas of potential manganese nodule production are also shown.

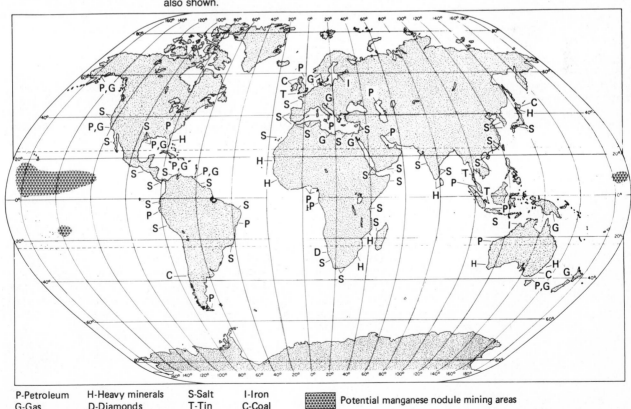

P-Petroleum   H-Heavy minerals   S-Salt   I-Iron
G-Gas   D-Diamonds   T-Tin   C-Coal   Potential manganese nodule mining areas

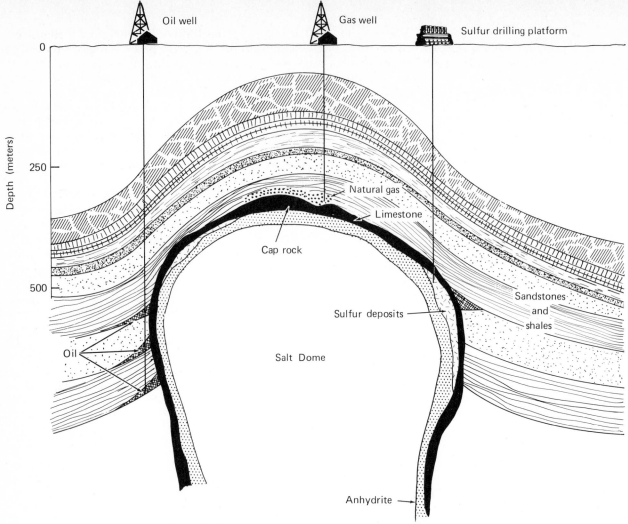

Oil well   Gas well   Sulfur drilling platform

Depth (meters)

0

250

500

Natural gas

Limestone

Cap rock

Oil

Sulfur deposits

Salt Dome

Sandstones
and
shales

Anhydrite

Figure 15-16

Schematic representation of a salt dome, showing deposits of petroleum, natural gas, and sulfur. The salt moving up through the sediment caused doming of the strata, producing traps for oil and gas. Sulfur was produced by bacteria breaking down anhydrite ($CaSO_4$), associated with the salt deposits.

Phosphorites, relatively insoluble deposits from which fertilizer is made, occur as nodules over the ocean floor, particularly on continental shelves. They are easily dredged and potentially valuable in areas short of phosphates. Phosphate content (P_2O_5) of the nodules tends to be relatively low, on the order of 20%, but many billions of tons are thought to be available, many times the world's production of phosphate (about 100 million tons per year) in the 1970s.

MINERALS FROM THE DEEP-OCEAN BOTTOM

Manganese nodules and deposits of massive sulfide minerals on the midocean ridges may become important sources of copper, nickel, cobalt, and other metals. Ignorance of their distribution and composition, lack of capability to produce them economically, and uncertainty about the legal status of the deep-ocean bottom have all delayed production of these potential resources. All three factors changed dramatically during the 1970s. Extensive exploration of the deep-ocean floor determined the composition and distribution of the nodules. Research submarines from the United States and France discovered sulfide minerals being formed by active hydrothermal vents on the midocean ridges. And United Nation Conference on the Law of the Sea helped clarify the legal status of the deep-ocean floor.

Manganese nodules (shown in Fig. 4-7) are potentially valuable, more for their copper, nickel, and cobalt contents than for the iron

438

439
minerals from the
deep-ocean bottom

and manganese that make up the bulk of the nodules. Potentially commercial nodules are concentrated in deep-ocean regions far from major sediment sources. The most interesting region is in the central Pacific, south of the Hawaiian Islands, an area of abundant nodules (see Figs. 4-11 and 15-15).

Slowly growing nodules have the highest copper–nickel–cobalt contents. The area of interest in the North Pacific covers about a million square kilometers. Nodules on the surface contain an estimated 5 billion tons of cobalt and 15 billion tons of nickel. The richest nodules lie at depths of about 5 kilometers. Thus special dredging techniques have been developed to "mine" them.

Two different mining techniques have been used in pilot-scale tests. One is an airlift technique in which compressed air is forced down a pipe to lift the nodules to the surface through a large-diameter pipe. The second technique utilizes a continuous bucket-line system to drag along the ocean bottom and recover the nodules in buckets as the line is hauled in (shown in Fig. 15-17). Both systems depend on underwater television to observe the mining processes because the waters are too deep for divers or commercially available submarines. Washing of the nodules and some processing may be done aboard ship. Processing plants ashore recover the metals from the nodules.

Potentially valuable metal-rich sulfide deposits associated with active midocean spreading centers have been discovered on the bottom of the Red Sea. These Red Sea muds contained 2% zinc and 0.7% copper, rich enough to be potentially exploitable. Other metal sulfide

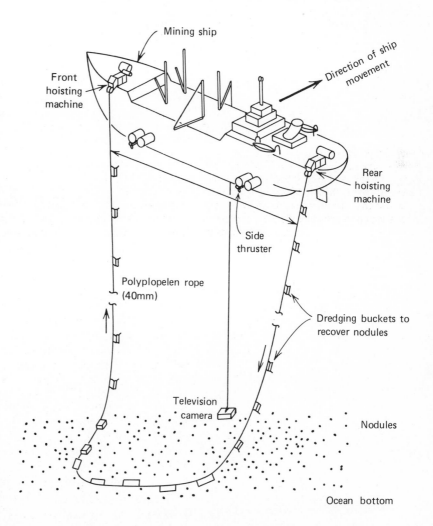

Figure 15-17

One technique to recover nodules from the deep-ocean floor uses buckets on a continuous cable. The buckets are dragged along the ocean bottom and filled with nodules and sediment before being hoisted back to the ship. (After Skinner and Turekian, 1973, p. 67.)

deposits containing high concentrations of zinc, copper, and iron have been recovered from the East Pacific Rise off Mexico. These deposits are associated with active or recently active hydrothermal vents. Similar deposits in ancient marine sediments are major copper producers on the island of Cyprus in the Mediterranean.

MAN-MADE ISLANDS AND OFFSHORE PORTS

The shallow ocean bottom provides space for urban and industrial growth. In many urban regions little space remains for expansion; so municipalities, utilities, and government agencies are considering the construction of facilities on continental shelves.

This seaward thrust is nothing new in the Netherlands. Between A.D. 1200 and 1950 the Dutch reclaimed about 1.6 million acres (6300 square kilometers) and reclamation of the shallow ocean bottom continues still (Fig. 15-18). Land is reclaimed by building dikes that enclose fields called *polders.* Some projects have been carried out on a monumental scale. In the 1930s the Zuider Zee, a large, shallow embayment, was cut off by a dike, changing it in a few years from a brackish estuary to the freshwater Ijssel Lake. Agricultural land was reclaimed from the former bay bottom. Land reclamation has also been done on a smaller scale in low-lying coastal areas around the North Sea. Parts of the lower Rhine estuary were shut off from the

Figure 15-18

Areas reclaimed from the North Sea, in the Netherlands. (After J. Van Venn, 1962. *Dredge, Drain, Reclaim! The Art of a Nation,* 5th ed. Martinus Nijhoff, The Hague.)

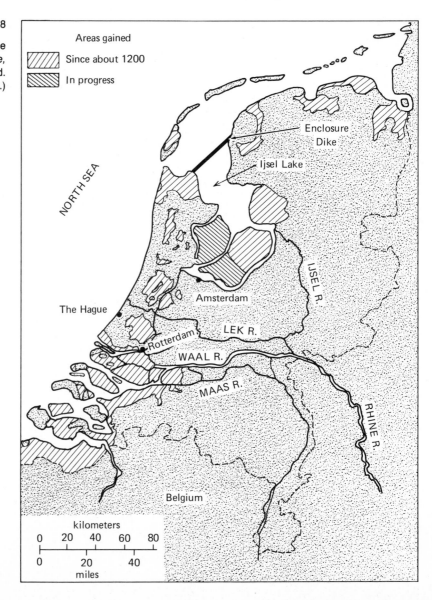

441
man-made islands
and offshore ports

North Sea in the 1950s and 1960s to prevent a repetition of the disastrous flooding accompanying the 1953 storm surge (Chapter 8).

Offshore port facilities in various parts of the world have been used for loading and unloading supertankers—tankers carrying hundreds of thousands of tons of petroleum and drawing up to 30 or 35 meters of water. Before the mid-1960s, tankers drew less than 15 meters so that they could move through the Suez Canal. In the 1970s closures of the canal and the savings in transportation cost provided by larger vessels made deep-draft ships more attractive. Unfortunately, such ships cannot operate in most estuaries without extensive and expensive dredging. But by locating loading and unloading facilities offshore, dredging is avoided. Offshore port facilities have been used for years in the Arabian Gulf and may be used in U.S. waters along the Atlantic and Gulf coasts.

Problems in obtaining both plant sites on land and access to adequate cooling waters have made offshore sites for large power plants especially attractive. Distance from shore reduces residents' objections to the disruptive appearance of a power plant on the coastal landscape and also reduces the exposure of the coastal population to potential radiation hazards caused by accidents at a nuclear plant (Fig. 15-19).

Offshore facilities can be constructed in several ways (Fig. 15-

Figure 15-19

Artist's model of an offshore nuclear power plant. In this scheme the nuclear reactors float on large barges. A semicircular breakwater protects the plant from wave action.

20). Simplest are tanker mooring and offloading facilities consisting of little more than a few pilings and moorings on which the vessel can tie. Oil is pumped ashore through a system of pipes to be stored and refined on land.

Other construction techniques suitable for building large offshore structures include

1. Dike and polder construction similar to techniques used in the Netherlands; costs range between $2000 and $30,000 per acre (1970 dollars).

Figure 15-20

Methods of constructing offshore facilities for industrial or other purposes.

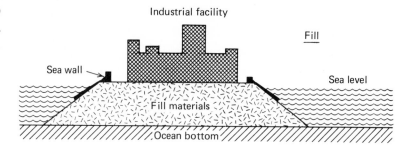

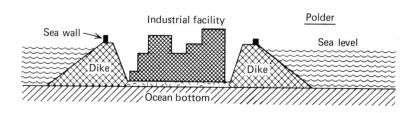

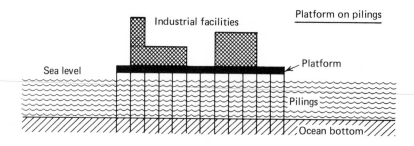

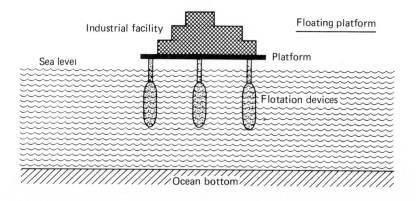

2. Conventional fill (including breakwater construction) in which the ocean bottom is built up by dumping fill materials; costs range between $8000 and $200,000 per acre.

3. Pile-supported deck or platform built on pilings; costs vary between $400,000 and $2.2 million per acre.

4. Floating platform, using flotation devices to support a platform; costs range between $1.3 million and $4.3 million per acre.

Regardless of which construction approach is used, costs of artificial islands are certain to be large. An offshore structure near New York City to be used as an airport was estimated to cost $1 billion to $7 billion in 1971. Smaller structures will be less expensive, but most will cost hundreds of millions. Despite their high costs, islands are likely to be built on the continental shelf. Many will probably be used for several purposes in order to reduce the costs for a single use.

WASTE DISPOSAL The ocean serves many little-known but vital functions. One of the most controversial is waste disposal. Near many coastal cities, the ocean is used to dilute and disperse a great variety of wastes, both liquid and solid (see Table 15-6). Urban wastes are often discharged by sewers into bays and harbors or into the coastal ocean directly from large pipes that take wastes offshore for disposal. The waters around discharge pipes dilute the wastes and currents disperse them in the surrounding coastal ocean.

TABLE 15-6

*Wastes Discharged into Estuaries and the Coastal Ocean**

SOURCES	WASTES DISCHARGED
Municipal storm sewers	Waste oils Street washings Raw sewage Suspended sediment
Municipal sewage treatment plants	Nutrients (phosphates and nitrates) Sewage sludges (solids from treatment)
Industrial wastes	Waste chemicals—e.g., acids, petrochemicals Waste oil
Runoff from agricultural lands	Nutrients from fertilizers Pesticides and herbicides Animal wastes
Electrical power plants	Waste heat Ash (from coal) Chemicals (corrosion-inhibiting, foam-suppressing, biocides)
Dredging operations and construction activities	Suspended sediment Nutrients from sediments
Petroleum production and exploration	Suspended sediment (drilling muds) Crude oil
Ships (commercial and recreational)	Untreated sewage Garbage Waste oil

*After M. G. Gross, 1980. *Oceanography,* 4th ed. Charles E. Merrill, Columbus, Ohio.

Coastal and estuarine circulation patterns do not always disperse wastes. Coastal currents tend to move materials along the coast and thereby inhibit movement of the wastes out into the open ocean where they could be more effectively diluted. Wastes that become associated with particles or taken up by organisms tend to be carried landward by bottom waters or in the estuarinelike circulation. Thus such nutrients as phosphates from detergents and nitrogen compounds from sewage, when discharged to bays and harbors, tend to remain near the point of discharge. This situation can stimulate the growth of phytoplankton, often of types not directly usable by organisms that live in the area. Problems arise if the plants are not eaten; they die and decompose, locally causing depletions of dissolved oxygen and fish kills, particularly in late summer.

Waste disposal in coastal waters interferes with shellfish production and sometimes with fisheries, especially for near-bottom fishes. It is an increasing problem on heavily populated, industrialized coasts, such as the U.S. Atlantic seaboard. Benthic organisms may be poisoned by sewage and chemicals (including pesticides) or suffocated by silt deposits. If not killed, they may be weakened and so be less able to resist disease and predation by natural enemies. A different problem arises when chemical wastes containing a high proportion of trace elements are taken up by shellfish. Silver, cadmium, chromium, and lead, for instance, have no known function in biological systems. In high concentrations, some of these elements are toxic to organisms. In the absence of any mechanism for excretion of such substances, they accumulate in plant and animal fats and proteins (Table 15-7). Such materials tend to be conserved in ecosystems because they are passed along in food chains. Some may substitute for elements essential to metabolic processes; cadmium, for example, substitutes for zinc in humans, resulting in impairment of fat metabolism. Elements incorporated in animal or plant skeletons are likely to have shorter lifetimes in an ecosystem and to be more quickly incorporated in sediments instead.

A more serious problem arises when filter-feeding clams and oysters ingest sewage-borne viruses and bacteria; because these organisms are often eaten raw and whole, including the digestive tract, they can spread hepatitis and gastrointestinal diseases. Waste disposal areas are closed to commercial shellfish production when judged unsafe by government inspectors, thereby disrupting the livelihood of oystermen and clamdiggers.

Bathing beaches are also closely monitored for evidence of sew-

TABLE 15-7

Enrichment of Various Elements in Shellfish*

ELEMENT	ENRICHMENT FACTORS		
	SCALLOP	OYSTER	MUSSEL
Ag	2300	18,700	330
Cd	2,260,000	318,000	100,000
Cr	200,000	60,000	320,000
Cu	3000	13,700	3000
Fe	291,500	68,200	196,000
Mn	55,500	4000	13,500
Mo	90	30	60
Ni	12,000	4000	14,000
Pb	5300	3300	4000
V	4500	1500	2500
Zn	28,000	110,300	9100

*After Horne, 1969.

age and closed for swimming if sewage-associated bacteria counts run too high. Thus many beaches close to cities are no longer available for swimming (Fig. 15-21).

Other, less obvious changes are also associated with waste disposal on continental shelves. The presence of sewage solids in the New York Bight and off southern California, for instance, has caused an increased incidence of diseases in bottom-dwelling fishes. In some cases, fins of affected fishes are eaten away, leaving the fish subject to easy capture by predators. Somewhat similar conditions are also observed to affect lobster and crabs taken from polluted areas.

Although sewage solids are a ubiquitous problem in urban coastal areas, they are by no means the only wastes discharged at sea. Dredging operations to build and maintain navigation facilities, extraction of ores, such as iron, titanium, and aluminum, and washing of sand and gravel are major sources of waste solids. Some of these sources rival or exceed natural processes as sources of sediment particles in coastal waters. Waste disposal operations in the New York metropolitan region are the largest single source of sedimentary materials to the entire mid-Atlantic area. And such disposal operations are quite widespread. In 1974 dredging produced about 100 million tons of sediment discharged on the continental shelf, about half from the Mississippi River alone. Virtually every major port requires dredging and much of the dredged material is disposed of at sea because of the lack of suitable sites on land.

Figure 15-21

Warnings are posted along the Potomac River waterfront, in Alexandria, Virginia. (Photograph courtesy U.S. Environmental Protection Agency, EPA-DOCUMERICA. Photographer, Erik Calonius.)

Other wastes discharged at sea include those carried in liquids. For example, desalination processes discussed earlier will produce brines that must be piped back to the ocean and with them the waste heat that was lost during the evaporation (or freezing) process. Steam electric power plants and refineries also discharge heated waters that can warm part of a bay or coastal area enough to cause changes in biological communities.

Highly radioactive wastes from the processing of nuclear fuels may someday be placed in holes drilled into the oceanic crust. The purpose is to place the wastes in inaccessible locations to ensure their isolation from humans for millions of years. The sites would be located far from land, far from the midocean ridges or subduction zones, and in the center of midocean gyres. The wastes would be sealed in containers and placed well below the sea floor.

Locations would be chosen to minimize exposure to humans if leaks did occur. A large fraction of any material that did leak would be retained by chemical reactions with the deep-ocean sediments. Any leakage into deep-ocean waters would take hundreds to thousands of years to reach the surface. Concentrations of the wastes would be reduced through mixing with large volumes of ocean waters.

REVIEW QUESTIONS

1. List the traditional uses made of the ocean. How is each changing? Is it increasing? Decreasing? Why?

2. Describe the typical sequence of events from the discovery of an underutilized fish stock to its final destruction as a commercial fishery.

3. Contrast the operations of mariculture with a traditional fishery.

4. List and briefly discuss the uses of some materials commercially produced from ocean water.

5. List the factors necessary for the accumulation of commercial quantities of petroleum. Discuss how the continental margin is a favorable area for formation of petroleum deposits.

6. What are some of the nonmetallic resources produced from surficial deposits on the continental shelf?

7. Describe why salt domes are favorable locations for oil and gas fields.

8. What are some of the materials that may be commercially produced from the deep-sea floor? Discuss some of the problems encountered in producing these materials on a commercial basis.

9. What are some of the problems involved in using the ocean water and the ocean floor for disposal of waste materials?

SUMMARY OUTLINE

Marine resources—fishing, defense, transportation; also fuels, metals, construction materials
    Control and management through international laws a positive factor
    Renewable resources—replenished naturally
    Nonrenewable resources—for instance, petroleum and metals ores

Fisheries—a renewable resource
    Man an important predator—60 million metric tons annual production
    Maximum yield probably 200 million tons, including species not now used
    Dynamics of fishery—aids in predicting yields
        Fishing increases population growth rate by reducing population size

Overfishing results if maximum sustainable yield is exceeded

Size of year class depends on environmental conditions, food supply

Pacific sardine fishery—replaced by anchovy off California

El Niño and Peruvian Anchovetta Fishery

Highly productive fishery—1960s

Overfishing and cessation of upwelling caused collapse of fishery

Worldwide effects of protein shortage

Heavy rains associated with El Niño cause flooding

Whaling—international regulation necessary

Mariculture—important in Orient; for instance, relocation of year class for fattening, enclosures to control conditions

Resources from seawater—salt, magnesium, bromine, water

Salt—by evaporation using solar energy

Metals—production costs usually exceed value of resource

Water—important in coastal areas; evaporation, distillation, freezing

Petroleum and natural gas—in porous sedimentary deposits, with organic matter

Major sources on continental shelf, in salt domes

Potential energy sources—renewable

Tidal power—requires large tidal range, suitable topography, energy storage

Waves—now used for buoys, whistles

Temperature differences—a use of solar energy

Geothermal power—steam or hot water from heat in ocean floor; used in Iceland

Minerals from continental margins

Sand, gravel, shell—demand from construction industry; for landfill

Heavy minerals (e.g., tin, titanium)—from erosion of ancient rocks, in valleys and beach deposits on continental shelf

Manganese nodules—valuable for copper, nickel, cobalt

Sulfur—produced from salt domes

Phosphorites—phosphate for fertilizer, not yet produced commercially

Minerals from deep-ocean floor

Manganese nodules—potential sources of copper, nickel, and cobalt

Especially promising areas in North Pacific

Massive sulfide deposits on midocean ridges are potential sources of copper and zinc

Man-made islands and offshore ports—pressure for land in coastal areas

Polders reclaimed as farmland in the Netherlands

Offshore port facilities—for deep-draft vessels—for instance, tankers

Sites for industry—power production, storage of materials

Four types: dike and polder, fill, piles supporting deck, floating platforms

Waste disposal—coastal, estuarine circulation retains wastes in coastal waters

Sewage and chemical wastes—damaging to fishes, shellfish, beaches

Dredging—may be the major sediment sources in a coastal area

Waste heat—carried in brines, coolants

Radioactive wastes buried in deep-ocean sediments

SELECTED REFERENCES

CUSHING, D. H. 1975. *Marine Ecology and Fisheries.* Cambridge University Press, Cambridge, England. 278 pp. Advanced treatment of fisheries, fish production, and marine food webs.

MERO, J. L. 1965. *The Mineral Resources of the Sea.* Elsevier, Amsterdam, 312 pp. Broad coverage of mineral resources.

SKINNER, B. J., AND K. K. TURKEKIAN. 1973. *Man and the Ocean.* Prentice-Hall, Englewood Cliffs, N.J. 149 pp.

Elementary discussion of ocean resources and ocean processes.

VAN VEEN, J. 1962. *Dredge, Drain, Reclaim! The Art of a Nation,* 5th ed. Martinus Nijhoff, The Hague, Netherlands. 200 pp. Elementary discussion of Dutch polder building.

WENK, EDWARD. 1969. The physical resources of the ocean. *Scientific American* 221(3): 166–176.

APPENDIX 1
Conversion Factors

EXPONENTIAL NOTATION

It is often necessary to use very large or very small numbers to describe the ocean or to make calculations about its processes. To simplify writing such numbers, scientists commonly indicate the number of zeros by *exponential notation,* indicating the powers of ten. Some examples:

1,000,000,000 =	10^9	(one billion)
1,000,000 =	10^6	(one million)
1,000 =	10^3	(one thousand)
100 =	10^2	(one hundred)
10 =	10^1	(ten)
1 =	10^0	(one)
0.1 =	10^{-1}	(one tenth)
0.01 =	10^{-2}	(one hundredth)
0.001 =	10^{-3}	(one thousandth)
0.000 001 =	10^{-6}	(one millionth)
0.000 000 001 =	10^{-9}	(one billionth)

Multiplication: To multiply exponential numbers (powers of ten), the exponents are added. For example, $10 \times 100 = 1000$, which is written exponentially as $10^1 \times 10^2 = 10^3$

Division: To divide exponential numbers, the exponent of the divisor is subtracted from the exponent of the dividend. For example, $100/10 = 10$ or written exponentially as $10^2/10^1 = 10^1$

UNITS OF MEASURE

length
1 *kilometer* (km) = 10^3 meters = 0.621 statute mile = 0.540 nautical mile

1 *meter* (m) = 10^2 centimeters = 39.4 inches = 3.28 feet = 1.09 yards = 0.547 fathom

1 *centimeter* (cm) = 10 millimeters = 0.394 inch = 10^4 microns

1 *micron* (μ) = 10^{-3} millimeter = 0.0000394 inch

area
1 *square centimeter* (cm²) = 0.155 square inch

1 *square meter* (m²) = 10.7 square feet

1 *square kilometer* (km²) = 0.386 square statute mile = 0.292 square nautical mile

volume
1 *cubic kilometer* (km³) = 10^9 cubic meters = 10^{15} cubic centimeters = 0.24 cubic statute mile

1 *cubic meter* (m³) = 10^6 cubic centimeters = 10^3 liters = 35.3 cubic feet = 264 U.S. gallons

1 *liter* (l) = 10^3 cubic centimeters = 1.06 quarts = 0.264 U.S. gallon

1 *cubic centimeter* (cm³) = 0.061 cubic inch

mass
1 *metric ton* = 10^6 grams = 2205 pounds

1 *kilogram* (kg) = 10^3 grams = 2.205 pounds

1 *gram* (g) = 0.035 ounce

time
1 day = 8.64×10^4 seconds (mean solar day)

1 year = 8765.8 hours = 3.156×10^7 seconds (mean solar year)

speed
1 *knot* (nautical mile per hour) = 1.15 statute miles per hour = 0.51 meter per second

1 *meter per second* (m/sec) = 2.24 statute miles per hour = 1.94 knots

1 *centimeter per second* (cm/sec) = 1.97 feet per minute = 0.033 feet per second

Conversion formulas

$$°C = \frac{°F - 32}{1.8}$$

$$°F = (1.8 \times °C) + 32$$

Conversion table

°C	°F
0	32
10	50
20	68
30	86
40	104
100	212

energy

1 *gram-calorie (cal)* = $\frac{1}{860}$ watt-hour = $\frac{1}{252}$ British thermal unit (Btu)

Graphs, Charts, and Maps

In any scientific discipline, data gathered from many sources are usually compiled and presented in graphic form at some point. Oceanography is no exception, although the wide variety of material obtained from studying the ocean poses some special problems. In this section we review some of the common means of graphically presenting data—graphs, profiles, maps, and diagrams—in an effort to show the uses as well as the limitations of each technique.

GRAPHS

Of the various techniques used to portray scientific data, *graphs* are perhaps the most widely used. A graph permits general aspects of the relationship between two properties to be understood at a glance. Furthermore, values of various properties can be measured from a graph.

In constructing a graph, data are often first organized into tables. For example, if we are studying the increase in boiling point of seawater with salinity changes, we may arrange our data as follows.

SALINITY (parts per thousand)	BOILING POINT INCREASE (°C)
4	0.06
12	0.19
20	0.31
28	0.44
36	0.57

From this table it is apparent that the boiling point of seawater increases with increased salinity. To see this more clearly, a graph can be drawn as in Fig. A2-1. Note that salinity is plotted on the *x* (horizontal) axis with values increasing to the right. Boiling point increase, in degree Celsius (°C), is plotted on the *y* (vertical) axis, with values increasing upward. We plot our experimental points and then draw a line through these points.

Generally a straight line is the best first estimate. In this case, it is a reasonably good approximation. For many other graphs, we may need to use more complicated curves; but even for complicated curves, a straight line is a reasonable estimate for small portions of the curve.

Note that the graph shows that the boiling point is raised as salinity increases. Also, you can estimate boiling point increases for salinities not studied here. For instance, for seawater with a salinity of 24 parts per thousand, we can interpolate a boiling point increase of 0.375°C; for a salinity of 40 parts per thousand, we extrapolate (or extend) the curve to indicate a boiling point increase of 0.63°C.

Figure A2-1

Graph of the increase in temperature of boiling and water salinity. The dots represent the experimental data from the table.

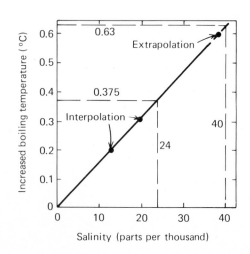

In reading a graph, always determine which property is plotted on each axis. Also, check both the scale intervals and the values at the origins. The appearance of the graph can be changed drastically by changing either; advertisers, for instance, often make use of graphs where the scales and origins are chosen to present their data in the most favorable light.

PROFILES

Profiles are used to show topography, either of ocean bottom or of land. An ocean-bottom profile can be considered as a vertical slice through the earth's surface. Such a profile can be drawn with no distortion—distances are equal vertically and horizontally. But imagine the problems involved in drawing a profile 10 centimeters long of the Atlantic Ocean between New York and London, a distance of 5500 kilometers but only 3.4 kilometers deep at the deepest point. A pencil line would be too thick to portray accurately the maximum relief; thus such a profile conveys no useful information.

To get around this problem, profiles—including those in this book—are usually distorted. Profiles showing oceanic features are typically distorted by factors of several hundred or several thousand. Consequently, even gently rolling hills look like impossibly rugged mountains. The effect of profile distortion can be seen rather dramatically when it is applied to the human profile, as in Fig. A2-2.

CONTOURS AND CONTOUR MAPS

Various means have been used to portray land forms or typography; the most useful employ contours in contour maps. Figure 2-21 is an example of a contour map. *Contour lines* connect points that are at equal elevations or at equal depths. Obviously not all elevations (or depths) can be connected by contour lines. Only certain ones at selected intervals are shown; otherwise the map would be solid black. The vertical interval represented by successive contour lines is called the *contour interval*.

To interpret a contour map, imagine the shoreline of a lake. The still-water surface is a horizontal plane, touching points of equal elevation along the shore. The shoreline is thus a contour line. If the water surface were controlled to fall by regular intervals, it would trace a series of contour lines, forming a contour map on the lake bottom (or hillside). Note that the shorelines formed at different lake levels do not cross one another; neither do contours on maps.

Contours reveal topography. For example, contours that are closed on a map (do not intersect at a boundary) indicate either a hill or a depression. To find out which, look to see if the elevation increases toward the closed contour(s). If so, you are looking at a hill. If the elevation decreases toward the closed contour, it is a depression. Contours around depressions are often marked by *hachures*—short lines on the contour pointing toward the depression. When a contour line crosses a valley or canyon, the contour line forms a V, pointing upstream.

Contours can also be used to depict properties other than elevation (or depth). For example, several maps in this book use contours to show distribution of properties, such as surface-ocean temperatures (see Figs. 6-6 and 6-7) and salinities (see Fig. 6-14). We would consider them to be temperature (or salinity) hills and valleys. High temperature corresponds to a hill, low temperature to a valley. Contours may also be used to show the distribution of temperature or salinity in a vertical section of the ocean.

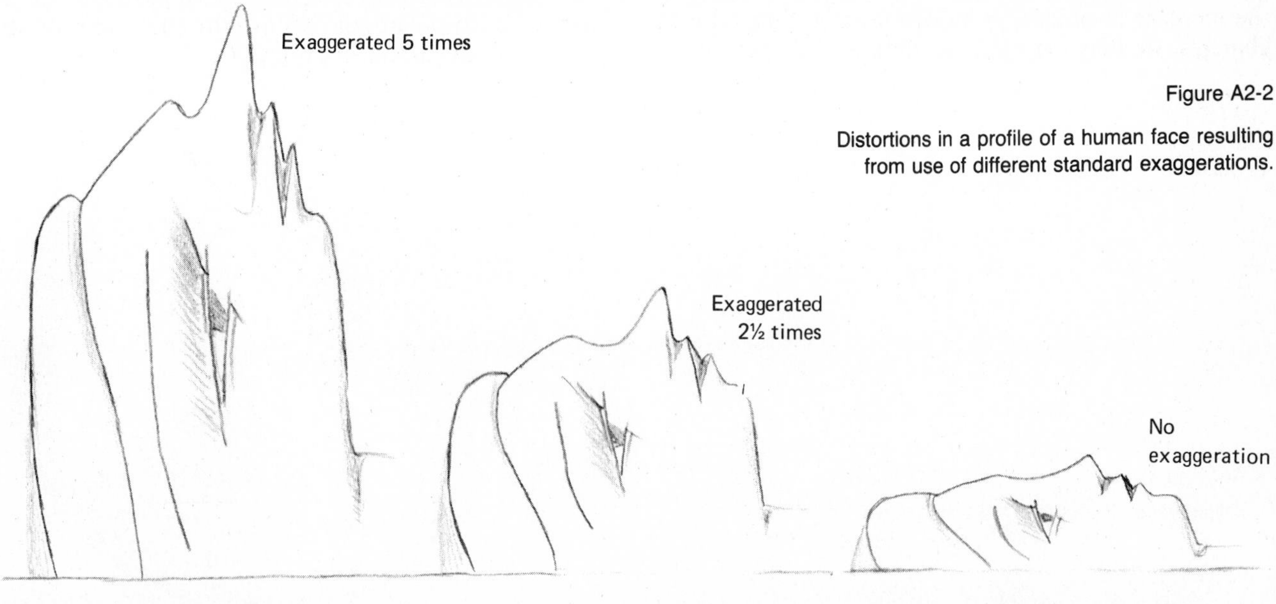

Exaggerated 5 times

Exaggerated 2½ times

No exaggeration

Figure A2-2

Distortions in a profile of a human face resulting from use of different standard exaggerations.

COORDINATES—LOCATIONS ON A MAP

A *coordinate* is a sort of address—a means of designating location. The most familiar coordinate system, found in many cities, is the network of regularly spaced, lettered or numbered streets crossing one another, usually at right angles, which enables us to locate a given address. This is an example of a *grid*. As soon as we determine how the streets and avenues are arranged, we can find Fifth Avenue and 42nd Street in New York or 16th and K Streets in Washington, D.C., even though we may never have been in those cities before.

A printed grid is used on large-scale maps to designate locations of features. In making maps of relatively small areas, it is simplest to assume that the world is flat. This works for areas extending up to 100 miles from a starting point. For larger areas, the earth's curvature must be considered.

Since the earth is round, we must use *spherical coordinates*—a grid fitted to a sphere. A small town or even a state has rather definite starting points for a grid—the edges or the center. But a sphere has no edges or corners; so we must designate those points where our numbering system is to begin.

Distances north or south are easiest to deal with. We can easily identify the earth's geographic poles, where the axis of rotation intersects the earth's surface. Using these points, it is fairly easy to draw a line circling the earth and equally distant from the North and South poles. This is the *equator*, which serves as our starting point to measure distances north and south. Going from the equator, the distance to the pole is divided into 90 equal parts *(degrees)*. The series of grid lines that circle the earth and connect points that are the same distance from the nearest pole are known as *parallels of latitude*.

To see how this works, imagine the earth with a section cut out, as in Fig. A2-3. Now look at the angle formed by the line connecting any point of interest with the earth's center and the line from the earth's center to a point directly south of that point, on the equator. This angle is a measure of the distance between the chosen point and the equator. The North Pole has a latitude of 90°N, Seattle is approximately 47°N, and Rio de Janeiro approximately 23°S.

Latitude was easily measured by early mariners. The angle between Polaris (the pole star) and the horizon provides a reasonably accurate measure of latitude. At the equator, the pole star is on the horizon (latitude 0°N). Midway to the North Pole (latitude 45°N), the pole star is 45° above the horizon. At the North Pole, Polaris is directly overhead. Although there is no star directly above the South Pole, the same principle holds except that a correction would be necessary to allow for the displacement from the South Pole of the star used.

Measuring east–west distances on the earth poses the problem—where do we start? The answer has been to establish an arbitrary starting point—the *prime meridian*—and to indicate distances as east or west of that meridian. Several prime meridians have been used by different nations, but today the Greenwich prime meridian is most commonly used. It passes through the famous observatory at Greenwich (a suburb southeast of London).

Longitude—distance east or west of the prime meridian—is indicated on a map by north–south lines, connecting points with equal angular separation from the prime meridian. They are called *meridians of longitude* and converge at the North and South poles. Longitude in degrees is measured by the size of the angle between the prime meridian and the meridian of longitude passing through the given point, as in

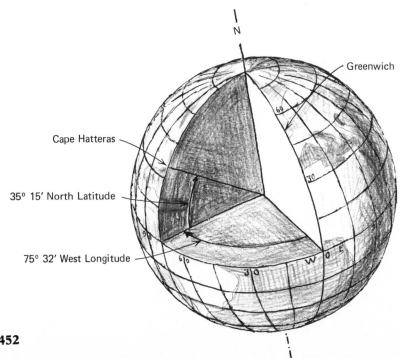

Figure A2-3

Latitude and longitude on the earth shown for
Cape Hatteras, North Carolina.

Cape Hatteras

35° 15′ North Latitude

75° 32′ West Longitude

Greenwich

Fig. A2-3. Going eastward from the prime meridian, longitude increases until we reach the middle of the Pacific Ocean, when we come to the 180° meridian. Going westward from Greenwich, longitude also increases until we reach the 180° meridian, which represents the juncture between the Eastern and Western hemispheres. Through much of the Pacific Ocean, the 180° meridian is also the location of the *international dateline*. This designation of the 180th meridian as the international dateline—where the "new day" begins—is no accident. Its position in the midst of the Pacific avoids the problem of adjacent cities being one day apart in time. This also explains, in part, the choice of the Greenwich Meridian as the Prime Meridian.

Longitude and time are intimately related. The earth turns on its axis once every 24 hours. Since it takes the earth 24 hours to make one complete turn (360°), we calculate that the earth turns 15° per hour. We use this relationship to find our relative position east or west of the prime meridian.

Each meridian of longitude is a *great circle*. If we sliced through the earth along one of the meridians of longitude, our cut would go through the center of the earth. Of the parallels of latitude, only the equator is a great circle. All the other parallels are *small circles*, for a plane (or slice) passing through them would not go through the center of the earth. Great circles are favored routes for ships or aircraft because a *great circle* route is the shortest distance between two points on a globe.

To study time and longitude, let us begin at local noon on the prime meridian, when the sun is directly overhead. One hour later the sun is directly over the meridian of a point 15° west of the prime meridian; 2 hours later it is over a meridian 30° west of the prime meridian. And 12 hours later (midnight) it is over the 180th meridian and the new day begins.

If we have an accurate clock keeping "Greenwich time" (the time on the Prime Meridian), we can determine our approximate longitude from the time of local noon, when the sun is highest in the sky. Assume that our clock read 2:00 P.M. Greenwich time at local noon. The 2-hour difference indicates that our position is 30° from the prime meridian. Since local noon is later than Greenwich, we know that we are west of Greenwich and our longitude is therefore 30°W. Another example—if our local noon occurs at 9:30 A.M. Greenwich time, we are 2.5 hours × 15° per hour = 37.5° east of the prime meridian; our longitude is thus 37.5°E.

Degrees—like hours—are divided into 60 parts known as *minutes*. Each minute is further divided into 60 *seconds*. Consequently, in the last example we would give our position as 37°30′E.

To determine longitude, therefore, a ship need only have accurate time. With modern electronic communications, this poses no problems. For centuries, however, seafarers had no means of keeping accurate time at sea. Not until the 1760s, when the first practical chronometers—accurate clocks for use abroad ship—were designed, was it possible for most ships' pilots to determine longitude. Even the Greek astronomer Ptolemy (A.D. 90–168) made maps with relatively accurate positions north and south. But he overestimated the length of the Mediterranean by 50%, an error that was not corrected until 1700.

Each map in this book shows latitude and longitude (usually at 20° intervals) to indicate the positions of the map features (see Fig. 2-6). The parallels of latitude may also be used to determine approximate distance on a map. Each degree of latitude equals approximately 60 nautical miles (69 statute miles or 111 kilometers). Each minute of latitude is approximately 1 nautical mile, or 1.85 kilometers. At the equator each minute of longitude is 1 nautical mile but decreases so that 1 minute of longitude is only 0.5 nautical mile at 60° north or south latitude and vanishes at the poles.

MAPS AND MAP PROJECTIONS

A *map* is a flat representation of the earth's surface. Symbols are used to depict surface features. Because the earth is a sphere, making a flat map distorts the shape or size of surface features. The only distortion-free map is a globe, but a globe is not practical for the study of relatively small areas and so maps are used almost exclusively in science.

In making a map, we would like to make the final product as useful as possible. In general, we would like a map to preserve the following properties of the earth's surface:

Equal area   each area on the map should be proportional to the area of the earth's surface it represents.

Shape   the general outlines of a large area shown on a map should approximate as nearly as possible the shape of the region portrayed. A map that preserves shape is said to be *conformal*.

Distance   a perfect map would permit distance to be measured accurately between any two points anywhere on the map. Many common maps, such as the Mercator projection, do not accurately portray distances in a simple way.

Direction   ideally, it would be possible to measure directions accurately anywhere on a map.

No map has all these properties; only a globe preserves size, shape, direction, and distance simultaneously.

A *map projection* takes the grid of latitude and longitude lines from a sphere and converts them into a grid on a flat surface. Sometimes the resulting grid is a simple rectangular one where longitude and latitude lines intersect at right angles. In other projections, latitude and longitude lines are complex curves that intersect at various angles.

TABLE A2-1

Characteristics of Various Map Projections

NAME OF PROJECTION (type)	DISTINCTIVE FEATURES	DESIRABLE FEATURES (UNDESIRABLE FEATURES)	USES
Mercator (cylindrical)	Horizontal parallels Vertical meridians	Compass directions are straight lines (Extreme distortion at high latitudes)	Navigation
Goode homolosine projection	Horizontal parallels Characteristic interruptions of outline	Equal area Little distortion of shape (Interruption of either continents or oceans)	Data presentation
Hoelzel's planisphere (modified)	Horizontal parallels Characteristic nearly oval outline	Shapes easily recognizable (Moderate scale distortion at high latitudes)	Index map Data presentation

With the network formed from the grid lines, the map is drawn by plotting points in the appropriate spot on the new projection. In this way, the various types of maps are prepared. We shall consider only a few of the many map projections that have been developed to serve specific functions (listed in Table A2-1).

Without a doubt, the *Mercator projection* is most familiar. Parallels of latitude and meridians of longitude are all straight lines, and cross at right angles. The outline shape is a square or rectangle, as shown in Fig. A2-4.

In its simplest form, the Mercator projection can be visualized as being made by a light inside a translucent globe projecting latitude and longitude onto a cylinder surrounding the globe. Even though the cylinder used for the projection is curved, it is easily made into the flat map desired.

The common Mercator projection in our example is most accurate within 15° of the equator and least accurate at the poles. Although shapes are well preserved by this projection, area is distorted, especially near the poles. For example, South America is in reality nine times the size of Greenland, but this fact is not obvious from common Mercator projections. The scale of a Mercator projection changes going away from the equator. If the reader uses the length of a degree of latitude as his scale, he can avoid serious error.

Another property of the Mercator projection useful to mariners is that a course of constant compass direction (a *rhumb line*) is a straight line on this projection. Although a rhumb course is not a great circle and thus not the shortest distance between any two points on the earth's surface, it is useful for navigation because a great circle course requires a constant changing of direction. A rhumb line is slightly longer but easier to navigate.

For world maps, the Mercator projection has distinct limitations; but for relatively small areas, such as navigation charts, the Mercator projection is without equal. Nearly all navigation charts used at sea are Mercator projections.

For use in this book, a *Hoelzel planisphere* (shown in Fig. A2-5) is used. This modified cylindrical projection shows the continents well, permitting most of the world's coastal regions to be easily recognized. Like the Mercator, this projection distorts areas near the poles. Instead of converging to a point at the poles, the meridians converge to a line that is only a fraction of the length of the equator. Because the projection used in this book splits the Pacific Ocean in the middle, it is not overly convenient for showing properties in the ocean.

A special projection (the *interrupted homolosine*, shown in Fig. A2-6) was developed by J. P. Goode in 1923 to show the ocean basins without any interruptions. In addition, the projection shows area equally. The continents are interrupted to show the ocean basins intact.

SELECTED REFERENCES

ALDRIDGE, B. G. 1968. *Mathematics for Physical Science.* Merrill Publishing Company, Columbus, Ohio. 137 pp. Elementary mathematical and data-plotting procedures.

COHEN, P. M. 1970. *Bathymetric Navigation and Charting.* United States Naval Institute, Annapolis, Md. 138 pp. Navigation through use of bathymetric data; nontechnical.

GREENWOOD, DAVID. 1964. *Mapping.* University of Chicago Press, Chicago. 289 pp. Elementary discussion of coordinates, maps, and map projections.

WILLIAMS, J. E. (Ed.). 1963. *World Atlas.* 2nd ed. Prentice-Hall, Englewood Cliffs, N.J. Includes graphical summary of different map projects. 96, 41 pp.

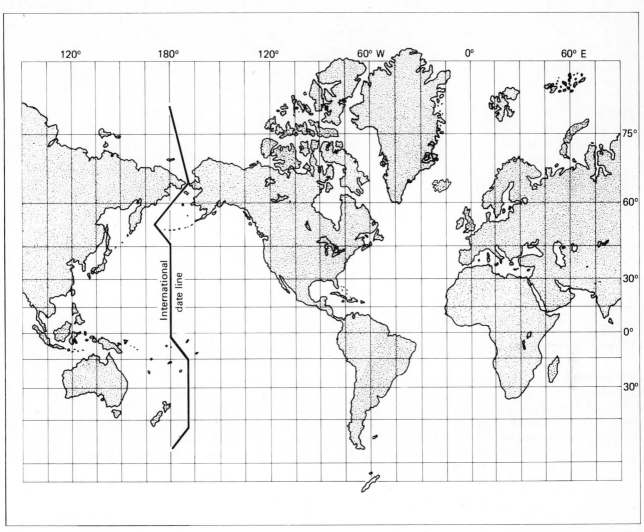

Figure A2-4

A mercator projection. Compare the shape of Greenland shown on this map with that in Figs. A2-5 and A2-6 to see the distortion at high latitudes.

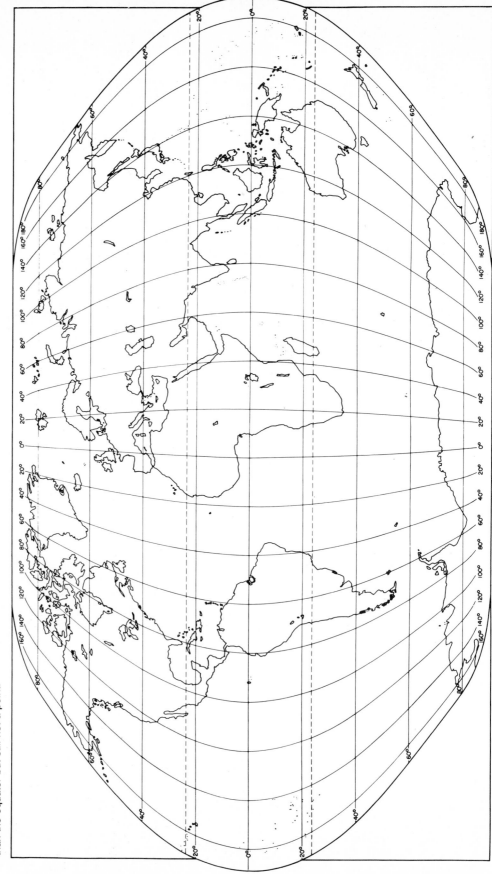

Figure A2-5

A Hoelzel planisphere. Note that the meridians of longitude converge to a line shorter than the equator but still not a point.

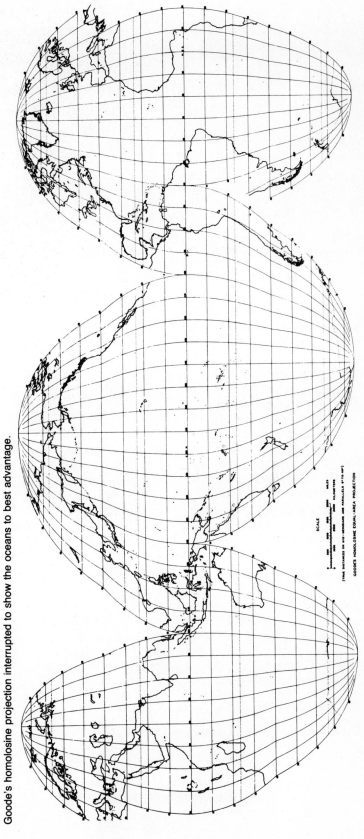

Figure A2-6
Goode's homolosine projection interrupted to show the oceans to best advantage.

Abyssal pertaining to the great depths of the ocean, generally below 3700 meters.

Albedo the ratio of radiation reflected by a body to the amount incident upon it, commonly expressed as a percentage.

Algae marine or freshwater plants, including phytoplankton and seaweeds.

Algal ridge the elevated margin of a windward reef built by actively growing calcareous algae.

Alternation of generations mode of development, characteristic of many coelenterates, in which a sexually reproducing generation gives rise (by union of egg and sperm) to an asexually reproducing form, from which new individuals arise by budding or simple division of the "parent" animal.

Amphidromic point the center of an amphidromic system; a nodal or no-tide point around which the crest of a standing wave rotates once in each tidal period.

Anaerobic condition in which oxygen is excluded, with the result that organisms that depend on the presence of oxygen cannot survive. Anaerobic bacteria can live under these conditions.

Andesite a type of volcanic rock intermediate in composition between granite and basalt, associated with partial melting of crust and mantle during subduction in the presence of some water. Andesitic mountain ranges are often associated with subduction zones.

Anion a negatively charged atom or radical.

Anoxic devoid of dissolved oxygen. See *Anaerobic*.

Antinode that part of a standing wave where the vertical motion is greatest and the horizontal velocities are least.

Aphotic zone that portion of the ocean where light is insufficient for plants to carry on photosynthesis.

Aquaculture the cultivation or propagation of water-dwelling organisms.

Archipelagic plain a gently sloping sea floor with a generally smooth surface, particularly found among groups of islands or seamounts.

Arrowworm see *Chaetognath*.

Arthropods animals with a segmented external skeleton of chitin or plates of calcium carbonate, and with jointed appendages—for example, a crab or an insect.

Asthenosphere upper zone of the earth's mantle, extending from the base of the lithosphere to depths of about 250 kilometers beneath continents and ocean basins; relatively weak, probably partially molten.

Atoll a ring-shaped organic reef that encloses a lagoon in which there is no preexisting land and that is surrounded by the open sea. Low sand islands may occur on the reef.

ATP adenosine triphosphate, a compound involved in energy transfer in metabolism, mainly formed during cell respiration.

Autotrophic nutrition that process by which an organism manufactures its own food from inorganic compounds.

Azoic devoid of life.

Backshore that part of a beach that is usually dry, being reached only by the highest tides; by ex-

*The glossary is taken in part after B. B. Baker, W. R. Deebel, and R. D. Geisenderfer (Eds.) *Glossary of Oceanographic Terms*, 2nd ed. U.S. Naval Oceanographic Office, Washington, D.C. (1966); and American Geological Institute, *Glossary of Geology and Related Sciences* and supplement, 2nd ed., Washington, D.C. (1960).

tension, a narrow strip of relatively flat coast bordering the sea.

Baleen (whalebone) horny material growing down from the upper jaw of large plankton-feeding (baleen) whales, which forms a strainer or filtering organ consisting of numerous plates with fringed edges.

Bank an elevation of the sea floor of a large area; a submerged plateau.

Bar an offshore ridge or mound that is submerged (at least at high tide), especially at the mouth of a river or estuary.

Barrier beach a bar parallel to the shore whose crest rises above high water.

Barrier island a detached portion of a barrier beach between two inlets.

Barrier reef a reef that is separated from a landmass by a lagoon, usually connected to the sea through passes (openings) in the reef.

Basalt a fine-grained igneous rock, black or greenish black, rich in iron, magnesium, and calcium.

Basin a depression of the sea floor more or less equidimensional in form and of variable extent.

Bathythermograph a device for obtaining a record of temperature at various depths in the ocean, usually within a few hundred meters of the ocean surface.

Baymouth bar a bar extending partially or entirely across the mouth of a bay.

Beach seaward limit of the shore (limits are marked approximately by the highest and lowest water levels).

Beach cusp one of a series of low mounds of beach material separated by troughs spaced at more or less regular intervals along the beach face.

Beach drift see *Littoral drift*.

Beach face this term is sometimes used to describe the highest portion of the foreshore, above the low tide terrace and exposed only to the action of wave uprush, not to the rise and fall of the tide.

Bed load see *Load*.

Benioff zone see *Subduction zone*.

Benthic that portion of the marine environment inhabited by marine organisms that live permanently in or on the bottom.

Benthos bottom-dwelling marine organisms.

Berm the low, nearly horizontal portion of a beach (backshore), having an abrupt fall and formed by the deposit of material by wave action. It marks the limit of ordinary high tides and waves.

Berm crest the seaward limit of a berm.

Bight a concavity in the coastline that forms a large, open bay.

Biogenous sediment sediment consisting of at least 30% by volume of particles derived from skeletal remains of organisms, which may contribute calcium carbonate, silica, or phosphate minerals to the sediment.

Biogeochemical cycles the paths by which elements essential to life circulate from the nonliving environment to living organisms and back again to the environment.

Bioluminescence the production of light without sensible heat by living organisms as a result of a chemical reaction either within certain cells or organs or in some form of secretion.

Biomass the amount of living matter per unit of water surface or volume expressed in weight units.

Bird-foot delta a delta formed by the outgrowth of pairs of natural levees, formed by the distributaries, making the digitate or bird-foot form.

Bivalves mollusks, generally sessile or burrowing into soft sediment, rock, wood, or other materials. Individuals possess a hinged shell and a hatchet-shaped foot, which is sometimes used in digging. Clams, oysters, and mussels belong to this group.

Black bottle (dark bottle) a container used in measuring respiratory activity of primary producers. The container is covered or coated to exclude light and thereby prevent photosynthetic activity.

Black mud (hydrogen sulfide mud) a dark, fine-grained sediment formed in poorly aerated bays, lagoons, and fjords. This sediment usually contains large quantities of decaying organic matter and iron sulfides and may exude hydrogen sulfide gas.

Bloom see *Plankton bloom*.

Bore see *Tidal bore*.

Bottom water the water mass at the deepest part of the water column. It is densest water that is permitted to occupy that position by the regional bottom topography.

Boundary currents northward- or southward-directed surface currents that flow parallel and close to the continental margins. They are caused by deflection of the prevailing eastward- and westward-flowing currents by the continental landmasses.

Brackish water water in which salinity values range from approximately 0.05 to 17.00 parts per thousand.

Breaker a wave breaking on the shore, over a reef, etc. Breakers may be roughly classified into four kinds, although the categories may overlap: *spilling breakers* break gradually over a considerable distance; *plunging breakers* tend to curl over and break with a crash; *surging breakers* peak up, but then, instead of spilling or plunging, they surge up on the beach face; *collapsing breakers* break in the middle or near the bottom of the wave rather than at the top.

Breakwater   a structure, usually rock or concrete, protecting a shore area, harbor, anchorage, or basin from waves.

Brine   water containing a higher concentration of dissolved salt than that of the ordinary ocean. Brine can be produced by the evaporation or freezing of seawater.

Calcareous algae   marine plants that form a hard external covering of calcium compounds.

Calorie   the amount of heat required to raise the temperature of 1 gram of water by 1°C (defined on basis of water's *specific heat*).

Canyon (submarine)   see *Submarine canyon.*

Capillary wave   a wave in which the primary restoring force is surface tension. A water wave in which the wavelength is less than 1.7 centimeters is considered a capillary wave.

Cation   a positively charged ion, tending to be deposited at the cathode in electrolysis or in a battery.

Celsius temperature   temperatures based on a scale in which water freezes at 0° and boils at 100° (at standard atmospheric pressure); also called *centigrade temperature.*

Cephalopods   benthic or swimming mollusks possessing a large head, large eyes, and a circle of arms or tentacles around the mouth; the shell is external, internal, or absent, and an ink sac usually is present. They include squids and octopus.

Chaetognath   small, elongate, transparent, wormlike animals, pelagic in all seas from the surface to great depths. They are abundant and may multiply rapidly into vast swarms. Some, especially *Sagitta,* have been identified as indicator species.

Change of state   a change in the physical form of a substance, as when a liquid changes to a solid due to cooling.

Chemosynthesis   carbon dioxide fixation (primary production) by certain bacteria in the absence of sunlight, using inorganic compounds, such as ammonia or methane, or elements, such as reduced iron, hydrogen, or sulfur.

Chitin   a nitrogenous carbohydrate derivative forming the skeletal substance in arthropods.

Chloride   an atom of chlorine in solution, bearing a single negative charge.

Chlorinity   a measure of the chloride content, by mass, of seawater (grams per kilogram, or per mille, of seawater).

Chlorophyll   a group of green pigments that occur chiefly in bodies called chloroplasts and that are active in photosynthesis.

Cilia   hairlike processes of cells, which beat rhythmically and cause locomotion of the cells or produce currents in water.

Clay   soil consisting of inorganic material, the grain of which have diameters smaller than 0.004 millimeter (or 4 microns).

Climate   The prevalent or characteristic meteorological conditions of a place or region—in contrast with *weather,* which is the state of the atmosphere at a particular time.

Climax community   end product of ecological succession in which there is a balance of production and consumption in the community and further change in species composition takes place only as a result of changes in environmental conditions or outside forces influencing the community.

Clupeoids   fish of the herring family, including sardine, anchovy, pilchard, and menhaden.

Coastal currents   currents flowing roughly parallel to the shore and constituting a relatively uniform drift in the water just seaward of the surf zone. These currents may be caused by tides or winds, or associated with the distribution of mass in coastal waters.

Coastal ocean   shallow portion of the ocean (in general, overlying the continental shelf), where the circulation pattern is strongly influenced by the bordering land.

Coastal plain   low-relief continental plain, adjacent to the ocean and extending inward to the first major change in terrain features.

Cobble   a rock fragment between 64 and 256 millimeters in diameter, larger than a pebble and smaller than a boulder; usually rounded or otherwise abraded.

Coccolithophores   microscopic, often abundant planktonic algae, the cells of which are surrounded by an envelope on which numerous small calcareous disks or rings (coccoliths) are embedded.

Coelenterates   a large, diverse group of simple animals possessing two cell layers and a digestive cavity with only one opening. This opening is surrounded by tentacles containing stinging cells. Some are sessile, some pelagic (medusae), and some undergo alternation of generations.

Compaction   the decrease in volume or thickness of a sediment under load through closer crowding of constituent particles and accompanied by decrease in porosity, increase in density, and squeezing out of water.

Compensation depth   the depth at which oxygen

production by photosynthesis equals that consumed by plant respiration during a 24-hour period.

Compressional wave a wave that causes alternate compression and expansion of the medium through which it passes.

Conduction transfer of energy through matter by internal particle or molecular motion without any external motion.

Conformal projection a map projection in which the angles around any point are correctly represented—for example, a Mercator projection.

Conservative property a property whose values do not change in the course of a particular, specified series of events or processes—for example, those properties of seawater, such as salinity, the concentrations of which are not affected by the presence or activity of living organisms but are affected by diffusion and currents.

Continental block a large landmass rising abruptly from the deep-ocean floor, including marginal regions that are shallowly submerged.

Continental climate a climate characterized by cold winters and hot summers in areas where the prevailing winds have traveled across large land areas.

Continental crust the thickened part of the crust forming continental blocks; typically about 35 kilometers thick, consisting primarily of granitic rocks.

Continental margin a zone separating the emergent continents from the deep-sea bottom; generally consists of the continental shelf, slope, and rise.

Continental rise a gentle slope with a generally smooth surface, rising toward the foot of the continental slope.

Continental shelf the sea floor adjacent to a continent, extending from the low-water line to the change in slope, usually about 180 meters depth, where continental shelf and continental slope join.

Continental slope a declivity from the outer edge of the continental shelf, extending from the break in slope to the deep-sea floor.

Contour line a line on a chart connecting points of equal value above or below a reference value; can portray elevation, temperature, salinity, or other values.

Convection vertical circulation resulting from density differences within a fluid, resulting in transport and mixing of the properties of that fluid.

Convergence area or zone where flow regimes come together or converge, usually resulting in sinking of surface water.

Convergence zone a band along which crustal area is lost. Colliding edges of crustal plates may be thickened, folded, or underthrust, or one plate may be subducted and destroyed.

Copepods minute shrimplike crustaceans, most species of which range between about 0.5 and 10 millimeters in length.

Coral the hard, calcareous skeletons of various sessile, colonial coelenterate animals, or the stony solidified mass of a number of such skeletons; also, the entire animal, a compound polyp that produces the skeleton.

Coral reef a complex ecological association of bottom-living and attached calcareous, shelled marine invertebrates forming fringing reefs, barrier reefs, or atolls.

Coralline algae red algae, having either a bushy or encrusting form and deposits of calcium carbonate either on the branches or as a crust on the substrate. Certain of them develop massive encrustations on coral reefs.

Core a vertical, cylindrical sample of sediments from which the nature and stratification of the bottom may be determined; also, the central zone of the earth.

Coriolis effect an apparent force acting on moving particles resulting from the earth's rotation. It causes moving particles to be deflected to the right in the Northern Hemisphere and to the left in the Southern Hemisphere; the deflection is proportional to the speed and latitude of the moving particle. The speed of the particle is unchanged by the apparent deflection.

Crinoids a group of echinoderms, most of which are attached by a long stalk to the bottom either permanently or when immature; species without stalks either swim or creep slowly about. Crinoids (such as sea lily, feather star) occur in shallow water as well as at great depths.

Critical depth the depth above which occurs the net effective plant production in the water column as a whole.

Crust of the earth the outer shell of the solid earth. Beneath the oceans, the outermost layer of crust is composed of unconsolidated sediment, the second layer of consolidated sediment and weathered lavas, and the inside layer probably of basaltic rocks. The crust varies in thickness from approximately 5 to 7 kilometers under the ocean basins to 35 kilometers under the continents. Its lower limit is defined to be the Mohorovičić discontinuity.

Crustaceans arthropods that breathe by means of gills or similar structures. The body is commonly covered by a hard shell or crust. The group includes barnacles, crabs, shrimps, and lobsters.

Ctenophores spherical, pear-shaped, or cylindrical animals of jellylike consistency ranging from less than 2 centimeters to about 1 meter in length.

The outer surface of the body bears eight rows of comblike structures.

Current ellipse a graphic representation of a rotary current in which the speed and direction of the current at different hours of the tide cycle are represented by vectors joined at one point. A line joining the extremities of the radius vectors forms a curve roughly approximating an ellipse.

Current meter any device for measuring and indicating speed or direction (usually both) of flowing water.

Current rose a graphic representation of currents for specified areas, utilizing arrows to show the direction toward which the prevailing current flows and the frequency (expressed as a percentage) of any given direction of flow.

Current tables tables that give daily predictions of the times, speeds, and directions of the currents.

Cypris larva the stage at which the young of barnacles attach to the substrate.

Daily (diurnal) inequality the difference in heights and durations of the two successive high waters or of the two successive low waters of each day; also, the difference in speed and direction of the two flood currents or the two ebb currents of each day.

Daily (diurnal) tide a tide having only one high water and one low water each tidal day.

Debris line a line near the limit of storm-wave uprush marking the landward limit of debris deposits.

Decomposers heterotrophic and chemoautotrophic organisms (chiefly bacteria and fungi) that break down nonliving matter, absorb some of the decomposition products, and release compounds usable in primary production.

Deep scattering layer a stratified population of organisms in ocean waters that causes scattering of sound as recorded on an echo sounder. Such layers may be from 50 to 200 meters thick. They occur less than 200 meters below the surface at night and several hundred meters below the surface during the day.

Deep-water waves water waves whose depth is greater than one-half the average wavelength.

Delta an alluvial deposit, roughly triangular or digitate in shape, formed at the mouth of a stream of tidal inlet or a river.

Demersal fishes those living near the bottom.

Density the mass per unit volume of a substance, usually expressed in grams per cubic centimeter. In centimeter–gram–second system, density is numerically equivalent to *specific gravity*.

Density current the flow (caused by density differences or gravity) of one current through, under, or over another; it retains its unmixed identity from the surrounding water because of density differences.

Deposit feeding removal of edible material from sediment or detritus either by ingesting material unselectively and excreting the unusable portion or by selectively ingesting discrete particles.

Depth of no motion the depth at which water is assumed to be motionless, used as a reference surface for computing geostrophic currents.

Desalination removal of water from seawater or brine.

Detritus any loose material produced directly from rock disintegration. *Organic detritus* consists of decomposition or disintegration products or dead organisms, including fecal material.

Diatoms microscopic phytoplankton organisms, possessing a wall of overlapping halves (valves) impregnated with silica.

Diffusion transfer of material (e.g., salt) or of a property (e.g., temperature) by eddies or molecular movement. Diffusion causes spreading or scattering of matter under the influence of a concentration gradient, with movement from the stronger to the weaker solution.

Dinoflagellates microscopic or minute organisms, which may possess characteristics of both plants (such as chlorophyll and cellulose plates) and animals (such as ingestion of food).

Discontinuity an abrupt change in a property, such as salinity or temperature, at a line or surface.

Dispersion the separation of a complex, surface gravity-wave disturbance into its component parts. Longer component parts of the wave travel faster than shorter ones and arrive at their destination first, as swell.

Distributary an outflowing branch of a river, which occurs characteristically on a delta.

Diurnal daily, especially pertaining to actions that are completed within approximately 24 hours and that recur every 24 hours.

Divergence horizontal flow of water in different directions from a common center or zone; upwelling is a common example of divergence.

Divergence zone a region along which crustal plates move apart and new lithospheric material solidifies from rising volcanic magma.

Doldrums a nautical term describing a belt of light, variable winds near the equator; an area of low atmospheric pressure.

Drift bottle a bottle released at sea for use in studying surface currents. It contains a card identifying the date and place of release and requesting the finder to return it, noting the date and place of recovery.

Drift current a slow-moving current, principally caused by wind, usually restricted to the waters above the pycnocline.

Drift net fishing net suspended in the water vertically so that drifting or swimming animals can be trapped or entangled in the mesh.

Dynamic topography the configuration formed by the geopotential difference (dynamic height) between a given surface and a reference surface, usually the layer of no motion. A contour map of dynamic "topography" may be used to estimate geostrophic currents.

Ebb current the movement of a tidal current away from shore or down a tidal stream.

Echinoderms principally benthic marine animals having calcareous plates with projecting spines forming a rigid or articulated skeleton or plates and spines embedded in the skin. They have radially symmetrical, usually five-rayed, bodies. They include the starfish, sea urchins, crinoids, and sea cucumbers.

Echiuroids unsegmented, burrowing marine worms.

Echo sounding determination of the depth of water by measuring the time interval between emission of a sonic or ultrasonic signal and the return of its echo from the bottom. The instrument used for this purpose is called an *echo sounder.*

Ecological efficiency ratio of the efficiency with which energy is transferred from one trophic level to the next.

Ecosystem ecological unit including both organisms and the nonliving environment, each influencing the properties of the other and both necessary for maintenance of life as it exists on earth.

Eddy a current of air, water, or any fluid, often on the side of a main current, especially one moving in a circle—in extreme cases, a whirlpool.

Eddy viscosity the turbulent transfer of momentum by eddies giving rise to an internal fluid friction, in a manner analogous to the action of molecular viscosity in laminar flow but taking place on a much larger scale.

Eelgrass a seed-bearing, grasslike marine plant that grows chiefly in sand or mud–sand bottoms and most abundantly in temperate water less than 10 meters deep.

Ekman spiral a theoretical representation of the currents resulting from a steady wind blowing across an ocean having unlimited depth and extent and uniform viscosity. Specifically, it would cause the surface layer to drift on an angle of 45° to the right of the wind direction in the Northern Hemisphere; water at successive depths would drift in directions more to the

right until, at some depth, the water would move in the direction opposite to the wind. Speed decreases with depth throughout the spiral. The net water transport is 90° to the right of the wind in the Northern Hemisphere.

Encrusting algae see *Coralline algae, Red algae.*

Epifauna animals that live at the water–substrate interface, either attached to the bottom or moving freely over it.

Epiphytes plants that grow attached to other plants.

Equal-area projection a map projection in which equal areas on the earth's surface are represented by equal areas on the map.

Equilibrium tide the hypothetical semidaily tide caused by gravitational attraction of the sun and moon on a frictionless, nonrotating entirely water-covered earth.

Estuary a semienclosed, tidal coastal body of saline water with free connection to the sea, commonly the lower end of a river.

Euphausids shrimplike, planktonic crustaceans, widely distributed in oceanic and coastal waters, especially in colder waters. They grow to 8 centimeters in length and many possess luminous organs.

Euphotic zone see *Photic zone.*

Fan a gently sloping, fan-shaped feature normally located near the lower termination of a canyon.

Fault a fracture or fracture zone in rock, along which one side has been displaced relative to the other.

Fauna the animal population of a particular location, region, or period.

Fermentation a type of respiration in which partial oxidation of organic compounds takes place in the absence of free oxygen—for example, the oxidation of glucose to alcohol by yeast.

Fetch an area of the sea surface over which seas are generated by a wind having a constant direction and speed, also called *generating area;* also, the length of the fetch area, measured in the direction of the wind in which the seas are generated.

Filter feeding filtering or trapping edible particles from seawater; a feeding mode typical of many zooplankters and other marine organisms of limited mobility.

Fjord a narrow, deep, steep-walled inlet of the ocean, formed either by the submergence of a mountainous coast or by entrance of the ocean into a deeply excavated glacial trough after the melting away of the glacier.

Flagellum a whiplike bit (process) of protoplasm that provides locomotion for a motile cell.

Flocculation process of aggregation into small lumps, especially with regard to soils and colloids.

Floe sea ice, either as a single unbroken piece or as many individual pieces covering an area of water.

Flood current the tidal current associated with the increase in the height of a tide. Flood currents generally set toward the shore or in the direction of the tide progression; sometimes erroneously called *flood tide.*

Flora the plant population of a particular location, region, or period.

Food chain simplification of a food web.

Food web the interrelated food relationships in an ecosystem, including its production, consumption, and decomposition, and the energy relationships among the organisms involved in the cycle.

Foraminifera benthic or planktonic protozoans possessing shells, usually of calcium carbonate.

Forced wave a wave generated and maintained by a continuous force, in contrast to a free wave, which continues to exist after the generating force has ceased to act.

Forereef the upper seaward face of a reef, extending above the lowest point of abundant living coral and coralline algae to the reef crest. This zone commonly includes a shelf, bench, or terrace that slopes to 15–30 meters, as well as the living, wave-breaking face of the reef. The terrace is an eroded surface or is veneered with organic growth. The living reef front above the terrace in some places is smooth and steep; in other places it is cut up by grooves separated by ridges that together have been called *spur-and-groove systems,* forming comb-tooth patterns.

Forerunner low, long-period swell that commonly precedes the main swell from a distant storm, especially a tropical cyclone.

Foreshore see *Low tide terrace.*

Foul to attach to or come to lie on the surface of submerged man-made or introduced objects, usually in large numbers or amounts, as barnacles on the hull of a ship or silt on a stationary object.

Fracture zone an extensive linear zone of unusually irregular topography of the ocean floor characterized by large seamounts, steepsided or asymmetrical ridges, troughs, or long, steep slopes.

Free wave any wave not acted on by any external force except for the initial force that created it.

Fringing reef a reef attached directly to the shore of an island or continental landmass. Its outer margin is submerged and often consists of algal limestone, coral rock, and living coral.

Fully developed sea the maximum height to which ocean waves can be generated by a given wind force blowing over sufficient fetch, regardless of duration, as a result of all possible wave components in the spectrum being present with their maximum amount of spectral energy.

Gastropods mollusks that possess a distinct head, generally with eyes and tentacles and a broad, flat foot and that are usually enclosed in a spiral shell.

Geostrophic current a current resulting from the balance between gravitational forces and the Coriolis effect.

Geothermal power power derived from heat energy that flows from the earth's crust, especially in volcanic areas.

Giant squids large cephalopods (length may be 15 meters or more) that inhabit mid-depths in oceanic regions but may come to the surface at night. They are eaten by sperm whales.

Gill a delicate, thin-walled structure, often an extension of the body wall, used for the exchange of gases in a water environment, and sometimes for excretion of wastes.

Glassworm see *Chactognath.*

Glacial marine sediment high-latitude, deep-ocean sediments that have been transported from the land to the ocean by means of glaciers or icebergs.

Glacier a mass of land ice, formed by recrystallization of old, compacted snow, flowing slowly from an area of accumulation to an area of sublimation, melting, or evaporation, where snow or ice is removed from the glacier surface.

Grab an instrument possessing jaws that seize a portion of the ocean bottom for retrieval and study.

Graded bedding a type of stratification in which each stratum displays a gradation in grain size from coarse below to fine above.

Gradient the rate of decrease (or increase) of one quantity with respect to another—for example, the rate of decrease of temperature with depth.

Granite a crystalline rock consisting essentially of alkali feldspar and quartz: in seismology, a rock in which the compressional-wave velocity varies approximately between 5.5 and 6.2 kilometers per second. *Granitic* is a textural term applied to coarse and medium-grained igneous rocks.

Gravel loose material ranging in size from 2 to 256 millimeters.

Gravity anomaly a disturbance in the earth's normal gravity field.

Gravity wave a wave whose velocity of propagation is controlled primarily by gravity. Water waves of length greater than 1.7 centimeters are considered gravity waves.

Great circle the intersection of the surface of a sphere and a plane through its center—for example, meridians of longitude and the equator are great circles on the earth's surface.

Greenhouse effect the result of the penetration of the atmosphere by comparatively short-wavelength solar radiation that is largely absorbed near and at the earth's surface whereas the relatively long-wavelength radiation emitted by the earth is partially absorbed by water vapor, carbon dioxide, and dust in the atmosphere, thus warming the atmosphere.

Groin a low, artificial, damlike structure of durable material placed so that it extends seaward from the land; used to slow littoral drift.

Group velocity the velocity with which a wave group travels. In deep water, it is equal to one-half the individual wave velocity.

Guyot a flat-topped submarine mountain or seamount.

Gyre a circular or spiral form, usually applied to a semiclosed current system.

Hadley cell a semiclosed system of vertical motions in the earth's atmosphere. Warm, moist air rises in equatorial regions, flows to midlatitudes (30°N, 30°S), where it sinks and returns to the equatorial zone as the trade winds.

Half-life the time required for the decay of one-half the atoms of a sample of a radioactive substance. Each radionuclide has a unique half-life, related to its rate of disintegration.

Halocline water layer with large vertical changes in salinity.

Headland a high, steep-faced point of land extending into the sea.

Heat budget the accounting for the total amount of the sun's heat received on the earth during any one year as being exactly equal to the total amount lost from the earth to radiation and reflection.

Heat capacity amount of heat required to raise the temperature of a substance by a given amount.

Herbivore an animal that feeds only on plant matter.

Heterotrophic nutrition that process by which an organism utilizes preformed organic compounds for its food.

High water the upper limit of the surface water level reached by the rising tide; also called high tide.

Higher high water the higher of the two high waters of any tidal day. The single high water occurring daily in a diurnal tide is considered a higher high water.

Higher low water the higher of two low waters occurring during a tidal day.

Holoplankton organisms whose life cycle is completely passed in the floating state.

Hurricane a cyclonic storm, usually of tropical origin, covering an extensive area and containing winds of 120 kilometers per hour or higher.

Hydrogen bond the relatively weak bond formed between adjacent molecules in liquid water, resulting from the mutual attraction of hydrogen and nearby atoms.

Hydrogenous sediment particles precipitated from solution in water.

Hydrography the measurement, description, or mapping of surface waters.

Hydroid the polyp form of coelenterate animals that exhibit alternation of generations. It is attached, often branching, and gives rise to the pelagic, medusa form by asexual budding.

Hydrologic cycle the composite cycle of water exchange, including change of state and vertical and horizontal transport, of the interchange of water among earth, atmosphere, and ocean.

Hydrophilic having the property of attracting water.

Hydrophobic not attractive to water; unwettable.

Hydrosphere the water portion of the earth, as distinguished from the solid part and the gaseous outer atmosphere. It consists of liquid water in sedimentary rocks, rivers, lakes, and oceans, as well as ice in sea ice and continental ice sheets.

Hydrothermal associated with high-temperature groundwaters—for example, the alteration or precipitation of minerals and mineral ores.

Hypsometric graph a representation of the respective elevations and depths of points on the earth's surface with reference to sea level.

Iceberg a large mass of detached land ice floating in the sea or stranded in shallow water.

Ice shelf a thick ice formation with a fairly level surface, formed along a polar coast and in shallow bays and inlets, where it is fastened to the shore and often reaches bottom.

Igneous rock formed by solidification of magma.

Infauna animals who live buried in soft sediment.

Insolation in general, solar radiation received at the earth's surface; also, the rate at which direct solar radiation is incident upon a unit horizontal surface at any point on or above the surface of the earth.

Instability a property of a system such that any disturbance grows larger instead of diminishing so that the system never returns to the original steady state; in oceanography, usually refers to the vertical displacement of a water parcel.

Interface surface separating two substances of different properties (such as different densities, salinities, or temperatures)—for example, the air–sea interface or the water–sediment interface.

Internal wave a wave that occurs within a fluid whose density changes with depth, either abruptly at a sharp surface of discontinuity (an interface) or gradually.

Interstitial water water contained in the pore spaces between the grains in rocks and sediments.

Intertidal zone (littoral zone) generally considered the zone between mean high-water and mean low-water levels.

Invertebrate animal lacking a backbone.

Ionic bond a linkage between two atoms, with a separation of electric charge on the two atoms; a linkage formed by the transfer or shift of electrons from one atom to another.

Iron–manganese nodules see *Manganese nodules*.

Island–arc system a group of islands, usually with a curving, archlike pattern and convex toward the open ocean, with a deep trench or trough on the convex side and usually enclosing a deep-sea basin on the concave side. Generally associated with volcanoes.

Isopods generally flattened marine crustaceans, mostly scavengers.

Isostasy the balance of portions of earth's crust, which rise or subside until their masses are in equilibrium relationship, "floating" on the denser mantle below. Isostatic theory implies that large continental masses are supported by "roots" of low-density crustal material.

Isotherm line connecting points of equal temperature.

Isotope one of several nuclides having the same number of protons in their nuclei and hence belonging to the same element but differing in the number of neutrons and therefore in mass number or energy content; also, a radionuclide or a preparation of an element with special isotopic composition, used principally as an isotopic tracer.

Jellyfish see *Medusa*.

Jet stream high-altitude swift air current.

Jetty in United States terminology, a structure built to influence tidal currents, to maintain channel

depths, or to protect the entrance to a harbor or river.

Knoll an elevation rising less than 1 kilometer from the ocean floor and of limited extent across the summit.

Lagoon a shallow sound, pond, or lake, generally separated from the open ocean.

Lamina a sediment or sedimentary-rock layer less than 1 centimeter thick, visually separable from the material above and below. *Lamination* refers to the alternation of such layers differing in grain size or composition.

Laminar flow a flow in which the fluid moves smoothly in streamlines, in parallel layers or sheets; a nonturbulent flow.

Langmuir circulation cellular circulation, with alternate left- and right-hand helical vortices, having axes in the direction of the wind, which is set up in the surface layer of a water body by winds exceeding 7 knots.

Lantern fishes (myctophids) small oceanic fishes that normally live at depths between a few hundred and a few thousand meters depth and characteristically have numerous small light organs on the sides of the body. Many undergo diurnal vertical migration.

Latent heat the heat released or absorbed per unit mass by a system undergoing a reversible change of state at a constant temperature and pressure.

Lava fluid rock, issuing from a volcano or fissure in the earth's surface, or the same material solidified by cooling.

Layer of no motion see *Depth of no motion*.

Levee an embankment bordering one or both sides of a seachannel or delta distributary.

Lithogenous sediment sediment composed primarily of mineral grains transported into the oceans from the continents.

Lithosphere the outer, solid portion of the earth; includes the crust of the earth.

Littoral see *Intertidal zone*.

Littoral drift sand moved parallel to the shore by wave and current action.

Load the quantity of sediment transported by a current. It includes the *suspended load* of small particles that float in suspension distributed through the whole body of the current and the bottom load or *bed load* of large particles that move along the bottom by rolling and sliding.

Longshore bar see *Bar*.

Longshore current a current located in the surf zone moving generally parallel to the shoreline; usually generated by waves breaking at an angle with the shoreline.

Low tide terrace   the zone that lies between the ordinary high- and low-water marks and is daily traversed by the oscillating water line as the tides rise and fall. This area, together with the vertical scarp that often occurs at its upper limit, is sometimes called the *foreshore,* which ends at the highest point of normal wave uprush.

Low water   the lowest limit of the surface-water level reached by the lowering tide; also called *low tide.*

Lower high water   the lower of two high tides occurring during a tidal day.

Lower low water   the lower of two low tides occurring during a tidal day.

Lunar tide   that part of the tide caused solely by the gravitational attraction of the moon as distinguished from that part caused by the gravitational attraction of the sun.

Magma   mobile rock material, generated within the earth, capable of intrusion and extrusion, from which igneous rock solidifies.

Manganese nodules   concretionary lumps of manganese and iron, containing copper and nickel; found widely scattered on the deep-ocean floor.

Mantle   the bulk of the earth, between crust and core, from about 40 to 3500 kilometers depth. Phase changes caused by increasing pressure with depth divide the mantle into concentric layers. Also, the tough, protective membrane possessed by all mollusks, within which water circulates.

Map projection   a method of representing part or all of the surface of a sphere, such as the earth, on a plane surface.

Marginal sea   semienclosed body of water adjacent to, widely open to, and connected with the ocean at the water surface but bounded at depth by submarine ridges—for example, the Yellow Sea.

Mariculture   see *Aquaculture.*

Marine humus   partially or entirely undecomposed organic matter in ocean sediment.

Marine snow   particles of organic detritus and living forms. The downward drift of these particles and living forms, especially in dense concentration, appears similar to a snowfall when viewed by underwater investigators.

Maritime climate   climate characterized by relatively little seasonal change, warm, moist winters, cool summers: result of prevailing winds blowing from ocean to land.

Marsh   an area of soft, wet land. Flat land periodically flooded by saltwater is called a *salt marsh.*

Maximum sustainable yield   the maximum yield of a fishery that can be sustained for many years without steady depletion of the stock.

Meander   a turn or winding of a current that may become detached from the main stream.

Mean sea level   the average height of the surface of the sea for all stages of the tide over a 19-year period, usually determined from hourly readings of tidal height.

Mean tidal range   the difference in height between mean high water and mean low water, measured in feet or meters.

Medusa (jellyfish)   any of various free-swimming coelenterates having a disk- or bell-shaped body of jellylike consistency. Many have long tentacles with stinging cells.

Meiobenthos   animals in the size range 100 to 500 microns that live between the grains in near-shore sediments, including many kinds of small invertebrates and larger protozoans.

Mélange   a large-scale formation containing diverse rock materials that were originally mixed and consolidated at great pressure, characteristic of compression or subduction zones.

Meridian (of longitude)   a great circle passing through the North and South poles. It connects points with an equal angular separation from the prime meridian.

Meroplankton   mainly organisms whose early developmental stages occur in the floating state; adults are benthic.

Metabolite   substance necessary to survival and growth of an organism; this includes vitamins and other materials required only in trace amounts, as well as the nutrients required for synthesis of living tissue.

Metamorphic rocks   formed in the solid state, below the earth's surface, in response to changes of temperature, pressure, or chemical environment.

Microbenthos   one-celled plants, animals, and bacteria living in or on the surface of bottom sediments.

Midocean ridge   a great median arch or sea-bottom swell extending the length of an ocean basin and roughly paralleling the continental margins.

Midocean rise   see *Oceanic rise.*

Mixed tide   type of tide in which a diurnal wave produces large inequalities in heights and/or durations of successive high and/or low waters. This term applies to the tides intermediate to those predominantly semidaily and those predominantly daily.

Mohorovičić discontinuity   the sharp discontinuity in composition between the outer layer of the earth (crust) and the inner layer (mantle).

Molecular diffusion   see *Diffusion.*

Mole   formula weight of a substance in grams—for example, 1 mole of H_2O weighs 18 grams (oxygen = 16; 2 hydrogen = 2).

Molecular viscosity   an internal resistance to flow in liquids arising from the attractive interactions between molecules.

Monsoons   a name for seasonal winds (derived from the Arabic *mausim,* meaning season), first applied to the winds over the Arabian Sea, which blow for 6 months from the northeast and the remaining 6 months from the southwest but subsequently extended to similar winds in other parts of the world.

Mud   detrital material consisting mostly of silt and clay-sized particles (less than 0.06 millimeter) but often containing varying amounts of sand and/or organic materials. It is also a general term applied to any fine-grained sediment whose exact size classification has not been determined.

Mysids   elongate crustaceans that usually are transparent (or nearly so) and benthic or deep living.

Nannoplankton   plankton whose length is less than 50 microns. Individuals will pass through most nets and usually are collected by centrifuging water samples.

Nansen bottle   a device used by oceanographers to obtain subsurface samples of seawater. The "bottle" is lowered by wire; its valves are open at both ends. It is then closed in situ by allowing a weight ("messenger") to slide down the wire and strike the reversing mechanism. This step causes the bottle to turn upside down, closing the valves and reversing the thermometers. Usually a series of bottles is lowered and then the reversal of each bottle releases another messenger to actuate the bottle beneath it.

Natural frequency   the characteristic frequency (number of vibrations or oscillations per unit time) of a body controlled by its physical characteristics (dimensions, density, etc.).

Nauplius   a limb-bearing early larval stage of many crustaceans.

Neap tide   lowest range of the tide, occurring near the times of the first and last quarter of the moon.

Nekton   those pelagic animals that are active swimmers, such as most of the adult squids, fishes, and marine mammals.

Net primary production   the total amount of organic matter produced by photosynthesis minus the amount consumed by the photosynthetic organisms in their own respiratory processes.

Nitrogen fixation   conversion of atmospheric nitrogen to oxides usable in primary food production.

Nodal line   a line in an oscillating area along which there is little or no rise and fall of the tide.

Nodal point   the no-tide point in an amphidromic region.

Node   that part of a standing wave where the vertical motion is least and the horizontal velocities are greatest.

Nonconservative property   a property whose values change in the course of a particular specified series of events or processes—for example, those properties of seawater, such as nutrient or dissolved oxygen concentrations, that are affected by biological or chemical processes.

Nonrenewable resource   one that is not replenished at a rate comparable to its rate of consumption.

Nuclide   a species of atom characterized by the constitution of its nucleus. The nuclear constitution is specified by the number of protons, number of neutrons, and energy content—or alternatively, by the atomic number, mass number, and atomic mass.

Nudibranchs (sea slugs)   gastropods in which the shell is entirely absent in the adult. The body bears projections that vary in color and complexity among the species.

Nutrient   in the ocean, any one of a number of inorganic or organic compounds or ions used primarily in the nutrition of primary producers. Nitrogen and phosphorus compounds are examples of *essential nutrients.*

Ocean basin   that part of the floor of the ocean that is more than about 2000 meters below sea level.

Oceanic crust   a mass of igneous material, approximately 7 kilometers thick, that lies under the ocean bottom.

Oceanic (midocean) rise   general term for the discontinuous ocean-bottom province that rises above the deep-ocean floor; area of crustal generation.

Ooze   a fine-grained deep-ocean sediment containing at least 30% (by volume) undissolved sand- or silt-sized, calcareous or siliceous skeletal remains of small marine organisms, the remainder usually being clay-sized material.

Open ocean   the part of the ocean that is seaward of the approximate edges of the continental shelves, usually more than 2 kilometers deep.

Ophiolitic suite   an assemblage of rocks on land, containing deep-sea sediments, submarine lavas, and oceanic crust, apparently thrust upward during crustal plate subduction.

Oxidation   loss of hydrogen or electrons; opposite of reduction.

Pack ice a rough, solid mass of broken ice floes forming a heavy obstruction and preventing navigation.

Pangaea single continent thought to have existed prior to Permian times. Fragments of it split apart about 200 million years ago to form the present continents.

Partial tide one of the harmonic components comprising the tide at any point. The periods of the partial tides are derived from various combinations of the angular velocities of earth, sun, and moon, relative to each other.

Patch reef a term for all isolated coral growths in lagoons of barrier and atolls. They vary in extent from expanses measuring several kilometers across to coral pillars or even mushroom-shaped growths consisting of a single colony.

Pelagic a primary division of the ocean that includes the whole mass of water. The division is made up of the coastal ocean (which includes water shallower than 200 meters) and the open ocean (which includes water deeper than 200 meters).

Pelagic deposits deep-ocean sediments that have accumulated by the settling out of the ocean on a particle-by-particle basis.

Photic zone (euphotic zone) the layer of a body of water that receives ample sunlight for the photosynthetic processes of plants; it is usually 80 meters deep or more, depending on angle of incidence of sunlight, length of day, and cloudiness.

Photosynthesis the manufacture of carbohydrate food from carbon dioxide and water in the presence of chlorophyll, by utilizing light energy and releasing oxygen.

Phytoplankton the plant forms of plankton.

Planetary winds see *Prevailing wind systems*.

Plankton passively drifting or weakly swimming organisms.

Plankton bloom an unusually high concentration of plankton (usually phytoplankton) in an area, caused either by an explosive or gradual multiplication of organisms and usually producing a discoloration of the sea surface.

Plate tectonics the theory that accounts for seismicity, mountain building, volcanism, and other manifestations of crustal plate movement with sea-floor spreading.

Polder a land area that has been reclaimed from the ocean bottom; often separated from the ocean by dikes, then drained.

Polychaetes segmented marine worms, some of which are tube worms; others are free-swimming worms.

Polyp an individual, sessile coelenterate.

Prevailing wind systems (planetary winds) large, relatively constant wind systems that result from the shape, inclination, revolution, and rotation of the planet. They are the northeast and southeast trade winds, the westerlies, and the polar easterlies.

Primary coastline a coastline shaped primarily by terrestrial forces rather than by wave action or other marine processes.

Primary productivity the amount of organic matter synthesized by organisms from inorganic substances in unit time in a unit volume of water or in a column of water of unit area cross section and extending from the surface to the bottom; also called *gross primary production*.

Prime meridian the meridian of longitude (0°), used as the reference for measurements of longitude, internationally recognized as the meridian of Greenwich, England.

Productivity see *Primary productivity*.

Profile a drawing showing a vertical section along a surveyed line.

Progressive wave a wave that is manifested by the progressive movement of the wave form.

Protozoa mostly microscopic, one-celled animals, constituting one of the largest populations in the ocean.

Pseudopod an extension of protoplasm that has no permanent shape but can be projected or withdrawn by the animal, for capturing food or for locomotion; characteristic of some protozoans.

Pteropods free-swimming gastropods in which the foot is modified into fins; both shelled and non-shelled forms exist. In some shallow areas, accumulated shells of these organisms form a type of bottom sediment called *pteropod ooze*.

Pycnocline a vertical density gradient in some layer of a body of water, positive with respect to depth and appreciably greater than the gradients above and below it; also, a layer in which such a gradient occurs.

Radioisotope any radioactive isotope of an element; also, a word loosely used as a synonym for *radionuclide*.

Radiolarians single-celled planktonic protozoans possessing a skeleton of siliceous spicules and radiating threadlike pseudopodia.

Radionuclide a synonym for *radioactive nuclide*.

Red algae reddish, filamentous, membranous, encrusting or complexly branched plants in which the color is imparted by the predominance of a red pigment over the other pigments present. Some are included among the coralline algae.

Red clay a more or less brown-to-red deep-sea deposit. It is the most finely divided clay material that is derived from the land and transported by ocean currents and winds, accumulating far from land and at great depths.

Red tide a red or reddish brown discoloration of surface waters most frequently in coastal regions, caused by concentrations of certain microscopic organisms, particularly dinoflagellates.

Reduction gain in electrons or in hydrogen; opposite of oxidation.

Reef an off-shore consolidated rock hazard to navigation with a least depth of 20 meters or less.

Reef flat (of a coral reef) a flat expanse of dead reef rock that is partly or entirely dry at low tide. Shallow pools, potholes, gullies, and patches of coral debris and sand are features of the reef flat. It is divisible into inner and outer portions.

Reflection the process whereby a surface or discontinuity turns back a portion of the incident radiation into the medium through which the radiation approached.

Refraction of water waves the process by which the direction of a wave moving in shallow water at an angle to the contours is changed, causing the wave crest to bend toward alignment with the underwater contours; also, the bending of wave crests by currents.

Relict sediment sediment having been deposited on the continental shelf by processes no longer active.

Renewable resource one that may be replenished at a rate comparable to its rate of consumption, either as a result of natural growth or by careful management of the resource.

Residence time time required for a flow of material to replace the amount of that material originally present in a given area. Assuming a steady flow, replacement time can be calculated for any substance, such as salt or water.

Respiration an oxidation-reduction process by which chemically bound energy in food is transformed into other kinds of energy on which certain processes in all living cells are dependent.

Reversing thermometer a mercury-in-glass thermometer that records temperature on being inverted and thereafter retains its reading until returned to the first position.

Reversing tidal current a tidal current that flows alternatively in approximately opposite directions, with a period of slack water at each reversal of direction. Reversing currents usually occur in rivers and straits where the flow is restricted. When the flow is toward the shore, the current is flooding; when in the opposite direction, it is ebbing.

Rift valley a trough formed by faulting in a divergence area.

Rill a small groove, furrow, or channel made in mud or sand on a beach by tiny streams following an outflowing tide.

Rip agitation of water caused by the meeting of currents or by a rapid current setting over an irregular bottom—for example, a *tide rip*.

Rip current a strong current, usually of short duration, flowing seaward from the shore. It usually appears as a visible band of agitated water and is the return movement of water piled up on the shore by incoming waves and wind.

Ripple a wave controlled to a significant degree by both surface tension and gravity.

Rise a long, broad elevation that rises gently and generally smoothly from the ocean bottom.

River-induced upwelling the upward movement of deeper water that occurs when seawater mixes with freshwater from a river—becoming less dense—and moves toward the surface.

Rotary current, tidal a tidally induced current that flows continually with the direction of flow, changing through all points of the compass during the tidal period. Rotary currents are usually found in the ocean where the direction of flow is not restricted by any barriers.

Sabellid see *Tube worm*.

Salinity a measure of the quantity of dissolved salts in seawater. Formally defined as the total amount of dissolved solids in seawater in parts per thousand (‰) by weight when all the carbonate has been converted to oxide, the bromide and iodide to chloride, and all organic matter is completely oxidized.

Salps transparent pelagic tunicates. The body is more or less cylindrical and possesses conspicuous ringlike muscle bands, the contraction of which propels the animal through the water.

Salt dome a generally cylindrical mass of salt, a kilometer or more in diameter, that rose to its present position through surrounding sediments that were originally deposited above it. Reservoirs in rocks around a dome may contain gas and oil.

Salt marsh see *Marsh*.

Saltwater wedge an intrusion in a tidal estuary of seawater in the form of a wedge characterized by a pronounced increase in salinity from surface to bottom.

Sand loose material that consists of grains ranging between 0.0625 and 2 millimeters in diameter.

Sargassum a brown alga characterized by a bushy form, substantial "holdfast" (rootlike structure) when attached, and a yellowish brown, greenish yellow, or orange color.

Scarp an elongated and comparatively steep slope of the ocean floor separating flat or gently sloping areas.

Scattering the dispersion of light when a beam strikes very small particles suspended in air or water. In light scattering there is, theoretically, no loss of intensity but only a redirection of light.

School a large number of one kind of fish or other aquatic animal swimming or feeding together.

Scour the downward and sideward erosion of a sediment bed by wave or current action.

Scyphozoans coelenterates in which the polyp or hydroid stage is minimized or insignificant and the medusoid stage is well developed. The true jellyfish belong to this group.

Sea waves generated or sustained by winds within their fetch, as opposed to *swell;* also, a subdivision of an ocean.

Seabed drifter a plastic float designed to drift in near-bottom currents; movements of the drifter indicate speed and direction of the current.

Sea breeze a light wind blowing toward the land, caused by unequal heating of land and water masses.

Seachannel a long, narrow, U- or V-shaped shallow depression of ocean floor, usually occurring on a gently sloping plain or fan.

Sea-floor spreading the mechanism by which oceanic crust is generated at divergence zones, such as midocean ridges, and adjacent crustal plates are moved apart as new material forms.

Seamount an elevation rising 900 meters or more from the ocean bottom.

Sea state (state of the sea) the numerical or written description of ocean-surface roughness.

Seawall a man-made structure of rock or concrete built along a portion of coast to prevent wave erosion of the beach.

Secondary coastline a coastline shaped primarily by marine forces, such as wave action, or by marine organisms.

Sediment particulate organic and inorganic matter that accumulates in a loose, unconsolidated form. It may be chemically precipitated from solution, secreted by organisms, or transported from land by air, ice, wind, or water and deposited.

Sedimentation the process of breakup and separation of particles from the parent rock, their transportation, deposition, and consolidation into another rock.

Seiche a standing-wave oscillation of an enclosed or semienclosed water body that continues, pendulum-fashion, after the cessation of the originating force, which may have been seismic, atmospheric, or wave induced; also, an oscillation of a fluid body in response to a disturbing force having the same frequency as the natural frequency of the fluid system.

Seismic pertaining to, characteristic of, or produced by earthquakes or earth vibrations.

Seismic reflection the measurements, and recording in wave form, of the travel time of acoustic energy reflected back to detectors from rock or sediment layers that have different elastic-wave velocities.

Seismic sea wave see *Tsunami.*

Semidaily (semidiurnal) tide a tide having a period or cycle of approximately one-half a tidal day; the predominating type of tide throughout the world is semidiurnal, with two high waters and two low waters each tidal day.

Semipermeable membrane a membrane through which a solvent, but not certain dissolved or colloidal substances, may pass.

Sensible heat the portion of energy exchanged between ocean and atmosphere that is utilized in changing the temperature of the medium into which it penetrates.

Sessile permanently attached, by a base of stalk; not free to move about.

Set (current direction) the direction toward which the current flows.

Shear wave a wave that causes particles in a medium to vibrate back and forth at right angles to the direction of propagation.

Shoreline the boundary between a body of water and the land at high tide (usually mean high water).

Significant wave height (characteristic wave height) the average height of the highest one-third of waves of a given wave group.

Sill shallow portion of the ocean floor that partially restricts water flow; may be either at the mouth of an inlet, fjord, or similar structure, or at the edge of an ocean basin—for example, the Bering Sill separates the Pacific and Arctic portions of the Atlantic Ocean.

Silt particles between sands and clays in size.

Sinking (downwelling) a downward movement of surface water generally caused by converging currents or as a result of a water mass becoming more dense than the surrounding water.

Siphonophores medusoid coelenterates, many of which are luminescent and some venomous. Some possess a gas-filled float. Some are colonial so that polyp and medusoid individuals function as a single individual.

Sipunculids wormlike marine animals, unsegmented, with the mouth surrounded by tentacles. The anterior (head) end can be withdrawn into the body. They are deposit feeders.

Slack water the state of a tidal current when its velocity is near zero; usually occurs when a reversing current changes its direction.

Slick area of quiescent water surface, usually elongated. Slicks may form patches or weblike nets where ripple activity is greatly reduced.

Slump the slippage or sliding of a mass of unconsolidated sediment down a submarine or subaqueous slope. Slumps occur frequently at the heads or along the sides of submarine canyons; triggered by any small or large earth shock, the sediment usually moves as a unit mass initially but often becomes a turbidity flow.

Solar tide the (partial) tide caused solely by the tide-producing forces of the sun.

Sounding the measurement of the depth of water beneath a ship.

Specific gravity the ratio of the density of a substance relative to the density of pure water at 4°C; in the centimeter–gram–second system, *density* and *specific gravity* may be used interchangeably.

Specific heat the quantity of heat required to raise the temperature of 1 gram of any substance by 1°C. The common unit is calories per gram per degree centigrade.

Sperm whale see *Toothed whales.*

Spicules crystals of newly formed sea ice; also, a minute, needlelike or multiradiate calcareous or siliceous body in sponges, radiolarians, some gastropods, and echinoderms.

Spit a small point of land projecting into a body of water from the shore.

Spring bloom the sudden proliferation of phytoplankton that occurs from the temperate zone to the poles whenever the critical depth (as determined by penetration of sunlight) exceeds the depth of the mixed, stable surface layer (as determined by the establishment of a pycnocline).

Spring tide tide of increased range that occurs about every 2 weeks when the moon is new or full.

Spur-and-groove structure see *Forereef.*

Stability the resistance to overturning or mixing in the water column, resulting from the presence of a positive density gradient.

Stand of the tide the interval at high or low water when there is no appreciable change in the height of the tide; its duration will depend on the range of the tide, being longer when the tidal range is small and shorter when the tidal range is large.

Standing crop the biomass of a population present at any given moment.

Standing wave a type of wave in which the surface of the water oscillates vertically between fixed points, called *nodes,* without progression. The points of maximum vertical rise and fall are called *antinodes.* At the nodes, the underlying water particles exhibit no vertical motion but maximum horizontal motion.

Steady state the absence of change with time.

Still-water level the level that the sea surface would assume in the absence of wind waves; not to be confused with mean sea level or half-level tide.

Storm surge (storm wave, storm tide, tidal wave) a rise or piling up of water against shore, produced by wind-stress and atmospheric-pressure differences in a storm.

Stratosphere that part of the earth's atmosphere between the troposphere and the upper layer (ionosphere).

Subduction zone an inclined plane descending away from a trench, separating a sinking oceanic plate from an overriding plate. A region of high seismicity, known as a *Benioff zone,* coincides with this plane.

Sublimation the transition of the solid phase of certain substances into the gaseous—and vice versa—without passing through the liquid phase. Water possesses this property; thus ice can change directly into water vapor or water vapor into ice.

Submarine canyon a V-shaped submarine depression of valley form with relatively steep slope and progressive deepening in a direction away from shore.

Subsurface current a current usually flowing below the pycnocline, generally at slower speeds and frequently in a different direction from the currents near the surface.

Subtropical high one of the semipermanent highs of the subtropical high-pressure belt.

Succession, ecological the orderly process of community change whereby communities replace one another in sequence.

Surf collective term for breakers; also, the wave activity in the area between the shoreline and the outermost limit of breakers.

Surf zone the area between the outermost breaker and the limit of wave uprush.

Surface-active agent substance, usually in solution, that can markedly change the surface or interfacial properties of the liquid fraction, even when present only in trace amounts.

Surface tension (surface energy, capillary forces, interfacial tension) a phenomenon peculiar to the surface of liquids, caused by a strong attraction toward the interior of the liquid acting on the liquid molecules in or near the surface in such a way as to reduce the surface area. An actual tension results, usually expressed in dynes per centimeter or ergs per square centimeter.

Surface zone (mixed zone) water layer above the pycnocline where wind waves and convection cause mixing of the water, resulting in more or less uniform temperature and salinity with depth.

Surge horizontal oscillation of water with comparatively short period accompanying a seiche (see also *Storm surge*).

Swallow float a tubular buoy, usually made of aluminum, that can be adjusted to remain at a selected density level to drift with the motion of that water mass. The float is tracked by listening devices, to determine current velocities. velocities.

Swash the rush of water up onto the beach following the breaking of a wave.

Swell ocean waves that have traveled out of their generating area.

Swimbladder a membranous sac of gases, lying in the body cavity between the vertebral column and the alimentary tract of certain fishes. It serves a hydrostatic function in most fishes that possess it; in some, it participates in sound production.

Symbiosis a relationship between two species in which one or both members benefit and neither is harmed.

Tablemount see *Guyot*.

Tectonic estuary an estuary occupying a basin formed as a result of mountain building.

Thermocline a vertical temperature gradient in some layer of a body of water, negative with respect to depth and appreciably greater than the gradients above and below it; also, a layer in which such a gradient occurs.

Thermohaline circulation vertical circulation induced by surface cooling, which causes convective overturning and consequent mixing.

Tidal bore a very rapid rise of the tide in which the advancing water forms an abrupt front, sometimes of considerable height, occurring in certain shallow estuaries where there is a large tidal range.

Tidal bulge (tidal crest) a long-period wave associated with the tide-producing forces of the moon and sun; identified with the rising and falling of the tide. The trough located between the two tidal bulges present at any given time on the earth is known as the *tidal trough*.

Tidal constituent one of the harmonic elements in a mathematical expression for the tide-generating force and in corresponding formulas for the tide or tidal current. Each constituent represents a periodic change or variation in the relative positions of the earth, moon, and sun.

Tidal current the alternating horizontal movement of water associated with the rise and fall of the tide caused by the astronomical tide-producing forces.

Tidal day the interval between two successive upper transits of the moon over a local meridian. The period of the *mean tidal day*, sometimes called a *lunar day*, is 24 hours, 50 minutes.

Tidal flats marshy or muddy areas which are covered and uncovered by the rise and fall of the tide; also called *tidal marshes*.

Tidal period the elapsed time between successive high or low waters.

Tidal range the difference in height between consecutive high and low waters.

Tidal trough see *Tidal bulge*.

Tide the periodic rise and fall of the earth's ocean and atmosphere that results from tide-producing forces of moon and sun acting on the rotating earth.

Tide curve a graphic presentation of the rise and fall of tide; time (in hours or days) is plotted against height of the tide.

Tide-producing forces the slight local difference between the gravitational attraction of two astronomical bodies and the centrifugal force that holds them apart. Gravitational attraction predominates at the surface point nearest to the other body while centrifugal "repulsion" predominates at the surface point farthest from the other body.

Tide pool depression in a rock within the intertidal zone, alternately submerged and exposed with water remaining inside, by the rise and fall of the tide.

Tide rip see *Rip*.

Tide tables tables that give daily predictions, usually a year in advance, of the times and heights of the tide.

Tide wave a long-period gravity wave that has its origin in the tide-producing force and that manifests itself in the rising and falling of the tide.

Tintinnids microscopic planktonic protozoans that possess a tubular or vase-shaped outer shell.

Tombolo an area of unconsolidated material deposited by wave or current action, that connects a rock, island and so on to the main shore or other body of land.

Toothed whales this group includes dolphins, porpoises, killer whales, and sperm whales.

Trade winds the wind system, occupying most of the tropics, which blows from the subtropical highs toward the equatorial trough; a major component of the general circulation of the atmosphere. The winds are northeasterly in the Northern Hemisphere and southeasterly in the Southern Hemisphere.

Transform fault a fault connecting the offset portions of a midocean ridge, along which crustal plates slide past each other.

Trawl a bag or funnel-shaped net for catching bottom fish by dragging along the bottom; also, a large net for catching zooplankton and fishes by towing in intermediate depths.

Trochophore The free-swimming pelagic stage of some segmented worms and mollusks.

Trophic level a successive stage of nourishment as represented by links of the food chain. Primary producers (phytoplankton) constitute the first trophic level, herbivorous zooplankton the second trophic level, and carnivorous organisms the third and higher trophic levels.

Tropics equatorial region between Tropic of Cancer and Tropic of Capricorn; climate found in the belt close to the equator; characteristics are daily variation in temperature exceeding seasonal variation and generally high rainfall.

Tropopause the upper limit of the troposphere.

Troposphere that portion of the atmosphere next to the earth's surface in which temperature generally rapidly decreases with altitude, clouds form, and convection is active. At middle latitudes the troposphere generally includes the first 10 to 12 kilometers above the earth's surface.

Tsunami (seismic sea wave) a long-period sea wave produced by a submarine earthquake or volcanic eruption. It may travel unnoticed across the ocean for thousands of miles from its point of origin and builds up to great heights over shallow water.

Tube worm any polychaete, chiefly the sabellids and related groups, that builds a calcareous or leathery tube on a submerged surface. Tube worms are notable fouling organisms.

Tunicates globular or cylindrical, often saclike animals, many of which are covered by a tough flexible material. Some are sessile; others are pelagic.

Turbidite turbidity-current deposit characterized by both vertically and horizontally graded bedding.

Turbidity reduced water clarity resulting from the presence of suspended matter. Water is considered turbid when its load of suspended matter is visibly conspicuous, but all waters contain some suspended matter and so are turbid to some degree.

Turbidity current a gravity current resulting from a density increase by suspended material.

Turbulent flow a flow characterized by irregular, random-velocity fluctuations.

Upwelling the process by which water rises from a lower to a higher depth, usually as a result of divergence and offshore currents.

van der Waals forces weak attractive forces between molecules that arise from interactions between the atomic nuclei of one molecule and the electrons of another molecule.

Veliger the planktonic larval second stage of many gastropods.

Viscosity internal resistance-to-flow property of fluids that enables them to support certain stresses and thus resist deformation for a finite time; arises from molecular and eddy effects.

Volcanic center ("hot spot") a locus of continuing volcanism due to long-term activity in the mantle. Sea-floor spreading moves the earth's crust with respect to such a center so that as a volcanic ridge or island chain is built at the surface, its orientation is governed by the direction of sea-floor motion.

Volcanic island island formed by the top of a volcano or solidified volcanic material, rising above the sea surface.

Water budget the accounting for the interchange of water substance among the earth, the atmosphere, and the ocean.

Water mass a body of water usually identified by its *T-S* curve or chemical content.

Water parcel water mass with a certain temperature and salinity, separated from surrounding waters by fairly sharp boundaries across which mixing occurs.

Wave a disturbance that moves through or over the surface of the ocean (or earth); wave speed depends on the properties of the medium.

Wave age the state of development of a wind-generated sea-surface wave, conveniently expressed by the ratio of wave speed to wind speed. Wind speed is usually measured at about 8 meters above still-water level.

Wave energy the capacity of a wave to do work. In a deep-water wave, about half the energy is kinetic energy, associated with water movement, and about half is potential energy, associated with the elevation of water above the still-water level in the crest or its depression below still-water level in the trough.

Wave group a series of waves in which the wave direction, wavelength, and wave height vary only slightly.

Wave height vertical distance between crest and preceding trough.

Wavelength horizontal distance between successive wave crests measured perpendicular to the crests.

Wave period the time required for two successive wave crests to pass a fixed point, such as a rock or an anchored buoy.

Wave spectrum a concept used to describe by mathematical function the distribution of wave energy (square of wave height) with wave frequency (1/period). The square of the wave height is related to the potential energy of the sea surface so that the spectrum can also be called the *energy spectrum*.

Wave steepness ratio of wave height to wavelength.

Wave train a series of waves from the same direction.

Wave velocity speed at which the individual wave form advances; also, a vector quantity that specifies the speed and direction with which a wave travels through a medium.

Weathering process of destruction or partial destruction of rock by thermal, chemical, and mechanical processes.

Wetland see *Marsh*.

Whitecap the white froth on crests of waves in a wind, caused by wind blowing the crest forward and over.

Wind-driven circulation surface-current system driven by the force of the winds.

Wind mixing mechanical stirring of water due to motion, induced by the surface wind.

"Wind tide" vertical rise in the water level on the leeward side of a body of water, caused by wind stress on the surface of the water.

Wind waves waves formed and growing in height under the influence of wind; loosely, any wave generated by wind.

Windrows rows of floating debris, aligned in the wind direction, formed on the surface of a lake or ocean by Langmuir circulation in the upper water layer.

Year class organisms of a particular species spawned during a single year or breeding season.

Zonation organization of a habitat into more or less parallel bands of distinctive plant and animal associations where conditions for survival are optimal.

Zone a layer that encompasses some defined feature, structure, or property in the ocean.

Zooplankton the animal forms of plankton.

INDEX

Cape Horn, 33
Cape May, New Jersey, 201, 301
Cape Mendocino, California, 263
Cape Verde Basin, 71
Cape Verde Islands, 68
Capillary waves, 217
Carbohydrate, 324, 333, 339
Carbon, organic (*see* Organic carbon;
    Organic matter)
Carbon:
    and plankton, 326, 327, 341
    in seawater, 123, 127, 196
Carbonate ion, 130–131
Carbonates (*see also* Calcium carbonate;
    Sand, carbonate; Sediment,
    carbonate):
    in seawater, 131
    shells, 131, 292, 362 (*see also*
        Sediment, carbonate; Shells in
        sediments)
Carbon dioxide:
    in the atmosphere, 57, 131, 143–145,
        150, 165
    and chemosynthesis, 403–404
    and early atmosphere, 79
    effects on climate, 165
    and photosynthesis, 78, 81, 130–131,
        333–335, 340, 356, 404, 407
    and respiration, 78–81, 130–131,
        333–334, 340, 341–344
    in seawater, 91, 129–131 (*see also*
        Carbonate ion)
    and zooxanthellae, 406
Carbon dioxide cycle, 130–131
Carbon fixation (*see* Chemosynthesis;
    Organic carbon, production of;
    Photosynthesis; Productivity)
Carbon-14, 18, 108, 196, 335
Carbonic Acid, 130–131
Carbon monoxide, 79
Cardisonia, 400
Caribbean coastal ocean region (U.S.),
    262–263
Caribbean Plate, 62, 68
Caribbean Sea, 23, 42, 43, 241, 248,
    262, 340, 374
Carlsberg Ridge, 66, 68
Carnivores and food web, 325, 331, 332,
    343, 347, 362, 366, 369, 371,
    373–375 (*see also* Predation)
Caroline Basin, 71
Carpathian Mountains, 68
Cascadia Channel, 41
Cassiterite, 436
Catalyst (ship), 17
Celebes Sea, 50
Celestite, 81
Cellulose, 322, 355, 356, 414
Centropages, 361
Cephalopods, 364, 370–372 (*see also*
    names of groups and individuals,
    e.g. *Cirrotheuthis; Octopus*)
Ceratius holboelli, 371
Cerbula gibba, 387
Ceylon (*see* Sri Lanka)

Chaetognatha, 362–363, 380, 408 (*see also*
    Arrowworms)
Chaetopterus, 401
Challenger Deep (Marianas Trench), 42
Challenger Deep-Sea Expedition, 5–8, 10,
    89, 357
Channel Islands, England, 386
Channels:
    continental shelf and detrital deposits,
        436
    deep sea, 47, 414
    in deltas (*see* Distributaries)
Chelating agents, 331
Chemical energy and photosynthesis,
    333–334
Chemosynthesis, 81, 403–404, 414
Chesapeake Bay, 263, 269, 276–279,
    290
    and marine organisms, 362, 379, 390,
        392, 406
    and tidal phenomena, 254–256,
        276–279
Chile, 290, 346, 434
Chile Rise, 68
Chimaera, 422
Chimneys, hydrothermal, 81
China, 75, 95, 422, 433, 435
China Plate, 62
Chinook salmon, 377
Chitin, 355, 356, 362, 366, 370, 407
Chloride ion, 122–124, 126–129
Chlorinity, 123–124
Chlorite, 87, 101–102
Chlorophyll, 333–334
Christmas Island, 7, 425
Chromis verator, 410
Chromite, 436
Chromium from marine deposits, 436,
    444
Chrysaora, 363, 390–391
Chukchi Sea, 340
Cilia, in marine organisms, 357, 363,
    366, 403, 405
Cirrotheuthis, 372
Cladocerans, 362
Clams, 367, 415
    commercial production, 423
    as deposit feeders, 396, 400–402
    sediment reworked by, 400–402
    and waste disposal, 444–445
Clay, red (*see* Red clay)
Clay minerals, 87, 101–107
Clays, 88, 97
    adsorption of surface active materials,
        396, 402
    and marine organisms, 364, 396–397
    settling velocity, 96–97, 106
    size of particles, 88, 97, 101, 305, 396
    transport, 96–98, 106, 305
Climate:
    effect of ocean, 18, 118, 164–165,
        172–173
Climatic changes:
    and fisheries, 424–426
    and sediment dating, 109

Climatic regions (zones), 170–172,
    174–175 (*see also* individual
    regions, e.g. Polar region;
    Tropical region)
    and sedimentation, 106–107
Climax community, 404
Clio, 362
Clouds, 57, 144–145, 153, 171, 270
    and heat budget, 144–145, 150
    and water budget, 151–155
Clupeoids, 374–375 (*see also* names of
    groups and individuals, e.g.
    Herring; *Sardina pilchardus*)
Clyde Sea, 105
Coal, 437
Coastal circulation:
    and boundary currents, 185–187, 262,
        281–284
    Northeast Pacific coastal ocean,
        281–284
    and upwelling (*see* Upwelling)
    and wastes, 443
    and winds, 184, 198, 281–284
Coastal currents, 185–187, 262, 265,
    280–284, 297, 377, 443
Coastal engineering, 316–318
Coastal modification, 315–318
Coastal ocean, 30–35, 165, 262–268,
    281–283, 324
    and bacteria, 339
    evolution, 268
    fisheries (*see* Fisheries and coastal
        ocean)
    and fishes, 377–388, 420–426
    and food, 331–333, 343, 355
    general features, 256–268
    and Gulf Stream System, 184–187,
        262
    and mixing processes, 161, 264–274,
        281–283, 327, 335–336, 346
    nutrients recycled in (*see* Nutrients in
        coastal waters)
    and organic carbon concentrations,
        338–340
    organic growth factors, 331
    and plankton, 355–356, 358, 360, 363
    and productivity, 267, 331, 334,
        344–346, 348, 355–356, 358, 374,
        384, 386, 424, 428
    and salinity, 124, 160–161, 191,
        262–263, 264–265, 268–270 (*see
        also* Salinity distribution,
        Northeast Pacific coastal ocean)
    and sediment transport, 47, 294, 297,
        299
    and temperature, 185–186, 191,
        262–265, 268, 276–278
    and tides, 263–264
    and turbidity, 142, 275
    and upwelling (*see* Upwelling)
    and waste disposal, 443–446
Coastal ocean regions (U.S.) (*see* United
    States and coastal ocean regions)
Coastal plain, 45, 46, 57, 103, 268, 270,
    276, 279, 308

Hwang-Ho River, 31, 95, 96, 291
Hychoides dianthus, 405
Hydration atmosphere (sheath), 128
Hydraulic dredge, 435–436
Hydraulic model, 318
Hydrogen, in ancient atmosphere and ocean, 79
  and chemosynthesis, 403–404
  as a fuel, 434
  and reducing power, 333, 404
  in seawater, 124, 127, 130
  in water molecule, 114–115
Hydrogen bomb testing, 165
Hydrogen bond, 115–117, 119–120, 123
Hydrogen chloride, 78
Hydrogen sulfide, 118, 130, 414, 424–425, 437
Hydrographers Canyon, 412
Hydroids, 390–392, 400, 404
Hydrological cycle, 151–152
Hydrophilic substances, 133–134
Hydrophobic substances, 134
Hydrophone, 60
Hydrosphere, 29, 57, 346
Hydrostatic pressure (*see* Pressure)
Hydrothermal circulation in oceanic crust, 61
Hydrothermal vents, 61, 80–81, 414
  and marine organisms, 321, 383
Hydroxides, in marine sediments, 87
Hyperiidae, 360, 376
Hypersaline lagoons, 269, 279–281
Hypsographic curve, 28, 30

I

Ice, 30 (*see also* Floe ice; Glaciers; Pack ice; Sea ice; Shelf ice)
  density of, 116–117, 121
  effect of heat on, 119–121
  structure, 116–117, 121
Ice Age, 20, 78, 172–175 (*see also* Pleistocene glacial period)
  and marine organisms, 172–173, 411
  and sea level, 45, 268
Icebergs, 104, 166–170
Icecap (terrestrial climate zone), 170
Iceland, 35, 39, 43, 68, 73, 196, 386, 433, 435
Ideal waves, 212–216
Idiacanthus, 369
Igneous rocks, 59
Ijssel Lake, Netherlands, 440
Illite, 87, 101–103
Ilmenite, 436
Incoming solar radiation (*see* Insolation)
India, 75–76, 346, 402, 434
  and monsoon circulation, 198–199
  and sediment discharge, 95–96, 304
  and tectonic processes, 75–76
  and upwelling, 266
Indian-Antarctic Basin, 71
Indian Basin, 33–35, 37, 71
  dimensions, 30, 35
  evolution of, 75–76

as a gulf, 28
sediment deposits, 99, 101–103 (*see also* Sedimentation of Indian Ocean Basin)
topography, 33, 35
trenches, 36, 42
Indian Central Water, 160
Indian Equatorial Water, 160
Indian Ocean, 31–35, 65
  boundaries, 31–35
  and currents, 171, 183, 185, 194–195, 198–199
  freshwater discharge, 31, 32
  and marine organisms, 402–403
  and nutrients, 327, 346
  oxygen distribution, 343
  and productivity, 345
  and salinity, 32, 154, 159, 160
  surface and drainage areas and average depth, 32
  and temperature, 32, 147–148, 158, 160, 194–195
  and tidal phenomena, 237
  and upwelling, 346
  volume of, 32–33
  water mass movements in, 158–160, 194–195
  waves, 222
Indian subcontinent, 75–76
Indonesia, 31, 43, 50, 74, 431
Indus River, 31, 35, 96
Industrial Revolution, 165
Industrial wastes (*see* Wastes, industrial)
Infauna, 396, 401, 414 (*see also* Meiobenthos; Microbenthos; and names of groups and individual organisms, e.g. Tube worms; *Urechis caupo*)
Infra-gravity waves, 217
Inlets (*see* Tidal inlets)
Insects and food web, 294–295
Insolation (*see also* Light in marine environment; Light in the ocean; Solar energy):
  and distribution of, 144–146, 175
  and food web, 325
  and heat budget, 143–145, 155, 157
Intergovernmental Oceanographic Commission (IOC), 20
Internal waves, 230–232, 274
International Council for the Exploration of the Sea (ICES), 12–13
International Council of Scientific Unions (ICSU), 17
International Date Line, 453
International Geophysical Year, 17
International Ice Patrol, 14
Intertidal region, and marine organisms, 238, 324, 386–390, 397
Intertidal zones, 386
Intertropical convergence zone, 149, 188
Intra-Coastal Canal, Laguna Madre, Texas, 281
Iodine, 123, 127, 132, 326
Ionic bond, 115, 128

Ion pairs, 129
Ipswich, England, 431
Iran, 72
Iron, in chemosynthesis, 403
  in earth's core, 56, 78
  and marine organisms, 326, 331, 403–404, 444
  in rocks, 56, 57
  in seawater, 78–79, 123, 127, 128, 135, 331
  in sedimentary rocks, 79, 81, 94–95, 106–107, 404
Iron-manganese nodules (*see* Manganese nodules)
Iron ore, 79, 437
Irrawaddy River, 31, 96, 276
Isaacs, John D., 353
Isias, 361
Island arc-trench systems, 36, 42–43, 49–50, 63–64, 68, 71, 73–76 (*see also* Volcanic islands, formation of)
Isopods, 360, 394–395, 412
Isostasy, 59, 66, 289

J

Jacks, 377, 408, 422
Jamaica, 310
James River, Virginia, 269, 276–278
Janthina, 363
Japan, 32, 50, 74, 81, 230, 346, 433
  and aquaculture, 428
  and fish production, 422
  and geothermal power, 433
Japan Sea, 31, 74
Japan Trench, 42, 68, 71
Jarmilla event, 61, 109
Java Sea, 50
Java Trench, 42, 48, 50, 68, 71
Jeanette (ship), 180
Jefferson, Thomas, 8
Jellyfish, 363–364, 408
Jet stream, 149, 173
Jetties, 212, 281, 290, 315–318
John F. Kennedy International Airport, New York, 316
Joint Oceanographic Institution for Deep Earth Sampling (JOIDES), 19
Julius Caesar, 238

K

Kalimonton (Borneo), 50, 431
Kamchatka Peninsula, 48, 68, 100, 346
Kane (ship), 413
Kaolinite, 87, 101–102
Kermadec Trench, 42, 51, 71
Key West, Florida, 290
Kidd, Lewis W., 353
Killer whale, 373
Killifish, 405
Kingfish, 377